VOLUME FOUR HUNDRED AND SIXTY-FIVE

METHODS IN ENZYMOLOGY

Liposomes, Part G

METHODS IN ENZYMOLOGY

VOLUME FOUR HUNDRED AND SIXTY-FIVE

METHODS IN ENZYMOLOGY

Liposomes, Part G

EDITED BY

NEJAT DÜZGÜNEŞ

Department of Microbiology
University of the Pacific Arthur A. Dugoni School of Dentistry
San Francisco, California, USA

ELSEVIER

AMSTERDAM • BOSTON • HEIDELBERG • LONDON
NEW YORK • OXFORD • PARIS • SAN DIEGO
SAN FRANCISCO • SINGAPORE • SYDNEY • TOKYO
Academic Press is an imprint of Elsevier

Academic Press is an imprint of Elsevier
525 B Street, Suite 1900, San Diego, CA 92101-4495, USA
30 Corporate Drive, Suite 400, Burlington, MA 01803, USA
32 Jamestown Road, London NW1 7BY, UK

First edition 2009

ISBN: 978-0-12-381379-4
ISSN: 0076-6879

Printed and bound in United States of America
09 10 11 12 10 9 8 7 6 5 4 3 2 1

Contents

Contributors

Theresa M. Allen
Department of Pharmacology, University of Alberta, Edmonton, Alberta, Canada

Maria José Alonso
Department of Pharmaceutical Technology, Faculty of Pharmacy, University of Santiago de Compostela, Santiago de Compostela, Spain

Luis A. Bagatolli
Membrane Biophysics and Biophotonics group/MEMPHYS Center of Biomembrane Physics, Department of Biochemistry and Molecular Biology, University of Southern Denmark, Odense M, Denmark

Krystof S. Bankiewicz
Department of Neurological Surgery, University of California San Francisco, San Francisco, California, USA

Pamela Becherini
Experimental Therapy Unit, Laboratory of Oncology, G. Gaslini Children's Hospital, Genoa, Italy

Vikram K. Bhatia
Bio-Nanotechnology Laboratory, Department of Neuroscience and Pharmacology and Nano-Science Center, University of Copenhagen, Copenhagen, Denmark

Chiara Brignole
Experimental Therapy Unit, Laboratory of Oncology, G. Gaslini Children's Hospital, Genoa, Italy

Irene Caffa
Experimental Therapy Unit, Laboratory of Oncology, G. Gaslini Children's Hospital, Genoa, Italy

Ana Cardoso
Department of Biochemistry, Faculty of Science and Technology, and Center for Neuroscience and Cell Biology, University of Coimbra, Coimbra, Portugal

Bridget Carragher
National Resource for Automated Molecular Microscopy, Department of Cell Biology, The Scripps Research Institute, La Jolla, California, USA

Edison L. S. Carvalho
Faculty of Pharmacy, Federal University of Rio de Janeiro, Rio de Janeiro, Brazil

Ian Cheong
The Ludwig Center for Cancer Genetics and Therapeutics, Johns Hopkins Kimmel Comprehensive Cancer Center, Baltimore, Maryland, USA

Angelo Corti
Division of Molecular Oncology, IIT Network Research Unit of Molecular Neuroscience, San Raffaele Scientific Institute, Milan, Italy

Nejat Düzgüneş
Department of Microbiology, University of the Pacific, Arthur A. Dugoni School of Dentistry, San Francisco, California, USA

Kai K. Ewert
Materials, Physics, and Molecular, Cellular and Developmental Biology Departments, University of California at Santa Barbara, Santa Barbara, California, USA

Daniel A. Fletcher
Department of Bioengineering, University of California, and Lawrence Berkeley National Laboratory, Berkeley, California, USA

John Forsayeth
Department of Neurological Surgery, University of California San Francisco, San Francisco, California, USA

Cristina Gagliani
Department of Experimental Medicine, Human Anatomy Section, and MicroSCBio Research Center, University of Genoa, Genoa, Italy; IFOM (FIRC Institute of Molecular Oncology), Milan, Italy

Electra Gizeli
Department of Biology, University of Crete, and Institute of Molecular Biology and Biotechnology, FORTH, Crete, Greece

Ana Grenha
Centre for Molecular and Structural Biomedicine (CBME), Institute for Biotechnology and Bioengineering (IBB), University of Algarve, Faro, Portugal

J. A. Helms
Division of Plastic and Reconstructive Surgery, Department of Surgery, Stanford University School of Medicine, Stanford, California, USA

Takayuki Ishii
Department of Medical Biochemistry, University of Shizuoka School of Pharmaceutical Sciences, Shizuoka, Japan

John Katsaras
Canadian Neutron Beam Centre, Steacie Institute for Molecular Sciences, Chalk River Laboratories, National Research Council Canada, Chalk River, and Department of Physics, Brock University, St. Catharines; Guelph-Waterloo Physics Institute and Biophysics Interdepartmental Group, University of Guelph, Guelph, Ontario, Canada

Michal T. Krauze
Department of Neurological Surgery, University of California San Francisco, San Francisco, California, USA

Norbert Kučerka
Department of Physical Chemistry of Drugs, Faculty of Pharmacy, Comenius University, Bratislava, Slovakia, and Canadian Neutron Beam Centre, Steacie Institute for Molecular Sciences, Chalk River Laboratories, National Research Council Canada, Chalk River, Ontario, Canada

Andreas H. Kunding
Bio-Nanotechnology Laboratory, Department of Neuroscience and Pharmacology and Nano-Science Center, University of Copenhagen, Copenhagen, Denmark

L. James Lee
NSF Nanoscale Science and Engineering Center (NSEC), and Department of Chemical and Biomolecular Engineering, The Ohio State University, Columbus, Ohio, USA

Robert J. Lee
Division of Pharmaceutics, College of Pharmacy, and NSF Nanoscale Science and Engineering Center (NSEC), The Ohio State University, Columbus, Ohio, USA

Jean Luc Lenormand
HumProTher Laboratory, TIMC-ThereX, UMR 5525 CNRS-UJF, Université Joseph Fourier, UFR de Médecine, Domaine de la Merci, La Tronche, France

P. Leucht
Division of Plastic and Reconstructive Surgery, Department of Surgery, Stanford University School of Medicine, Stanford, California, USA

Thomas H. Li
Department of Mechanical Engineering, University of California, Berkeley, California, USA

Lavinia Liguori
HumProTher Laboratory, TIMC-ThereX, UMR 5525 CNRS-UJF, Université Joseph Fourier, UFR de Médecine, Domaine de la Merci, La Tronche, France

Maria C. Pedroso de Lima
Center for Neuroscience and Cell Biology, and Department of Biochemistry, Faculty of Science and Technology, University of Coimbra, Coimbra, Portugal

Christina Lohr
Bio-Nanotechnology Laboratory, Department of Neuroscience and Pharmacology and Nano-Science Center, University of Copenhagen, Copenhagen, Denmark

Monica Loi
Experimental Therapy Unit, Laboratory of Oncology, G. Gaslini Children's Hospital, Genoa, Italy

Renato Longhi
Istituto di Chimica del Riconoscimento Molecolare, Consiglio Nazionale delle Ricerche, Milan, Italy

Philippe Méléard
CNRS UMR 6226, ENSCR, Avenue du Général Leclerc, Cedex, France

Danilo Marimpietri
Experimental Therapy Unit, Laboratory of Oncology, G. Gaslini Children's Hospital, Genoa, Italy

Kathryn Melzak
Department of Chemistry, University of British Columbia, Vancouver, British Columbia, Canada

João Nuno Moreira
Laboratory of Pharmaceutical Technology, Faculty of Pharmacy, and Center for Neuroscience and Cell Biology, University of Coimbra, Coimbra, Portugal

N. Morrell
Division of Plastic and Reconstructive Surgery, Department of Surgery, Stanford University School of Medicine, Stanford, California, USA

Hiroyuki Nakamura
Department of Chemistry, Faculty of Science, Gakushuin University, Toshima-ku, Tokyo, Japan

Mu-Ping Nieh
Canadian Neutron Beam Centre, Steacie Institute for Molecular Sciences, Chalk River Laboratories, National Research Council Canada, Chalk River, Ontario, Canada

Naoto Oku
Department of Medical Biochemistry, University of Shizuoka School of Pharmaceutical Sciences, Shizuoka, Japan

Gabriella Pagnan
Experimental Therapy Unit, Laboratory of Oncology, G. Gaslini Children's Hospital, Genoa, Italy

Daniela Di Paolo
Experimental Therapy Unit, Laboratory of Oncology, G. Gaslini Children's Hospital, Genoa, Italy

Fabio Pastorino
Experimental Therapy Unit, Laboratory of Oncology, G. Gaslini Children's Hospital, Genoa, Italy

Mirco Ponzoni
Experimental Therapy Unit, Laboratory of Oncology, G. Gaslini Children's Hospital, Genoa, Italy

Tanja Pott
CNRS UMR 6226, ENSCR, Avenue du Général Leclerc, Cedex, France

Clinton S. Potter
National Resource for Automated Molecular Microscopy, Department of Cell Biology, The Scripps Research Institute, La Jolla, California, USA

Joel Quispe
National Resource for Automated Molecular Microscopy, Department of Cell Biology, The Scripps Research Institute, La Jolla, California, USA

Carmen Remuñán-López
Department of Pharmaceutical Technology, Faculty of Pharmacy, University of Santiago de Compostela, Santiago de Compostela, Spain

S. M. Rooker
Division of Plastic and Reconstructive Surgery, Department of Surgery, Stanford University School of Medicine, Stanford, California, USA

Cyrus R. Safinya
Materials, Physics, and Molecular, Cellular and Developmental Biology Departments, University of California at Santa Barbara, Santa Barbara, California, USA

Hiromi Sakai
Research Institute for Science and Engineering, Waseda University, 3-4-1 Okubo, Shinjuku, Tokyo, Japan, and Waseda Bioscience Research Institute in Singapore, 11 Biopolis Way, Helios, Singapore

Begoña Seijo
Department of Pharmaceutical Technology, Faculty of Pharmacy, University of Santiago de Compostela, Santiago de Compostela, Spain

D. Simanovskii
Hansen Experimental Physics Laboratory (HEPL), Stanford University School of Medicine, Stanford, California, USA

Keitaro Sou
Research Institute for Science and Engineering, Waseda University, 3-4-1 Okubo, Shinjuku, Tokyo, Japan, and Waseda Bioscience Research Institute in Singapore, 11 Biopolis Way, Helios, Singapore

Jeanne C. Stachowiak
Sandia National Laboratories, Livermore, California, USA

Dimitrios Stamou
Bio-Nanotechnology Laboratory, Department of Neuroscience and Pharmacology and Nano-Science Center, University of Copenhagen, Copenhagen, Denmark

Carlo Tacchetti
Department of Experimental Medicine, Human Anatomy Section, and MicroSCBio Research Center, University of Genoa, Genoa, Italy; IFOM (FIRC Institute of Molecular Oncology), Milan, Italy

Sara Trabulo
Center for Neuroscience and Cell Biology, and Department of Biochemistry, Faculty of Science and Technology, University of Coimbra, Coimbra, Portugal

Achilleas Tsortos
Institute of Molecular Biology and Biotechnology, FORTH, and Department of Biology, University of Crete, Crete, Greece

Eishun Tsuchida
Research Institute for Science and Engineering, Waseda University, 3-4-1 Okubo, Shinjuku, Tokyo, Japan, and Waseda Bioscience Research Institute in Singapore, 11 Biopolis Way, Helios, Singapore

Kanta Tsumoto
Liposome Engineering Laboratory Inc., and Graduate School of Engineering, Mie University, Tsu, Mie, Japan

Patrick Vermette
Research Centre on Aging, Institut Universitaire de Gériatrie de Sherbrooke, and Laboratoire de Bioingénierie et de Biophysique de l'Université de Sherbrooke, Department of Chemical and Biotechnological Engineering, Université de Sherbrooke, Sherbrooke, Québec, Canada

Dali Yin
Department of Neurological Surgery, University of California San Francisco, San Francisco, California, USA

Tetsuro Yoshimura
Liposome Engineering Laboratory Inc., and Graduate School of Engineering, Mie University, Tsu, Mie; Nagoya Industrial Science Research Institute, Nagoya, Aichi, Japan

Bo Yu
NSF Nanoscale Science and Engineering Center (NSEC), and Department of Chemical and Biomolecular Engineering, The Ohio State University, Columbus, Ohio, USA

L. Zhao
Division of Plastic and Reconstructive Surgery, Department of Surgery, Stanford University School of Medicine, Stanford, California, USA

Shibin Zhou
The Ludwig Center for Cancer Genetics and Therapeutics, Johns Hopkins Kimmel Comprehensive Cancer Center, Baltimore, Maryland, USA

Alexandra Zidovska
Materials, Physics, and Molecular, Cellular and Developmental Biology Departments, University of California at Santa Barbara, Santa Barbara, California, USA

Alessia Zorzoli
Experimental Therapy Unit, Laboratory of Oncology, G. Gaslini Children's Hospital, Genoa, Italy

Preface

Previous Methods in Enzymology volumes on Liposomes have described methods of liposome preparation and the physicochemical characterization of liposomes (volume 367), and the use of liposomes in biochemistry, molecular cell biology (volume 372), immunology, diagnostics, gene delivery and gene therapy (volume 373). Methods involved in the production and application of antibody- or ligand-targeted liposomes, environment-sensitive liposomes, and liposomal oligonucleotides were provided in volume 387, as were methods for studying the *in vivo* fate of liposomes. Finally, volume 391 presented methods in liposomal anti-cancer, antibacterial, antifungal and antiviral agents, miscellaneous liposomal therapies and electron microscopy of liposomes. The latter volume also included a short introductory chapter on "The Origin of Liposomes: Alec Bangham at Babraham".

This new volume includes topics focusing on advances in liposome formation and characterization, and recent developments in the use of liposomes in therapeutics.

I hope that these chapters will be helpful to graduate students, postdoctoral fellows, research associates and established scientists initiating projects on liposomes, or shifting the focus of their research. Although the chapters are not written in "protocol" format, they describe the experimental methods in sufficient detail that can be adopted readily for the reader's project. In addition, the chapters provide the perspective of the authors on the field, as well as examples of results obtained with the described methods.

I would like to thank all the colleagues who graciously contributed to this volume with relatively short notice, Associate Editor Tara Hoey, Editorial Services Manager Delsy Retchagar, and Project Manager Priya Kumaraguruparan of Elsevier for their help in preparing this volume, and Shirley Light, formerly of Academic Press, for her initiation of the Liposomes volumes about a decade ago. I would also like to express my gratitude to my supportive and loving family, my wife, Diana, and my children, Avery and Maxine.

I would like to dedicate this book to the memory of my father, Orhan Düzgüneş, formerly Professor of Genetics and Statistics at the University of Ankara, and my mother, Zeliha Düzgüneş, formerly Professor of Entomology at the same university. I still appreciate their enthusiasm for their fields of study and their dedication to scientific research and teaching.

Nejat Düzgüneş

Methods in Enzymology

Volume XVI. Fast Reactions
Edited by Kenneth Kustin

Volume XVII. Metabolism of Amino Acids and Amines (Parts A and B)
Edited by Herbert Tabor and Celia White Tabor

Volume XVIII. Vitamins and Coenzymes (Parts A, B, and C)
Edited by Donald B. McCormick and Lemuel D. Wright

Volume XIX. Proteolytic Enzymes
Edited by Gertrude E. Perlmann and Laszlo Lorand

Volume XX. Nucleic Acids and Protein Synthesis (Part C)
Edited by Kivie Moldave and Lawrence Grossman

Volume XXI. Nucleic Acids (Part D)
Edited by Lawrence Grossman and Kivie Moldave

Volume XXII. Enzyme Purification and Related Techniques
Edited by William B. Jakoby

Volume XXIII. Photosynthesis (Part A)
Edited by Anthony San Pietro

Volume XXIV. Photosynthesis and Nitrogen Fixation (Part B)
Edited by Anthony San Pietro

Volume XXV. Enzyme Structure (Part B)
Edited by C. H. W. Hirs and Serge N. Timasheff

Volume XXVI. Enzyme Structure (Part C)
Edited by C. H. W. Hirs and Serge N. Timasheff

Volume XXVII. Enzyme Structure (Part D)
Edited by C. H. W. Hirs and Serge N. Timasheff

Volume XXVIII. Complex Carbohydrates (Part B)
Edited by Victor Ginsburg

Volume XXIX. Nucleic Acids and Protein Synthesis (Part E)
Edited by Lawrence Grossman and Kivie Moldave

Volume XXX. Nucleic Acids and Protein Synthesis (Part F)
Edited by Kivie Moldave and Lawrence Grossman

Volume XXXI. Biomembranes (Part A)
Edited by Sidney Fleischer and Lester Packer

Volume XXXII. Biomembranes (Part B)
Edited by Sidney Fleischer and Lester Packer

Volume XXXIII. Cumulative Subject Index Volumes I-XXX
Edited by Martha G. Dennis and Edward A. Dennis

Volume XXXIV. Affinity Techniques (Enzyme Purification: Part B)
Edited by William B. Jakoby and Meir Wilchek

Volume 89. Carbohydrate Metabolism (Part D)
Edited by Willis A. Wood

Volume 90. Carbohydrate Metabolism (Part E)
Edited by Willis A. Wood

Volume 91. Enzyme Structure (Part I)
Edited by C. H. W. Hirs and Serge N. Timasheff

Volume 92. Immunochemical Techniques (Part E: Monoclonal Antibodies and General Immunoassay Methods)
Edited by John J. Langone and Helen Van Vunakis

Volume 93. Immunochemical Techniques (Part F: Conventional Antibodies, Fc Receptors, and Cytotoxicity)
Edited by John J. Langone and Helen Van Vunakis

Volume 94. Polyamines
Edited by Herbert Tabor and Celia White Tabor

Volume 95. Cumulative Subject Index Volumes 61–74, 76–80
Edited by Edward A. Dennis and Martha G. Dennis

Volume 96. Biomembranes [Part J: Membrane Biogenesis: Assembly and Targeting (General Methods; Eukaryotes)]
Edited by Sidney Fleischer and Becca Fleischer

Volume 97. Biomembranes [Part K: Membrane Biogenesis: Assembly and Targeting (Prokaryotes, Mitochondria, and Chloroplasts)]
Edited by Sidney Fleischer and Becca Fleischer

Volume 98. Biomembranes (Part L: Membrane Biogenesis: Processing and Recycling)
Edited by Sidney Fleischer and Becca Fleischer

Volume 99. Hormone Action (Part F: Protein Kinases)
Edited by Jackie D. Corbin and Joel G. Hardman

Volume 100. Recombinant DNA (Part B)
Edited by Ray Wu, Lawrence Grossman, and Kivie Moldave

Volume 101. Recombinant DNA (Part C)
Edited by Ray Wu, Lawrence Grossman, and Kivie Moldave

Volume 102. Hormone Action (Part G: Calmodulin and Calcium-Binding Proteins)
Edited by Anthony R. Means and Bert W. O'Malley

Volume 103. Hormone Action (Part H: Neuroendocrine Peptides)
Edited by P. Michael Conn

Volume 104. Enzyme Purification and Related Techniques (Part C)
Edited by William B. Jakoby

VOLUME 123. Vitamins and Coenzymes (Part H)
Edited by FRANK CHYTIL AND DONALD B. MCCORMICK

VOLUME 124. Hormone Action (Part J: Neuroendocrine Peptides)
Edited by P. MICHAEL CONN

VOLUME 125. Biomembranes (Part M: Transport in Bacteria, Mitochondria, and Chloroplasts: General Approaches and Transport Systems)
Edited by SIDNEY FLEISCHER AND BECCA FLEISCHER

VOLUME 126. Biomembranes (Part N: Transport in Bacteria, Mitochondria, and Chloroplasts: Protonmotive Force)
Edited by SIDNEY FLEISCHER AND BECCA FLEISCHER

VOLUME 127. Biomembranes (Part O: Protons and Water: Structure and Translocation)
Edited by LESTER PACKER

VOLUME 128. Plasma Lipoproteins (Part A: Preparation, Structure, and Molecular Biology)
Edited by JERE P. SEGREST AND JOHN J. ALBERS

VOLUME 129. Plasma Lipoproteins (Part B: Characterization, Cell Biology, and Metabolism)
Edited by JOHN J. ALBERS AND JERE P. SEGREST

VOLUME 130. Enzyme Structure (Part K)
Edited by C. H. W. HIRS AND SERGE N. TIMASHEFF

VOLUME 131. Enzyme Structure (Part L)
Edited by C. H. W. HIRS AND SERGE N. TIMASHEFF

VOLUME 132. Immunochemical Techniques (Part J: Phagocytosis and Cell-Mediated Cytotoxicity)
Edited by GIOVANNI DI SABATO AND JOHANNES EVERSE

VOLUME 133. Bioluminescence and Chemiluminescence (Part B)
Edited by MARLENE DELUCA AND WILLIAM D. MCELROY

VOLUME 134. Structural and Contractile Proteins (Part C: The Contractile Apparatus and the Cytoskeleton)
Edited by RICHARD B. VALLEE

VOLUME 135. Immobilized Enzymes and Cells (Part B)
Edited by KLAUS MOSBACH

VOLUME 136. Immobilized Enzymes and Cells (Part C)
Edited by KLAUS MOSBACH

VOLUME 137. Immobilized Enzymes and Cells (Part D)
Edited by KLAUS MOSBACH

VOLUME 138. Complex Carbohydrates (Part E)
Edited by VICTOR GINSBURG

Volume 157. Biomembranes (Part Q: ATP-Driven Pumps and Related Transport: Calcium, Proton, and Potassium Pumps)
Edited by Sidney Fleischer and Becca Fleischer

Volume 158. Metalloproteins (Part A)
Edited by James F. Riordan and Bert L. Vallee

Volume 159. Initiation and Termination of Cyclic Nucleotide Action
Edited by Jackie D. Corbin and Roger A. Johnson

Volume 160. Biomass (Part A: Cellulose and Hemicellulose)
Edited by Willis A. Wood and Scott T. Kellogg

Volume 161. Biomass (Part B: Lignin, Pectin, and Chitin)
Edited by Willis A. Wood and Scott T. Kellogg

Volume 162. Immunochemical Techniques (Part L: Chemotaxis and Inflammation)
Edited by Giovanni Di Sabato

Volume 163. Immunochemical Techniques (Part M: Chemotaxis and Inflammation)
Edited by Giovanni Di Sabato

Volume 164. Ribosomes
Edited by Harry F. Noller, Jr., and Kivie Moldave

Volume 165. Microbial Toxins: Tools for Enzymology
Edited by Sidney Harshman

Volume 166. Branched-Chain Amino Acids
Edited by Robert Harris and John R. Sokatch

Volume 167. Cyanobacteria
Edited by Lester Packer and Alexander N. Glazer

Volume 168. Hormone Action (Part K: Neuroendocrine Peptides)
Edited by P. Michael Conn

Volume 169. Platelets: Receptors, Adhesion, Secretion (Part A)
Edited by Jacek Hawiger

Volume 170. Nucleosomes
Edited by Paul M. Wassarman and Roger D. Kornberg

Volume 171. Biomembranes (Part R: Transport Theory: Cells and Model Membranes)
Edited by Sidney Fleischer and Becca Fleischer

Volume 172. Biomembranes (Part S: Transport: Membrane Isolation and Characterization)
Edited by Sidney Fleischer and Becca Fleischer

Volume 173. Biomembranes [Part T: Cellular and Subcellular Transport: Eukaryotic (Nonepithelial) Cells]
Edited by Sidney Fleischer and Becca Fleischer

Volume 174. Biomembranes [Part U: Cellular and Subcellular Transport: Eukaryotic (Nonepithelial) Cells]
Edited by Sidney Fleischer and Becca Fleischer

Volume 175. Cumulative Subject Index Volumes 135–139, 141–167

Volume 176. Nuclear Magnetic Resonance (Part A: Spectral Techniques and Dynamics)
Edited by Norman J. Oppenheimer and Thomas L. James

Volume 177. Nuclear Magnetic Resonance (Part B: Structure and Mechanism)
Edited by Norman J. Oppenheimer and Thomas L. James

Volume 178. Antibodies, Antigens, and Molecular Mimicry
Edited by John J. Langone

Volume 179. Complex Carbohydrates (Part F)
Edited by Victor Ginsburg

Volume 180. RNA Processing (Part A: General Methods)
Edited by James E. Dahlberg and John N. Abelson

Volume 181. RNA Processing (Part B: Specific Methods)
Edited by James E. Dahlberg and John N. Abelson

Volume 182. Guide to Protein Purification
Edited by Murray P. Deutscher

Volume 183. Molecular Evolution: Computer Analysis of Protein and Nucleic Acid Sequences
Edited by Russell F. Doolittle

Volume 184. Avidin-Biotin Technology
Edited by Meir Wilchek and Edward A. Bayer

Volume 185. Gene Expression Technology
Edited by David V. Goeddel

Volume 186. Oxygen Radicals in Biological Systems (Part B: Oxygen Radicals and Antioxidants)
Edited by Lester Packer and Alexander N. Glazer

Volume 187. Arachidonate Related Lipid Mediators
Edited by Robert C. Murphy and Frank A. Fitzpatrick

Volume 188. Hydrocarbons and Methylotrophy
Edited by Mary E. Lidstrom

Volume 189. Retinoids (Part A: Molecular and Metabolic Aspects)
Edited by Lester Packer

Volume 190. Retinoids (Part B: Cell Differentiation and Clinical Applications)
Edited by Lester Packer

Volume 191. Biomembranes (Part V: Cellular and Subcellular Transport: Epithelial Cells)
Edited by Sidney Fleischer and Becca Fleischer

Volume 192. Biomembranes (Part W: Cellular and Subcellular Transport: Epithelial Cells)
Edited by Sidney Fleischer and Becca Fleischer

Volume 193. Mass Spectrometry
Edited by James A. McCloskey

Volume 194. Guide to Yeast Genetics and Molecular Biology
Edited by Christine Guthrie and Gerald R. Fink

Volume 195. Adenylyl Cyclase, G Proteins, and Guanylyl Cyclase
Edited by Roger A. Johnson and Jackie D. Corbin

Volume 196. Molecular Motors and the Cytoskeleton
Edited by Richard B. Vallee

Volume 197. Phospholipases
Edited by Edward A. Dennis

Volume 198. Peptide Growth Factors (Part C)
Edited by David Barnes, J. P. Mather, and Gordon H. Sato

Volume 199. Cumulative Subject Index Volumes 168–174, 176–194

Volume 200. Protein Phosphorylation (Part A: Protein Kinases: Assays, Purification, Antibodies, Functional Analysis, Cloning, and Expression)
Edited by Tony Hunter and Bartholomew M. Sefton

Volume 201. Protein Phosphorylation (Part B: Analysis of Protein Phosphorylation, Protein Kinase Inhibitors, and Protein Phosphatases)
Edited by Tony Hunter and Bartholomew M. Sefton

Volume 202. Molecular Design and Modeling: Concepts and Applications (Part A: Proteins, Peptides, and Enzymes)
Edited by John J. Langone

Volume 203. Molecular Design and Modeling: Concepts and Applications (Part B: Antibodies and Antigens, Nucleic Acids, Polysaccharides, and Drugs)
Edited by John J. Langone

Volume 204. Bacterial Genetic Systems
Edited by Jeffrey H. Miller

Volume 205. Metallobiochemistry (Part B: Metallothionein and Related Molecules)
Edited by James F. Riordan and Bert L. Vallee

Volume 206. Cytochrome P450
Edited by Michael R. Waterman and Eric F. Johnson

Volume 207. Ion Channels
Edited by Bernardo Rudy and Linda E. Iverson

Volume 208. Protein–DNA Interactions
Edited by Robert T. Sauer

Volume 209. Phospholipid Biosynthesis
Edited by Edward A. Dennis and Dennis E. Vance

Volume 210. Numerical Computer Methods
Edited by Ludwig Brand and Michael L. Johnson

Volume 211. DNA Structures (Part A: Synthesis and Physical Analysis of DNA)
Edited by David M. J. Lilley and James E. Dahlberg

Volume 212. DNA Structures (Part B: Chemical and Electrophoretic Analysis of DNA)
Edited by David M. J. Lilley and James E. Dahlberg

Volume 213. Carotenoids (Part A: Chemistry, Separation, Quantitation, and Antioxidation)
Edited by Lester Packer

Volume 214. Carotenoids (Part B: Metabolism, Genetics, and Biosynthesis)
Edited by Lester Packer

Volume 215. Platelets: Receptors, Adhesion, Secretion (Part B)
Edited by Jacek J. Hawiger

Volume 216. Recombinant DNA (Part G)
Edited by Ray Wu

Volume 217. Recombinant DNA (Part H)
Edited by Ray Wu

Volume 218. Recombinant DNA (Part I)
Edited by Ray Wu

Volume 219. Reconstitution of Intracellular Transport
Edited by James E. Rothman

Volume 220. Membrane Fusion Techniques (Part A)
Edited by Nejat Düzgüneş

Volume 221. Membrane Fusion Techniques (Part B)
Edited by Nejat Düzgüneş

Volume 222. Proteolytic Enzymes in Coagulation, Fibrinolysis, and Complement Activation (Part A: Mammalian Blood Coagulation Factors and Inhibitors)
Edited by Laszlo Lorand and Kenneth G. Mann

Volume 223. Proteolytic Enzymes in Coagulation, Fibrinolysis, and Complement Activation (Part B: Complement Activation, Fibrinolysis, and Nonmammalian Blood Coagulation Factors)
Edited by Laszlo Lorand and Kenneth G. Mann

Volume 224. Molecular Evolution: Producing the Biochemical Data
Edited by Elizabeth Anne Zimmer, Thomas J. White, Rebecca L. Cann, and Allan C. Wilson

Volume 225. Guide to Techniques in Mouse Development
Edited by Paul M. Wassarman and Melvin L. DePamphilis

Volume 226. Metallobiochemistry (Part C: Spectroscopic and Physical Methods for Probing Metal Ion Environments in Metalloenzymes and Metalloproteins)
Edited by James F. Riordan and Bert L. Vallee

Volume 227. Metallobiochemistry (Part D: Physical and Spectroscopic Methods for Probing Metal Ion Environments in Metalloproteins)
Edited by James F. Riordan and Bert L. Vallee

Volume 228. Aqueous Two-Phase Systems
Edited by Harry Walter and Göte Johansson

Volume 229. Cumulative Subject Index Volumes 195–198, 200–227

Volume 230. Guide to Techniques in Glycobiology
Edited by William J. Lennarz and Gerald W. Hart

Volume 231. Hemoglobins (Part B: Biochemical and Analytical Methods)
Edited by Johannes Everse, Kim D. Vandegriff, and Robert M. Winslow

Volume 232. Hemoglobins (Part C: Biophysical Methods)
Edited by Johannes Everse, Kim D. Vandegriff, and Robert M. Winslow

Volume 233. Oxygen Radicals in Biological Systems (Part C)
Edited by Lester Packer

Volume 234. Oxygen Radicals in Biological Systems (Part D)
Edited by Lester Packer

Volume 235. Bacterial Pathogenesis (Part A: Identification and Regulation of Virulence Factors)
Edited by Virginia L. Clark and Patrik M. Bavoil

Volume 236. Bacterial Pathogenesis (Part B: Integration of Pathogenic Bacteria with Host Cells)
Edited by Virginia L. Clark and Patrik M. Bavoil

Volume 237. Heterotrimeric G Proteins
Edited by Ravi Iyengar

Volume 238. Heterotrimeric G-Protein Effectors
Edited by Ravi Iyengar

Volume 257. Small GTPases and Their Regulators (Part C: Proteins Involved in Transport)
Edited by W. E. Balch, Channing J. Der, and Alan Hall

Volume 258. Redox-Active Amino Acids in Biology
Edited by Judith P. Klinman

Volume 259. Energetics of Biological Macromolecules
Edited by Michael L. Johnson and Gary K. Ackers

Volume 260. Mitochondrial Biogenesis and Genetics (Part A)
Edited by Giuseppe M. Attardi and Anne Chomyn

Volume 261. Nuclear Magnetic Resonance and Nucleic Acids
Edited by Thomas L. James

Volume 262. DNA Replication
Edited by Judith L. Campbell

Volume 263. Plasma Lipoproteins (Part C: Quantitation)
Edited by William A. Bradley, Sandra H. Gianturco, and Jere P. Segrest

Volume 264. Mitochondrial Biogenesis and Genetics (Part B)
Edited by Giuseppe M. Attardi and Anne Chomyn

Volume 265. Cumulative Subject Index Volumes 228, 230–262

Volume 266. Computer Methods for Macromolecular Sequence Analysis
Edited by Russell F. Doolittle

Volume 267. Combinatorial Chemistry
Edited by John N. Abelson

Volume 268. Nitric Oxide (Part A: Sources and Detection of NO; NO Synthase)
Edited by Lester Packer

Volume 269. Nitric Oxide (Part B: Physiological and Pathological Processes)
Edited by Lester Packer

Volume 270. High Resolution Separation and Analysis of Biological Macromolecules (Part A: Fundamentals)
Edited by Barry L. Karger and William S. Hancock

Volume 271. High Resolution Separation and Analysis of Biological Macromolecules (Part B: Applications)
Edited by Barry L. Karger and William S. Hancock

Volume 272. Cytochrome P450 (Part B)
Edited by Eric F. Johnson and Michael R. Waterman

Volume 273. RNA Polymerase and Associated Factors (Part A)
Edited by Sankar Adhya

Volume 274. RNA Polymerase and Associated Factors (Part B)
Edited by Sankar Adhya

VOLUME 294. Ion Channels (Part C)
Edited by P. MICHAEL CONN

VOLUME 295. Energetics of Biological Macromolecules (Part B)
Edited by GARY K. ACKERS AND MICHAEL L. JOHNSON

VOLUME 296. Neurotransmitter Transporters
Edited by SUSAN G. AMARA

VOLUME 297. Photosynthesis: Molecular Biology of Energy Capture
Edited by LEE MCINTOSH

VOLUME 298. Molecular Motors and the Cytoskeleton (Part B)
Edited by RICHARD B. VALLEE

VOLUME 299. Oxidants and Antioxidants (Part A)
Edited by LESTER PACKER

VOLUME 300. Oxidants and Antioxidants (Part B)
Edited by LESTER PACKER

VOLUME 301. Nitric Oxide: Biological and Antioxidant Activities (Part C)
Edited by LESTER PACKER

VOLUME 302. Green Fluorescent Protein
Edited by P. MICHAEL CONN

VOLUME 303. cDNA Preparation and Display
Edited by SHERMAN M. WEISSMAN

VOLUME 304. Chromatin
Edited by PAUL M. WASSARMAN AND ALAN P. WOLFFE

VOLUME 305. Bioluminescence and Chemiluminescence (Part C)
Edited by THOMAS O. BALDWIN AND MIRIAM M. ZIEGLER

VOLUME 306. Expression of Recombinant Genes in Eukaryotic Systems
Edited by JOSEPH C. GLORIOSO AND MARTIN C. SCHMIDT

VOLUME 307. Confocal Microscopy
Edited by P. MICHAEL CONN

VOLUME 308. Enzyme Kinetics and Mechanism (Part E: Energetics of Enzyme Catalysis)
Edited by DANIEL L. PURICH AND VERN L. SCHRAMM

VOLUME 309. Amyloid, Prions, and Other Protein Aggregates
Edited by RONALD WETZEL

VOLUME 310. Biofilms
Edited by RON J. DOYLE

VOLUME 311. Sphingolipid Metabolism and Cell Signaling (Part A)
Edited by ALFRED H. MERRILL, JR., AND YUSUF A. HANNUN

Volume 436. Globins and Other Nitric Oxide-Reactive Protiens (Part A)
Edited by Robert K. Poole

Volume 437. Globins and Other Nitric Oxide-Reactive Protiens (Part B)
Edited by Robert K. Poole

Volume 438. Small GTPases in Disease (Part A)
Edited by William E. Balch, Channing J. Der, and Alan Hall

Volume 439. Small GTPases in Disease (Part B)
Edited by William E. Balch, Channing J. Der, and Alan Hall

Volume 440. Nitric Oxide, Part F Oxidative and Nitrosative Stress in Redox Regulation of Cell Signaling
Edited by Enrique Cadenas and Lester Packer

Volume 441. Nitric Oxide, Part G Oxidative and Nitrosative Stress in Redox Regulation of Cell Signaling
Edited by Enrique Cadenas and Lester Packer

Volume 442. Programmed Cell Death, General Principles for Studying Cell Death (Part A)
Edited by Roya Khosravi-Far, Zahra Zakeri, Richard A. Lockshin, and Mauro Piacentini

Volume 443. Angiogenesis: *In Vitro* Systems
Edited by David A. Cheresh

Volume 444. Angiogenesis: *In Vivo* Systems (Part A)
Edited by David A. Cheresh

Volume 445. Angiogenesis: *In Vivo* Systems (Part B)
Edited by David A. Cheresh

Volume 446. Programmed Cell Death, The Biology and Therapeutic Implications of Cell Death (Part B)
Edited by Roya Khosravi-Far, Zahra Zakeri, Richard A. Lockshin, and Mauro Piacentini

Volume 447. RNA Turnover in Bacteria, Archaea and Organelles
Edited by Lynne E. Maquat and Cecilia M. Arraiano

Volume 448. RNA Turnover in Eukaryotes: Nucleases, Pathways and Analysis of mRNA Decay
Edited by Lynne E. Maquat and Megerditch Kiledjian

Volume 449. RNA Turnover in Eukaryotes: Analysis of Specialized and Quality Control RNA Decay Pathways
Edited by Lynne E. Maquat and Megerditch Kiledjian

Volume 450. Fluorescence Spectroscopy
Edited by Ludwig Brand and Michael L. Johnson

SECTION ONE

ADVANCES IN LIPOSOME FORMATION AND CHARACTERIZATION

CHAPTER ONE

Spontaneously Formed Unilamellar Vesicles

Mu-Ping Nieh,* Norbert Kučerka,*,† *and* John Katsaras*,‡,§

Contents

Abstract

Mixtures of long- and short-chain phospholipids can spontaneously form uniform unilamellar vesicles (ULVs) with diameters 50 nm (polydispersities of <0.3) or less. The morphology of these ULVs has mainly been characterized using small angle neutron scattering (SANS), a technique highly suited for the study of hydrogenous materials. Once formed, these ULVs have turned out to be highly stable and show great promise as imaging and therapeutic carriers.

* Canadian Neutron Beam Centre, Steacie Institute for Molecular Sciences, Chalk River Laboratories, National Research Council Canada, Chalk River, Ontario, Canada
† Department of Physical Chemistry of Drugs, Faculty of Pharmacy, Comenius University, Bratislava, Slovakia
‡ Department of Physics, Brock University, St. Catharines, Ontario, Canada
§ Guelph-Waterloo Physics Institute and Biophysics Interdepartmental Group, University of Guelph, Guelph, Ontario, Canada

Methods in Enzymology, Volume 465

ISSN 0076-6879, DOI: 10.1016/S0076-6879(09)65001-1

1. Introduction

Liposomes composed of phospholipids have proven to be highly effective in encapsulating therapeutic and diagnostic molecules (Hofheinz *et al.*, 2005; Zhang *et al.*, 2008), and for targeting specific sites of disease when functionalized with antibodies (Emerich and Thanos, 2007; Simone *et al.*, 2009; Torchilin, 2007). Liposomes possess many advantages including high loading capacities, the ability to entrap both hydrophilic and hydrophobic molecules, and biocompatibility (Lasic, 1998). The permeability of a phospholipid bilayer to small molecules is maximal at its phase-transition temperature, T_M (i.e., gel-to-liquid crystalline L_α phase), a desired condition for the release of encapsulated materials (Hays *et al.*, 2001; Inoue, 1974; Papahadjopoulos *et al.*, 1973; Yatvin *et al.*, 1978). There have been many studies devoted to controlling the transition temperature of liposomes through additives, such as cholesterol (Kraske and Mountcastle, 2001; Papahadjopoulos *et al.*, 1972; Sujatha and Mishra, 1998) and alcohols (McIntosh *et al.*, 1983; Rowe, 1983, 1985; Simon and McIntosh, 1984), to name a few.

Some zwitterionic phospholipids have a T_M within the physiological range of temperatures, but they naturally form large multilamellar vesicles (MLVs). Compared to unilamellar vesicles (ULVs), when introduced inside the human body they generally exhibit relatively short circulation half-lives, a property that can be controlled, for example, by varying the ULV size (Juliano and Stamp, 1975; Nabar and Nadkarni, 1998; Van Borssum Waalkes *et al.*, 1993; Zou *et al.*, 1995). ULVs have traditionally been formed from MLVs being sonicated or extruded through ceramic filters of specific pore size. Generally speaking, ULVs produced by sonication are polydisperse (different sizes), while extrusion methods, although capable of producing uniform size ULVs, is a labor intensive method that can be problematic for the production of small ULVs (<50 nm diameter) in quantities desired by industry, as the ceramic filters are easily fouled. As the size of ULVs is dictated by the filter's pore size, the smallest size ULVs that can routinely be produced by extrusion methods is ~40 nm in diameter. It is well known that ULVs larger than 50 nm in diameter have short circulation half-lives and tend to quickly accumulate in the liver and spleen (Allen *et al.*, 1989; Gregoriadis, 1995; Oku, 1999). Spontaneously formed ULVs provide the possibility of resolving some, or all of the aforementioned issues.

Self-assembled ULVs made of surfactants were first reported nearly two decades ago (Kaler *et al.*, 1989; Safran *et al.*, 1990; Talmon *et al.*, 1983). ULVs have been produced by mixing cationic and anionic surfactants (Iampietro and Kaler, 1999; Kaler *et al.*, 1992; Marques, 2000; Yatcilla *et al.*, 1996) through dilution (Caria and Khan, 1996; Demé *et al.*, 2002;

O'Connor *et al.*, 1997; Villeneuve *et al.*, 1999), and rapid changes in temperature (Lesieur *et al.*, 2000; Nieh *et al.*, 2001, 2004). However, the fact that the size of these ULVs was directly related to surfactant concentration (Bergström and Pedersen, 2000; Bergström *et al.*, 1999; Egelhaaf and Schurtenberger, 1999; Leng *et al.*, 2003; Oberdisse and Porte, 1997; Schurtenberger *et al.*, 1985) implied that leakage could be problematic during the fusion or fission of these ULVs. The use of these concentration-dependent ULVs is thus limited to situations where the environment, is, for the most part, reasonably stable. It was not until recently that we reported on a self-assembled ULV system composed entirely of phospholipids and whose size was independent of lipid concentration (Nieh *et al.*, 2003, 2004), greatly expanding the possible applications of ULVs. Over the years, we have conducted small angle neutron scattering (SANS) experiments to characterize the size and polydispersity of spontaneously forming ULVs, and found their average size to be less than 50 nm in diameter, with corresponding polydispersities of less than 0.3. It was also determined that their morphology is robust and not easily altered by additives (Nieh *et al.*, 2006). As a result, they can readily accommodate a range of amphiphilic molecules, making them highly desirable as contrast imaging and drug delivery vehicles. Importantly, the size of self-assembled ULVs can be controlled through the judicious use of long- to short-chain lipid mixing ratios, bilayer rigidity (e.g., inclusion of cholesterol), and charge density.

2. Preparation of Spontaneously Forming ULVs

Generally speaking, spontaneously formed ULVs are composed of neutral long-chain [e.g., dimyristoyl phosphatidylcholine (di-14:0, DMPC); however, ditridecanoyl (di-13:0, DTPC) or dipalmitoyl phosphatidylcholine (di-16:0, DPPC) can also be used] and short-chain [e.g., dihexanoyl phosphatidylcholine (di-6:0, DHPC)] lipids, and a long-chain charged lipid (e.g., dimyristoyl phosphatidylglycerol, DMPG). A mixture of DMPC/DHPC/DMPG (e.g., molar ratio of 3.2:1:0.04) dissolves in water (or appropriate buffer) with an initial total lipid concentration of 20 wt% through successive vortexing and temperature cycling between 4 and 60 °C, until the solution is transparent at 4 °C. The transparent solution can then be progressively diluted to 10, 5, and finally 2.0 wt% total lipid concentration, keeping in mind to vortex and temperature cycle at each lipid concentration. It is important that the 2 wt% sample be kept at 4 °C prior to any further dilution, and then only diluted with 4 °C water. The morphology at these low-total lipid concentrations is one of the bilayered micelles (commonly known as bicelles), which transform into small monodisperse ULVs at temperatures >40 °C, as was determined by small angle neutron scattering (SANS) (Nieh *et al.*, 2003, 2004).

3. Characterization of ULVs

ULVs are best characterized using a combination of dynamic light scattering (DLS) and small angle X-ray or neutron scattering (SAXS or SANS). For example, DLS can provide information regarding the hydrodynamic radius, R_H, which in addition, includes all of the water molecules attached to the ULV, and is calculated from the diffusion of ULVs using the Stokes–Einstein relation, keeping in mind that it is possible for a small molecule to have a larger hydrodynamic radius than a large molecule if it is surrounded by a greater number of solvent molecules (Nieh *et al.*, 2004). However, the structure of ULVs is more precisely obtained through SANS (Nieh *et al.*, 2001, 2002) or SAXS (Lesieur *et al.*, 2000; Weiss *et al.*, 2005). Normally, SANS (or SAXS) data are plotted as a function of scattered intensity, I, versus scattering vector, q [$=(4\pi/\lambda)\sin(\theta/2)$, where λ and θ are the wavelength and the scattering angle, respectively]. The probing range of length scales thus lies between $2\pi/q_{\mathrm{min}}$ and $2\pi/q_{\mathrm{max}}$, where q_{min} and q_{max} are the attainable minimum and maximum q values, in the case of SANS typically between 0.003 and 0.5 Å^{-1} (i.e., between 2000 and 10 Å). With this sensitivity, SANS is ideally suited in determining the diameter (not the hydrodynamic radius as obtained from DLS) and shell thickness of ULVs.

Figure 1.1 illustrates typical SANS data arising from a low-polydispersity ULV sample using the 30 m SANS instrument located at the National Institute of Standards and Technology (NIST, USA) Center for Neutron

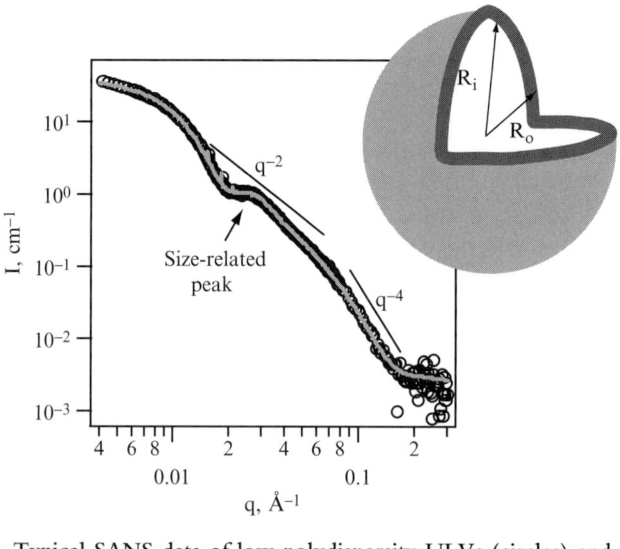

Figure 1.1 Typical SANS data of low-polydispersity ULVs (circles) and the best-fit curve (gray line) using the depicted spherical shell model.

Research (NCNR). The data are best fitted through the use of a polydisperse spherical shell model, as described in Eq. (1.1) (Feigin and Svergun, 1987) using the IGOR Pro software®. The data fitting procedure was developed by the NCNR (Kline, 2006).

SANS scattered intensity can be written as follows:

$$I_{\text{vesicle}}(q) = \frac{\phi_{\text{vesicle}}}{V_{\text{vesicle}}} \int_0^{\infty} f(r) A_{\text{o}}^2(qr)\,\mathrm{d}r, \tag{1.1}$$

where ϕ_{vesicle} and V_{vesicle} are the total volume fraction of ULVs, and the total volume occupied by an individual ULV, respectively. The amplitude of the form factor (spherical shell model), $A_{\text{o}}(qr)$, is given as

$$A_{\text{o}}(qr) = \frac{4\pi(\rho_{\text{lipid}} - \rho_{\text{solvent}})}{q^3}\left[\left(\sin q\frac{R_{\text{i}}}{R_{\text{o}}}r - \sin qr\right) - qr\left(\frac{R_{\text{i}}}{R_{\text{o}}}\cos q\frac{R_{\text{i}}}{R_{\text{o}}}r - \cos qr\right)\right]$$

$$f(r) = \frac{(p^{-2/p^2})(r/\langle R_{\text{o}}\rangle)^{(1-p^2)/p^2}\,\mathrm{e}^{-r/(p^2\langle R_{\text{o}}\rangle)}}{\langle R_{\text{o}}\rangle \Gamma(1/p^2)},$$

where R_{i}, R_{o}, ρ_{lipid} and ρ_{solvent} are the inner and outer ULV radii, and the coherent neutron scattering length densities of the lipid and the solvent, respectively. $f(r)$ is the Schulz distribution function describing the size distribution of ULV radii and p is the polydispersity of R_{o}, which is defined as $\sigma/\langle R_{\text{o}}\rangle$, where σ^2 is the variance of R_{o}. $\langle R_{\text{o}}\rangle$ represents the average R_{o}. The scattering pattern from small, uniform ULVs generally exhibits the following two features when plotted on a log–log graph (Fig. 1.1):

1. SANS data at q less than $\sim$0.01 (Å^{-1}) reveal the average ULV size (also known as the "Guinier" regime), while data at q greater than $\sim$0.01 (Å^{-1})—excluding oscillations—follow a q^{-2} dependence, indicative of a planar structure (i.e., the ULV's shell). At the high q regime ($q > 0.1$ Å^{-1}), the intensity exhibits a q^{-4} dependence corresponding to the interfacial scattering from the lipid bilayer (Porod's law).
2. The broad peak appearing along the SANS curve ($q \sim 0.25$ Å^{-1} in Fig. 1.1) is a measure of ULV size, while the number of oscillations is a good indicator of the range of ULV sizes (i.e., more oscillations translate to a narrower size distribution or low polydispersity).

4. ULV Stability

One important characteristic of a system's suitability as a potential drug delivery carrier is its stability under variable conditions, for example, changes in concentration and/or temperature. As mentioned, although surfactant-based self-assembled ULVs can exhibit low polydispersities (Oberdisse and Porte, 1997; Schurtenberger *et al.*, 1985), their size is for the most part, concentration dependent, with ULV fusion and fission taking place with changes in total surfactant concentration. Table 1.1 shows that spontaneously formed ULVs are extremely stable when they are diluted at a temperature of 45 °C and exhibit polydispersities ranging from 0.15 to 0.23 (Nieh *et al.*, 2003)—it is believed that the small decrease in ULV radii after 2 weeks is due to lipid degradation.

5. Parameters Affecting ULVs

5.1. The path of formation

Phase diagrams of DMPC/DMPG/DHPC phospholipid mixtures have previously been constructed (Katsaras *et al.*, 2005; Nieh *et al.*, 2004; Yue *et al.*, 2005), as shown in Figure 1.2. From these phase diagrams it is clear that ULVs exist at temperatures $\geq$35 °C and at lipid concentrations $\leq$1.25 wt%. Moreover, it was determined that low-polydispersity ULVs could only be formed from low temperature monodisperse bicelles, whereby DMPC undergoes a gel to the liquid crystalline (L_α) phase transition. Importantly, it is believed that bicelles dictate ULV size (discussed in a later section)—diluting a high-concentration lipid mixture at a temperature beyond DMPC's T_M results in polydisperse ULVs (Nieh *et al.*, 2005).

5.2. Charge density

Charge density plays an important role in the formation of spontaneously formed ULVs. Figure 1.3 shows SANS data from three lipid mixtures with different charge densities, R (defined as [DMPG]/[DMPC]) of 0, 0.01, and 0.67, respectively. ULVs were found to form at $R = 0.01$. In the case of the neutral system, the spherical shell model was able to describe most of data except for the peak at $\sim$0.1 Å^{-1}, a signature of DMPC MLVs (Hui and He, 1983; Janiak *et al.*, 1976), thus implying the coexistence of ULVs and MLVs. As for the highly charged lipid mixture ($R = 0.067$), the bilayered micelle morphology persisted throughout the temperature range studied (Nieh *et al.*, 2002).

These results demonstrate that there is an optimal charge density for the formation of ULVs, and that the different morphologies are highly dependent on the delicate balance of interactions between membrane fluctuations and the various Coulombic repulsive, van der Waal, and hydration forces (Pozo-Navas *et al.*, 2003).

Table 1.1 ULV $\langle R_o \rangle$ and polydispersities as a function of lipid concentration at 45 °C

Lipid concentration	1.0 wt%	0.5 wt%			0.25 wt%
Duration	5 h	5 h	4 days	2 weeks	5 h
$\langle R_o \rangle$ (Å)	378 ± 20	378 ± 15	382 ± 25	332 ± 20	363 ± 15
Polydispersity (p)	0.23 ± 0.10	0.16 ± 0.05	0.15 ± 0.05	0.19 ± 0.08	0.22 ± 0.08

1 wt% samples were diluted at room temperature.

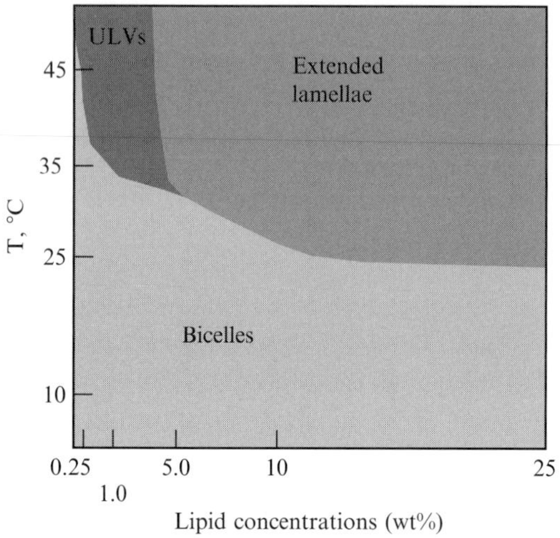

Figure 1.2 Structural phase diagram constructed by SANS. (See Color Insert.)

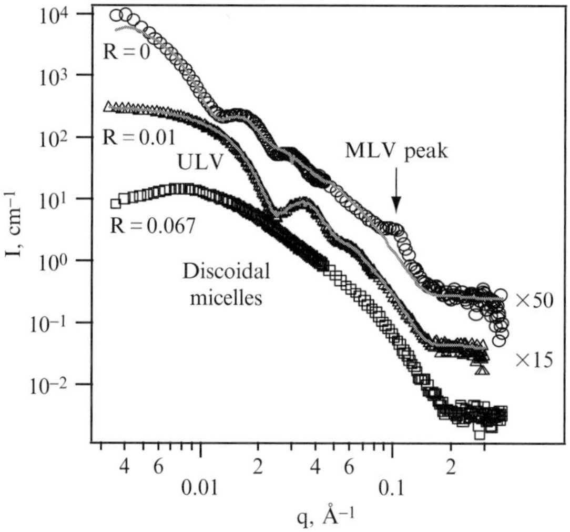

Figure 1.3 SANS data from mixtures of 0.25 wt% DMPC/DHPC/DMPG at 45 °C, and *R* ([DMPG]/[DMPC]) equal 0 (circles), 0.01 (triangles), and 0.067 (squares). The gray lines are best-fits to the data using the spherical shell model form factor.

5.3. The effect of long- to short-chain lipid molar ratios on morphology

The effect of Q (=[DMPC]/[DHPC]) on ULV size has been studied by Nieh *et al.* (to be published). Over a Q range between 2.5 and 4, all ULVs exhibited relatively low polydispersities (i.e., $0.23 \leq p \leq 0.31$), with the

Table 1.2 $\langle R_i \rangle$ and ULV polydispersities as a function of Q

Long- to short-chain lipid molar ratio (Q)	$\langle R_i \rangle$ (Å)	Polydispersity (p)
2.5	190	0.31
3.0	168	0.28
3.5	133	0.26
4.0	117	0.23

average ULV R_i decreasing from 190 to 117 Å (Table 1.2). This Q dependence of ULV radii is believed to be highly correlated with bicelle size, from which ULVs are formed. At higher Q values, because there are insufficient amounts of DHPC coating the bicelle rim, bicelles fold into ULVs earlier, resulting in much smaller ULVs.

5.4. Initial lipid concentration

As was mentioned, concentration-independent spontaneously formed ULVs were diluted at room temperature; however, their final size was found to vary depending on the path of formation. For example, it is known that ULV size strongly depends on initial total lipid concentration at low temperatures where bicelles are present. Increasing temperature resulted in the formation of ULVs with $\langle R_i \rangle$ of ~180 and ~80 Å for total lipid concentrations of 1.0 and 0.1 wt%, respectively (Nieh *et al.*, 2008). Once formed, the size of these ULVs was not affected by further dilutions (Nieh *et al.*, 2003). An important note is that bicelle size (i.e., bicelle concentration) dictates ULV size.

5.5. Chain length of the long-chain lipid

DMPC is the most commonly used long-chain phospholipid in preparing spontaneously formed ULVs. However, the longer chain lipid DPPC or the shorter DTPC can be used instead of DMPC to form low-polydispersity ULVs. The average ULV size for $Q = 3.2$ and 0.1 wt% lipid concentration decreased slightly as a function of increasing chain length, that is, $\langle R_i \rangle = 112$, 97, and 95 Å in the case of DTPC, DMPC, and DPPC, respectively (to be published). This dependence can be rationalized in the manner by which bicelles fold onto themselves to form ULVs (see later section).

5.6. Membrane rigidity

It is well known that cholesterol increases membrane rigidity (Boggs and Hsia, 1972; Mendelsohn, 1972). Cholesterol was used to modify the rigidity of bicelles from which highly stable ULVs can be obtained up to a cholesterol concentration of 20 mol%. Preliminary SANS data also indicates that

Table 1.3 Parameters affecting the size of self-assembled ULVs

Parameters	Effects
Path	Polydispersity
Charge density	Lamellarity
Q, long- to short-chain molar ratio	−−
Initial lipid concentration	++
Chain length of long-chain lipid	−
Membrane rigidity	+

The symbols "+" and "−" represent increasing or decreasing ULV size, respectively. Their numbers are an indicator of their intensity.

$\langle R_i \rangle$ of spontaneously formed ULVs increases with increasing cholesterol content (to be published). Table 1.3 summarizes the various parameters studied. It seems that initial lipid concentration and [DMPC]/[DHPC] molar ratio have the most profound effect on ULV size, affecting $\langle R_i \rangle$ by more than 100%. However, besides the parameters listed in Table 1.3, another important factor that may affect ULV size is the solution's salinity (Yue *et al.*, 2005).

6. Mechanism of ULV Formation

There are a few theories purporting to explain the formation of spontaneously formed ULVs. These include a negative modulus of Gaussian curvature as a result of charged bilayers (Winterhalter and Helfrich, 1992), a nonzero spontaneous curvature from the unequal distribution of the two molecular species making up the inner and outer bilayer leaflets (Safran *et al.*, 1991), a calculation based on a molecular thermodynamics model (Yuet and Blankschtein, 1996), and a kinetically trapped model of disk-like micelles transforming into vesicles (Leng *et al.*, 2003). However, experimental support for these various theories is mostly based on particular systems, where formation mechanisms can vary greatly. For the most part, the results discussed in this chapter are best described by the model proposed by Fromherz (1983), whereby the transformation of bicelles to ULVs is driven by the line tension at the bicelle's rim (shown in Fig. 1.4).

Bicelles at low temperature are thought to have the long-chain DMPC lipid residing in their planar region and the short-chain DHPC lipid predominantly located at their rim—thus minimizing the energy penalty arising from the high curvature at the disk's rim and possible exposure of DMPC's hydrophobic chains to water. Segregation between the two lipid species is most likely the direct result of their immiscibility, as DMPC is in the gel phase and DHPC is in the L_α phase. Around 24 °C, DMPC's acyl

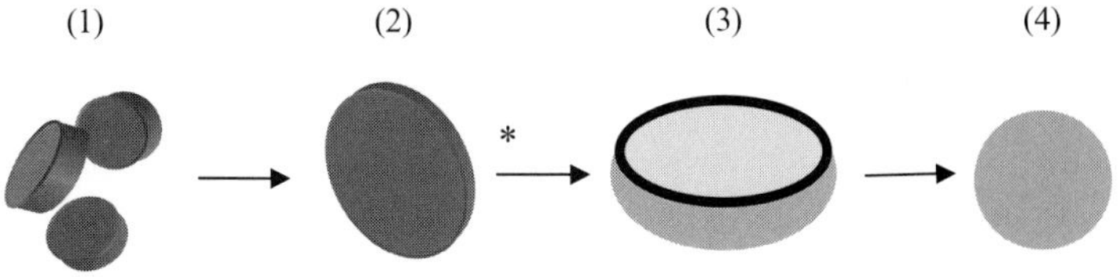

Figure 1.4 Proposed mechanism for the formation of monodisperse spontaneously formed ULVs from bicelles. For the most part, it is believed that DHPC molecules populate the bicelle's rim, while DMPC are found in the bicelle's planar region, see Fromherz (1983).

chains melt resulting in its increased miscibility with DHPC. As a result, some of the DHPC partitions into the bicelle's DMPC-rich planar region causing the rim's line tension to increase. At some point, when enough DHPC has partitioned into the bicelle's planar region, in order to prevent the bicelle's hydrophobic core from becoming exposed to water, the bicelles fuse forming larger bicelles. At some point (yet to be understood), the larger bicelles begin to fold, at first forming a bowl and eventually a hollow sphere (ULV).

The proposed model suggests that the ULV size-determining stage is when the larger bicelles begin to fold. This implies that a more flexible membrane should yield smaller ULVs consistent with what was observed in the case of cholesterol-doped systems. However, membrane flexibility is not the only factor influencing ULV size. Another key factor is bicelle stability, whereby more stable bicelles can have more time to interact and coalesce with neighboring bicelles. This is reflected by the fact that larger ULVs were formed at lower Q values, as bicelles are more stable with higher amounts of DHPC. Nevertheless, the fact that different long-chain lipids resulted in a small decrease in ULV size—as their hydrocarbon chain length was increased (i.e., DTPC, DMPC, and DPPC)—can be rationalized as follows: Longer hydrocarbon chains translate into stiffer bilayers and higher transition temperatures stabilizing the bicelle morphology. On the other hand, in the case of thicker bilayers (i.e., longer chain lipids), more DHPC is required to stabilize the bicelles; therefore, for a given concentration of DHPC the bicelles cannot grow as large and fold into small ULVs. It therefore seems that these competing effects practically cancel each other out, resulting in no dramatic change to the size of ULVs.

7. Encapsulation and Controlled Release Mechanism of Spontaneously Formed ULVs

The encapsulation and leakage properties of spontaneously formed ULVs were studied using fluorescence spectroscopy (Nieh *et al.*, 2008). Two ULV mixtures (0.2 and 1.0 wt%) were prepared at 4 °C in an aqueous solution containing 12.5 m*M* of the fluorescent probe 8-aminonaphthalene-1,3,8-trissulfonic disodium salt (ANTS), and 45 mM of the fluorescence quencher *p*-xylene-bis-pyridinium bromide (DPX). If both the fluorescent probe and quencher are in close proximity (i.e., confined within an individual ULV), there is practically no fluorescence. After encapsulation, freely floating fluorescent probes and quenchers were removed from the solution at 45 °C using a PD-10 size-exclusion chromatography (SEC) column containing Sephadex^{TM} G-25. Triton X-100, an agent commonly used to compromise vesicle structure was added to the eluted ULVs, forcing them to release their contents and the resultant fluorescence was monitored. When ULVs leak or break apart—reducing the probability of the quencher interacting with the fluorescent probe—fluorescence intensity dramatically increases. Figure 1.5 shows the fluorescence response of 1 wt% ULVs at 10 and 45 °C upon the addition of Triton X-100. From the 45 °C data it is clear that ULVs (as determined by SANS) encapsulate both fluorescent probe and quencher; this is not, however, the case for bicelles (10 °C).

The 0.2-wt% ULV sample at 45 °C (Fig. 1.6) behaves in a similar manner (Fig. 1.5). However, unlike the 1 wt% sample, some degree of encapsulation is retained, even at 10 °C. This result is supported by SANS data, whereby it was observed that at low lipid concentrations ULV do not form bilayered micelles. Instead, they deform into ellipsoidal vesicles (Nieh *et al.*, 2005), thus maintaining a certain degree of encapsulating capability.

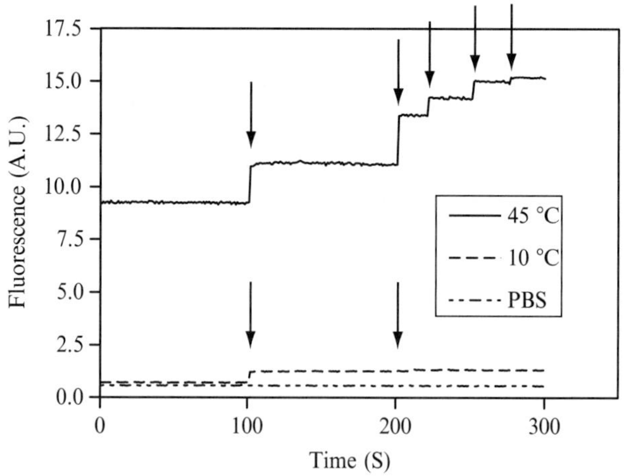

Figure 1.5 Fluorescence response of a 1.0 wt% DMPC/DHPC/DMPG mixture at 45 °C (solid line) and 10 °C (dashed line), and PBS buffer only (broken line). The arrows represent the addition of Triton X-100. Increase in fluorescence intensity is the result of ULVs (45 °C) being compromised, thus releasing fluorescent probe and quencher molecules. For further details, see Nieh *et al.* (2008).

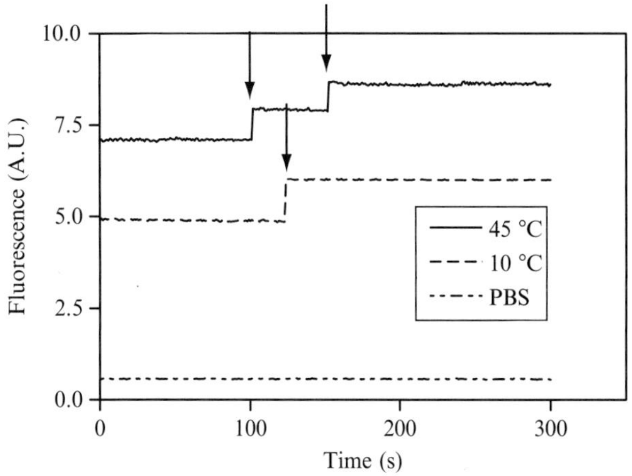

Figure 1.6 Fluorescent response of a 0.2 wt% DMPC/DHPC/DMPG mixture at 45 °C (solid line) and 10 °C (dashed line), and PBS buffer only (broken line). The arrows represent the addition of Triton X-100. Increase in fluorescence intensity is due to ULVs (45 °C) or ellipsoidal vesicles (10 °C) being compromised, thus releasing fluorescent probe and quencher molecules. For further details see Nieh *et al.* (2008).

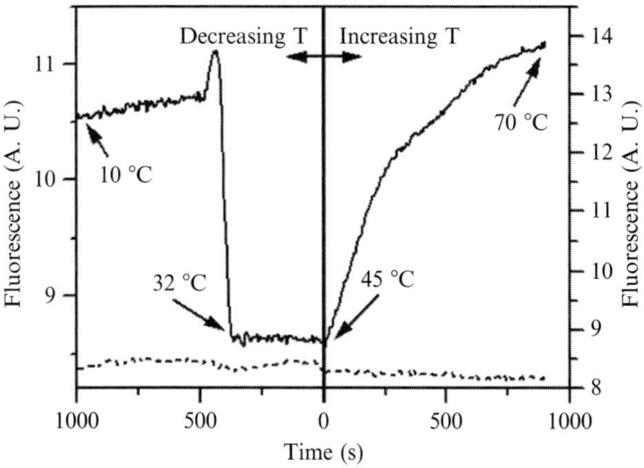

Figure 1.7 Fluorescence response of 1.0 wt% ULVs (solid line) and PBS buffer only (broken line), upon heating and cooling. Two ULV samples were prepared at 45 °C (0 s), one was cooled, while the other was heated.

It is also known that self-assembled ULVs composed of DMPC/DHPC/DMPG exhibit two temperature-dependent release mechanisms. Figure 1.7 shows that between 32 and 45 °C, ULVs are highly stable. However, beyond 45 °C they begin to leak and continue to do so at temperatures up to 70 °C, albeit at a different rate. On the other hand, at a temperature

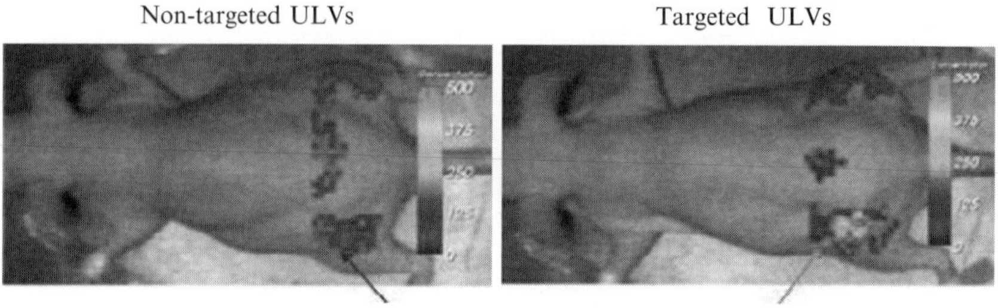

Figure 1.8 Imaging payloads delivered to a xenograft tumor using antibody-functionalized ULVs loaded with Gd, Cy5.5, and the C225 antibody. The Cy5.5 signal is predominant in the tumor where functionalized ULVs were used. (See Color Insert.)

below 32 °C, ULVs transform into bicelles, releasing their contents. At elevated temperatures, it is thought that ULVs begin to leak at an increasing rate because of lipid flip-flop taking place between the inner and outer bilayer leaflets.

8. Application

Recently, spontaneously formed ULVs have been reformulated to target and image disease (Abulrob *et al.* 2009). The ULVs were modified as follows: (a) functionalized PEGylated lipids (i.e., distearoyl PC-polyethylene glycol (PEG)-maleimide) was incorporated, onto which a targeting molecule (i.e., C225 single domain antibody) was attached; (b) DMPC was replaced with a magnetic resonance imaging (MRI) contrast agent [i.e., gadolinium (Gd)-diethylenetriamine pentaacetic acid (DTPA)-2-benzoxazolinone (BOA)]; and (c) di-18:1 dodecanylamine phosphatidylethanolamine was added to conjugate the near-infrared probe, Cy 5.5.

For the *in vivo* study, the human glioblastoma cell line (U87MG) that expresses the epidermal growth factor receptor (EGFR) was injected into the flank of nude mice and allowed to develop a tumor over a period of 10 days. Disease targeted and nontargeted ULVs were then injected into animals, which were imaged 24 h after injection. Significant accumulation of Cy5.5 in the xenograft tumor was detected only in the case of the targeted ULV formulation (Fig. 1.8).

9. Concluding Remarks

Treatment of disease can involve surgery and/or therapies, including radiation, chemo and hormonal and biological therapies. In the course of these treatments, healthy tissues can be damaged, resulting in unwanted side effects. However, the use of drugs specifically targeted to a disease can minimize the toxic side effects associated with many conventional therapies (e.g., chemotherapy). Low polydispersity, spontaneously formed ULVs show great promise in enhancing the efficacy of various medical imaging techniques (e.g., MRI and positron emission tomography) and drug treatments (e.g., cancers and diseases of the brain). The features that make them attractive for commercialization are as follows: (1) they are inexpensive, made up exclusively of low-cost phospholipids; (2) their size and polydispersity can be controlled; (3) they are highly stable, providing long shelf-life and extended circulation half-lives when inserted into the body; and (4) the ULV preparation is easily adaptable to industrial scale production. The latter two features are distinct advantages over lipid-based ULVs produced by traditional extrusion and sonication methods. Preliminary *in vivo* experiments show that these ULVs can be fabricated to both target and image diseased tissues.

ACKNOWLEDGEMENT

NK would like to acknowledge funding from the Advanced Foods and Materials Network, part of the Networks of Centres of Excellence.

REFERENCES

Abulrob, A., Stanimirovic, D., Iqbal, U., Nieh, M.-P., and Katsaras, J. (2009). *Single domain antibody-targeted carrier for contrast agents and drug delivery agents* US Provisional Patent 61/228,117.

Allen, T. M., Hansen, C., and Rutledge, J. (1989). Liposomes with prolonged circulation times: Factors affecting uptake by reticuloendothelial and other tissues. *Biochim. Biophys. Acta* **981,** 27–35.

Bergstrom, M., and Pedersen, J. S. (2000). A small-angle neutron scattering study of surfactant aggregates formed in aqueous mixtures of sodium dodecyl sulfate and didodecyldimethylammonium bromide. *J. Phys. Chem. B* **104,** 4155–4163.

Bergstrom, M., Pedersen, J. S., Schurtenberg, P., and Egelhaaf, S. U. (1999). Small-angle neutron scattering (SANS) study of vesicles and lamellar sheets formed from mixtures of an anionic and a cationic surfactant. *J. Phys. Chem. B* **103,** 9888–9897.

Boggs, J. M., and Hsia, J. C. (1972). Effect of cholesterol and water on the rigidity and order of phosphatidylcholine bilayers. *Biochim. Biophys. Acta* **290,** 32–42.

Caria, A., and Khan, A. (1996). Phase behavior of catanionic surfactant mixtures: Sodium bis (2-ethylhexyl)sulfosuccinate-didodecyldimethylammonium bromide–water system. *Langmuir* **12,** 6282–6290.

Demé, B., Dubois, M., Gulik-Krzywicki, T., and Zemb, T. (2002). Giant collective fluctuations of charged membranes at the lamellar-to-vesicle unbinding transition. 1. Characterization of a new lipid morphology by SANS, SAXS, and electron microscopy. *Langmuir* **18,** 997–1004.

Egelhaaf, S. U., and Schurtenberger, P. (1999). Micelle-to-vesicle transition: A time-resolved structural study. *Phys. Rev. Lett.* **82,** 2804–2807.

Emerich, D. F., and Thanos, C. G. (2007). Targeted nanoparticle-based drug delivery and diagnosis. *J. Drug Target.* **15,** 163–183.

Feigin, L. A., and Svergun, D. I. (1987). Determination of the integral parameters of particles. *In* Structure Analysis by Small Angle X-Ray and Neutron Scattering pp. 59–105. Plenum Press, New York.

Fromherz, P. (1983). Lipid-vesicle structure: Size control by edge-active agents. *Chem. Phys. Lett.* **94,** 259–266.

Gregoriadis, G. (1995). Engineering liposomes for drug delivery. *Trends Biotechnol.* **13,** 527–537.

Hays, L. M., Crowe, J. H., Wolkers, W., and Rudenko, S. (2001). Factors affecting leakage of trapped solutes from phospholipid vesicles during thermotropic phase transitions. *Cryobiology* **42,** 88–102.

Hofheinz, R.-D., Gnad-Vogt, S. U., Beyer, U., and Hochhaus, A. (2005). Liposomal encapsulated anti-cancer drugs. *Anticancer Drugs* **16,** 691–707.

Hui, S. W., and He, N. B. (1983). Molecular organization in cholesterol-lecithin bilayers by X-ray and electron diffraction measurements. *Biochemistry* **22,** 1159–1164.

Iampietro, D. J., and Kaler, E. W. (1999). Phase behavior and microstructure of aqueous mixtures of cetyltrimethylammonium bromide and sodium perfluorohexanoate. *Langmuir* **15,** 8590–8601.

Inoue, K. (1974). Permeability properties of liposomes prepared from dipalmitoyllecithin, dimyristoyllecithin, egg lecithin, rat liver lecithin and beef brain sphingomyelin. *Biochim. Biophys. Acta* **339,** 390–402.

Janiak, M. J., Small, D. M., and Shipley, G. G. (1976). Nature of the thermal pretransition of synthetic phospholipids: Dimyristoyl- and dipalmitoyllecithin. *Biochemistry* **15,** 4575–4580.

Juliano, R. L., and Stamp, D. (1975). The effect of particle size and charge on the clearance rates of liposomes and liposome encapsulated drugs. *Biochem. Biophys. Res. Commun.* **63,** 651–658.

Kaler, E. W., Murthy, A. K., Rodrigues, B. E., and Zasadzinski, J. A. N. (1989). Spontaneous vesicle formation in aqueous mixtures of single-tailed surfactant. *Science* **245,** 1371–1374.

Kaler, E. W., Herrington, K. L., Murthy, A. K., and Zasadzinski, J. A. N. (1992). Phase behavior and structures of mixtures of anionic and cationic surfactants. *J. Phys. Chem.* **96,** 6698–6707.

Katsaras, J., Harroun, T. A., Pencer, J., and Nieh, M.-P. (2005). "Bicellar" lipid mixtures as used in biochemical and biophysical studies. *Naturwissenschaften* **92,** 355–366.

Kline, S. R. (2006). Reduction and analysis of SANS and USANS data using IGOR Pro. *J. Appl. Crystallogr.* **39,** 895–900.

Kraske, W. V., and Mountcastle, D. B. (2001). Effects of cholesterol and temperature on the permeability of dimyristoylphosphatidylcholine bilayers near the chain melting phase transition. *Biochim. Biophys. Acta* **1514,** 159–164.

Lasic, D. D. (1998). Novel application of liposomes. *Trends Biotechnol.* **16,** 307–321.

Leng, J., Egelhaaf, S. U., and Cates, M. E. (2003). Kinetics of micelle-to-vesicle transition: Aqueous lecithin-bile salt mixtures. *Biophys. J.* **85,** 1624–1646.

Lesieur, P., Kiselev, M. A., Barsukov, L. I., and Lombardo, D. (2000). Temperature-induced micelle to vesicle transition: Kinetic effects in DMPC/NaC system. *J. Appl. Crystallogr.* **33,** 623–627.

Marques, E. F. (2000). Size and stability of catanionic vesicles: Effects of formation path, sonication and aging. *Langmuir* **16,** 4798–4807.

McIntosh, T. J., McDaniel, R. V., and Simon, S. A. (1983). Induction of an interdigitated gel phase in fully hydrated phosphatidylcholine bilayers. *Biochim. Biophys. Acta* **731,** 109–114.

Mendelsohn, R. (1972). Laser-Raman spectroscopic study of egg lecithin and egg lecithin-cholesterol mixtures. *Biochim Biophys. Acta* **290,** 15–21.

Nabar, S. J., and Nadkarni, G. D. (1998). Effect of size and charge of liposomes on biodistribution of encapsuled ^{99m}Tc-DTPA in rats. *Indian J. Pharmacol.* **30,** 199–202.

Nieh, M.-P., Glinka, C. J., Krueger, S., Prosser, R. S., and Katsaras, J. (2001). SANS study of structural phases of magnetically alignable lanthanide-doped phospholipid mixtures. *Langmuir* **17,** 2629–2638.

Nieh, M.-P., Glinka, C. J., Krueger, S., Prosser, S. R., and Katsaras, J. (2002). SANS study on the effect of lanthanide ions and charged lipids on the morphology of phospholipid mixtures. *Biophys. J.* **82,** 2487–2498.

Nieh, M.-P., Harroun, T. A., Raghunathan, V. A., Glinka, C. J., and Katsaras, J. (2003). Concentration-independent spontaneously forming biomimetic vesicles. *Phys. Rev. Lett.* **91,** 1581051–1581054.

Nieh, M.-P., Harroun, T. A., Raghunathan, V. A., Glinka, C. J., and Katsaras, J. (2004). Spontaneously formed monodisperse biomimetic unilamellar vesicles: The effect of charge, dilution, and time. *Biophys. J.* **86,** 2615–2629.

Nieh, M.-P., Raghunathan, V. A., Kline, S. R., Harroun, T. A., Huang, C.-Y., Pencer, J., and Katsaras, J. (2005). Spontaneously formed unilamellar vesicles with path-dependent size distribution. *Langmuir* **21,** 6656–6661.

Nieh, M.-P., Pencer, J., Katsaras, J., and Qi, X. (2006). Spontaneously forming ellipsoidal phospholipid unilamellar vesicles and their interactions with helical domains of saposin C. *Langmuir* **22,** 11028–11033.

Nieh, M.-P., Katsaras, J., and Qi, X. (2008). Controlled release mechanisms of spontaneously forming unilamellar vesicles. *Biochim. Biophys. Acta* **1778,** 1467–1471.

Oberdisse, J., and Porte, G. (1997). Size of microvesicles from charged surfactant bilayers: Neutron scattering data compared to an electrostatic model. *Phys. Rev. E* **56,** 1965–1975.

O'Connor, A. J., Hatton, T. A., and Bose, A. (1997). Dynamics of micelle-vesicle transitions in aqueous anionic/catinonic surfactant mixtures. *Langmuir* **13,** 6931–6940.

Oku, N. (1999). Delivery of contrast agents for positron emission tomography imaging by liposomes. *Adv. Drug. Deliv. Rev.* **37,** 53–61.

Papahadjopoulos, D., Nir, S., and Ohki, S. (1972). Permeability properties of phospholipid membranes: Effect of cholesterol and temperature. *Biochim. Biophys. Acta* **266,** 561–583.

Papahadjopoulos, D., Jacobson, K., Nir, S., and Isac, T. (1973). Phase transitions in phospholipid vesicles: Fluorescent polarization and permeability measurements concerning the effect of temperature and cholesterol. *Biochim. Biophys. Acta* **311,** 330–348.

Pozo-Navas, B., Raghunathan, V. A., Katsaras, J., Rappolt, M., Lohner, K., and Pabst, G. (2003). Discontinuous unbinding of lipid bilayers. *Phys. Rev. Lett.* **91,** 028101.

Rowe, E. S. (1983). Lipid chain length and temperature dependence of ethanol-phosphatidylcholine interactions. *Biochemistry* **22,** 3299–3305.

Rowe, E. S. (1985). Thermodynamic reversibility of phase transitions specific effects of alcohols on phosphatidylcholine. *Biochem. Biophys. Acta* **813,** 321–330.

Safran, S. A., Pincus, P., and Andelman, D. (1990). Theory of spontaneous vesicle formation in surfactant mixtures. *Science* **248,** 354–356.

Safran, S. A., Pincus, P., Andelman, D., and MacKintosh, F. C. (1991). Stability and phase behavior of mixed surfactant vesicles. *Phys. Rev. A* **43,** 1071–1078.

Schurtenburger, P., Mazer, N., and Kanzig, W. (1985). Micelle-to-vesicle transition in aqueous solutions of bile salt and phosphatidylcholine. *J. Phys. Chem.* **89,** 1042–1049.

Simon, S. A., and McIntosh, T. J. (1984). Interdigitated hydrocarbon chain packing causes the biphasic transition behavior in lipid/alcohol suspensions. *Biochim. Biophys. Acta* **773,** 169–172.

Simone, E., Ding, B.-S., and Muzykantov, V. (2009). Targeted delivery of therapeutics to endothelium. *Cell Tissue Res.* **335,** 283–300.

Sujatha, J., and Mishra, A. K. (1998). Phase transitions in phospholipid vesicles: Excited state prototropism of 1-naphthol as a novel probe concept. *Langmuir* **14,** 2256–2262.

Talmon, Y., Evans, D. F., and Ninham, B. W. (1983). Spontaneous vesicles formed from hydroxide surfactants: Evidence from electron microscopy. *Science* **221,** 1047–1048.

Torchilin, V. P. (2007). Targeted pharmaceutical nanocarriers for cancer therapy and imaging. *AAPS J.* **9,** E128–E147.

Van Borssum Waalkes, M., Kuipers, F., Havinga, R., and Scherphof, G. L. (1993). Conversion of liposomal 5-fluoro-2′-deoxyuridine and its dipalmitoyl derivative to bile acid conjugates of α-fluoro-β-alanine and their excretion into rat bile. *Biochim. Biophys. Acta* **1176,** 43–50.

Villeneuve, M., Kaneshina, S., Imae, T., and Aratono, M. (1999). Vesicle-micelle equilibrium of anionic and cationic surfactant mixture studied by surface tension. *Langmuir* **15,** 2029–2036.

Weiss, T. M., Narayanan, T., Wolf, C., Gradzielski, M., Panine, P., Finet, S., and Helsby, W. I. (2005). Dynamics of self-assembly of unilamellar vesicles. *Phys. Rev. Lett.* **94,** 038303.

Winterhalter, M., and Helfrich, W. (1992). Bending elasticity of electrically charged bilayers: Coupled monolayers, neutral surfaces and balancing stresses. *J. Phys. Chem.* **96,** 327–330.

Yatcilla, M. T., Herrington, K. L., Brasher, L. L., Kaler, E. W., Chiruvolu, S., and Zasadzinski, J. A. N. (1996). Phase behavior of aqueous mixtures of cetyltrimethylammonium bromide (CTAB) and sodium octyl sulfate (SOS). *J. Phys. Chem.* **100,** 5874–5879.

Yatvin, M. B., Weinstein, J. N., Dennis, W. H., and Blumenthal, R. (1978). Design of liposomes for enhanced local release of drugs by hyperthermia. *Science* **202,** 1290–1293.

Yue, B., Huang, C.-Y., Nieh, M.-P., Glinka, C. J., and Katsaras, J. (2005). Highly stable phospholipid unilamellar vesicles from spontaneous vesiculation: A DLS and SANS study. *J. Phys. Chem. B* **109,** 609–616.

Yuet, P. K., and Blankschtein, D. (1996). Molecular-thermodynamic modeling of mixed cationic/anionic vesicles. *Langmuir* **12,** 3802–3818.

Zhang, L., Gu, F. X., Chan, J. M., Wang, A. Z., Langer, R. S., and Farokhzad, O. C. (2008). Nanoparticles in medicine: Therapeutic applications and developments. *Clin. Pharmacol. Ther.* **83,** 761–769.

Zou, Y., Lin, Y.-H., Reddy, S., Priebe, W., and Perez-Soler, R. (1995). Effect of vesicle size and lipid composition on the in vivo tumor selectivity and toxicity of the non-cross-resistant anthracycline annamycin incorporated in liposomes. *Int. J. Cancer* **61,** 666–671.

CHAPTER TWO

Use of Acoustic Sensors to Probe the Mechanical Properties of Liposomes

Kathryn Melzak,[*,1] Achilleas Tsortos,[†,‡] *and* Electra Gizeli[†,‡]

Contents

[*] Department of Chemistry, University of British Columbia, Vancouver, British Columbia, Canada
[†] Department of Biology, University of Crete, Crete, Greece
[‡] Institute of Molecular Biology and Biotechnology, FORTH, Crete, Greece
[1] Current address: Centro de Investigación Cooperativa en Biomateriales, Parque Tecnológico de San Sebastián, San Sebastián, Spain

Methods in Enzymology, Volume 465
ISSN 0076-6879, DOI: 10.1016/S0076-6879(09)65002-3

Abstract

Acoustic sensors probe the response of a thin layer to the mechanical displacement associated with an acoustic wave. Acoustic measurements provide two simultaneous time-resolved signals; one signal is related to the velocity or frequency of the acoustic wave and is mainly a function of adsorbed mass, while the second signal, related to the oscillation amplitude, is associated with energy dissipation and is a function of the viscoelastic properties of the adsorbed layer. The methods described in this chapter explore the relationship between the acoustic measurements of adsorbed liposomes and the mechanical properties of the lipid bilayer. This is carried out using a well-characterized model system consisting of liposomes prepared from an unsaturated phospholipid and a range of mole fractions of cholesterol. Real-time acoustic measurements are shown to be sensitive to changes in the liposome cholesterol content, regardless of the mode of attachment of the liposome to the device surface. This sensitivity is not due to changes in the density of the bilayer, or to changes in the extent of liposome–surface interactions, thus leaving the mechanical properties of the bilayer as the feature that is probably being measured. Some mechanisms by which the acoustic response could be generated are suggested in this chapter.

1. Introduction

Liposomes are small vesicles consisting of one or more layers of a lipid bilayer surrounding an aqueous core. These vesicles will undergo a deformation when subjected to an external force; the extent of this deformation will be determined by the mechanical properties of the vesicles. There are several points to consider when investigating the use of acoustic sensors to probe the mechanical properties of attached liposomes: these include the nature of the deformations that the acoustic devices exert on the liposomes, the model system to use when interpreting the data and the biological relevance of the information that can be obtained. We will discuss these points briefly in this section and then in more detail later in the chapter.

The acoustic devices that have been employed here are mainly a Love wave device, in which a shear wave travels between two pairs of electrodes on the device surface, as well as a quartz crystal microbalance (QCM) which supports a bulk shear acoustic wave. The principle of the acoustic measurements is that the propagation of the acoustic wave through the solid medium of the sensor is affected by changes in the adjacent medium that contains the analyte of interest.

For simple, well-defined systems, there is a simple and well-defined acoustic response. If an elastic layer such as a metal is deposited on the device surface, then there will be a linear relationship between the mass on the surface and the change in the wave velocity; adsorbed mass on the device surface can thus be determined by monitoring the wave velocity (Ballantine *et al.*, 1997). If a purely viscous solution is placed in contact with the device, a simple relationship may be obtained between the viscosity–density product of the solution, and the drop in the efficiency of the signal transmission. Viscoelastic layers such as polymers will exhibit both a viscous response, which is a deformation or flow in response to applied stress, and an elastic response, which is a tendency to return to the original shape after applied stress. Such layers are described by a complex shear modulus $G = G' + jG''$, where G' is the storage modulus and G'' is the loss modulus, measuring energy dissipation (Lucklum *et al.*, 1997).

In the present case, the analyte is a layer of adsorbed or otherwise attached liposomes, as shown in Fig. 2.1. Due to the fact that liposomes are discrete particles, and at a molecular level are neither continuous nor isotropic, the created layer cannot be considered continuous/homogeneous, but heterogeneous (Tellechea *et al.*, 2009). The force that is applied at one point in the layer of adsorbed liposomes may thus not elicit the same response as the same force when it is applied at a different location or

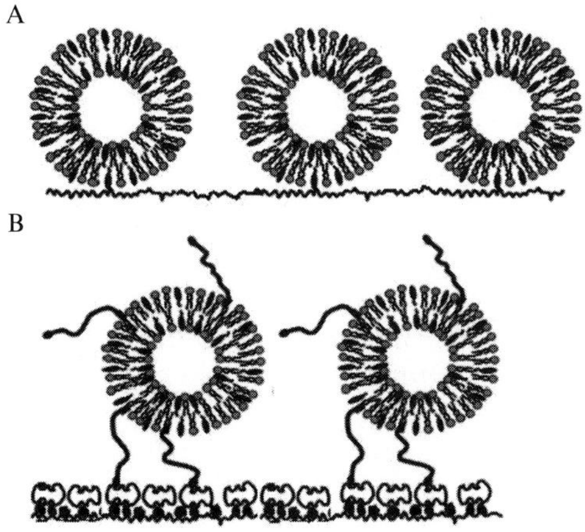

Figure 2.1 Layers of liposomes attached to the surface of the acoustic devices in two different ways. (A) POPC/cholesterol liposomes adsorb directly onto the surface of a device that has been modified with a layer of silicate gel. (B) The device is modified with a layer of gold, biotinylated BSA (black dots) and then streptavidin (white); the streptavidin provides binding sites for biotinylated liposomes prepared from POPC, cholesterol, and biotin-PEG-PE. Acoustic measurements were carried out with liposomes attached via both the methods shown here.

orientation. When the liposome layer is on the surface of the acoustic device, this discontinuous and anisotropic system will be subjected to a combination of different applied forces, as is discussed in Section 2. It therefore becomes clear that it would be difficult to develop a model that describes all aspects of the liposome interaction with acoustic devices. Continuum formulations have been described as being unsuitable for characterizing the molecular structures of biomaterials (Bao and Suresh, 2003), although this depends on the scales involved (Zhu *et al.*, 2000). This does not, however, mean that it is not possible to use acoustic measurements to obtain useful information about liposomes, although it does limit the approach that must be taken for interpretation of data.

Here, the primary interest is in the mechanical properties of the lipid bilayer that forms the boundary of the liposomes, and that can serve as a model for the lipid bilayer component of cell membranes. The mechanical properties of the lipid bilayer are of interest due to their effects on the biological properties of cells; one example of this is in the regulation of the osmotic balance of cells, where stretching of the lipid bilayer due to osmotic stress will activate ion channels, leading to a relief of the stress (Kung, 2005; Sukharev *et al.*, 1997). Since the activation of these ion channels is controlled solely through stretching the lipid bilayer, the resistance of the lipid bilayer toward stretching is of interest. A second example of the importance of the bilayer properties is their low resistance to shear, which permits red blood cells to pass through capillaries smaller than the cell diameter without damaging the cells. A third membrane mechanical property is the membrane resistance to bending, which has been associated with membrane fusion (Chernomordik *et al.*, 1995). The relationship of cell membrane cholesterol content, that is, membrane fluidity, and the cytoskeleton organization and cell function has also been shown (Kwik *et al.*, 2003).

To explore the relationship between the acoustic measurements of adsorbed liposomes and mechanical properties of the lipid bilayer, well-characterized liposome suspensions have been applied for which the mechanical properties varied in a known manner (Melzak and Gizeli, 2009; Melzak *et al.*, 2008). The systems employed here are an adsorbed layer of unilamellar liposomes prepared from 1-Oleoyl-2-palmitoyl-*sn*-glycero-3-phosphocholine (POPC) and cholesterol, in various ratios, as well as *Escherichia coli*-like liposomes. Addition of cholesterol increases the resistance of the bilayer to stretching and bending in a concentration-dependent manner. *E. coli*-like liposomes are used as a more physiologically relevant system; their acoustic response upon surface binding is compared to that of proteins and cells.

Some additional analysis of data from acoustic measurements of liposomes is performed by making comparisons to models that have been derived for simpler systems. This would permit us to say if the liposome layers were behaving more like a liquid or a solid. Consideration of the data in terms of existing models has the potential to provide some new insights into the acoustic response.

Although there are already established techniques for determining the stretching (Needham and Nunn, 1990) and bending moduli (Evans and Rawicz, 1990) of lipid bilayers, providing absolute values of their mechanical properties, acoustic measurements do have two advantages: they are simple to carry out, and they are suitable for making kinetic measurements. Acoustic experiments could be readily automated, in contrast to the other technically demanding existing techniques. Data can be acquired as a function of time, making the measurements suitable for kinetic analysis; this could be of interest for characterizing the interaction of lipid bilayers with pharmaceuticals or other substances with the potential to affect the bilayer mechanical properties. The kinetic analysis is demonstrated here by monitoring the removal of cholesterol from the adsorbed liposomes.

2. Acoustic Measurements

2.1. Love wave acoustic device

Acoustic waves are regular elastic deformations that propagate in solids. Surface acoustic waves (SAWs) can be generated via the piezoelectric effect, by application of an alternating current to interdigital electrodes placed on the surface of a piezoelectric substrate. When a voltage is applied across a piezoelectric material, this gets deformed in a direction determined by the crystal orientation. Application of an alternating current results in an oscillating stress, thus generating the acoustic wave. The nature of the acoustic wave generated will be a function of the orientation and thickness of the piezoelectric substrate. Devices that operate in a shear mode, with displacements in the plane of the device surface, are less damped in aqueous medium and are therefore suitable for analysis of biological samples (Ballantine *et al.*, 1997). The acoustic device employed in these experiments is known as a Love wave device. Love waves are produced in acoustic sensors when a waveguide layer is added to a SAW device that supports a shear wave (Gizeli, 2000; Martin *et al.*, 2004). Addition of the waveguide layer has the effect of confining the acoustic energy close to the surface that is exposed to the analyte, thus increasing the sensitivity of the device (Fig. 2.2).

Two of the measurement techniques suitable for characterizing acoustic waves involve network analyzers and oscillator circuits. Network analyzers can apply a range of frequencies of alternating current to an acoustic device and can determine the frequency-dependent device characteristics, including the change in phase and the power levels of the reflected or transmitted wave, relative to a reference signal. The device employed in the experiments presented here is a two-port device in which the acoustic wave is

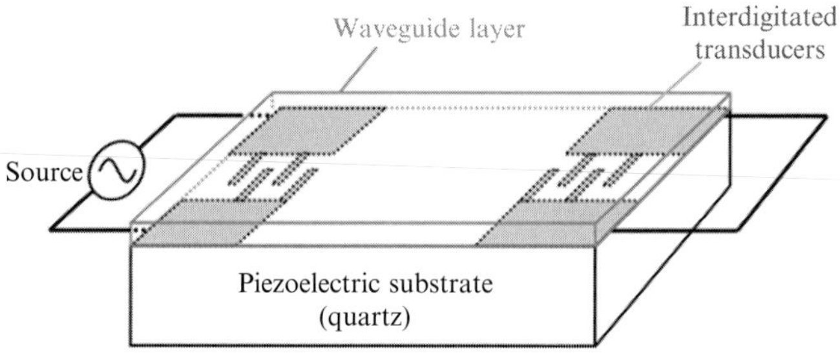

Figure 2.2 Love wave device with a waveguide layer on top of a piezoelectric substrate with suitably placed electrodes. The electrodes are not drawn to scale: there are 40 repeat units at each end of the device.

generated at one set of electrodes and travels across the device before being converted back to an electrical signal at a second set of electrodes. The device is attached to a network analyzer via two coaxial cables. The transmitted power is measured in decibels; for the 50 Ω terminated output of the network analyzer, the default value for applied power is 1 mW, or 0.225 V, so that the transmitted power levels are in fact measured in dBm, or dB relative to 1 mW, with values that can be converted to watts or volts. The output values for the Love wave devices employed in these experiments are typically in the range of 3–6 mV. The peak shape for the transmitted power as a function of applied frequency is determined by addition of the signals from the repeating pairs of interdigital electrodes. The peak shape will thus be determined by the device production process and is not suitable for extraction of viscoelastic parameters.

To monitor interactions that occur as a function of time on the device surface, data can be collected at one frequency, measuring the insertion loss in dB and the change in phase in degrees. The change in phase is directly proportional to adsorbed mass for continuous elastic films such as metal layers deposited on the device surface. The insertion loss is a measure of the extent to which energy is dissipated as the wave travels across the device and is a function of viscous or viscoelastic losses.

2.2. Quartz crystal microbalance

Shear acoustic waves can also be generated using an acoustic device based on a thickness shear mode (TSM) resonator geometry (Rodahl *et al.*, 1995). This device consists of a piezoelectric substrate of a thickness of 0.5 mm or less, with two electrodes in the form of uniform gold patches deposited on the opposite sides of the substrate. The corresponding measured quantities are the frequency, f (instead of phase) and energy dissipation, D (instead of amplitude). The mass of a solid, elastic layer deposited on the device surface is linearly related to frequency change, based on the known Sauerbery equation (Ballantine *et al.*, 1997); such a film will result in zero dissipation. Generally, the interpretation of the data obtained with the QCM is similar to that described for the Love wave device. QCM devices are commercially available together with the measuring unit.

Liquid in contact with either of the above acoustic devices will couple with the oscillating surface (Fig. 2.3). It is commonly assumed that the thickness of the coupled liquid layer is given by $\delta = (2\eta/\rho\omega)^{1/2}$, where the outer boundary of the layer is taken to be the point where the wave amplitude has decayed to $1/e$ of its initial value. For a 103-MHz Love wave operating in water, δ is calculated to be 55 nm and for a QCM operating at 35 MHz $\delta = 95$ nm.

If the liquid above the acoustic sensor behaves as a Newtonian fluid, then the relative change in phase and amplitude of the acoustic wave are proportional to $(\eta\rho)^{1/2}$, where η is the dynamic viscosity and ρ is the solution density (Kanazawa and Gordon, 1985; Mitsakakis *et al.*, 2009; Ricco and Martin, 1987). At higher viscosities, the relaxation time of the liquid becomes significant. The relaxation time is also significant at low viscosities for solutes with high molecular weight.

2.3. Concentration-dependent effects of cholesterol

In these experiments, we have used different ratios of cholesterol to phospholipid in order to obtain liposomes having lipid bilayers with different mechanical properties. We shall first describe how the bilayer mechanical properties are affected by cholesterol and then summarize some other effects of cholesterol that have the potential to affect the acoustic experiments.

Cholesterol increases the resistance of lipid layers to changes in area within the plane of the lipid layer. This has been measured by manipulation of individual liposomes, with the force required to change the area being measured directly (Needham and Nunn, 1990), and by using a Langmuir trough to determine the compressibility of monolayers (Li *et al.*, 2001). The resistance to changes in area is described by the elastic area expansion

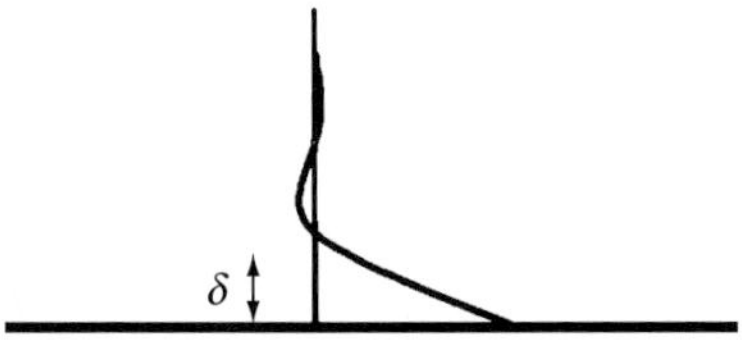

Figure 2.3 Transverse displacement of the sheared liquid boundary layer adjacent to the surface of the acoustic device. The effective thickness δ of the entrained fluid layer is a function of the viscosity η and density of the fluid and the frequency ω of the acoustic wave.

modulus K, in units of force per distance. Over the concentration range of 0–43% cholesterol, vesicle bilayers composed of stearoyloleoyl phosphatidylcholine (SOPC) and cholesterol showed a threefold increase in the expansion modulus, with the greater part of this increase occurring at values greater than 28 mol% cholesterol (Needham and Nunn, 1990). The POPC employed in the experiments described here differs from SOPC by having two carbons less on the saturated fatty acyl chain.

The resistance of bilayers to bending is predicted by models to be linearly proportional to the area compressibility modulus K_a, and proportional to the square of the layer thickness (Marsh, 2006). Addition of cholesterol is thus predicted to increase the resistance to bending; the bending modulus k_c can be measured directly (Evans and Rawicz, 1990) and a summary of measurements from different sources indicates that cholesterol increases k_c (Marsh, 2006).

Cholesterol will also increase the density of layers formed from lipids having one unsaturated fatty acid chain. Density here refers to mass of lipid per area covered by the lipid layer; potential changes in layer thickness are disregarded. The layer density has been measured for cholesterol-lipid monolayers formed at the air–water interface of a Langmuir trough, where the area per molecule can be readily calculated. As the mole fraction of cholesterol increases from 0 to 0.1, 0.2, and 0.3, the density of the layer increases from 1.5 ng/mm^2 to 1.62, 1.75, and 1.85 ng/mm^2 (calculated from area per molecule presented by Kauffman *et al.* (2000), knowing the average molecular weight). It may be noted that addition of cholesterol changes the average area per lipid molecule, but has very little effect on the average area per phospholipid; changing the cholesterol content will therefore have minimal effect on the liposome diameter.

The density of the lipid layer that comprises the outer shell of the liposomes will thus be increased by cholesterol. The effect of cholesterol on the density of the whole liposome is, however, minimal. The radius of the liposomes was found to be 70 nm, by light scattering measurements. A lipid layer with a thickness of 5 nm would thus constitute 20% of the volume of the liposomes; changing the cholesterol content of this layer from 0% to 30% would only change the density of the whole liposomes by 3%. To obtain this number, the liposomes are considered as hollow spheres with a known radius (70 nm) and a known shell thickness (5 nm). The proportion of the total volume occupied by the bilayer and by the aqueous core can thus be determined; the aqueous core was assumed to have a constant density while the density of the bilayer shell as a function of cholesterol content was estimated as described above.

The density of the layer adjacent to the surface of the acoustic device will affect the transmission of the acoustic wave. Increasing the cholesterol content of the liposomes will cause a change in density that occurs simultaneous to the change in the bilayer mechanical properties; if we want to demonstrate that the observed response is due specifically to the bilayer mechanical properties, it is necessary to devise suitable controls. This is done by carrying out surface plasmon resonance (SPR) measurements on the same samples that were applied to the surface of the acoustic device. The SPR measurements are only going to be affected by the change in optical density of the adsorbed layer, and will not be affected by the mechanical properties.

Increasing the cholesterol content will also lead to separation of the lipid bilayer into lateral domains. For the POPC/cholesterol system at ambient temperature (21–26 °C, in the case of our experiments) and without applied pressure, lipid bilayers will exist as a mix of liquid ordered (lo) and liquid disordered (ld) states, depending on the mole fraction of cholesterol present (de Almeida *et al.*, 2003). Fluorescence depolarization measurements show that there is only the ld phase present up to 12% cholesterol, ld and lo phases are both present over the range between 12% and 46% cholesterol, and that only the lo phase is present at cholesterol concentrations above 46%.

In the experiments described here, the mole fraction of cholesterol is varied from 0 to 0.35. Liposomes prepared over this concentration range of cholesterol would be expected to have varying extents of separation into liquid ordered and liquid disordered states. In the majority of our experiments, the liposomes are adsorbed directly onto the modified surface of the acoustic devices. The adsorption process could potentially be affected by the inhomogeneities that develop at higher mole fractions of cholesterol; it is therefore necessary to produce controls to demonstrate that the mode of attachment of the liposomes is not affecting the acoustic response. This is achieved by making measurements on systems in which the liposomes were attached to the device surface by a long tether, as shown in Fig. 2.1B.

3. Experimental Procedures

3.1. Materials

POPC and phosphate-buffered saline (PBS) tablets (0.01 *M* phosphate, 0.0027 *M* potassium chloride, and 0.137 *M* sodium chloride, pH 7.4) are purchased from Sigma. 1,2-Distearoyl-*sn*-glycero-3-phosphoethanolamine-*N*-[biotinyl(polethylene glycol) 2000] (ammonium salt) (biotin-PEG-PE) and *E. coli*-like liposomes with a company-reported composition of ~67% phosphatidylethanolamine, ~23.2% phosphatidylglycerol, and ~9.8% cardiolipin are obtained from Avanti. Chloroform is obtained from Merck and glycerol from Biomol. Medium molecular weight poly(methyl methacrylate) (PMMA) and 2-ethoxyethyl acetate (2-EEA) are purchased from Aldrich. Hydrochloric acid (HCl) and Triton X-100 detergent are obtained from Merck, and tetraethyl orthosilicate (TEOS) and β-cyclodextrin (βCD) from Fluka. Sputter-coating of a gold layer is carried out with a gold foil target (99.99%) from BAL-TEC. Streptavidin and avidin are purchased from Sigma, neutravidin from Pierce and biotinylated BSA from Fluka, for deposition of liposomes via a PEG linker. All aqueous solutions are prepared with water purified by distillation followed by deionization with a Barnstead NANOpure filtration system. Water-soluble cholesterol (WSC) refers to cholesterol that has been prepared with an excess of methyl-β-cyclodextrin and is purchased from Sigma.

3.2. Love wave device

The acoustic wave device (Fig. 2.2) is prepared by photolithographic patterning of gold interdigitated electrodes (IDTs) on a piezoelectric quartz substrate at the Southampton Electronics Centre (Southampton, UK) using single-crystal y-cut, z-propagating 0.5 mm thick quartz, with a 200-nm gold overlayer and a 10-nm chromium adhesion layer. Each IDT consists of 40 pairs of split fingers with a periodicity of 45 μm, thus defining the acoustic wavelength. A waveguide layer is applied by spin-coating a 17% solution of PMMA in 2-EEA at 4000 rpm. The PMMA-coated devices are heated to 195 °C for 2 h to promote solvent evaporation. The operating frequency of the acoustic waveguide device is 103 MHz. The waveguide layer can be readily cleaned off in acetone for recycling of the devices.

3.3. Instrumentation and experimental setup for Love wave experiments

A Hewlett-Packard 4195A network analyzer is used to measure the amplitude A (insertion loss) and phase Ph of the output electrical signal with respect to a reference signal. The amplitude is reported in decibels, which are 10 times the logarithm of the power ratio of the input and output signals. Data are collected on a PC using LabView software. A Perspex flow cell with a silicone rubber gasket is used to hold the solution in place over a region of the device between the IDTs, exposing an area of approximately 0.12 cm^2. Experiments are carried out with a constant flow of solution over the surface of the device, at 0.1 ml/min unless otherwise specified. During experiments, a 3-MHz interval around the main Love mode is scanned every 43 s to monitor the signal. Data are collected at a fixed frequency so that one data point is collected every 43 s. Experiments are carried out at ambient temperature which is 26 $\pm$ 1 °C for the experiments with the silicate-coated devices and 21 $\pm$ 1 °C for experiments with the gold-coated devices and the liposomes attached via a PEG linker. The change in phase and the change in amplitude are normalized to zero with respect to buffer so that the value plotted indicates the relative change over the course of each experiment, written as ΔA (for the change in amplitude) for the insertion loss and ΔPh for the change in phase.

3.4. QCM device

A QCM-D instrument (Qsense-D300, Sweden) is used and data are collected at 35 MHz frequency using the software provided by the company. The flow of molecules over the device surface is facilitated with a pump at a flow rate of 20 μl/min. Reported changes in dissipation and frequency are normalized to zero with respect to buffer. Experiments are carried out at a constant temperature of 25 °C, controlled by a Peltier element.

3.5. Liposome preparation

Unilamellar lipid vesicles are prepared by extrusion using an Avestin Liposofast Basic extrusion apparatus. Vesicles are made from POPC and cholesterol in varying mole ratios, or from POPC and cholesterol with 2.4 mol% biotin-PEG-PE. Stock solutions (20 mg/ml) of lipids are made up in chloroform and mixed to give the appropriate mole ratio. The chloroform is removed by evaporation under a stream of nitrogen for 1 h. The dried lipids are resuspended in PBS (pH 7.4) at a total lipid concentration of 2 mg/ml and then extruded 25 times through a 50-nm pore membrane (Avestin). The radius of gyration (R_g) and the hydrodynamic radius (R_h) of POPC liposomes are determined by static and dynamic light scattering measurements using an ALV-5000 goniometer/correlation setup with a Nd:YAG laser at 532 nm. The R_g extraction is from a Guinier plot and R_h is calculated based on the Stokes–Einstein relation, with values for D being calculated from a stretched exponential fit. The values for R_g and R_h are found to be 67 and 74 nm, respectively; the ratio R_g/R_h is thus greater than the 0.774 value for solid spheres, as would be expected for the hollow liposomes. Due to the close agreement between the two values measured here, an average value of 70 nm is used for the radius of the liposomes in subsequent calculations. *E. coli*-like liposomes (0.02–0.5 mg/ml) used in QCM experiments are prepared in 0.5 M HEPES buffer (pH 7.4, 150 mM NaCl) by extrusion through a 50-nm pore filter membrane (Avestin).

3.6. Adsorption of the liposomes to the surface of the acoustic devices

The acoustic devices are modified with a silicate gel that induces adsorption of the POPC/cholesterol liposomes, forming a layer shown schematically in Fig. 2.1A. Although POPC has a net neutral charge, it is zwitterionic, with the positively charged choline group situated so that it can interact with the negatively charged silanol groups on the silicate. A silicate gel is prepared by condensation from TEOS in the presence of acid (Melzak and Gizeli, 2002). The TEOS (0.25 g) is added to a 1.5-ml microcentrifuge tube and 5.8 *M* HCl is added to give a final volume of 1 ml. The 2-phase mixture is left for 1 min before being vortexed at 1400 rpm for 2 min, so that the TEOS is hydrolyzed and one phase is formed. The hydrolyzed TEOS mixture is left at room temperature until sufficient cross-linking has occurred that a bubble of air injected into the mix remained in place (this occurrs at approximately 20 min after the start of the mixing process, and about 1.5 min after the mixture is sufficiently solid to remain in place when the tube is inverted). The TEOS mixture is then placed on the PMMA-coated acoustic device for 2 min before being rinsed off with water. The surface roughness of the silicate is approximately 3 nm, as measured by atomic force microscopy (Digital Instruments Nanoscope IIIa instrument, operating in tapping mode with Veeco tips).

A continuous flow of PBS is pumped over the surface of the modified Love wave devices using a peristaltic pump. For the POPC/cholesterol liposomes, vesicle suspensions at a total lipid content of 0.4 mg/ml are pumped over the surface of the silicate-modified device for 10 min. The device is then rinsed with PBS before the surface is regenerated with a rinse with 0.1% (w/v) Triton X-100 in PBS. Following a PBS rinse of 15 min, the device surface is returned to its initial state so that an additional cycle of liposome deposition can be started. Multiple depositions can therefore be carried out in one experiment. The change in phase and amplitude observed for multiple depositions of POPC liposomes during the course of one experiment remain constant, with a standard deviation of 5% in the change in signal amplitude, ΔA, observed during liposome deposition and 2% in the change in phase, ΔPh.

E. coli-like liposomes are adsorbed on a protein (neutravidin)-coated gold QCM device surface followed by buffer rinse.

3.7. Formation of a tethered liposome layer

POPC liposomes are also attached to the surface of the acoustic device via tethers, forming a layer as shown schematically in Fig. 2.1B. This is done with liposomes that incorporated a biotin-terminated PEG, as described above; the biotin coupled specifically to streptavidin that is deposited via a multistep procedure on the device surface. The initial step in the process is to deposit a gold layer on one PMMA-coated device. A 20-nm gold layer is deposited on the region between the IDTs by sputter-coating with a Bal-Tec SCD 050 sputter-coater. The gold layer is etched immediately prior to the acoustic experiments to ensure a clean surface. Devices are reused by etching to clean off adsorbed sample and the top layer of gold; the rate of etching is calculated and the gold is replaced by sputter-coating to maintain the 20 nm thickness.

Initially, a continuous flow of PBS is pumped over a freshly coated gold surface of an acoustic device. A layer of biotinylated BSA is deposited on the gold (0.25 mg/ml), followed by buffer rinse and deposition of a layer of streptavidin (0.025 mg/ml). After buffer rinse, liposomes containing biotin-PEG-PE are deposited on the surface at a concentration of 0.4 mg/ml lipid. The device is then rinsed with PBS until the signal has equilibrated. Due to the nature of the bond between the liposomes and the surface, the devices cannot in this case be regenerated with a detergent rinse; instead the surface is regenerated by etching in an argon plasma followed by recoating with gold.

3.8. Addition of β-cyclodextrin and water-soluble cholesterol

The cholesterol concentration of liposomes adsorbed to the surface of the acoustic device is modified *in situ* by addition of βCD or WSC. The βCD has a cyclic structure with a hydrophilic exterior that permits solubility in water and a hydrophobic interior cavity that will take up hydrophobic molecules of suitable size. Addition of a large excess of βCD will therefore remove cholesterol from the adsorbed liposomes. In contrast, WSC can be prepared by addition of a sufficient excess of a βCD to cholesterol; the cholesterol can then leave the βCD cavity and insert into the liposomes.

Solutions of βCD at 3 mg/ml are added to liposome layers on the silicate-coated acoustic devices after the liposomes have been rinsed with buffer for 15–20 min. The βCD is added for 30 min, followed by a 5-min rinse and addition of detergent to regenerate the surface. During the course of one experiment, multiple cycles of liposome deposition and cholesterol removal via βCD can be carried out. The initial liposome composition is varied over the range of 0–35 mol%; this makes it possible to determine the effect of initial cholesterol concentration on one modified surface. Analogous experiments are also carried out with WSC added in place of the βCD. For liposomes attached to gold-coated devices via a PEG linker, the device surface cannot be regenerated. Experiments are carried out with 35% cholesterol and with controls at 0% cholesterol.

3.9. Surface plasmon resonance

Surface plasmon resonance (SPR) experiments are carried out with a Reichert SR7000 instrument and gold-coated devices from Xantec. The gold-coated slides used for the SPR are modified with a 1% PMMA solution and then with the same silicate modification procedure described above. The SPR slides and acoustic devices are modified in pairs to improve the reproducibility of the silicate modification, so that the SPR results can be compared more readily to the acoustic data. The SPR measurements are sensitive to the change in optical density of the region immediately adjacent to the surface of the gold slide, and are thus generally assumed to be proportional to changes in the density in this region. Buffer contained within the liposomes has the same refractive index as the surrounding medium. The SPR measurements monitor the change in refractive index as the buffer adjacent to the slide is replaced with liposomes; these measurements will therefore be detecting only the lipid component of the liposomes.

4. Change in ΔPh and Δ*A* for Adsorbed Vesicles as a Function of Cholesterol Content

The cholesterol content is varied in two ways, firstly by depositing liposomes prepared with different initial mole fractions of cholesterol (Fig. 2.4) and secondly by changing the cholesterol content *in situ*, either by removal (Fig. 2.5) or by insertion using βCD (Fig. 2.6).

The results show clearly that liposomes with a greater cholesterol content are associated with a greater change in signal (Fig. 2.4). The pattern observed for Δ*A* is similar to that observed for ΔPh. The response on addition of βCD is similar regardless of the mode of attachment of the liposomes (Fig. 2.5); this implies that the acoustic sensor is responding to the intrinsic properties of the liposomes and that the liposome–surface interactions are not the dominant factor governing the response.

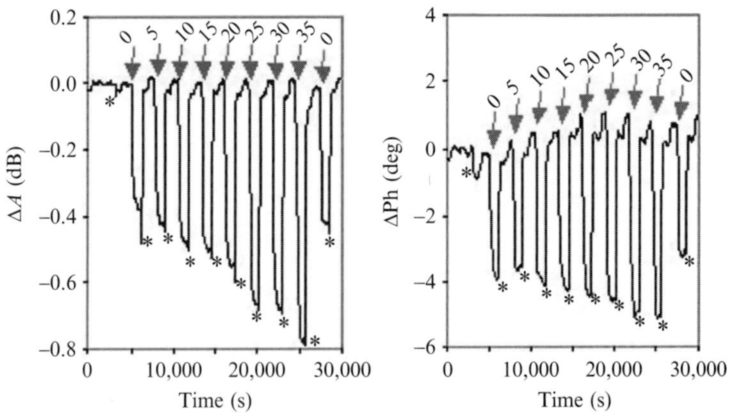

Figure 2.4 Acoustic response for sequential depositions of liposomes with increasing concentrations of cholesterol on a silicate-modified device. The mol% of cholesterol is indicated for each deposition. The device was regenerated between depositions by addition of detergent at the points marked by an asterisk (*).

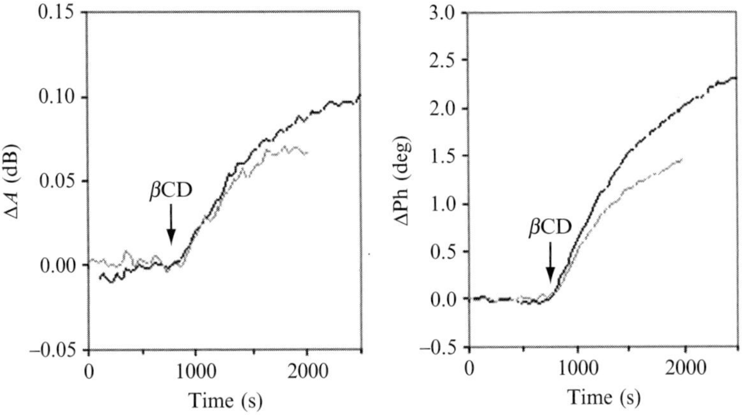

Figure 2.5 Net change in signal on removal of cholesterol from liposomes with 35 mol% initial cholesterol, by addition of βCD at the time indicated; the gray line represents liposomes attached via a PEG tether and the black line represents liposomes adsorbed to the device surface.

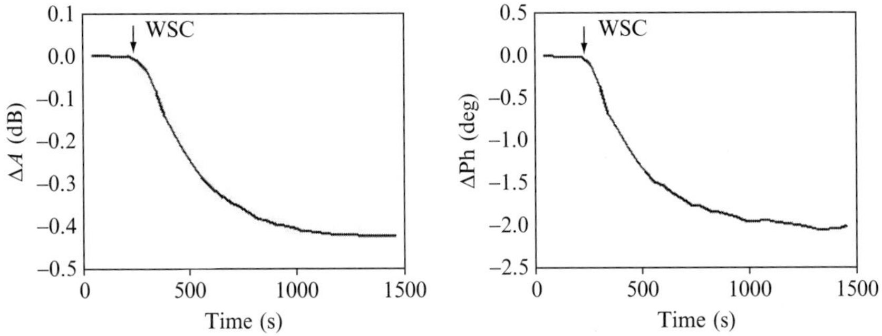

Figure 2.6 Increasing the cholesterol content *in situ*: net change in signal on insertion of cholesterol into liposomes (with no initial cholesterol) by addition of 1.84 mg/ml WSC at the time indicated.

The extent to which the acoustic signals change as a function of cholesterol is summarized in Fig. 2.7, which also shows how the SPR response changes. The SPR response for the deposited 30% cholesterol liposomes is 24% greater than that for the 0% cholesterol liposomes; this is less than that observed for either of the acoustic signals, but is similar to the 23% increase in the monolayer density that has been measured previously for the POPC/cholesterol system with a Langmuir trough (data calculated from Kauffman *et al.*, 2000). The results in Fig. 2.7 show that the acoustic signals are not reflecting solely on the density of the lipid layer; other possibilities are discussed below.

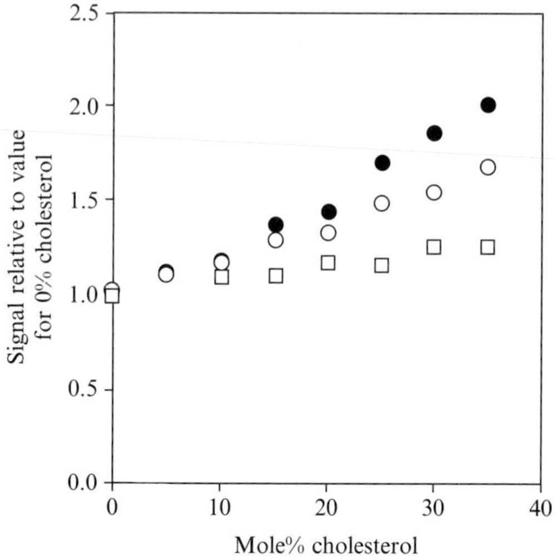

Figure 2.7 Change in acoustic and SPR responses with increasing mol% cholesterol, using the first depositions of 0 mol% in each experiment as an internal standard for normalization (ΔA, black circles; ΔPh, open circles; SPR, squares).

5. Comparison of the Results to Various Model Systems

5.1. Do the liposomes behave like rigid spheres?

If the liposomes were completely rigid and did not deform, then they would move synchronously with the oscillating surface of the acoustic device. This would lead to a change in phase directly proportional to the adsorbed mass, with a proportionality constant of 10 deg cm$^2/\mu$g, as determined by deposition of sputter-coated gold (data not shown). The adsorbed mass would include all the mass oscillating synchronously with the device; for rigid liposomes, this would include all the water enclosed within the liposome. This case can be eliminated based on two reasons: it does not provide a means of explaining the extent of the increase of ΔPh with increasing mole fractions of cholesterol and it would also lead to a much greater change in phase on liposome deposition than that observed here. Given the mass sensitivity for this device and a liposome diameter of 140 nm, the observed phase change of 3–6 deg is only compatible with the treatment of liposomes as rigid spheres if the liposomes only occupy 6% of the available surface area. This possibility can be eliminated based on fluorescence microscopy observations with labeled liposomes (Melzak and Gizeli, 2002). In addition, the observed energy dissipation (Fig. 2.4) is not compatible with the deposition of rigid mass. We can therefore say that the liposomes are not rigid, and that the adsorbed layer does not move synchronously with the device surface.

5.2. Acoustic behavior of liposomes in comparison to glycerol, proteins, and cells

Figure 2.4 indicates that energy dissipation is significant and increases with increasing cholesterol content in the lipid bilayer. In order to put things in perspective and decide on the viscous or viscoelastic behavior of attached liposomes, their acoustic response was compared to that of other model systems. A useful way to represent acoustic data and discriminate between elastic mass and viscous liquid loading is by plotting the observed amplitude or dissipation changes as a function of the corresponding phase or frequency drop (Tsortos *et al.*, 2008a,b).

Figure 2.8 shows the case for four different systems: two proteins, avidin and neutravidin, *E. coli*-like liposomes, LG2 Lymphoblastoid human cells as well as glycerol. As can be seen, liposomes appear to have an intermediate acoustic behavior of that of purely viscous glycerol and semirigid proteins adsorbed on the gold sensor surface (the "Sauerbrey limit" of zero energy dissipation). Interestingly, LG2 cells exhibit a purely viscous response nearly overlapping with that of glycerol. This can be attributed to the large difference in the size of the two systems, with liposomes lying completely

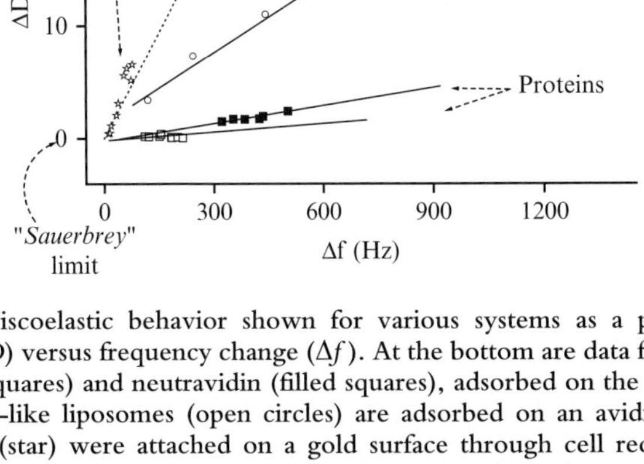

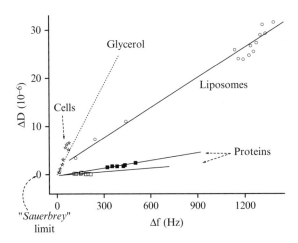

Figure 2.8 Viscoelastic behavior shown for various systems as a plot of energy dissipation (ΔD) versus frequency change (Δf). At the bottom are data for the proteins avidin (open squares) and neutravidin (filled squares), adsorbed on the gold electrode surface. *E. coli*-like liposomes (open circles) are adsorbed on an avidin-coated gold surface. Cells (star) were attached on a gold surface through cell receptor (HLA)/ antibody (anti-HLA) interactions (Saitakis *et al.*, 2008). Glycerol was applied on bare gold. All data were obtained with a QCM-D device at 35 MHz.

within the 95 nm penetration depth of the 35 MHz QCM device, while large cells (15 μm) being only partly probed by the wave.

Having established from Fig. 2.8, the viscoelastic behavior of surface-attached liposomes of interest is also the observation that the increase of cholesterol content results in larger dissipation (Fig. 2.4). This effect can be explained by stating that the cholesterol is increasing the effective viscosity of the layer by making the liposome shell more rigid, consistent with bulk viscosity measurements on gels with entrapped liposomes having different bilayer rigidities (Mourtas *et al.*, 2008).

6. Can the Response Be Explained in Terms of Variable Slip at the Surface?

Another possible factor that may be considered here is slip due to a lack of complete coupling between the oscillating surface of the acoustic device and the adjacent medium. A nonslip boundary condition assumes that the displacements in the liquid layer immediately adjacent to the surface will be the same as those at the surface of the solid. This may not be a valid assumption for shear-mode acoustic sensors that operate with high-surface accelerations (Duncan-Hewitt and Thompson, 1992; Watts *et al.*, 1990). The effect of variable coupling between the device surface and the adsorbed liposomes was investigated by attaching the liposomes to the device surface via two different mechanisms. Liposomes adsorbed directly to the silicate-modified surface of the acoustic device may indeed be subject to slip, but this should not be a relevant factor for liposomes attached via a PEG linker. The similarity of the responses shown in Fig. 2.5 implies that the response is not due to variable amounts of slip of the liposomes on the acoustic device. There is, however, still the question of whether there could be variable extent of slip of the water past the surface of the liposomes.

7. Displacements Associated with the Liposomes

The acoustic signal changes associated with the deposition of liposomes indicate that the adsorbed liposomes do not move synchronously with the substrate surface as rigid spheres. If we assume, however, that there is no slip, then the portion of the liposome in contact with the acoustic device would move synchronously with the surface, while the remainder of the liposome does not. This would lead to distortion of the adsorbed liposomes; if we assume that the liposomes maintain a constant surface area, then the bilayer shell would be sheared and bent. The resistance of the bilayer shell toward bending would be affected by the presence of cholesterol, and we thus have a mechanism by which the acoustic device can probe the mechanical properties of the bilayer.

Alternatively, we could consider the liposomes in terms of their interaction with water, both inside and outside the vesicles. Although the liposomes do not behave like rigid spheres, they will entrap some of the enclosed water; an increase in the cholesterol content of the bilayer shell may act to increase the extent to which this occurs. This would give the liposomes a greater effective mass, thus increasing the acoustic response. Concomitantly, the aqueous medium outside the liposomes may also be considered: it will couple to the oscillating acoustic device, forming an entrained fluid layer with an oscillation amplitude that decays with increasing distance away from the device surface. Increasing the cholesterol/ rigidity of the liposome will increase the extent to which the fluid layer moves (higher viscosity) relative to the adsorbed vesicles (Mourtas *et al.*, 2008). In a similar manner, increased viscosity (in particular intrinsic viscosity) has been offered as an explanation for tethered polymers (DNA) where dissipation also increases (Tsortos *et al.*, 2008a,b).

8. Concluding Remarks

We have demonstrated experimentally that the acoustic measurements are sensitive to changes in the cholesterol content of adsorbed or tethered liposomes. This sensitivity is not due to changes in the density of the bilayer, or to changes in the extent of liposome–surface interactions. The cholesterol-induced changes in the bilayer mechanical properties are thus left as the prime candidate for the feature being detected by the acoustic measurements. From the comparison of cell-like liposomes with cells and proteins, the acoustic behavior of the liposomes is that of viscoelastic material when attached at the surface of the sensor.

Acknowledgments

We thank Mr. M. Saitakis and Ms. A. Prigipaki for providing data on cells and liposomes. We also thank the Human Frontier Science Program (HFSP) and the Greek General Secretariat of Research and Technology (GSRT) for financially supporting this work.

References

Ballantine, D. S., White, R. M., Martin, S. J., Ricco, A. J., Zellers, E. T., Frye, G. C., and Wohltjen, H. (1997). Acoustic Wave Sensors. Academic Press, San Diego, CA.

Bao, G., and Suresh, S. (2003). Cell and molecular mechanics of biological material. *Nat. Mater.* **2,** 715–725.

Chernomordik, L., Kozlov, M. M., and Zimmerberg, J. (1995). Lipids in biological membrane fusion. *J. Membr. Biol.* **146,** 1–14.

de Almeida, R. F. M., Fedorov, A., and Prieto, M. (2003). Sphingomyelin/phosphatidylcholine/cholesterol phase diagram: Boundaries and composition of lipid rafts. *Biophys. J.* **85,** 2406–2416.

Duncan-Hewitt, W. C., and Thompson, M. (1992). Four-layer slip theory for the acoustic shear wave sensor in liquids incorporating interfacial slip and liquid structure. *Anal. Chem.* **64,** 94–105.

Evans, E., and Rawicz, W. (1990). Entropy-driven tension and bending elasticity in condensed-fluid membranes. *Phys. Rev. Lett.* **64,** 2094–2097.

Gizeli, E. (2000). Study of the sensitivity of the acoustic waveguide sensor. *Anal. Chem.* **72,** 5967–5972.

Kanazawa, K. K., and Gordon, J. G. (1985). The oscillation frequency of a quartz resonator in contact with a liquid. *Analyt. Chim. Acta* **175,** 99–105.

Kauffman, J. M., Westerman, P. W., and Carey, M. C. (2000). Fluorocholesterols, in contrast to hydroxycholesterols, exhibit interfacial properties similar to cholesterol. *J. Lipid Res.* **41,** 991–1003.

Kung, C. (2005). A possible unifying principle for mechanosensation. *Nature* **436,** 647–654.

Kwik, J., Boyle, S., Fooksman, D., Margolis, L., Sheetz, M. P., and Edidin, M. (2003). Membrane cholesterol, lateral mobility, and the phosphatidylinositol 4,5-bisphosphate-dependent organization of cell actin. *Proc. Natl. Acad. Sci. USA* **100,** 13964–13969.

Li, X. M., Momsen, M. M., Smaby, J. M., Brockman, H. L., and Brown, R. E. (2001). Cholesterol decreases the interfacial elasticity and detergent solubility of sphingomyelins. *Biochemistry* **40,** 5954–5963.

Lucklum, R., Behling, C., Cernosek, R. W., and Martin, S. J. (1997). Determination of complex shear modulus with thickness shear mode resonators. *J. Phys. D: Appl. Phys.* **30,** 346–356.

Marsh, D. (2006). Elastic curvature constants of lipid monolayers and bilayers. *Chem. Phys. Lipids* **144,** 146–159.

Martin, F., Newton, M. I., McHale, G., Melzak, K. A., and Gizeli, E. (2004). Pulse mode shear horizontal-surface acoustic wave (SH-SAW) system for liquid based sensing applications. *Biosens. Bioelectron.* **19,** 627–632.

Melzak, K. A., and Gizeli, E. (2002). A silicate gel for promoting deposition of lipid bilayers. *J. Colloid Interface Sci.* **246,** 21–28.

Melzak, K. A., and Gizeli, E. (2009). Relative activity of cholesterol in OPPC/cholesterol/sphingomyelin mixtures measured with an acoustic sensor. *Analyst* **134,** 609–614.

Melzak, K. A., Bender, F., Tsortos, A., and Gizeli, E. (2008). Probing mechanical properties of liposomes using acoustic sensors. *Langmuir* **24,** 9172–9180.

Mitsakakis, K., Tsortos, A., Kondoh, J., and Gizeli, E. (2009). Parametric study of SH-SAW device response to various types of surface perturbations. *Sens. Actuators B Chem.* **138,** 408–416.

Mourtas, S., Haikou, M., Theodoropoulou, M., Tsakiroglou, C., and Antimisiaris, S. G. (2008). The effect of added liposomes on the rheological properties of a hydrogel: A systematic study. *J. Colloid Interface Sci.* **317,** 611–619.

Needham, D., and Nunn, R. S. (1990). Elastic-deformation and failure of lipid bilayer-membranes containing cholesterol. *Biophys. J.* **58,** 997–1009.

Ricco, A. J., and Martin, S. J. (1987). Acoustic wave viscosity sensor. *Appl. Phys. Lett.* **50,** 1474–1476.

Rodahl, M., Hook, F., Krozer, A., Brzezinski, P., and Kasemo, B. (1995). Quartz crystal microbalance setup for frequency and Q-factor measurements in gaseous and liquid environments. *Rev. Sci. Instrum.* **66,** 3924–3930.

Saitakis, M., Dellaporta, A., and Gizeli, E. (2008). Measurement of two-dimensional binding constants between cell-bound major histocompatibility complex and immobilized antibodies with an acoustic biosensor. *Biophys. J.* **95,** 4963–4971.

Sukharev, S. I., Blount, P., Martinac, B., and Kung, C. (1997). Mechanosensitive channels of *Escherichia coli*: The MscL gene, protein, and activities. *Annu. Rev. Physiol.* **59,** 633–657.

Tellechea, E., Johannsmann, D., Steinmetz, N. F., Richter, R. P., and Reviakine, I. (2009). Model-independent analysis of QCM data on colloidal particle adsorption. *Langmuir* **25,** 5177–5184.

Tsortos, A., Papadakis, G., Mitsakakis, K., Melzak, K. A., and Gizeli, E. (2008a). Quantitative determination of size and shape of surface-bound DNA using an acoustic wave sensor. *Biophys. J.* **94,** 2706–2715.

Tsortos, A., Papadakis, G., and Gizeli, E. (2008b). Shear acoustic wave biosensor for detecting DNA intrinsic viscosity and conformation: A study with QCM-D. *Biosens. Biolectron.* **24,** 836–841.

Watts, E. T., Krim, J., and Widom, A. (1990). Experimental observations of interfacial slippage at the boundary of molecularly thin films with gold substrates. *Phys. Rev. B: Condens. Matter* **41,** 3466–3472.

Zhu, C., Bao, G., and Wang, N. (2000). Cell mechanics: Mechanical response, cell adhesion, and molecular deformation. *Annu. Rev. Biomed. Eng.* **2,** 189–2000.

CHAPTER THREE

Liposome Characterization by Quartz Crystal Microbalance Measurements and Atomic Force Microscopy

Patrick Vermette[*,†]

Contents

Abstract

This chapter reviews liposome characterization by quartz crystal microbalance (QCM) measurements and atomic force microscopy (AFM). In many studies, AFM imaging is simply used to image liposomes with resolution often that does not allow morphological analysis. Although liposome size can be obtained by processing AFM images, it is found that liposomes flatten upon surface adsorption or immobilization. Liposome stability and stiffness have been characterized by using AFM imaging or AFM force measurements, although the latter method, using a microsphere attached on the AFM cantilever, seems more appropriate to limit liposome damage and to obtain more quantitative analysis, such as the Young's modulus. Investigation of liposome layers by QCM revealed that liposomes can be detected from a combined analysis of frequency and bandwidth shifts. However, QCM by itself provides only limited information on liposomes. QCM can be used to assess the presence of a layer and also to discriminate between rigid and viscoelastic ones. Liposome properties have been derived

* Laboratoire de Bioingénierie et de Biophysique de l'Université de Sherbrooke, Department of Chemical and Biotechnological Engineering, Université de Sherbrooke, Sherbrooke, Québec, Canada
† Research Centre on Aging, Institut Universitaire de Gériatrie de Sherbrooke, Sherbrooke, Québec, Canada

Methods in Enzymology, Volume 465
ISSN 0076-6879, DOI: 10.1016/S0076-6879(09)65003-5

from QCM curves, but often this requires making hypotheses that are difficult to assess. AFM and QCM analyses need to be combined with other techniques to provide complementary information.

Abbreviations

DMPC	dimyristoyl-*sn*-glycero-3-phosphocholine
DMPG	1,2-dimyristoyl-*sn*-glycero-3-phospho-*rac*-91-glycerol
DOPA	1,2-dioleoyl-*sn*-glycero-3-phosphate
DOPC	1,2-dioleoyl-*sn*-glycero-3-phosphocholine
DOPE	1,2-dioleoyl-*sn*-glycero-3-phosphoethanolamine
DOPS	dioleoyl-phosphatidylserine
DOTAP	1,2-dioleoyl-3-trimethylammonium propane
DPPC	dipalmitoyl-*sn*-glycero-3-phosphocholine
DSPC	distearoyl-*sn*-glycero-3-phosphocholine
DSPE	distearoyl-phosphatidylethanolamine
DSPE-PEG-carbonate	poly(ethylene glycol)-α-disteroylphosphatidylethanolamine-ω-benzotriazole carbonate
PA	phosphatidic acid
PC	phosphatidylcholine
PE	phosphatidylethanolamine
PEG	poly(ethylene glycol)
PEO	poly(ethylene oxide)
PG	phosphatidyl-glycerol
POPC	1-palmitoyl-2-oleoyl-*sn*-glycero-3-phosphocholine
POPE	1-palmitoyl-2-oleoyl-*sn*-glycero-3-phosphoethanolamine
POPG	1-palmitoyl-2-oleoyl-*sn*-glycero-3-phosphatidyl-DL-glycerol
PS	phosphatidylserine

1. Introduction

The interactions between lipid vesicles or liposomes and solid surfaces find application in several fields. Supported phospholipid bilayers can be made by exposing solid surfaces to suspensions of lipid vesicles, providing models of biological membranes (Plant, 1999). Supported phospholipid bilayers can be relatively well defined but this depends on the fabrication technique. There may be problems associated with their use, primarily due to the decreased fluidity caused by the underlying rigid substrate (Dahmen-Levison *et al.*, 1998; Jung *et al.*, 2000). Thus, layers of intact liposomes have been proposed as an alternative approach to investigate biological membranes. Such usage hinges, however, on the ability to bind liposomes intact on solid surfaces, which is contrary to the notion that liposomes are always prone to destabilization when in contact with a surface.

Many articles dealing with the adsorption of liposomes on surfaces report some interactions between the liposomes and the surfaces. These interactions can induce stresses on lipid membranes resulting in deformation, flattening, and even rupture of the liposomes (Lipowsky and Seifert, 1991). Any modification of the membrane curvature can induce a variation in its behavior, which can even lead to membrane fusion and/or rupture, as it is often seen in vesicle adsorption experiments (Lipowsky and Seifert, 1991; Plant, 1999; Reviakine and Brisson, 2000). The driving forces for liposome deformation are a competition between the membrane bending and adhesion energies (Seifert and Lipowsky, 1990). Many groups have investigated the mechanisms behind liposome adsorption, fusion, and rupture on surfaces with the aim to produce supported phospholipid bilayers (Plant, 1999; Radler *et al.*, 1995; Reviakine and Brisson, 2000; Steinem *et al.*, 1996). In spite of the practical importance of vesicle disruption upon contact with solid supports, little is known about the mechanisms and kinetics of supported phospholipid bilayer formation from a liposome suspension.

From a different perspective, surface-bound liposomes have been suggested as a possible method to locally deliver drugs in the vicinity of, for example, biomedical devices (Danion *et al.*, 2007a,b; Vermette *et al.*, 2002). When considering liposomes as drug delivery systems, it is necessary to maintain their structural integrity up to the intended end use, to allow the sustained release of the encapsulated therapeutic agent for the required time. Other applications of surface-immobilized liposomes involve immunodiagnosis (Lee *et al.*, 2005, 2008) and coating of capillaries in capillary electrophoresis (Mei *et al.*, 2008).

Methods to immobilize liposomes on solid surfaces or into solid, or gel scaffolds have been reviewed by Brochu *et al.* (2004). For instance, immobilization of intact liposomes can be done with naked vesicles (i.e., vesicles with no polymer protective shell or biomolecules) but only under given experimental conditions and with proper combination of surfaces and liposomes (Keller and Kasemo, 1998; Reimhult *et al.*, 2002, 2003). Attachment of intact liposomes requires a strategy to avoid the rupture of the vesicles. Methods have been investigated to fix intact liposomes onto solid surfaces and into 3-D matrices (Brochu *et al.*, 2004). These include (i) steric entrapment (Glavas-Dodov *et al.*, 2002; Pleyer *et al.*, 1994; Ruel-Gariepy *et al.*,

2002; Weiner *et al.*, 1985), (ii) attachment by hydrophobic interactions (Hara *et al.*, 2000), (iii) covalent binding (Khaleque *et al.*, 2003; Mao *et al.*, 2003), (iv) electrostatic attraction (Lee *et al.*, 2008), and (v) specific binding (Vermette *et al.*, 2002; Yang *et al.*, 1998), to name the most common.

This chapter reviews the use of atomic force microscopy (AFM) and quartz crystal microbalance (QCM) measurements to study intact liposomes or lipid vesicles. These techniques can yield valuable information on liposomes; but at the same time, because of the popularity of the techniques, data derived from AFM and QCM have often been overinterpreted. Here, liposome properties that can be gathered by these two instruments are highlighted. Advantages and pitfalls associated with their use to characterize liposomes are also addressed.

The information obtained from the scientific literature relative to the interactions of liposomes with solid surfaces is often confusing, and sometimes contradictory. Discrepancies reported in the literature may perhaps be attributable to variability in the liposome formulation, the method of liposome preparation, and/or the surfaces used to adsorb or immobilize liposomes. Also, considering that many articles are often very technical and descriptive, that is, providing very little scientific insight, this chapter will try to avoid the often phenomenological analysis conveyed in some QCM and AFM studies of liposomes.

2. Liposomes Assessed by AFM

2.1. The AFM technique

AFM involves the use of a cantilever bearing a tip at its end that can be used to scan or probe surfaces. When the tip approaches a solid surface, forces between the tip and the sample lead to a deflection of the cantilever. AFM can be used either to image samples or to obtain force profiles between the AFM cantilever tip and a sample. For imaging, the AFM can be operated in a number of modes, depending on the application. Imaging can be carried out in contact mode, Tapping mode, and noncontact mode (Digital Instruments, Veeco Metrology Group, 2000). Contact mode and Tapping mode are the most common methods that have been used to image liposomes. In contact mode, the cantilever tip is scanned across the surface while monitoring the change in cantilever deflection. A feedback loop maintains a constant "setpoint" deflection between the cantilever and the surface while moving the scanner across the scanned surface area. The force between the tip and the surface is kept constant during this operation. In the Tapping mode, the cantilever tip is scanned across the solid surface while oscillating. A feedback loop maintains a constant "setpoint" oscillation amplitude while moving the scanner across the scanned surface area. In traditional AFM force measurements, the AFM tip is extended toward and retracted from a solid surface, while cantilever deflection is monitored as a function of the piezoelectric displacement.

Thus, AFM can be used either to image liposomes or to carry out force measurements to extract some information on their rigidity and stability. Liposome characterization using these two AFM methods is reviewed in the following subsections.

2.2. AFM imaging of liposomes: Topology and size analyses

In many studies, AFM is simply used to visualize liposomes or detect their presence on surfaces (Casals *et al.*, 2003; Colas *et al.*, 2007; Estes and Mayer, 2005; Garbuzenko *et al.*, 2009; Jung *et al.*, 2005, 2006; Kanno *et al.*, 2002; Kim *et al.*, 2005; Kleemann *et al.*, 2007; Lee *et al.*, 2005; Mei *et al.*, 2008; Paclet *et al.*, 2000; Ruozi *et al.*, 2005; Silva *et al.*, 2008).

For example, AFM imaging in Tapping mode (in liquid) of PC liposomes adsorbed on glass showed that the topology of actin-containing liposomes was dependent on the actin concentration and liposome size (Li and Palmer, 2004). Liposomes encapsulating monomeric actin or filamentous actin had different structures (Palmer *et al.*, 2003). The structure can be associated with a change in the liposome stiffness, as AFM force measurements with bare cantilever tips revealed that liposomes encapsulating actin were stiffer than control liposomes (Zhang *et al.*, 2007). AFM imaging in air was used to confirm the difference between linear and circular DNA-liposome complexes (von Groll *et al.*, 2006). Another example of AFM imaging involved carboxylic acid-terminated self-assembled monolayers (SAM) that were exposed to DMPC or DMPC:DMPG (4:1) liposomes having multimodal size populations (Mechler *et al.*, 2009). The combination of AFM imaging and QCM analysis revealed that lipid mono- and multilayers, as well as vesicles, were found on the surfaces. Liposomes (PC:DSPE-PEG-carbonate:cholesterol) were covalently attached on amine-bearing surfaces (Lunelli *et al.*, 2005). In a rare case, X-ray photoelectron spectroscopy (XPS) was used to ascertain the presence and chemical composition of the layers. AFM imaging in air was used to inspect the liposome topography (Lunelli *et al.*, 2005).

In some studies, liposome sizes were obtained by AFM imaging (Colas *et al.*, 2007; Kleemann *et al.*, 2007). However, the liposome size derived from AFM images can be different from that measured using dynamic light scattering (Colas *et al.*, 2007).

AFM imaging (contact or Tapping mode) and force measurements (with bare cantilever tips) were conducted on egg-PC liposomes adsorbed on mica (Liang *et al.*, 2004). The liposome sizes as measured by light scattering and AFM imaging did not concur. Although the presented AFM images were of good quality, a constant size of ca. 40–60 nm was observed by AFM sectional analysis, regardless of the cholesterol concentration, while light scattering showed that the liposome average diameters increased as the cholesterol concentration increased (Liang *et al.*, 2004). This is a rare study that compares liposome sizes obtained by AFM and light scattering. Perhaps, the discrepancy was partly due to the fact that liposomes were made in 20 m*M* NaCl solution, and AFM imaging was carried out in pure water while light scattering experiments were conducted in 20 m*M* NaCl. However, such a condition should expose liposomes only to a small osmotic stress as the NaCl concentration was low. This finding is in line with a study showing that AFM imaging did not match photon correlation spectroscopy and hydrodynamic chromatography for the particle size measurements of lipid nanocapsules (Yegin and Lamprecht, 2006). In another study, AFM imaging (contact mode) in aqueous solutions was used to calculate the width-to-height ratio of PC vesicles adsorbed on glass surfaces (Schönherr *et al.*, 2004). The critical rupture radius of the vesicles was estimated to be larger than 250 nm (Schönherr *et al.*, 2004). The vesicles were found to flatten upon adsorption with a width-to-height ratio of approximately 5 (Schönherr *et al.*, 2004). Flattening is perhaps one of the main reasons why liposome sizes obtained by AFM imaging can differ from that measured by light scattering.

On the other hand, average diameters of liposomes composed of DPPC: cholesterol:PE-PEG, DSPE-cyanur, or PE-PEG-cyanur measured by photon correlation spectroscopy and by AFM were in agreement (Bakowsky *et al.*, 2008). Considering that liposomes maintain their integrity for only a few minutes after deposition, others suggested that if AFM images were obtained within 10 min after liposome deposition, sizes obtained from processing AFM images were comparable with those measured by photon correlation spectroscopy (Ruozi *et al.*, 2005).

There are also some issues associated with the use of photon correlation spectroscopy to gather particle size distribution. Dilute particle concentrations are required to ensure that multiple scattering and particle-particle interactions are negligible. I also recommend using volume-weighted size distributions, which give a more representative scattered intensity; the intensity-weighted size distributions are more biased toward large particles, while the number-weighted distributions are more influenced by small particles (Berne and Pecora, 1976). Also, its use is more problematic with broad particle size distribution and with nonspherical particles (Berne and Pecora, 1976; Ostrowsky, 1993). In photon correlation spectroscopy measurements, small particles are weakly weighted (Berne and Pecora, 1976; Ostrowsky, 1993). By contrast, a small number of large particles can result in important variation in size distribution (McCracken and Sammons, 1987). Furthermore, amphiphilic structures such as liposomes should be considered as soft or fluid-like, with molecules in constant thermal motion within each aggregate. Unlike monodisperse particles, most liposomes do not have well-define size or shape (at the molecular level) but rather a distribution about a mean value (Israelachvili, 1992). In most commercial photon correlation spectroscopy instruments, intensity data are converted to the number distributions by assuming a solid sphere. A liposome is not a solid sphere, but rather a hollow particle with an aqueous medium entrapped within a bilayer shell. When the intensity data for lecithin lipid vesicles was treated using a hollow-sphere model, the mean number distribution diameter was 10% larger than that given by the instrument using a model of a solid sphere (McCracken and Sammons, 1987). Moreover, it may be relevant to collect data at a few scattering angles to help lower the measured polydispersity of the liposome populations.

Using AFM, Shibata-Seki *et al.* (1996) imaged in liquid human IgG-bearing liposomes adsorbed onto antihuman IgG-covered substrates. AFM imaging of the liposomes showed balloon-like structures. The authors reported that the quality (contrast and/or reproducibility) of the AFM images depended both on the type of cantilever tips and on the load forces at which samples were scanned (Shibata-Seki *et al.*, 1996). While specific binding via IgG/anti-IgG interaction would allow the binding of liposomes on surfaces, the authors should have investigated if liposomes were attracted to the substrates via nonspecific (physisorptive) interaction forces. Furthermore, the method by which liposomes were "decorated" with IgG was not described.

Liposome immobilization using the specific interaction between NeutrAvidin^{TM} and biotin has been proposed by Vermette *et al.* (2002) to fabricate biomedical device coatings to locally deliver drugs. AFM images showed intact liposomes, but their surface density was below monolayer packing. A low leakage rate was observed, probably caused by the PEGylated lipids in the membrane of the liposomes (Vermette *et al.*, 2002). In a subsequent study, liposomes containing PEG-biotinylated lipids were docked at high density onto NeutrAvidin^{TM} surface immobilized on contact lenses (Danion *et al.*, 2007b). In a different investigation but with a similar liposome immobilization technique, AFM Tapping mode imaging revealed high density of intact vesicles (Tarasova *et al.*, 2008). The root-mean-squared (rms) roughness of surfaces increased from 1.5 nm for NeutrAvidin^{TM}-coated surfaces to 10 nm for surfaces with liposomes. From AFM images, the average height and width of individual surface-bound liposomes were 46 and 132 nm, respectively. Again, this illustrates that liposomes flatten upon interaction with surfaces.

Some authors suggest that sample preparation for AFM is easy and fast and that the technique allows liposomes to preserve their native state. These statements should be interpreted with great care. In my opinion, AFM imaging often does not provide high-resolution analyses over the morphology and structure of vesicles, as stated by others (Mozafari *et al.*, 2005; Ruozi *et al.*, 2007). Papers presenting AFM images of liposomes with the required resolution to obtain morphological information are hard to come by. Although electron microscopy methods can require chemical fixation, information on the morphology of liposomal structures has been obtained. Electron microscopy is a traditional way to get structural information and has been a valuable tool to study lipid aggregate structures, in some cases with a resolution in the nanometer range (Cevc and Seddon, 1993). Furthermore, electron microscopy in the cryo mode has been used to study lipid vesicles (Almgren *et al.*, 2000; Frederik *et al.*, 1991). For example, the freeze-fracture technique enables the characterization of phase separations (Hope *et al.*, 1989; Verkleij and Ververgaert, 1978). However, while preparing liposome samples for electron microscopy investigations, care must be taken to avoid osmotic stress, as this could significantly alter the morphology of liposomes. Temperature gradients caused by insufficiently rapid cryofixation are another potential source of problems. As for AFM, in every case, a sufficient number of images should be collected to ensure reproducibility.

Too often liposome imaging by AFM is carried out in a dehydrated state, that is, in air. Given the fact that liposomes are composed of amphiphilic molecules and are dynamic hydrated structures, it is surprising that many researchers limit themselves to perform experiments in air. This is perhaps due to the fact that it is easier to image in air and that resolution can be higher in air than in a liquid environment for such dynamic assemblies.

2.3. AFM force measurements: Characterization of liposome stiffness and stability

In some applications of surface-bound liposomes, it is required that liposomes remain intact upon surface attachment. This can be done by protecting liposomes with, for example, poly(ethylene glycol) (PEG)-bearing lipids. The use of PEGylated-lipids in liposome preparation increases their stability and circulation time in the bloodstream. For example, the absence of DSPC-PEG in DSPC:cholesterol extruded liposomes resulted in liposomes that aggregated into large clusters of deformed liposomes (Tardieu *et al.*, 1973). With the use of small concentrations of PEG-lipids, liposomes became spherical and appeared to be better separated.

AFM imaging can provide some information on the stiffness or stability of liposomes. For example, the incorporation of PEGylated lipids into liposomes resulted in round and spherical structures in AFM Tapping mode imaging, while irregular structures were observed with plain liposomes (DSPC:cholesterol) (Anabousi *et al.*, 2005). AFM images in air by Tapping mode revealed that liposomes with DSPE-PEG were less spread on mica than liposomes without PEG lipids (Carmo *et al.*, 2008).

In an elegant study, the rigidity of liposomes composed of either egg-PC, DMPC, or DSPC with cholesterol and stearylamine was investigated by comparing the change in the ratio of the height of the particles adsorbed on mica obtained by AFM imaging against the particle diameter measured by photon correlation spectroscopy (Nakano *et al.*, 2008). The liposomes made with the phospholipid having the highest phase transition temperature, that is, DSPC, showed the highest height/particle diameter ratio. The rigidity of liposomes decreased as the particle size increased. The authors also found that as the cholesterol content increased, the rigidity of the vesicles decreased. The effect of cholesterol on liposome stiffness and physical and biological stability has been well documented using other well-accepted techniques to investigate lipid systems. For example, the addition of cholesterol (30-50 mol%) into DSPC large unilamellar vesicles (LUV) resulted in stable vesicles in blood circulation ($t_{1/2} = 5\text{-}6$ h) and without cholesterol, pure DSPC vesicles were cleared rapidly ($t_{1/2} < 2$ min) (Semple *et al.*, 1995). It is often reported that cholesterol prevents the drastic change in membrane fluidity during the gel-liquid-crystalline phase transition by increasing the fluidity in the gel phase and by decreasing it in the liquid-crystalline phase (Hayakawa *et al.*, 1998). Indeed, cholesterol can have a stabilizing, and in some systems, a fluidizing effect, on lipid bilayers. As reported by Shin and Freed (1989a,b), electron spin resonance revealed that cholesterol increased ordering in the liquid crystalline state. On the other hand, in the gel phase, cholesterol induced local disorder. It was also found that cholesterol tends to aggregate. In fact, cholesterol-rich regions were more solid-like due to the dense packing. In any lipid bilayer membrane, there is an upper limit of cholesterol concentration than can be accommodated within the membrane; excess cholesterol will precipitate as crystals (Huang *et al.*, 1999). Furthermore, when lipids are in the gel state, cholesterol had a dose-dependent effect on the permeability (Raffy and Teissie, 1999). Besides modulating the membrane fluidity, cholesterol affects the polymorphism of lipid membranes. For example, it was shown that cholesterol destabilized lamellar structure and promoted the hexagonal H_{II} structure for PC:PE mixtures, depending upon the PC/PE ratio, the ratio of cholesterol to lipid and the temperature (Cullis and De Kruijff, 1978; Tilcock *et al.*, 1982).

While AFM imaging provides some insight on liposome stiffness and/or stability, AFM force measurements appear better suited to extract more quantitative information. For example, compared with pure egg-PC liposomes, the Young's modulus obtained by AFM force measurements of cholesterol-containing liposomes increased (Liang *et al.*, 2004).

AFM imaging and force measurements revealed that Pluronic® with longer poly(ethylene oxide) (PEO) chain length resulted in more stable Pluronic®-modified egg-PC liposomes (Liang *et al.*, 2005). The enhanced stabilization, as revealed by the bending modulus, can be attributed to a shell-like protective effect over the liposomes and/or an increased rigidity of the vesicle bilayers as a result of the poly(propylene oxide) (PPO) block incorporation (Liang *et al.*, 2005).

We have used AFM force measurements using a colloidal probe combined with the monitoring of a fluorescent dye release to characterize surface-bound liposomes (Vermette *et al.*, 2004). After liposome binding using the NeutrAvidin^{TM}-biotin link, approaching the surface with a colloidal probe glued on an AFM cantilever showed considerable compression, consistent with what is expected from intact, deformable liposomes, but not lipid bilayers. Plastic deformation suggestive of liposome disruption on compression was not detected. The kinetics of fluorescent dye release also revealed that intact liposomes had been attached on surfaces. Blocking surface-docked NeutrAvidin^{TM} with excess biotin in solution prior to exposure to liposomes showed that the attachment of liposomes was dependent largely, but not exclusively, on the specific biotin-NeutrAvidin^{TM} affinity binding, with some nonspecific physisorption. Chemisorption (i.e. sharing/exchange of electrons) and physisorption (i.e. no sharing/exchange of electrons) are the two categories of interactions occurring in protein-surface investigations. In fact, the level of liposome adsorption on preblocked NeutrAvidin^{TM} surfaces was much lower than for unblocked surfaces, but not zero. Consecutive addition of further NeutrAvidin^{TM} and liposome layers allowed fabrication of multilayers, as seen in AFM compressibility and fluorescent dye release assays (Vermette *et al.*, 2004).

In the continuation of the previous work, force-distance curves were obtained by AFM using colloidal probes over liposome layers and used to calculate Young's moduli by using the Hertz's contact theory (Brochu and Vermette, 2008). Young's moduli of 40 and 8 kPa have been obtained for one and three layers of intact liposomes, respectively. Compression work performed by the colloidal probe to compress these liposome layers has also been calculated.

AFM force measurements, combined with a model of contact theory, have been used to estimate Young's moduli (E) of different materials (Radmacher *et al.*, 1995; Weisenhorn *et al.*, 1993). By gathering Young's moduli of liposomes, some properties such as their structural stability could be better estimated.

Young's moduli can be extracted from AFM force curves by fitting data with a model of contact. Models often encountered are Hertz (1881), Johnson-Kendall-Roberts (JKR) (Johnson *et al.*, 1971), and Derjaguin-Muller-Toporov (DMT) (Derjaguin *et al.*, 1980; Muller *et al.*, 1980, 1983). Viscoelastic properties of hydrated layers can be derived by analyzing the Brownian motion of AFM cantilever in contact with the surface (Benmouna and Johannsmann, 2004), but this method also requires some assumptions. The JKR and DMT models are based on the Hertz's theory but they treat adhesive forces between the probe and the sample during contact. When no adhesive forces are involved, they return to the Hertz's model. Since the original Hertz model (Hertz, 1881) involves the contact between a sphere and a plane surface, variations of Hertz's theory have been suggested for systems involving other geometries of contact (Sneddon, 1965). For example, a modified Hertz model has been developed by Dimitriadis *et al.* (2002) to treat the effect of the sample thickness. To calculate Young's moduli using a corrected Hertz model (e.g. Dimitriadis *et al.*'s model), the sample thickness needs to be specified. Estimating the thickness of hydrated layers is a challenging task. We have tested the effect of the estimated thickness of the liposome layers on the calculated Young's modulus, and found that this estimation has a considerable impact (Brochu and Vermette, 2008). Also, the Poisson ratio needs to be provided; a ratio of 0.5 is often assumed for liposome layers since samples are considered incompressible, as for most biological samples (Dimitriadis *et al.* 2002). Finally, sample indentation needs to be measured, because there is an indentation in soft samples and this is also not an easy task. Indentation is defined as the difference between the piezo displacement and the cantilever deflection. This indentation can be calculated from the difference between the piezo position, considering that the deflection of the cantilever on a hard sample (e.g. a smooth glass surface) and on a soft sample (e.g. liposomes) is the same. Indentation must be calculated from AFM force curves with no force (e.g., no jump-to-contact) interfering between the AFM tip and the sample. To calculate the indentation, the contact point has to be known. Because it is difficult to identify the contact point, the latter should be estimated, along with the Young's modulus during calculations to obtain the best fit to the model (Brochu and Vermette; 2008; Dimitriadis *et al.*, 2002). Indentation over a certain percentage (usually 10%) of the total sample thickness should be used for the calculation to limit the influence of the underlying (often rigid) substrate.

Discrepancies of Young's moduli of intact liposomes immobilized on surfaces are observed in the scientific literature (see Table 3.1). These can be explained by the techniques used to produce these complex layers, which are often different from one study to another, by the liposome composition, and/or by the absence of verification to prove that liposomes remain intact upon surface immobilization. Table 3.1 lists examples of Young's moduli of different liposomal systems. The implications of the Young's modulus in liposome systems have not been investigated. Nevertheless, it can be hypothesized that the stability and release rate of encapsulated molecules can be correlated to the Young's modulus of liposomes, which the latter can depend on liposome composition and size. Also, in some applications of surface-bound liposomes (e.g. coating for contact lenses) it can be important to match liposome mechanical properties to those of the contacting cell layers/tissues.

It is interesting to compare Young's moduli of liposomes to those of living cells. From the more abundant literature on mechanical properties of cells, it can be observed that Young's moduli of cells vary from one study to another. Also, different cell types can have different moduli. Furthermore, the location

Table 3.1 Young's moduli of liposomes obtained by AFM force measurements

Liposome types	Young elastic modulus	Model	References
Biotinylated-PEG liposomes (ca. 110 nm) on NeutrAvidin™ surfaces	One liposome layer: 40 ± 20 kPa Three liposome layers: 8 ± 2 kPa	Hertz's model with AFM cantilever tips bearing a microsphere	Brochu and Vermette (2008)
• Egg-PC liposomes • Egg-PC liposomes:cholesterol (85:15) • Egg-PC liposomes:cholesterol (80:20) • Egg-PC liposomes:cholesterol (70:30) • Egg-PC liposomes:cholesterol (50:50) Numbers in brackets represent molar ratios	1.97 ± 0.75 MPa 12.07 ± 1.53 MPa 10.77 ± 0.64 MPa 10.4 ± 4.06 MPa 13.0 ± 2.97 MPa	Hertz's model with bare tips	Liang *et al.* (2004)
• DOPC:DOPS vesicles with cisplatin • DOPC:DOPS vesicles without cisplatin	880 kPa 450 kPa	Hertz-Sneddon's model with bare AFM tips	Ramachandran *et al.* (2006)
DOPC:DOPE liposomes	6.3 ± 0.6 MPa for control liposomes without actin filaments 26.3 ± 15.4 MPa for artificial cell-like structure	Hertz's model with bare tips	Zhang *et al.* (2007)

on cells where Young's moduli are measured needs to be specified (Chouinard *et al.*, 2008). For example, the Hertz model was used to fit AFM colloidal probe force measurements obtained over human epithelial cells to extract Young's moduli (Berdyyeva *et al.*, 2005). Values of 14 and 37 kPa were measured over the nucleus and cytoplasm of young cells. For comparison, Young's moduli of gelatin gel, at different pH, ranged between 1 and 9 kPa (Benmouna and Johannsmann, 2004). Polyacrylamide gels with various cross-linker concentrations showed E ranging between 1 and 8 kPa (Engler *et al.*, 2004).

In an excellent study, liposomes (DSPC/cholesterol/DSPE-PEG-biotin) were bound on solid surfaces bearing NeutrAvidin™ docked on a biotin-PEG-NHS coating produced under cloud point conditions. AFM force measurements between a microsphere attached to the cantilever and surface-bound liposomes resulted in repulsive forces for all tested approach velocities (Tarasova *et al.*, 2008). These AFM force measurements revealed that the surface-bound liposomes were stable and were barely affected by the application of normal and lateral forces. The interaction between albumin-coated AFM probes and the surface-bound liposomes resulted in low magnitude adhesive forces when compared to the other tested surfaces, revealing that surface-bound liposomes can provide antifouling properties. This is in agreement with QCM experiments showing that surface-bound liposomes were able to resist adsorption from fetal bovine serum (Brochu and Vermette, 2007). The term anti-fouling or low-fouling relates to the resistance of a surface towards protein adsorption, which subsequently affects or dictates cell adhesion to the surface. Friction analysis of surface-bound liposomes using AFM was also carried out (Tarasova *et al.*, 2008). Although some results reported in Tarasova *et al.*'s study cannot be fully explained, due to the complexity of the systems and the experiments, this work should pave the way for further investigations of liposomes by the proposed methodology.

AFM force measurements with bare cantilever tips revealed that liposomes containing cisplatin (a chemotherapy drug) were found to be stiffer and more stable than empty liposomes (Ramachandran *et al.*, 2006). AFM force measurements with bare cantilever tips were used to detect the interaction forces between the C2A domain of synaptotagmin I and liposomes (Park *et al.*, 2006). The C2A domain of synaptotagmin I binds to negatively charged phospholipids (Park *et al.*, 2006). In this study, the C2A domain of synaptotagmin I was claimed to be immobilized on gold-coated cantilevers (Park *et al.*, 2006), although no proof of such immobilization was provided. Also, liposome size distribution was not reported and the method of liposome immobilization on a Biacore sensor chip remains elusive.

AFM force measurements on surface-bound liposomes carried out with bare cantilevers can create artifacts. As noted by some authors (Liang *et al.*, 2004, 2005), retraction curves appear to involve resealing of portions of lipid bilayers, adhesion forces between the tip and the vesicles, and elongation of lipid structures. Essentially, it is possible for bare AFM cantilever tips to damage the liposomes. A colloidal probe is recommended to conduct AFM force measurements, as liposomes do not appear to break in such conditions.

2.4. Examples of protocols to carry out AFM measurements

The following guidelines provide the general instructions to carry out AFM experiments in imaging and force modes. The sequence of some steps can be modified.

Procedures prior to start either AFM imaging or AFM force measurements:

- Substrates are prepared and cleaned for liposome immobilization. Substrates should be molecularly smooth. Examples of smooth substrates are freshly cleaved mica, silicon wafers and some borosilicate plates. Substrate roughness should be assessed (and not assumed) prior to AFM experiments in order to determine if the substrate can be used for liposome immobilization. A root-mean-squared (rms) roughness of the substrate equal or below 0.5 nm is recommended.
- Liposomes are immobilized. An example of liposome immobilization protocol is provided elsewhere (Brochu and Vermette, 2008). Layers of liposomes should always be maintained in a hydrated state and in iso-osmolar solutions (compared to the osmolarity of their inner contents) to avoid liposome rupturing or structure changes.

Procedures for AFM imaging:

- The AFM mode of operation is selected (e.g. Tapping mode or contact mode). An adequate probe is selected as a function of the type of experiments and samples. Before mounting, the probe is cleaned thoroughly in surfactant, followed by rinses in water and then in ethanol, finishing with an ozone/UV treatment. Tweezers should be used to manipulate the probes. Samples are mounted in a liquid environment, using clean tweezers to manipulate the samples. All aqueous solutions that will come in contact with the liposome-covered surfaces should be filtered through membranes of 0.2 μm or smaller pore diameter. The optical head/scanner is mounted, depending on the AFM instrument setup used. The system is allowed to reach thermal equilibrium (at least one hour). The laser is aligned by adjusting the photodiode signal, and the tip-sample approach is performed. In the Tapping mode, the cantilever frequency is tuned. Depending on the imaging mode, the initial scan parameters are set (scan size, integral and proportional gains, setpoint, scan rate, etc.). After the initial engagement, the scan parameters are readjusted. Images are acquired on different spots on the same substrate to verify sample homogeneity, and on different samples of the same conditions.

Procedures for AFM force measurements:

- The force mode is selected. The probe is cleaned and mounted, as described above for AFM imaging. For colloidal probe force measurements, silica microspheres can be attached to the cantilevers. The spring constant of the cantilevers needs to be measured (I suggest using Cleveland *et al.*'s method (Cleveland *et al.*, 1993)) and specified prior to analyse AFM force curves. The samples are mounted in a liquid environment. All aquesous solutions should be filtered through membranes of 0.2 μm or smaller pore diameter. The samples should be firmly fixed. Particular attention should be devoted to this, especially with the Bioscope-like setup. The optical head/scanner is mounted, depending on the AFM instrument setup used, and at least one hour is allowed to reach thermal equilibrium. The laser is aligned by adjusting the photodiode signal, and the tip-sample approach is performed. The scan parameters (e.g. z-scan start, z-scan size, setpoint, etc.) are defined. The AFM force curves are processed and analyzed.
- Transformation of deflection versus z-position curves needs to be done to adequately compare force curves; to do so, zero-distance and zero-force positions need to be specified. The hard wall can be difficult to define for some samples. Also, the cantilever deflection needs to be transformed into force values, provided that the cantilever spring constant has been measured with sufficient accuracy. To analyse indentable samples such as liposome layers, I suggest making force-vs-indentation graphs. Processing AFM force curves can be done by a computer program, which can be made freely available upon request to me. Note that not many commercial computer programs are available to analyse AFM force curves.

3. QCM Measurements of Intact Liposomes

The principles involved in QCM measurements have been well described by Janshoff *et al.* (2000) and Bottom (1982). Briefly, QCM can be depicted as mechanical transducers for sensing chemical and biological reactions or events that transform the mass of a deposited layer into an electrical signal. The work by Nomura and Okuhara (1982) showed that QCM can be operated in a liquid phase. As the excellent review provided in Wikipedia (http://en.wikipedia.org/wiki/QCM, July 15th, 2009) and in other papers (Bottom, 1982; Johannsmann, 1999; Wegener *et al.*, 2001), in some QCM systems, frequency (f) shifts are monitored along with the bandwidth (w). The bandwidth is related to energy loss in the system. Parameters other than w are used: the Q-factor (quality factor) is given by $Q = f/w$, the dissipation (also referred to as damping), D, which is the inverse of Q ($D = Q^{-1} = w/f$), and the half-band-half-width, Γ (Γ = $w/2$). The delta *f* is proportional to the deposited mass if the resonator is operated in air or vacuum, or if the layer is as rigid as the quartz. The mass sensitivity of the resonator is maximal in its center and decreases toward the electrode edges. In water, the decay length of a 5-MHz quartz is estimated at 250 nm (Janshoff *et al.*, 2000). The Sauerbrey equation (Sauerbrey, 1959) can be used to estimate film masses, but it appears to be not valid in some cases, such as for thick films, viscous liquids, and viscoelastic layers. Mechanical models have been transformed into equivalent electrical circuits allowing the description of the oscillation of thin layers. Also, models have been proposed to evaluate properties of more complex layers such as viscoelastic films. Analyzing and comparing the bandwidth allows the verification of the applicability of the Sauerbrey equation and, this is sufficient. Trying to disentangle layer properties such as film thickness, viscosity, and elasticity from frequency and bandwidth shifts is not warranted. Also, some parameters can influence the resonant frequency of the quartz crystal. These include the conductance and permittivity of the solution, the surface roughness and change in the surface hydrophilicity, the electrochemical double layer, only to name a few (Janshoff *et al.*, 2000). These factors are, however, often overlooked.

Examples involving the use of QCM to study liposomes are presented below and some QCM results on liposomes are listed in Table 3.2.

The adsorption of egg-PC liposomes (diameter of approximately 20 nm) resulted in larger frequency and dissipation shifts on hydrophilic substrates than on hydrophobic ones (Rodahl *et al.*, 1997). Methyl-terminated and activated carboxythiol SAM were exposed to POPC/POPE liposomes (Morita *et al.*, 2006). Larger frequency shifts were observed on the activated carboxythiol SAM, although significant adsorption was detected on the methyl-terminated SAM, indicating nonspecific liposome adsorption (Morita *et al.*, 2006). Also, frequency shifts increased as the liposome diameter increased. Immobilization of liposome multilayers was accomplished using streptavidin–biotin binding repeated several times; liposomes-bearing biotinylated lipids were first immobilized by amino-coupling on the SAM and then alternatively exposed to streptavidin and more biotinylated liposomes. Frequency shifts of ca. 2000 Hz were measured for such multilayers (Morita *et al.*, 2006).

The adsorption of small unilamellar vesicles (SUV, egg-PC) with a mean radius of 12.5 nm on gold with alkane thiol, SiO_2, and oxidized gold was studied by QCM (Keller and Kasemo, 1998). From frequency and dissipation changes, the authors reported that liposome adsorption resulted in monolayers, bilayers, and intact vesicles on gold with alkane thiol, SiO_2, and oxidized gold, respectively. In further studies from the same research group, adsorption on SiO_2 of SUV (radius of ca. 12.5 nm) composed of egg-PC was followed by QCM and surface plasmon resonance (Keller *et al.*, 2000). The authors postulated that lipid bilayers were formed. Masses

Table 3.2 Liposome characterization by QCM measurements

Liposome types	Frequency shifts[a]	Dissipation[b]	References
• DPPC:cholesterol (2:1) • DPPC:cholesterol:PE-PEG • ConA-modified liposomes (DPPC:cholesterol:DSPE-cyanur) • ConA-modified liposomes (DPPC:cholesterol:DSPE-PEG-cyanur)	124 ± 18 Hz ca. 56 Hz ca. 825 Hz ca. 360 Hz	NA	Bakowsky *et al.* (2008)
Biotinylated-PEG liposomes (ca. 110 nm) on NeutrAvidin™ surfaces	ca. 200 ± 50 Hz	85 ± 20 Hz[c]	Brochu and Vermette (2007)
SUV POPC on gold	ca. 109-128 Hz	ca. 6-10 × 10^{-6}	Cho *et al.* (2007)
PE:PC:PS + DNA-cholesterol (28 nm)	ca. 100 Hz	8 × 10^{-6}	Granéli *et al.* (2004)
Carbohydrate-functionalized liposomes on lectin-bearing surfaces	700 Hz	NA	Hildebrand *et al.* (2002)
DPPC liposomes on polymer surfaces bearing phosphorylcholine groups	ca. 180-990 Hz	NA	Iwasaki *et al.* (1997)
Egg-PC (radius of 12.5 nm)	On Au + alkane thiol: 13 Hz On SiO_2: 26 Hz On oxidized Au: 90 Hz	ca. 0 ca. 0 ca. 3 × 10^{-6}	Keller and Kasemo (1998)
SUV (12.5 nm of radius) from egg-PC adsorbed on SiO_2 (hypothesized to form bilayers)	ca. 30 Hz (mass: ca. 500 ng/cm^2)	Close to zero	Keller *et al.* (2000)
DOPC:DOPA liposomes	95 ± 10 Hz	4.5 × 10^{-6}	Kohli *et al.* (2006)
Multilayers of POPC:POPG:cholesterol (with or without poly-D-lysine) vesicles (200 nm)	ca. 30 Hz	Depended on liposomes	Michel *et al.* (2004)

(*continued*)

Table 3.2 (*continued*)

Liposome types	Frequency shifts[a]	Dissipation[b]	References
• POPC/POPE liposomes on activated (carbodiimide) carboxythiol SAM • POPC/POPE liposomes on methyl-terminated thiol	ca. 500 Hz ca. 300 Hz	NA	Morita *et al.* (2006)
• DOPC:DOPS (1:1) • DOPC:DOPS (4:1) • DOTAP Vesicles were adsorbed on SiO_2	50 Hz 25 Hz 20.5 Hz	2.5×10^{-6} 0.1×10^{-6} 0.2×10^{-6}	Richter *et al.* (2003)
Egg-PC liposomes (diameter of 20 nm)	On hydrophobic surface: 13 Hz On hydrophilic surface: 87 Hz	ca. 0.4×10^{-6} ca. 3×10^{-6}	Rodahl *et al.* (1997)
Cholesterol-cDNA POPC vesicles extruded through 50 nm filters on DNA surfaces	ca. 100 Hz	ca. 10×10^{-6}	Städler *et al.* (2004)
• POPC/PG • POPC/PS • POPC/PA Vesicles extruded in 100 nm filters	ca. 30 Hz ca. 60 Hz with Ca^{2+}: ca. 100 Hz without Ca^{2+}: ca. 35 Hz	2 Ω[d] 15 Ω 100 Ω 0 Ω	Viitala *et al.* (2007)
Egg yolk phospholipids + antibody	ca. 385 Hz	NA	Yun *et al.* (1998)

[a] In some studies, frequency shifts are normalized to the overtone, while some omit to specify this information.
[b] Or other parameters related to viscous loss.
[c] Correspond to half-band-half-width.
[d] Correspond to resistance changes.
NA, not available; ConA, concanavalin A; SUV, small unilamellar vesicles.

calculated from surface plasmon resonance and QCM curves were different. The difference was attributed to the total mass of water trapped at the surface by the lipids. The authors stated that the calculated mass of the adsorbed lipid bilayer on SiO_2 from surface plasmon resonance curves would be independent of the morphology of the layer. It is, however, challenging to obtain the refractive index of such complex coatings; the refractive index is needed to calculate the optical thickness of a coating from surface plasmon resonance measurements, and small changes of this refractive index can have a significant effect on the calculated optical thickness. From the same research group, the interaction of liposomes with Au and SiO_2 was studied by QCM and surface plasmon resonance (Reimhult *et al.*, 2006). Upon interaction with the surfaces, POPC liposomes (mean diameter of 55 nm) formed vesicle layers and bilayers on Au and SiO_2, respectively. In this study, the liposome adsorption on Au resulted in a calculated QCM mass using the Voigt–Kelvin modeling of approximately 2700 ng/cm^2, while liposome adsorption on SiO_2 resulted in a calculated QCM mass of approximately 425 ng/cm^2 (Reimhult *et al.*, 2006). The dissipation was used to discriminate between vesicles and supported phospholipid bilayers, with corresponding values of 12×10^{-6} and 0, respectively (Reimhult *et al.*, 2006). AFM imaging (soft contact mode) was used to follow vesicle adsorption (mean diameter of 55 nm) on SiO_2 in a buffer solution. From these AFM images, the authors concluded that surface coverage by vesicles increased as the exposure time increased, and for the longest exposure time, bilayer patches were observed (Reimhult *et al.*, 2006), although it can be hypothesized that these patches could also have been the bare underlying SiO_2 substrate. It should be noted that AFM imaging does not allow discriminating chemical composition across a scanned area.

Formation of supported vesicle layers and supported lipid bilayers was studied by Brisson's group (Richter *et al.*, 2003). QCM and AFM imaging were combined to gather information on the mechanisms of layer formation. Interestingly, they used PEG-modified AFM tips for AFM imaging to reduce tip-sample interactions (Richter *et al.*, 2003). The AFM images were of good quality and were used to gather information on the organization of surface-bound liposomes and lipids. The AFM tip induced some local rupture of vesicles and supported lipid bilayer formation was observed (Richter *et al.*, 2003). According to this study, the tendency to form supported lipid bilayers and the quality of formed bilayers varied considerably. It is important to note that QCM and AFM analyses of such complex events should be interpreted with caution.

DPPC liposome adsorption on a polymer surface with phosphorylcholine groups was studied by QCM and AFM (Iwasaki *et al.*, 1997). The adsorption of liposomes on the surfaces was identified by frequency shifts in QCM measurements. AFM imaging confirmed the presence of vesicles on the surfaces by the change in roughness of the substrates. Interestingly, the authors used QCM to study the effect of temperature on the resonance frequency change of the polymer-coated resonator exposed to DPPC liposomes. Although it is a risky procedure because the QCM signal is by itself sensitive to temperature (Bottom, 1982), it was concluded that the resonance frequency changes observed on some surfaces were due to the phase transition of the DPPC liposomes. The temperature dependence of QCM measurements on liposomes should be further validated.

QCM was successfully used to demonstrate the specific binding of immunoliposomes to antigen-bearing surfaces (Yun *et al.*, 1998). The frequency change increased as the concentration of antibody used to prepare liposomes increased to reach saturation.

Liebau *et al.* (1998) investigated the adhesion of receptor-coupled liposomes using QCM measurements. In this study, a supported bilayer with glycolipid ligands was transferred onto a quartz surface and subsequently incubated in a lectin concanavalin A-bearing liposome suspension. Specific interactions could be differentiated from nonspecific liposome attachment, but the latter could not be eliminated. In another work, QCM was also used to study the adsorption of carbohydrate-functionalized liposomes onto surfaces bearing lectins (Hildebrand *et al.*, 2002). Specific versus nonspecific liposome adsorption was clearly discriminated. However, no test was carried out to verify whether or not liposomes remained intact upon contact with the surfaces. The adsorption of concanavalin A-modified liposomes of different compositions on glycolipid bearing supported lipid bilayers was investigated in detail (Bakowsky *et al.*, 2008). In this excellent study, steric protection of the liposomes by PE-PEG lipids led to a decrease in nonspecific adhesion compared to plain liposomes composed of DPPC:cholesterol. The direct linkage of concanavalin A to the liposome surface resulted in a higher decrease in frequency compared to PEGylated lectin-modified liposomes. However, PEGylated lectin-modified liposomes suffered from a very low level of nonspecific adhesion. AFM imaging revealed that non-PEGylated-lectin-modified liposomes formed a packed lipid layer, and liposome coalescence was also observed, while PEGylated lectin-modified liposomes appeared more round-shaped (Bakowsky *et al.*, 2008). These findings imply that the use of a PEG spacer arm (or perhaps also other low-fouling molecules) is necessary to limit non-specific interactions between functional liposomes and surfaces.

The specific adhesion of unilamellar vesicles with an average diameter of 100 nm on functionalized surfaces mediated by avidin–biotin binding has been investigated (Pignataro *et al.*, 2000). AFM imaging and QCM analysis were used to study the binding of liposomes made of DPPC and DHPE-X-biotin to avidin-coated surfaces. (X represents a spacer arm between the phospholipid and biotin) QCM monitoring of the adhesion of the biotin-bearing vesicles to avidin-coated surfaces revealed an increased shift in resonance frequency with increasing biotin concentration up to 10 mol%

DHPE-X-biotin, indicating that the liposomes were "docking" onto the avidin molecules. Increasing the biotin–lipid concentration (up to 30 mol%) resulted in a decrease in the height of the surface-bound liposomes, up to the point where vesicle rupture was observed (Pignataro *et al.*, 2000). The liposome height measured by AFM decreased as the applied load force by the cantilever increased.

By plotting the change of damping in function of the frequency change in QCM measurements, Höpfner *et al.* (2008) studied the surface interaction of several types of liposomes with surfaces. From this damping analysis, the authors reported that soy-PC liposomes seemed deformed when adsorbed on unmodified gold, while liposomes made of soy-PC and PE-PEG were less deformed. PEG seems to stabilize the liposomes. Soy-PC liposomes adsorbed on hydrophobic-modified gold seemed not deformed, while DOTAP liposomes adsorbed on negatively charged gold were flattened. A higher lipid-biotin concentration in PC liposomes resulted in drastic deformation of these liposomes bound to avidin-modified sensors. Also, the use of PEGylated-lipids in biotinylated PC liposomes diminished liposome interaction with the avidin-modified surfaces (Höpfner *et al.*, 2008). The addition of lipid-PEG in the PC liposomes decreased liposome deformation. The addition of lipids with PEG-grafted biotin resulted in a stronger binding ability, but also to more flattening. Although this study is interesting, only QCM measurements were used to derive the liposome behavior upon interaction with the solid surfaces.

DNA-tagged lipid vesicles were hybridized on surfaces bearing complementary DNA, and the immobilization was followed by QCM (Granéli *et al.*, 2004; Städler *et al.*, 2004). Dissipation and frequency shifts were reported, although no experiments were carried out to verify the extent of the nonspecific adsorption of the liposomes.

Intact liposomes have been attached on solid surfaces by a NeutrAvidin™-biotin link (Brochu and Vermette, 2007). Combined with XPS and measurements of the release of fluorescent probes encapsulated in liposomes, QCM measurements were used to follow the buildup of the different layers in real time and *in situ*, to show that biotinylated liposomes stay intact upon surface attachment on NeutrAvidin™-coated surfaces, and had viscoelastic behavior. Frequency and half-band-half-width shifts were associated with large standard deviations for the attachment of layers of intact liposomes.

Although the under layers used in the surface immobilization of liposomes have reproducible surface properties (composition, thickness, and structure), a small difference in the surface characteristics of one or more of the interlayers can be amplified in the QCM signals. This can result in batch-to-batch variations in QCM measurements owing to the high sensitivity of the method. If comparisons between the QCM signals of the buildup of the different layers is only done on one "representative" sample, significant shifts between the different layers could be obtained (Brochu and Vermette, 2007). Often, only "representative" QCM data have been reported in the literature and perhaps compared with over-enthusiasm. To obtain meaningful QCM data and interpretation, frequency and half-band-half-width shifts on the same sample can be compared, but should be combined with a thorough statistical analysis to compare QCM results obtained from different samples.

The thickness of adsorbed SUV made of POPC estimated using the Voigt–Voinova model was compared to those estimated using the Sauerbrey equation (Cho *et al.*, 2007). The Voigt–Voinova model estimated the thickness of the adsorbed vesicles to be 22 nm, while the Sauerbrey thicknesses for different overtones ranged between 18 and 20 nm. Although the authors stated that Sauerbrey underestimates the thickness of the layers, these differences were very small, considering the precision of the QCM instrument and the numerous hypotheses made to apply the Voigt-based model. Both thicknesses remain estimations of the real thickness.

QCM was combined with fluorescence recovery after pattern photobleaching (FRAPP) to study liposome (DOPC:DOPA:fluorescent probe) adsorption on patterned polyelectrolyte multilayers (Kohli *et al.*, 2006). FRAPP was used to determine if the adsorbed liposomes remained intact upon adsorption, based on the mobility of the fluorescently tagged lipids. It was found that liposomes remained on some surfaces by matching the mobile fraction of the fluorescently-tagged lipids calculated by FRAPP with larger frequency and dissipation shifts measured by QCM. This is an excellent combination of characterization techniques to assess if liposomes remain intact upon adsorption on solid surfaces.

Some authors have tried to disentangle properties of surface-bound liposomes from the equivalent circuit analysis of the QCM measurements (Viitala *et al.*, 2007). Some fundamental information on the formed vesicle layers need to be estimated to extract any properties. As shown by Viitala *et al.* (2007), QCM measurements can be used to discriminate between rigid and viscoelastic layers formed upon liposome adsorption from frequency shifts and changes in resistance. This level of analysis is sufficient, from my viewpoint, and further interpretation of QCM signals requires much speculation. The rationale behind macroscopic viscoelasticity measurements examined by mechanical modeling, for example, using Maxwell and Kelvin–Voigt models, has been applied to QCM data interpretation. This can be a risky procedure, however, as viscoelastic properties of macroscopic bodies are obtained under more defined and standardized practices. In my opinion, QCM cannot alone be used to obtain structural or morphological information on surface-bound liposomes. For example, liposome deformation upon adsorption cannot be explicitly determined from QCM measurements despite what many authors have hypothesized.

4. Surface-Bound Liposomes in Perspective and General Conclusions

The interaction between intact liposomes and solid supports is of interest for several fields, including the development of biomedical device coatings to locally deliver drugs, for immunodiagnosis and electrophoresis. One of the common features of these applications is that, to fulfill their functions, liposomes must remain intact upon their immobilization on solid surfaces. Liposomes protected by PEGylated lipids seem to be more stable than naked liposomes. Nevertheless, strategies need to be further developed or validated to improve liposome stability on surfaces, and examples of such strategies are briefly presented below.

Michel *et al.* (2004) described a method to fabricate more stable liposome layers by embedding into polyelectrolyte multilayers either unmodified LUV (diameter of approx. 200 nm) made of POPC, POPG, and cholesterol, or LUV with the same composition, but modified with poly-(D-lysine). The poly-(D-lysine) was shown to adsorb on the surface of the negatively charged unmodified LUV. Multilayers were built by the alternate deposition of polyanions and polycations. QCM and AFM imaging (in air with contact mode) were used to ascertain the presence of stable lipid vesicles on the surfaces. AFM imaging was used to compare the topography of the different systems, clearly indicating the presence of liposomal structures (Michel *et al.*, 2004). In a subsequent study from the same group, surface-immobilized liposomes were used as vehicles to support mineralization (Michel *et al.*, 2006). AFM imaging (in air with the Tapping mode) was used to follow the formation of calcium phosphate crystals inside surface-immobilized liposomes. This method appears to be a relatively easy way of immobilizing liposomes on solid surfaces to enhance the loading capacity for drug release applications.

A class of vesicular nanoparticles, called Cerasomes, has been described (Katagiri *et al.*, 2002) with the aim to provide more stable vesicles. A Cerasome is an organic-inorganic hybrid composed of a liposomal membrane with a ceramic surface. Cerasome multilayers were deposited on solid substrates and imaged by AFM in Tapping mode. The Cerasome particles were clearly seen. Cerasomes seem more stable than other lipid vesicles because their surface is covered with a siloxane network to prevent collapse and fusion of the vesicles. This class of vesicles could find applications as coatings for biomedical implants for drug release.

Methods to produce arrays of intact liposomes have been reported (Kalyankar *et al.*, 2006). Arrays of intact surface-bound liposomes could find applications in the high throughput screening of biomolecule-membrane interactions.

In summary, AFM in the imaging mode allows detection of individual liposomes. The image quality varies from one study to another. On one hand, the quality of AFM images depends on parameters related to the instrument and its operation. Such parameters include the type of scanner, the imaging mode, the type of cantilever, the scan size, the scan speed, the applied vertical and lateral forces, the drive amplitude, and the operator. On the other hand, it depends on parameters related to the samples such as sample preparation, method of liposome attachment or adsorption, substrate used to support liposomes, liposome composition, liposome preparation method, and liposome size distribution. It is necessary to consider these parameters that might affect the observed liposome morphology. In some studies, the substrate used to support liposome for AFM imaging is not even mentioned nor the conditions used (air vs aqueous environment).

Furthermore, extracting liposome sizes from AFM imaging experiments can lead to substantial errors. Liposomes can flatten upon deposition on a solid substrate. AFM imaging underestimates the height of surface-bound liposomes and overestimates the width owing to tip broadening effects. Although some drawbacks are associated with the use of light scattering to assess liposome sizes, it is advisable to use dynamic light scattering to obtain the size distribution of liposomes which would be used in a suspension formulation and not as surface coatings.

Unfortunately, with the current state of knowledge, it is not possible—nor useful—to recommend a particular "toolkit" of parameters most appropriate to investigate liposomes on solid surfaces by AFM imaging, as no general tendency can be found from my analysis of the scientific literature. Tapping mode imaging is often claimed to limit liposome damage as opposed to contact mode operation. An experienced operator is required to operate AFM because the images obtained from complex structures such as liposomes can be difficult to elucidate.

AFM force measurements should be used to probe the stability and compression of layers of liposomes. The use colloidal probes (i.e. cantilevers bearing microspheres) is recommended, and not bare cantilevers. Colloidal probe measurements with a microsphere attached to AFM cantilevers can probe material properties with a more defined geometry. It can also minimize sample damage, considering the high pressure involved with the use of bare cantilevers.

Although the QCM is a surface-sensitive technique to carry out surface adsorption experiments, firm conclusions on the stability of surface-immobilized liposomes cannot be drawn using this technique alone. It appears necessary to carry out independent stability experiments (e.g. by measuring the release of a fluorescent probe from the surface-bound liposomes) to support the statement that liposomes remain intact upon immobilization or adsorption on solid supports. These experiments are essential even with the use of QCM with dissipation monitoring. Raw data from QCM measurements only provide changes in frequency and dissipation, and the latter can be expressed in different forms.

When dealing with lipid systems such as liposomes, lipid polymorphism is an important parameter to consider. Model lipid membranes can exist in a number of different phases. Over the years, a wide range of analytical techniques has been applied to study the physicochemical properties of lipids, lipid membranes and liposomes. These techniques should not be left behind.

For example, electron microscopy techniques such as scanning electron microscopy (SEM) and transmission electron microscopy (TEM) have provided valuable information on the morphology of liposomes and other lipid systems. Freeze-fracture and cryo-TEM techniques have provided images of lipid systems with high resolution. Also, X-ray methods, nuclear magnetic resonance (NMR), and electron spin resonance (ESR) have been used to identify lipid structures. Furthermore, in lipid membranes a wide range of modes of molecular motion have been identified (Blume, 1993). The characteristic time constants for these motions range from 10^{-14} s for molecular vibrations to hours or days for the so-called "lipid flip-flop," that is, the transbilayer exchange of lipid molecules. No single technique can study time constants for molecular motions over such a large timeframe. Various vibrational spectroscopy methods have been applied to characterize the lipid polymorphic behavior. Infrared (IR) spectroscopy, for example, has been used to characterize lipid polymorphism. Moreover, differential scanning calorimetry (DSC) played an important role to investigate lipid phase transitions.

In conclusion, there is a need to perform a multitechnique analysis. AFM and QCM can only provide limited information on surface-bound or surface-adsorbed liposomes, as highlighted above. AFM and QCM do not and cannot replace traditional techniques that have provided useful information on many lipid systems. Surfaces used to support liposome adsorption or immobilization should be well characterized in order to allow proper interpretation of results. The presence of layers should be confirmed by XPS for example, and not assumed. Also, when dealing with specific liposome immobilization, using avidin–biotin linkage for example, the level of non-specific adsorption should be assessed. Each technique has particular strengths and limitations, and for a detailed study of liposome properties it is necessary to apply a range of methods, and to appreciate their potential and limitations when interpreting data. This strongly applies to AFM and QCM measurements, which have too often been over-interpreted.

REFERENCES

Almgren, M., Edwards, K., and Karlsson, G. (2000). Cryo transmission electron microscopy of liposomes and related structures. *Colloids Surf. A* **174,** 3–21.

Anabousi, S., Laue, M., Lehr, C. M., Bakowsky, U., and Ehrhardt, C. (2005). Assessing transferrin modification of liposomes by atomic force microscopy and transmission electron microscopy. *Eur. J. Pharm. Biopharm.* **60,** 295–303.

Bakowsky, H., Richter, T., Kneuer, C., Hoekstra, D., Rothe, U., Bendas, G., Ehrhardt, C., and Bakowsky, U. (2008). Adhesion characteristics and stability assessment of lectin-modified liposomes for site-specific drug delivery. *Biochim. Biophys. Acta* **1778,** 242–249.

Benmouna, F., and Johannsmann, D. (2004). Viscoelasticity of gelatin surfaces probed by AFM noise analysis. *Langmuir* **20,** 188–193.

Berdyyeva, T. K., Woodworth, C. D., and Sokolov, I. (2005). Human epithelial cells increase their rigidity with ageing *in vitro*: Direct measurements. *Phys. Med. Biol.* **50,** 81–92.

Berne, B., and Pecora, R. (1976). Dynamic Light Scattering. John Wiley & Sons, New York.

Blume, A. (1993). Dynamic properties. *In* "Phospholipids Handbook," (G Cevc, ed.), pp. 455–509. Marcel Dekker Inc, New York.

Bottom, V. E. (1982). *Introduction to Quartz Crystal Unit Design.* Toronto, Van Nostrand Reinhold Company.

Brochu, H., and Vermette, P. (2007). Liposome layers characterized by quartz crystal microbalance measurements and multirelease delivery. *Langmuir* **23,** 7679–7686.

Brochu, H., and Vermette, P. (2008). Young's moduli of surface-bound liposomes by atomic force microscopy force measurements. *Langmuir* **24,** 2009–2014.

Brochu, H., Polidori, A., Pucci, B., and Vermette, P. (2004). Drug delivery systems using immobilized intact liposomes: A comparative and critical review. *Curr. Drug Deliv.* **1,** 299–312.

Carmo, V. A., De Oliveira, M. C., Reis, E. C., Guimarães, T. M., Vilela, J. M., Andrade, M. S., Michalick, M. S., and Cardoso, V. N. (2008). Physicochemical characterization and study of *in vitro* interactions of pH-sensitive liposomes with the complement system. *J. Liposome Res.* **18,** 59–70.

Casals, E., Verdaguer, A., Tonda, R., Galán, A., Escolar, G., and Estelrich, J. (2003). Atomic force microscopy of liposomes bearing fibrinogen. *Bioconjug. Chem.* **14,** 593–600.

Cevc, G., and Seddon, J. (1993). Physical characterization. *In* "Phospholipids Handbook," (G Cevc, ed.), pp. 351–401. Marcel Dekker Inc, New York.

Cho, N. J., Kanazawa, K. K., Glenn, J. S., and Frank, C. W. (2007). Employing two different quartz crystal microbalance models to study changes in viscoelastic behavior upon transformation of lipid vesicles to a bilayer on a gold surface. *Anal. Chem.* **79,** 7027–7035.

Chouinard, J. A., Grenier, G., Khalil, A., and Vermette, P. (2008). Oxidized-LDL induce morphological changes and increase stiffness of endothelial cells. *Exp. Cell Res.* **314,** 3007–3016.

Cleveland, J. P., Manne, S., Bocek, D., and Hansma, P. K. (1993). A nondestructive method for determining the spring constant of cantilevers for scanning force microscopy. *Rev. Sci. Instrum.* **64,** 403–405.

Colas, J. C., Shi, W., Rao, V. S., Omri, A., Mozafari, M. R., and Singh, H. (2007). Microscopical investigations of nisin-loaded nanoliposomes prepared by Mozafari method and their bacterial targeting. *Micron* **38,** 841–847.

Cullis, P., and De Kruijff, B. (1978). Polymorphic phase behaviour of lipid mixtures as detected by ^{31}P NMR. Evidence that cholesterol may destabilize bilayer structure in membrane systems containing phosphatidylethanolamine. *Biochim. Biophys. Acta* **507,** 207–218.

Dahmen-Levison, U., Brezesinski, G., and Mohwald, H. (1998). Specific adsorption of PLA_2 at monolayers. *Thin Solid Films* **327,** 616–620.

Danion, A., Arsenault, I., and Vermette, P. (2007a). Antibacterial activity of contact lenses bearing surface-immobilized layers of intact liposomes loaded with levofloxacin. *J. Pharm. Sci.* **96,** 2350–2363.

Danion, A., Brochu, H., Martin, Y., and Vermette, P. (2007b). Fabrication and characterization of contact lenses bearing surface-immobilized layers of intact liposomes. *J. Biomed. Mater. Res. A* **82,** 41–51.

Derjaguin, B. V., Muller, V. M., and Toporov, Y. P. (1980). On different approaches to the contact mechanics. *J. Colloid Interface Sci.* **73,** 293–294.

Digital Instruments, Veeco Metrology Group (2000). *Scanning probe microscopy training notebook. Version 3.0.* Santa Barbara, CA 93117, USA.

Dimitriadis, E. K., Horkay, F., Maresca, J., Kachar, B., and Chadwick, R. S. (2002). Determination of elastic moduli of thin layers of soft material using the atomic force microscope. *Biophys. J.* **82,** 2798–2810.

Engler, A. J., Richert, L., Wong, J. Y., Picart, C., and Discher, D. E. (2004). Surface probe measurements of the elasticity of sectioned tissue, thin gels and polyelectrolyte multilayer films: Correlations between substrate stiffness and cell adhesion. *Surf. Sci.* **570,** 142–154.

Estes, D. J., and Mayer, M. (2005). Electroformation of giant liposomes from spin-coated films of lipids. *Colloids Surf. B Biointerfaces* **42,** 115–123.

Frederik, P. M., Burger, K. N., Stuart, M. C., and Verkleij, A. J. (1991). Lipid polymorphism as observed by cryo-electron microscopy. *Biochim. Biophys. Acta* **1062,** 133–141.

Garbuzenko, O. B., Saad, M., Betigeri, S., Zhang, M., Vetcher, A. A., Soldatenkov, V. A., Reimer, D. C., Pozharov, V. P., and Minko, T. (2009). Intratracheal versus intravenous liposomal delivery of siRNA, antisense oligonucleotides and anticancer drug. *Pharm. Res.* **26,** 382–394.

Glavas-Dodov, M., Goracinova, K., Mladenovska, K., and Fredro-Kumbaradzi, E. (2002). Release profile of lidocaine HCl from topical liposomal gel formulation. *Int. J. Pharm.* **242,** 381–384.

Granéli, A., Rydström, J., Kasemo, B., and Höök, F. (2004). Utilizing adsorbed proteoliposomes trapped in a non-ruptured state on SiO_2 for amplified detection of membrane proteins. *Biosens. Bioelectron.* **20,** 498–504.

Hara, M., Yuan, H. Q., Miyake, M., Iijima, S., Yang, Q., and Miyake, J. (2000). Amphiphilic polymer–liposome interaction: A novel immobilization technique for liposome on gel surface. *Mater. Sci. Eng. Biomim. Mater. Sens. Syst.* **13,** 117–121.

Hayakawa, E., Naganuma, M., Mukasa, K., Shimozawa, T., and Araiso, T. (1998). Change of motion and localization of cholesterol molecule during L_α-H_{II} transition. *Biophys. J.* **74,** 892–898.

Hertz, H. (1881). On the contact of elastic solids. *J. Reine Angew. Math.* **92,** 156–171.

Hildebrand, A., Schaedlich, A., Rothe, U., and Neubert, R. H. (2002). Sensing specific adhesion of liposomal and micellar systems with attached carbohydrate recognition structures at lectin surfaces. *J. Colloid Interface Sci.* **249,** 274–281.

Hope, M., Wong, K., and Cullis, P. (1989). Freeze-fracture of lipids and model membrane systems. *J. Electron Microsc. Tech.* **13,** 277–287.

Höpfner, M., Rothe, U., and Bendas, G. (2008). Biosensor-based evaluation of liposomal behavior in the target binding process. *J. Liposome Res.* **18,** 71–82.

Huang, J., Buboltz, J. T., and Feigenson, G. W. (1999). Maximum solubility of cholesterol in phosphatidylcholine and phosphatidylethanolamine bilayers. *Biochim. Biophys. Acta* **1417,** 89–100.

Israelachvili, J. N. (1992). Intermolecular and Surface Forces. 2nd edn. New York, Academic Press.

Iwasaki, Y., Tanaka, S., Hara, M., Ishihara, K., and Nakabayashi, N. (1997). Stabilization of liposomes attached to polymer surfaces having phosphorylcholine groups. *J. Colloid Interface Sci.* **192,** 432–439.

Janshoff, A., Galla, H.-J., and Steinem, C. (2000). Piezoelectric mass-sensing devices as biosensors—An alternative to optical biosensors? *Angew. Chem. Int. Ed. Engl.* **39,** 4004–4032.

Johannsmann, D. (1999). Viscoelastic analysis of organic thin films on quartz resonators. *Macromol. Chem. Phys.* **200,** 501–516.

Johnson, K. L., Kendall, K., and Roberts, A. D. (1971). Surface energy and the contact of elastic solids. *Proc. R. Soc. Lond. A Math. Phys. Sci.* **324,** 301–313.

Jung, L. S., Shumaker-Parry, J. S., Campbell, C. T., Yee, S. S., and Gelb, M. H. (2000). Quantification of tight binding to surface-immobilized phospholipid vesicles using surface plasmon resonance: Binding constant of phospholipase A_2. *J. Am. Chem. Soc.* **122,** 4177–4184.

Jung, H. S., Kim, J. M., Park, J. W., Lee, H. Y., and Kawai, T. (2005). Amperometric immunosensor for direct detection based upon functional lipid vesicles immobilized on nanowell array electrode. *Langmuir* **21,** 6025–6029.

Jung, H., Kim, J., Park, J., Lee, S., Lee, H., Kuboi, R., and Kawai, T. (2006). Atomic force microscopy observation of highly arrayed phospholipid bilayer vesicle on a gold surface. *J. Biosci. Bioeng.* **102,** 28–33.

Kalyankar, N. D., Sharma, M. K., Vaidya, S. V., Calhoun, D., Maldarelli, C., Couzis, A., and Gilchrist, L. (2006). Arraying of intact liposomes into chemically functionalized microwells. *Langmuir* **22,** 5403–5411.

Kanno, T., Yamada, T., Iwabuki, H., Tanaka, H., Kuroda, S., Tanizawa, K., and Kawai, T. (2002). Size distribution measurement of vesicles by atomic force microscopy. *Anal. Biochem.* **309,** 196–199.

Katagiri, K., Hamasaki, R., Ariga, K., and Kikuchi, J. (2002). Layer-by-layer self-assembled polyelectrolyte multilayers with embedded liposomes: Immobilized submicronic reactors for mineralization. *J. Am. Chem. Soc.* **124,** 7892–7893.

Keller, C., and Kasemo, B. (1998). Surface specific kinetics of lipid vesicle adsorption measured with a quartz crystal microbalance. *Biophys. J.* **75,** 1397–1402.

Keller, C. A., Glasmastar, K., Zhdanov, V. P., and Kasemo, B. (2000). Formation of supported membranes from vesicles. *Phys. Rev. Lett.* **84,** 5443–5446.

Khaleque, M. A., Okumura, Y., Yabushita, S., and Mitani, M. (2003). Liposome immobilization on polymer gel particles by *in situ* formation of covalent linkages. *Chem. Lett.* **32,** 416–417.

Kim, J. M., Jung, H. S., Park, J. W., Yukimasa, T., Oka, H., Lee, H. Y., and Kawai, T. (2005). Spontaneous immobilization of liposomes on electron-beam exposed resist surfaces. *J. Am. Chem. Soc.* **127,** 2358–2362.

Kleemann, E., Schmehl, T., Gessler, T., Bakowsky, U., Kissel, T., and Seeger, W. (2007). Iloprost-containing liposomes for aerosol application in pulmonary arterial hypertension: Formulation aspects and stability. *Pharm. Res.* **24,** 277–287.

Kohli, N., Vaidya, S., Ofoli, R. Y., Worden, R. M., and Lee, I. (2006). Arrays of lipid bilayers and liposomes on patterned polyelectrolyte templates. *J. Colloid Interface Sci.* **301,** 461–469.

Lee, H. Y., Jung, H. S., Fujikawa, K., Park, J. W., Kim, J. M., Yukimasa, T., Sugihara, H., and Kawai, T. (2005). New antibody immobilization method via functional liposome layer for specific protein assays. *Biosens. Bioelectron.* **21,** 833–838.

Lee, H. Y., Lee, B. K., Park, J. W., Jung, H. S., Kim, J. M., and Kawai, T. (2008). Self-organized functional lipid vesicle array for sensitive immunoassay chip. *Ultramicroscopy* **108,** 1325–1327.

Li, S., and Palmer, A. F. (2004). Structure of small actin-containing liposomes probed by atomic force microscopy: Effect of actin concentration & liposome size. *Langmuir* **20,** 7917–7925.

Liang, X., Mao, G., and Ng, K. Y. (2004). Mechanical properties and stability measurement of cholesterol-containing liposome on mica by atomic force microscopy. *J. Colloid Interface Sci.* **278,** 53–62.

Liang, X., Mao, G., and Ng, K. Y. (2005). Effect of chain lengths of PEO–PPO–PEO on small unilamellar liposome morphology and stability: An AFM investigation. *J. Colloid Interface Sci.* **285,** 360–372.

Liebau, M., Bendas, G., Rothe, U., and Neubert, R. H. H. (1998). Adhesive interactions of liposomes with supported planar bilayers on QCM as a new adhesion model. *Sens. Actuators B Chem.* **47,** 239–245.

Lipowsky, R., and Seifert, U. (1991). Adhesion of vesicles and membranes. *Mol. Cryst. Liq. Cryst.* **202,** 17–25.

Lunelli, L., Pasquardini, L., Pederzolli, C., Vanzetti, L., and Anderle, M. (2005). Covalently anchored lipid structures on amine-enriched polystyrene. *Langmuir* **21,** 8338–8343.

Mao, X. Q., Kong, L., Li, X., Guo, B. C., and Zou, H. F. (2003). Unilamellar liposomes covalently coupled on silica gel for liquid chromatography. *Anal. Bioanal. Chem.* **375,** 550–555.

McCracken, M., and Sammons, M. (1987). Sizing of a vesicle drug formulation by quasi-elastic light scattering and comparison with electron microscopy and ultracentrifugation. *J. Pharm. Sci.* **76,** 56–59.

Mechler, A., Praporski, S., Piantavigna, S., Heaton, S. M., Hall, K. N., Aguilar, M. I., and Martin, L. L. (2009). Structure and homogeneity of pseudo-physiological phospholipid bilayers and their deposition characteristics on carboxylic acid terminated self-assembled monolayers. *Biomaterials* **30,** 682–689.

Mei, J., Xu, J. R., Xiao, Y. X., Liao, X. Y., Qiu, G. F., and Feng, Y. Q. (2008). A novel covalent coupling method for coating of capillaries with liposomes in capillary electrophoresis. *Electrophoresis* **29,** 3825–3833.

Michel, M., Vautier, D., Voegel, J. C., Schaaf, P., and Ball, V. (2004). Layer by layer self-assembled polyelectrolyte multilayers with embedded phospholipid vesicles. *Langmuir* **20,** 4835–4839.

Michel, M., Arntz, Y., Fleith, G., Toquant, J., Haikel, Y., Voegel, J. C., Schaaf, P., and Ball, V. (2006). Layer-by-layer self-assembled polyelectrolyte multilayers with embedded liposomes: Immobilized submicronic reactors for mineralization. *Langmuir* **22,** 2358–2364.

Morita, S., Nukui, M., and Kuboi, R. (2006). Immobilization of liposomes onto quartz crystal microbalance to detect interaction between liposomes and proteins. *J. Colloid Interface Sci.* **298,** 672–678.

Mozafari, M. R., Reed, C. J., Rostron, C., and Hasirci, V. (2005). A review of scanning probe microscopy investigations of liposome-DNA complexes. *J. Liposome Res.* **15,** 93–107.

Muller, V. M., Yushchenko, V. S., and Derjaguin, B. V. (1980). On the influence of molecular forces on the deformation of an elastic sphere and its sticking to a rigid plane. *J. Colloid Interface Sci.* **77,** 91–101.

Muller, V. M., Derjaguin, B. V., and Toporov, Y. P. (1983). On two methods of calculation of the force of sticking of an elastic sphere to a rigid plane. *Colloids Surf. A* **7,** 251–259.

Nakano, K., Tozuka, Y., Yamamoto, H., Kawashima, Y., and Takeuchi, H. (2008). A novel method for measuring rigidity of submicron-size liposomes with atomic force microscopy. *Int. J. Pharm.* **355,** 203–209.

Nomura, T., and Okuhara, M. (1982). Frequency shifts of piezoelectric quartz crystals immersed in organic liquids. *Anal. Chim. Acta* **142,** 281–284.

Ostrowsky, N. (1993). Liposome size measurements by photon-correlation spectroscopy. *Chem. Phys. Lipids* **64,** 45–56.

Paclet, M. H., Coleman, A. W., Vergnaud, S., and Morel, F. (2000). P67-phox-mediated NADPH oxidase assembly: Imaging of cytochrome b558 liposomes by atomic force microscopy. *Biochemistry* **39,** 9302–9310.

Palmer, A. F., Wingert, P., and Nickels, J. (2003). Atomic force microscopy and light scattering of small unilamellar actin-containing liposomes. *Biophys. J.* **85,** 1233–1247.

Park, J. H., Kwon, E. Y., Jung, H. I., and Kim, D. E. (2006). Direct force measurement of the interaction between liposome and the C2A domain of synaptotagmin I using atomic force microscopy. *Biotechnol. Lett.* **28,** 505–509.

Pignataro, B., Steinem, C., Galla, H. J., Fuchs, H., and Janshoff, A. (2000). Specific adhesion of vesicles monitored by scanning force microscopy and quartz crystal microbalance. *Biophys. J.* **78,** 487–498.

Plant, A. (1999). Supported hybrid bilayer membranes as rugged cell membrane mimics. *Langmuir* **15,** 5128–5135.

Pleyer, U., Elkins, B., Ruckert, D., Lutz, S., Grammer, J., Chou, J., Schmidt, K. H., and Mondino, B. J. (1994). Ocular absorption of cyclosporine A from liposomes incorporated into collagen shields. *Curr. Eye Res.* **13,** 177–181.

Radler, J., Strey, H., and Sackmann, E. (1995). Phenomenology and kinetics of lipid bilayer spreading on hydrophilic surfaces. *Langmuir* **11,** 4539–4548.

Radmacher, M., Fritz, M., and Hansma, P. (1995). Imaging soft samples with the atomic force microscope: Gelatin in water and propanol. *Biophys. J.* **69,** 264–270.

Raffy, S., and Teissie, J. (1999). Control of lipid membrane stability by cholesterol content. *Biophys. J.* **76,** 2072–2080.

Ramachandran, S., Quist, A. P., Kumar, S., and Lal, R. (2006). Cisplatin nanoliposomes for cancer therapy: AFM and fluorescence imaging of cisplatin encapsulation, stability, cellular uptake, and toxicity. *Langmuir* **22,** 8156–8162.

Reimhult, E., Höök, F., and Kasemo, B. (2002). Vesicle adsorption on SiO_2 and TiO_2: Dependence on vesicle size. *J. Chem. Phys.* **117,** 7401–7404.

Reimhult, E., Höök, F., and Kasemo, B. (2003). Intact vesicle adsorption and supported biomembrane formation from vesicles in solution: Influence of surface chemistry, vesicle size, temperature, and osmotic pressure. *Langmuir* **19,** 1681–1691.

Reimhult, E., Zäch, M., Höök, F., and Kasemo, B. (2006). A multitechnique study of liposome adsorption on Au and lipid bilayer formation on SiO_2. *Langmuir* **22,** 3313–3319.

Reviakine, I., and Brisson, A. (2000). Formation of supported phospholipid bilayers from unilamellar vesicles investigated by atomic force microscopy. *Langmuir* **16,** 1806–1815.

Richter, R., Mukhopadhyay, A., and Brisson, A. (2003). Pathways of lipid vesicle deposition on solid surfaces: A combined QCM-D and AFM study. *Biophys. J.* **85,** 3035–3047.

Rodahl, M., Hook, F., Fredriksson, C., Keller, C. A., Krozer, A., Brzezinski, P., Voinova, M., and Kasemo, B. (1997). Simultaneous frequency and dissipation factor QCM measurements of biomolecular adsorption and cell adhesion. *Faraday Discuss.* **107,** 229–246.

Ruel-Gariepy, E., Leclair, G., Hildgen, P., Gupta, A., and Leroux, J. C. (2002). Thermosensitive chitosan-based hydrogel containing liposomes for the delivery of hydrophilic molecules. *J. Control. Release* **82,** 373–383.

Ruozi, B., Battini, R., Tosi, G., Forni, F., and Vandelli, M. A. (2005). Liposome–oligonucleotides interaction for *in vitro* uptake by COS I and HaCaT cells. *J. Drug Target* **13,** 295–304.

Ruozi, B., Tosi, G., Leo, E., and Vandelli, M. A. (2007). Application of atomic force microscopy to characterize liposomes as drug and gene carriers. *Talanta* **73,** 12–22.

Sauerbrey, G. (1959). Verwendung von Schwingquarzen zur Wägung dünner Schichten und zur Mikrowägung. *Zeitschrift für Physik* **155,** 206–222.

Schönherr, H., Johnson, J. M., Lenz, P., Frank, C. W., and Boxer, S. G. (2004). Vesicle adsorption and lipid bilayer formation on glass studied by atomic force microscopy. *Langmuir* **20,** 11600–11606.

Seifert, U., and Lipowsky, R. (1990). Adhesion of vesicles. *Phys. Rev. A* **42,** 4768–4772.

Semple, S., Chonn, A., and Cullis, P. (1995). Influence of cholesterol on the association of plasma proteins with liposomes. *Biochemistry* **35,** 2521–2525.

Shibata-Seki, T., Masai, J., Tagawa, T., Sorin, T., and Kondo, S. (1996). *In-situ* atomic force microscopy study of lipid vesicles adsorbed on a substrate. *Thin Solid Films* **273,** 297–303.

Shin, Y., and Freed, J. (1989a). Dynamic imaging of lateral diffusion by electron spin resonance and study of rotational dynamics in model membrane. Effect of cholesterols. *Biophys. J.* **55,** 537–550.

Shin, Y., and Freed, J. (1989b). Thermodynamics of phosphatidylcholine-cholesterol mixed model membranes in the liquid crystalline state studied by the orientational order parameter. *Biophys. J.* **56,** 1093–1100.

Silva, L. P., Leite, J. R., Brand, G. D., Regis, W. B., Tedesco, A. C., Azevedo, R. B., Freitas, S. M., and Bloch, C., Jr. (2008). Dermaseptins from *Phyllomedusa oreades* and *Phyllomedusa distincta*: Liposomes fusion and/or lysis investigated by fluorescence and atomic force microscopy. *Comp. Biochem. Physiol. A Mol. Integr. Physiol.* **151,** 329–335.

Sneddon, I. N. (1965). The relation between load and penetration in the axisymmetric Boussinesq problem for a punch of arbitrary profile. *Int. J. Eng. Sci.* **3,** 47–57.

Städler, B., Falconnet, D., Pfeiffer, I., Höök, F., and Vörös, J. (2004). Micropatterning of DNA-tagged vesicles. *Langmuir* **20,** 11348–11354.

Steinem, C., Janshoff, A., Ulrich, W.-P., Sieber, M., and Galla, H.-J. (1996). Impedance analysis of supported lipid bilayer membranes: A scrutiny of different preparation techniques. *Biochim. Biophys. Acta* **1279,** 169–180.

Tarasova, A., Griesser, H. J., and Meagher, L. (2008). AFM study of the stability of a dense affinity-bound liposome layer. *Langmuir* **24,** 7371–7377.

Tardieu, A., Luzzati, V., and Reman, F. (1973). Structure and polymorphism of the hydrocarbon chains of lipids: A study of lecithin–water phases. *J. Mol. Biol.* **75,** 711–733.

Tilcock, C., Bally, M., Farren, S., and Cullis, P. (1982). Influence of cholesterol on the structural preferences of dioleoylphosphatidylethanolamine–dioleoylphosphatidylcholine systems: A phosphorus-31 and deuterium nuclear magnetic resonance study. *Biochemistry* **21,** 4596–4601.

Verkleij, A., and Ververgaert, P. (1978). Freeze-fracture morphology of biological membranes. *Biochim. Biophys. Acta* **515,** 303–327.

Vermette, P., Meagher, L., Gagnon, E., Griesser, H. J., and Doillon, C. J. (2002). Immobilized liposome layers for drug delivery applications: Inhibition of angiogenesis. *J. Control. Release* **80,** 179–195.

Vermette, P., Griesser, H. J., Kambouris, P., and Meagher, L. (2004). Characterization of surface-immobilized layers of intact liposomes. *Biomacromolecules* **5,** 1496–1502.

Viitala, T., Hautala, J. T., Vuorinen, J., and Wiedmer, S. K. (2007). Structure of anionic phospholipid coatings on silica by dissipative quartz crystal microbalance. *Langmuir* **23,** 609–618.

von Groll, A., Levin, Y., Barbosa, M. C., and Ravazzolo, A. P. (2006). Linear DNA low efficiency transfection by liposome can be improved by the use of cationic lipid as charge neutralizer. *Biotechnol. Prog.* **22,** 1220–1224.

Wegener, J., Janshoff, A., and Steinem, C. (2001). The quartz crystal microbalance as a novel means to study cell-substrate interactions *in situ*. *Cell Biochem. Biophys.* **34,** 121–151.

Weiner, A. L., Carpenter-Green, S. S., Soehngen, E. C., Lenk, R. P., and Popescu, M. C. (1985). Liposome-collagen gel matrix: A novel sustained drug delivery system. *J. Pharm. Sci.* **74,** 922–925.

Weisenhorn, A., Khorsandi, M., Kasas, S., Gotzos, V., and Butt, H.-J. (1993). Deformation and height anomaly of soft surfaces studied with an AFM. *Nanotechnology* **4,** 106–113.

Yang, Q., Liu, X. Y., Ajiki, S., Hara, M., Lundahl, P., and Miyake, J. (1998). Avidin-biotin immobilization of unilamellar liposomes in gel beads for chromatographic analysis of drug-membrane partitioning. *J. Chromatogr. B Biomed. Sci. Appl.* **707,** 131–141.

Yegin, B. A., and Lamprecht, A. (2006). Lipid nanocapsule size analysis by hydrodynamic chromatography and photon correlation spectroscopy. *Int. J. Pharm.* **320,** 165–170.

Yun, K., Kobatake, E., Haruyama, T., Laukkanen, M. L., Keinänen, K., and Aizawa, M. (1998). Use of a quartz crystal microbalance to monitor immunoliposome–antigen interaction. *Anal. Chem.* **70,** 260–264.

Zhang, Y., Cheng, C. M., Cusick, B., and LeDuc, P. R. (2007). Chemically encapsulated structural elements for probing the mechanical responses of biologically inspired systems. *Langmuir* **23,** 8129–8134.

CHAPTER FOUR

Mixing Solutions in Inkjet Formed Vesicles

Thomas H. Li,* Jeanne C. Stachowiak,† *and* Daniel A. Fletcher‡,§,1

Contents

Abstract

Controlling the contents of liposomes and vesicles is essential for their use in medicine, biotechnology, and basic research. Cargos such as proteins, DNA, and RNA are of growing interest for therapeutic applications as well as for fundamental studies of cellular organization and function, but controlled encapsulation and mixing of biomolecules within vesicles has been a challenge. Recently, microfluidic encapsulation has been shown to efficiently load arbitrary solutions of biomolecules into unilamellar vesicles. This method utilizes a piezoelectrically driven liquid jet to deform a planar bilayer and form a vesicle, with the fluid vortex formed by the jet mixing the solution in the jet with the

* Department of Mechanical Engineering, University of California, Berkeley, California, USA
† Sandia National Laboratories, Livermore, California, USA
‡ Department of Bioengineering, University of California, Berkeley, California, USA
§ Lawrence Berkeley National Laboratory, Berkeley, California, USA

1 Corresponding author

Methods in Enzymology, Volume 465

ISSN 0076-6879, DOI: 10.1016/S0076-6879(09)65004-7

surrounding solution. Here, we describe the equipment and protocol used for loading mixtures within unilamellar vesicles by microfluidic encapsulation, and we measure the encapsulated fraction to be 79 ± 5% using a falling vesicle technique. Additionally, we find that the presence of a continuous flow from the nozzle and changes in actuation voltage polarity do not significantly affect the encapsulated fraction. These results help to guide current applications and future development of this microfluidic encapsulation technique for forming and loading unilamellar vesicles.

1. Introduction

Successful clinical applications of lipid-encapsulated drugs over the past two decades have fueled interest in developing liposomes containing a mixture of biomolecular components that are capable of multiple therapeutic and diagnostic functions. Current applications of liposomes, which are typically 50–200 nm in diameter, include drug delivery (Düzgüneş *et al.*, 1995, 2005; Konduri *et al.*, 2005; Salem *et al.*, 2005; Weissig *et al.*, 2006) and diagnostic imaging (Lasic and Papahadjopoulos, 1995; Martina *et al.*, 2005), as well as cosmetics (Ramon *et al.*, 2005). The lipid bilayer membrane boundary of liposomes keeps their contents concentrated and shielded from exposure as they pass through the body. Additionally, modifications of the membrane can be made to improve drug bioavailability and reduce side effects in some clinical applications. When encapsulated in polyethylene glycol-coated liposomes, the anticancer drug Doxorubicin, has been shown to achieve much better performance in tumor targeting (Gabizon *et al.*, 1994) and reduced cardio toxicity (Safra *et al.*, 2000) compared to the unencapsulated form. Additional modification of the vesicle's bilayer with antibodies, aptamers, or ligands allows encapsulated drugs to be targeted for delivery to sites within the body (Debbage, 2009; Dos Santos *et al.*, 2007; Gantert *et al.*, 2009). Further functionality can be achieved by adding probes that facilitate medical imaging. Liposomes labeled with radionuclides can be imaged using single-photon emission computed tomography and magnetic resonance imaging (Zielhuis *et al.*, 2006), and radiolabeled liposomes containing anticancer therapeutics offer the ability to actively evaluate liposomal drug delivery (Elbayoumi *et al.*, 2007). The opportunity to combine multiple capabilities to address specific clinical needs makes liposomes attractive therapeutic and diagnostic vehicles. Despite the promise of multifunctional liposomes, encapsulating proteins, DNA and RNA, and other high-molecular-weight molecules and inorganic particles within lipid bilayer membrane compartments has been a challenge.

Over decades of liposome research, many methods have been developed to produce synthetic vesicles. One simple method to form vesicles is to dry a lipid film from organic solvent onto the bottom of a flask and rehydrate the film over the course of hours with an aqueous solution. Slight agitation is used to aid in the budding process (Reeves and Dowben, 1969). The resulting vesicles are multilamellar and vary in size. Based on the process, this technique is often referred to as swelling or hydration. Extrusion is a technique used to produce unilamellar vesicles of controlled size, which, involves taking a population of multilamellar vesicles and passing them repeatedly through a filter (Olson *et al.*, 1979). As the filter pore size is decreased, the vesicle population approaches that of the pore size. Below a pore diameter of 0.2 μm, extruded vesicles are uniform in size distribution. With the electroformation technique, lipids are dried onto a surface and immersed in a buffer solution. An alternating electric field is applied across the lipid film to produce a population of giant unilamellar vesicles with a heterogeneous size distribution (Angelova and Dimitrov, 1986). The electroinjection technique uses electroporation to facilitate the insertion of a small diameter tip into a liposome followed by the injection of the desired contents (Karlsson *et al.*, 2000). Although this method is impressive in its capability to control the vesicle contents, it is highly limited in throughput. In reverse-phase evaporation, aqueous solution is added to a suspension of lipid in an organic solvent. The solution is sonicated and the solvent is evaporated, leaving large unilamellar vesicles (Szoka and Papahadjopoulos, 1978). Some of these techniques can be combined to produce a desired population of vesicles, although combining techniques compounds the limitations of each method. For example, any vesicle formation process which uses extrusion to form vesicles of a uniform size distribution will be limited to sizes below that produced by a 0.2-μm pore. The ideal vesicle formation technique would offer uniform size distributions comparable to extruded vesicles, unilamellarity comparable to electroformed vesicles, and precise control over contents comparable to electroinjection without the respective limitations.

In 2008, we demonstrated that a liquid jet generated by a piezoelectrically actuated plunger–syringe system can be used to deform a planar lipid bilayer to make unilamellar vesicles loaded with contents of unrestricted size (Stachowiak *et al.*, 2008). More recently, we demonstrated that high-throughput vesicle formation and control of vesicle size (10–400 μm) can be achieved with an inkjet system (Stachowiak *et al.*, 2009). Figure 4.1 shows images of the formation process taken from a high-speed video. This technique is unique in that it offers the ability to form unilamellar vesicles with a high encapsulation efficiency and homogeneous size distribution at a high throughput rate (up to 200 Hz). Additionally, the fluid jet that forms the vesicle entrains the surrounding fluid in the process, which allows the user to simultaneously mix two fluids inside the vesicle during formation in order to trigger reactions or vary solution concentrations within the vesicle. Vesicles on the scale of micrometers to tens of

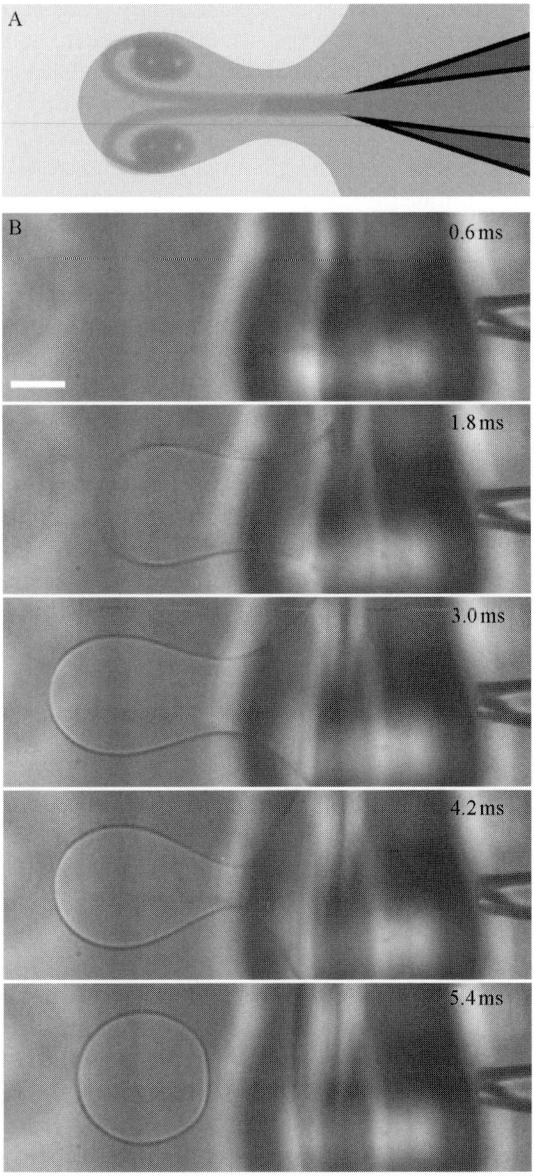

Figure 4.1 Inkjet-based microfluidic encapsulation to form unilamellar vesicles. (A) Diagram of the fluid jet and vortex ring deforming a planar lipid bilayer. (B) High-speed image sequence of the vesicle formation process. Conditions used to form vesicles are 20 pulses of a 35 V trapezoidal signal. A 200 mOsm sucrose solution is used in the inkjet and surrounding droplet to make a vesicle of 100% sucrose which is formed into a solution of 200 mOsm glucose. Scale bar represents 100 μm.

micrometers can be used directly for cell-like reconstitution, in which the spatial organization and reaction dynamics of cellular structures such as those formed by the cytoskeleton can be studied. To obtain liposomes on the scale of 100 nm, the vesicles can be sonicated or extruded as noted earlier, with contents expected to be similar to those of the original vesicle (Sugiura *et al.*, 2008). We have previously demonstrated the unilamellarity of the vesicles formed with our technique by inserting membrane pores and confirming the existence of membrane tubes during vesicle formation (Stachowiak *et al.*, 2008).

As microfluidic encapsulation matures as a technique for vesicle and liposome formation, there is a need to understand the efficiency at which two solutions—one in the jet and one surrounding the jet—can be encapsulated within a vesicle. Control over encapsulation efficiency allows one to accurately initiate a chemical reaction between the two solutions within the vesicle at the time of formation. Additionally, control over fluid mixing offers real-time management of the encapsulation fraction, allowing for convenient variation in concentration during an experiment, which is important for biological reconstitutions and small volume reactions. In this chapter, we investigate the vesicle encapsulation fraction, defined as the fraction of vesicle volume originating from the fluid surrounding the jet, using the falling drop method applied to vesicles. Our results on quantifying and controlling the encapsulated fraction are important for improving the reliability and usefulness of the microfluidic encapsulation technique.

2. Unilamellar Vesicle Formation by Microfluidic Encapsulation

In this section we describe the equipment and protocol used for generating vesicles by microfluidic encapsulation.

2.1. Inkjet vortex ring generator

Our microfluidic encapsulation technique for unilamellar vesicle formation is based on the ability to precisely form and control a liquid jet. We have chosen a single inkjet drop-on-demand device as a means of generating the liquid jet and tuning vortex dynamics. This inkjet device consists of a glass capillary surrounded by a cylindrical piezoelectric sleeve. Excitation of the piezoelectric actuator, by application of a voltage pulse, forces the piezoelectric sleeve to contract and expand around the glass capillary causing acoustic compression and rarefaction waves to propagate laterally along the nozzle. In typical inkjet usage, the excitation signal is tuned such that constructive and destructive interference among the waves leads to a compression wave with large enough amplitude to overcome the surface tension at the orifice and cause the ejection of a fluid droplet from the inkjet nozzle (Bogy and Talke, 1984). In our implementation, the inkjet is submerged in a miscible fluid such that there is no surface tension at the orifice, and therefore fluid ejection may be achieved by a smaller compression amplitude and is less dependent on optimizing the interference of waves. The bulk of the inkjet device used in our experiment is manufactured and assembled by Microfab Technologies, while the nozzle of the inkjet is manufactured in-house with an orifice diameter of ~10 μm using a P-97 Flaming/Brown micropipette puller (Sutter Instruments) and an MFG-3 Microforge (MicroData Instrument).

Through software controls, one can design the voltage profile used to excite the piezoelectric actuator. In many drop-on-demand applications, the voltage profile consists of a positive amplitude, or upright, trapezoidal profile followed by a negative amplitude, or inverted, trapezoidal profile. We have found that using a full upright and inverted trapezoidal profile provided power beyond what is needed to form vesicles, and thus we limited the voltage profile to an upright trapezoidal shape for simplicity. Adequate ejection of fluid to form a vesicle generally requires the application of numerous pulses of the trapezoidal waveform (Stachowiak *et al.*, 2009). Parameters for vesicle formation vary depending on the distance between the inkjet tip and bilayer, the viscosity of the solutions, and the condition of the lipid in the bilayer. Typical parameters for the experiments reported in this manuscript are a trapezoidal voltage profile with a 3-μs rise time from 0 to 35 V, a dwell time of 30 μs, and a fall time of 3 μs from 35 to 0 V, repeated for 20 pulses at a frequency of 20 kHz.

2.2. Microscopy of vesicle formation

Our microfluidic encapsulation system is built on an Axiovert 200 microscope base (Carl Zeiss), which was modified to facilitate vesicle production. The original microscope stage was replaced by a custom stage made to accommodate the jetting device. The stage is composed of parts purchased from Thorlabs or machined in house. The custom stage uses two sets of three axis translation stages. The first set of translation stages is used to position the entire system (inkjet device, formation chamber, and fixtures) relative to the microscope base for viewing, while the second set of stages is used to insert and position the inkjet device relative to the vesicle formation chamber as shown in Figure 4.2.

A high-speed camera is used to visualize the vesicle formation process. In our experiments, we use a monochrome Photron 1024PCI high-speed camera controlled through Photron Fastcam Viewer software. For typical vesicle formation processes, the full 1024×1024 pixel CCD area is usually reduced to a region-of-interest of approximately 600×200 pixels to facilitate high frame rate recording. The camera is triggered from a TTL signal

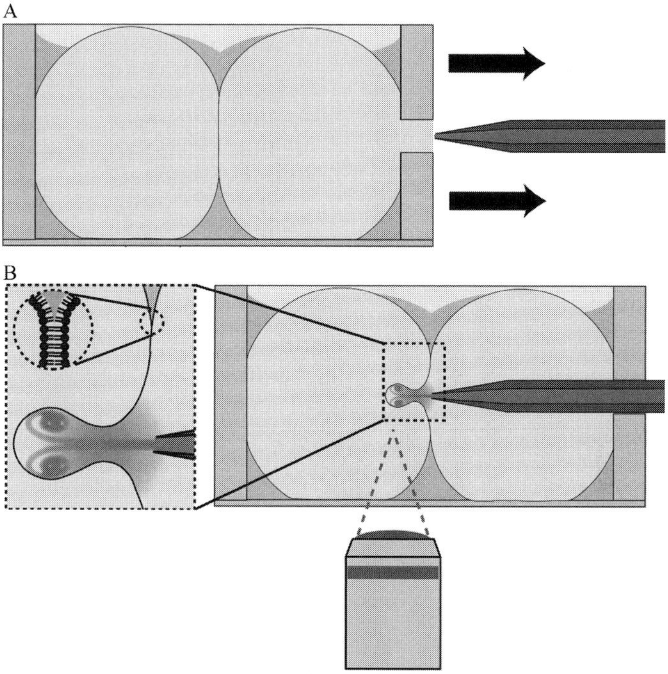

Figure 4.2 Illustration of the experimental setup used for making vesicles by microfluidic encapsulation. (A) A lipid bilayer is formed between two aqueous droplets in a chamber independent from the inkjet. (B) Inkjet nozzle is inserted into the chamber holding the bilayer and steered within range of the bilayer to form unilamellar vesicles by moving the chamber. This alignment procedure occurs on a microscope where the formation process can be recorded. Fluid flows continuously from the inkjet orifice to maintain the solution concentration inside the nozzle.

generated at the instant that the inkjet device is triggered. For most situations, a frame rate of approximately 5000 fps is used, which yields roughly 25 frames during the formation process. Generally, a 5× objective is used, which yields a field-of-view of 2 mm×0.6 mm when limiting the region-of-interest to 600×200 pixels.

In addition to the high-speed camera, a low-magnification side-view camera (Micropublisher 3.3 RTV, Qimaging) controlled through QCapture software has been added to the microscope. While the high-speed camera images from the underside of the system to record an image plane parallel to the table surface, the side-view camera is mounted to image the system on a plane perpendicular to the plane of the table surface. The main use of the side-view camera is to align the inkjet system and the vesicle chamber along the vertical direction. The side-view camera uses a 4× objective (Edmund Optics) and images a 4 cm × 3 cm field-of-view. A 45° mirror along the imaging path further reduces the field-of-view to 3 cm × 3 cm.

2.3. Nozzle flow system

In addition to the high-speed jet formed by the piezoelectric actuator, a continuous slow nozzle flow is used to prevent diffusion of outside fluid into the nozzle of the inkjet. Positive pressure inside the nozzle helps to minimize clogs in the nozzle due to particles in the solution surrounding the inkjet. If the solution inside the inkjet is different than the surrounding solution, the flow can be used to manipulate the solution concentration around the tip orifice. To enable continuous nozzle flow, the rear of the inkjet is attached by a Luer-Lock fitting to a 1-ml disposable syringe (Becton Dickinson). A motorized CMA-12PP linear actuator (Newport) connected to the plunger of the syringe, allows the user to drive fluid from the syringe through the inkjet and out the front of the nozzle prior to inkjet actuation and vesicle formation. The CMA-12PP actuator used for this setup is controlled through SMC100 software. During typical usage the actuator is run at a rate of 0.0003 mm/s which corresponds to a volumetric flow rate of 0.019 ml/h when used with a 1-ml BD syringe. The flow velocity out of the inkjet nozzle due to this flow system is calculated to be $\sim$66 mm/s.

3. Determination of Encapsulation Fraction within Vesicles

In this section, we describe a method for determining the encapsulation fraction within vesicles formed by microfluidic encapsulation when different solutions are used in and around the jet.

3.1. Falling vesicle method

To initiate chemical reactions and reconstitute biological machinery within the vesicles, it is important to know and control the ratio of two solutions mixed within the vesicle. The density of a droplet has been determined previously by the falling drop method (Sims, 1954) and, more recently, this method has been used to calculate the volume of oil present in double emulsions formed by jets (Funakoshi *et al.*, 2007). We use the falling drop method with vesicles to determine vesicle density due to the entrainment and mixing of two solutions during vesicle formation. The density of the solution within the vesicle is a linear combination of the solution jetted by the inkjet and the surrounding solution that is entrained into the jet vortex before the vesicle is formed. By jetting one solution of known density (sucrose) through a second solution (glucose) of known density, the fraction of each fluid within the vesicle can be determined.

To measure the fall time of the vesicle the plane of focus of the microscope objective is set to the plane of the inkjet and then lowered a

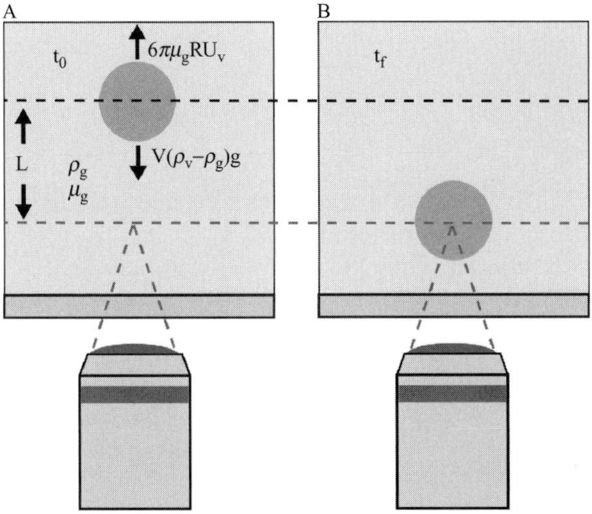

Figure 4.3 Illustration of experimental setup of the falling vesicle method used to estimate the encapsulation fraction of vesicles. (A) Focus of the microscope objective is offset a given distance prior to vesicle formation. (B) The fall time of the vesicle is used to estimate the density of vesicles. Glucose and sucrose solutions of known density and viscosities are used, and the fraction of each solution encapsulated in the vesicle is determined.

recorded distance downward such that vesicles would be formed out of focus and fall a determined distance into focus, as illustrated in Figure 4.3. The fall distance is established by focusing on the midplane of the inkjet tip then lowering the objective a distance of 0.5 mm calibrated in air. The fall distance is then corrected using Snell's law to account for the refractive index of the glucose and sucrose solutions. Using the high-speed camera with a time stamp on each frame, the fall time of the vesicle can be determined. Acceleration time is assumed to be negligibly small (~0.01 s) and the terminal velocity of the vesicle is determined from the recorded distance and fall time. By excluding data with Reynolds number greater than 0.1, we limit the flow regime to fully laminar where Stokes law has been shown to be highly accurate (Rhodes, 1998). The vesicles are assumed to remain perfectly spherical during the falling process. This assumption is evaluated by calculating the Bond number (*Bo*), the ratio of body forces to surface tension, as shown in the following equation:

$$Bo = \frac{g\Delta\rho R^2}{\gamma} \tag{4.1}$$

In our experiments, a vesicle filled completely with sucrose yields a Bond number of ~0.0004, which fulfills the requirement of being significantly smaller than unity. Therefore, the spherical approximation is valid and the velocity of the vesicle can be described by the Stokes equation, shown in the following equation:

$$U_s = \frac{2g\Delta\rho R^2}{9\mu} \tag{4.2}$$

where $\Delta\rho$ is the difference in density between the surrounding and internal fluid. The density of the surrounding fluid can be measured beforehand, leaving only the density of the fluid within the vesicle to be solved. In our case, the fluid surrounding the vesicle and the fluid entrained by the jet vortex are both 200 mOsm glucose. The resulting encapsulation fraction of glucose in the vesicle is given by

$$E_g = 1 - \frac{9U_s\mu_g}{2R^2g(\rho_s - \rho_g)} \tag{4.3}$$

where E_g is the fraction of glucose encapsulated within the vesicle.

3.2. Entrainment by vortex rings

The entrainment process occurs before the pinch off of the vesicle, during the formation of the vortex ring. Studies on single vortex ring entrainment have been performed by Gharib *et al.* (1998), who used a piston-cylinder vortex generator to determine the limiting stroke-to-diameter ratio, beyond which a vortex ring would pinch off from the trailing jet. This limiting ratio of approximately 4 was termed the formation number. Beyond this number the translational vortex velocity exceeds that of the trailing jet and fluid from the jet is no longer delivered to the vortex ring (Shusser and Gharib, 2000). Later, Dabiri and Gharib (2004) demonstrated that a vortex ring's entrainment fraction of surrounding fluid increased monotonically until the formation number was reached and pinch off occurred. The bulk of the entrainment occurs during this time and entrainment values of 30–40% were measured at and beyond pinch off. Additionally, Dabiri and Gharib were able to increase the entrainment fraction to 65% by delaying vortex pinch off beyond the formation time through the use of bulk counter flow.

In contrast to the piston-cylinder generator used in single vortex studies, our inkjet system produces a series of vortex rings traveling along the same axis (Stachowiak *et al.*, 2009). In the case of multiple vortex rings, Stachowiak *et al.* observed that the rearward rings in the trailing jet catch up to the lead vortex ring and combine to form a single, large composite vortex ring, a phenomenon that has been documented before with two vortex rings (Maxworthy, 1972). An effective stroke length for the first vortex generated by our inkjet can be estimated by multiplying the exit velocity by the period of a single voltage pulse. Using our high-speed camera at 30,000 frames/s, the exit velocity was estimated to be approximately 1.1 m/s, and duration of each voltage pulse was 50 μs based on a pulse frequency of 20 kHz. With an orifice diameter of $\sim$10 μm, we found the stroke-to-diameter ratio of the first vortex ring to be $\sim$5.5, which exceeds the formation number. Using this estimation one would expect that the first vortex ring produced by our system would contain 30–40% of the surrounding fluid if no subsequent rings were formed. Catching up of the rearward rings to the first vortex ring may be analogous to delaying the pinch off of the lead vortex ring in that fluid from the trailing jet continues to be delivered to the first vortex ring beyond the formation time. In this way, the entrainment fraction of the final composite vortex ring would have an expected value greater than 30–40% of surrounding fluid. Since this vortex ring goes on to deform the lipid bilayer until it collapses into a vesicle encapsulating the vortex, one would expect the entrapment fraction of the vesicle to be correlated to the entrainment fraction of the composite vortex ring. We carried out the experiments described below to test this entrainment prediction for vesicles formed by microfluidic encapsulation.

3.3. Solution preparation

DPHPC (1,2-diphytanoyl-*sn*-glycero-3-phosphocholine) lipid suspended in decane ($C_{10}H_{22}$) is used to form the planar bilayer. The lipid (Avanti Polar Lipids) comes dissolved in chloroform ($CHCl_3$) and sealed under argon atmosphere. In preparation for usage, a 25-mg vial of lipid is opened and divided into two glass test tubes, and the bulk of the chloroform is evaporated off under a gentle stream of nitrogen gas for approximately 20 min. The two test tubes are then placed into a vacuum chamber for 1 h to finalize chloroform removal. After drying, 0.5 ml of filtered decane (0.22 μm pore) is added to each test tube to produce a final concentration of 25 mg/ml of DPHPC suspended in decane.

A sucrose solution is used as the solution in the inkjet. A 200-mOsm sucrose solution is initially prepared by dissolving 3.42 g of crystalline sucrose (Fisher Scientific) into filtered deionized water for a final volume of 50 ml. The concentration is checked with an Osmette II osmometer (Precision Instruments). Before each experiment the sucrose solution is refiltered through a 0.22-μm pore filter to minimize particulates.

For the following experiments, a glucose solution is used as the medium surrounding the vesicles and as the solution entrained by the jet. A 200-mOsm solution is made by mixing 1.80 g of crystalline glucose with deionized, filtered water to a final volume of 50 ml. The concentration is checked using an osmometer and the solution is also filtered before usage.

3.4. Effect of planar bilayer age on oil exclusion in vesicles

To test the accuracy of the falling drop method for estimating encapsulation fraction, vesicles containing only sucrose solution are formed. This is done by loading the inkjet with a 200-mOsm sucrose solution and surrounding the inkjet in the same solution. Vesicles are made using a trapezoidal voltage profile with a 3 μs linear rise time, 30 μs dwell time at 35 V, and a fall time of 3 μs back to 0 V. Each vesicle is made by applying this voltage profile 20 times at a frequency of 20 kHz to create 20 pulses that coalesce into a single vortex ring. The vesicles are formed into a less dense 200 mOsm glucose solution and begin sinking immediately. The fall time of the vesicle is then recorded and used to determine whether the falling drop method could accurately give the expected settling rate for a sucrose-filled vesicle, equivalent to an encapsulation fraction of glucose of exactly zero.

Estimating the encapsulation fraction using vesicles formed within 30 min of forming the planar lipid bilayer led to large variability in the results. Figure 4.4 shows the variability in estimated glucose encapsulation fraction decreases ~30 min after the bilayer formation. Before the 30 min mark, the mean fraction of glucose entrapped was estimated to be 0.11 with a standard deviation of 0.19. Vesicles formed after the 30 min mark had a mean and standard deviation of 0.05 and 0.12, respectively. The most likely explanation for the time dependence is the presence of oil lenses during the formation and growth of the bilayer in the first 30 min. Using second harmonic generation microscopy, Ries *et al.* observed a similar time scale for the incorporation, growth, and exclusion of solvent in black lipid membranes, noting a growth dependence on the substrate used to support the black lipid membranes (Ries *et al.*, 2004). Figure 4.5 shows oil lenses

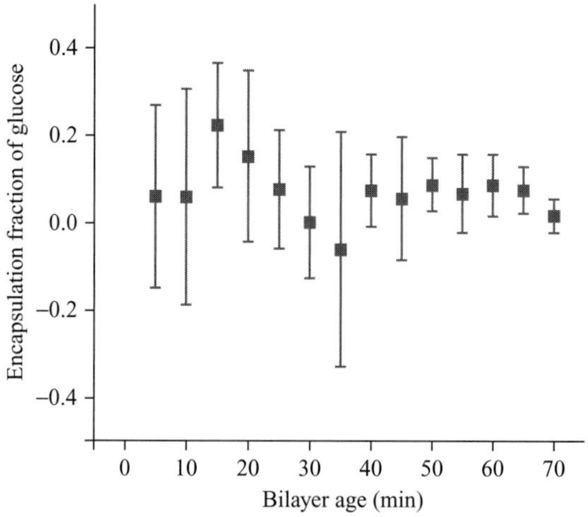

Figure 4.4 Experimental demonstration of the falling vesicle method with 0% glucose (100% sucrose) vesicles. The accuracy of the falling vesicle method was tested by measuring the falling rate of sucrose-filled vesicles in glucose and confirming that the calculated encapsulation fraction of glucose was zero. Early variability in the data suggests the presence of oil lenses in vesicles interfering with falling drop estimations for times <30 min.

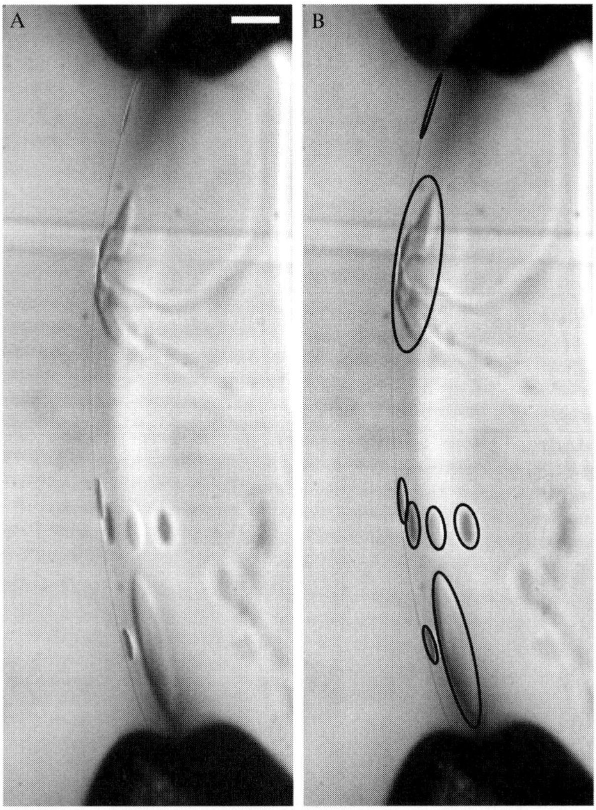

Figure 4.5 Image of oil lenses suspended in the lipid bilayer shortly after formation. The presence of oil lenses supports the observation of variability in the falling vesicle data at early times and suggests experiments should wait ~30 min after the bilayer is formed before making vesicles to ensure oil is excluded from the bilayer. Oil lenses in the image on the left are circled in the image on the right. Scale bar represents 100 um.

suspended in the bilayer. The presence of decane in the bilayer can result in oil within the vesicle membrane that obscures encapsulation estimations early in the bilayer lifetime.

3.5. Effect of continuous flow on encapsulation fraction

To test the effect of vortex entrainment on encapsulation fraction, 200 mOsm glucose solution is used to surround the inkjet and 200 mOsm sucrose solution is loaded inside the inkjet. The vesicles are formed into 200 mOsm glucose solution on the opposite side of the lipid bilayer. Vesicles are formed using the same voltage parameters described in Section 3.4. The nozzle flow system is used to slowly flow 200 mOsm sucrose solution through the inkjet. The purpose of the flow in this situation is to alter the local concentration of fluid surrounding the inkjet from pure glucose to a mix of glucose and sucrose. One might expect that the higher flow rate would lead to a higher local concentration of sucrose surrounding the inkjet tip and result in a lower encapsulation fraction of glucose. We experimentally measured encapsulation fraction of glucose at three flow rates, 0.006, 0.019, and 0.042 ml/h, as shown in Figure 4.6.

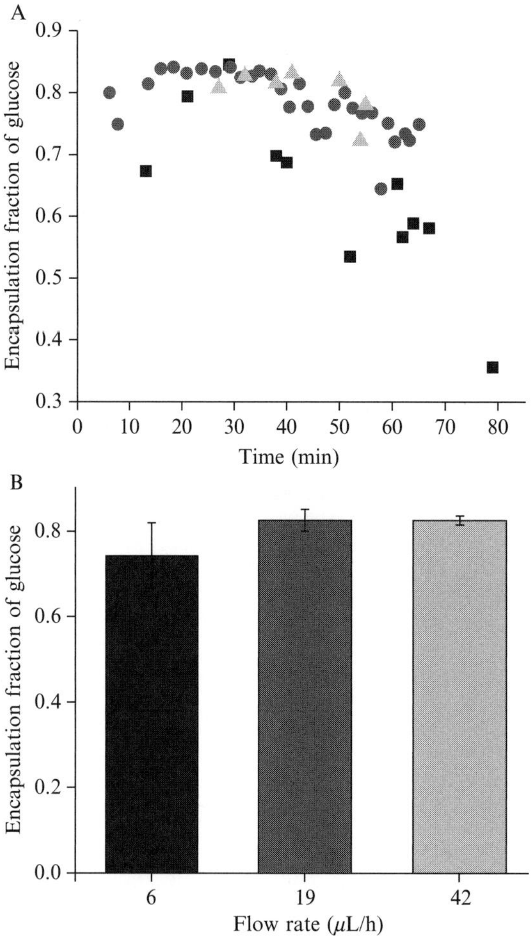

Figure 4.6 Experimentally determined glucose encapsulation fraction by the falling vesicle method comparing three different nozzle flow rates, with (A) encapsulation fraction over time and (B) average encapsulation fraction for each flow rate during the first 40 min. Vesicles are formed by jetting sucrose solution through glucose. The lack of a trend suggests that continuous flow through the nozzle does not significantly affect encapsulation fraction in vesicles. (See Color Insert.)

The lowest flow rate of 0.006 ml/h resulted in vesicles with a mean glucose encapsulation fraction of 0.64 and standard deviation of 0.13. Vesicles formed in the presence of 0.019 ml/h flow had a mean glucose encapsulation fraction of 0.79 with a standard deviation of 0.05. The highest flow rate resulted in vesicles with a mean glucose encapsulation fraction of 0.80 and a standard deviation of 0.04. At 95% confidence level, the set of vesicles made with a 0.019-ml/h flow rate were not statistically different from the vesicles made using a 0.042 ml/h flow rate. The lower encapsulation fraction of glucose in vesicles formed using the 0.006 ml/h flow suggests that other factors may be influencing encapsulation fraction, such as the separation distance between the bilayer and inkjet tip. At higher flow rates the fluid volume added around the inkjet is large enough to bend the bilayer away from the inkjet tip, requiring the inkjet to be repositioned to maintain an approximately constant separation distance. Agitating the fluid surrounding the inkjet tip would lower local concentration of sucrose, leading to high glucose encapsulation corresponding to higher flow.

3.6. Effect of inverting voltage polarity on encapsulation fraction

One explanation for the observed low encapsulation of fluid from the jet (sucrose) compared to fluid from surrounding the jet (glucose) is that no net flow from within the inkjet nozzle is actually expelled during the generation of the liquid jet that deforms the bilayer membrane during vesicle formation. This is possible because the inkjet generates the jet acoustically rather than through slug flow. Since expansion and contraction of the piezoelectric actuator forms acoustic waves in pairs, each compression wave has a corresponding rarefaction wave of equal amplitude. This equal and opposite pairing leads to oscillation of fluid in and out of the inkjet device that delivers momentum to drive the jet but provides no net fluid from within the inkjet nozzle to the jet itself. If this hypothesis is correct, and the jet that forms the vesicle is generated by fluid oscillations rather than slug flow from the inkjet, different voltage signals that produce the same oscillations should yield similar encapsulation fraction. More specifically, a positive voltage profile should lead to vesicles with encapsulation fractions comparable to those created by the equivalent negative, or inverted, voltage profile.

As described earlier, the inkjet is driven by repeat application of a trapezoidal voltage trace which rises from 0 to 35 V. Under these conditions, acoustic waves are generated by repeat expansion of the diameter of the inkjet's capillary tube. The same inkjet device can generate acoustic waves by contracting the capillary diameter using an inverted trapezoidal voltage signal which falls from 0 to -35 V. All other parameters including the timing and number of pulses applied remained the same for this experiment.

As in the previous experiment, the fluid within the inkjet was 200 mOsm sucrose, while the fluid surrounding the inkjet and the newly formed vesicles was 200 mOsm glucose. The flow rate was set to 0.019 ml/h. Using the inverted voltage trace, the resulting vesicles had a mean glucose encapsulation fraction of 0.73 and a standard deviation of 0.07. At a 95% confidence level, the encapsulation fraction of vesicles produced using the positive trapezoidal trace and vesicles produced using negative voltage trace were not statistically different, as shown in Figure 4.7.

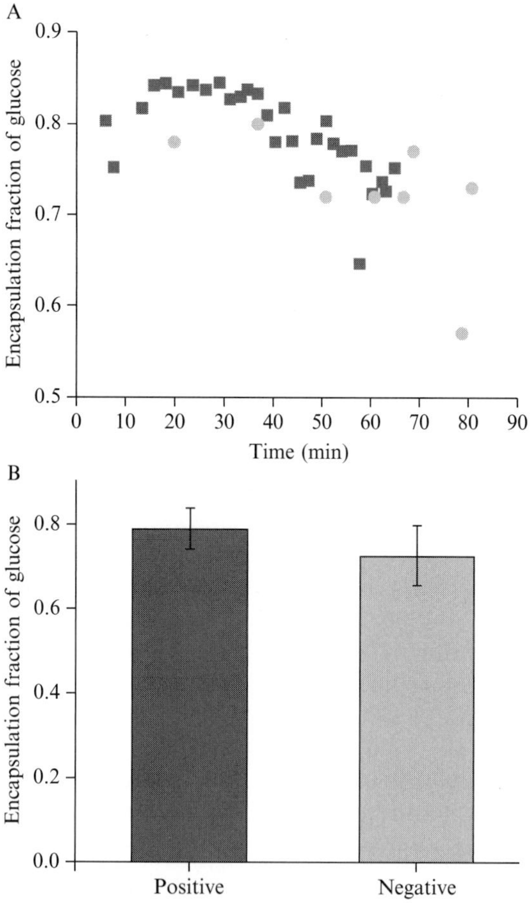

Figure 4.7 Experimentally determined glucose encapsulation fraction by the falling vesicle method comparing vesicles made by an inkjet driven by a positive voltage trace and a negative voltage trace, with (A) encapsulation fraction over time and (B) average encapsulation fraction for both polarities over the entire time. No statistical difference at 95% confidence level suggests no difference in the encapsulation process for different polarities.

This result provides insight into the inkjet driven vortex formation process. In the case of the positive trapezoidal signal, the increase in voltage leads to a rarefaction wave reaching the tip first. With no surface tension to counteract the wave, a small amount of fluid immediately outside of the orifice will be pulled into the inkjet and subsequently ejected when the voltage decreases back to zero, causing a compression wave to propagate to the orifice. In the case of a negative trapezoidal signal, a compression wave reaches the tip first, ejecting a small volume of liquid immediately inside of the orifice, and the rarefaction wave that follows draws liquid in from immediately outside the orifice. With repeat pulses the oscillations generated by the positive profile should be identical to the oscillations generated by the inverted profile. The findings of this experiment suggest that there are no differences in fluid mixing that occur from driving the inkjet with a positive voltage trace rather than a negative trace, which is likely an effect of the jet being generated acoustically rather than through bulk flow.

3.7. Discussion

The main advantage of estimating the encapsulation fraction using the falling drop method is simplicity. Given the described vesicle formation system, a viscometer, and a scale, the encapsulation fraction of vesicles can be quickly estimated. This method can be done with no modifications to the vesicle formation device, and therefore it can be performed as a precursor to another assay with minimal additional effort. Furthermore, encapsulation fraction can be estimated using a single vesicle which offers convenience over bulk estimation methods (Oku *et al.*, 1982).

Several fundamental assumptions are made in the application of Stokes law that can limit the accuracy of the falling method. The fluid surrounding the vesicle is assumed to be infinite in volume to have no wall effects, based on the experiment using purely sucrose-filled vesicles this approximation seems acceptable for the current chamber design; however, future vesicle chambers may evolve toward a more compact microfluidic design, in which case the infinite fluid assumption may not be satisfied. Stokes drag applies to a solid sphere while the Rybczynski–Hadmard equation offers a more accurate prediction which accounts for circulation for a falling fluid sphere under the same condition (Lamb, 1932). The drag on a vesicle should fall between the two results as a vesicle should have reduced inner fluid circulation compared to free fluid droplet. Experiments using vesicles filled with only 200 mOsm sucrose suggest that Stokes law as a more appropriate estimation of the drag on a vesicle than the Rybczynski–Hadmard equation. Application of Stokes law requires low Reynolds and Bond numbers, which also limits the solution combinations that can be used in this method; large differences in densities greatly reduce the applicability of Stokes law. In situations where greater accuracy is required, encapsulation estimations can potentially be done using quantitative fluorescence on individual vesicles, though establishing a calibration standard is a challenge.

To generalize the encapsulation results to other experimental situations, one must consider the vortex ring. When a vortex is created, the formation number is a function of the vortex generator parameters rather than the fluid properties. Therefore, one could expect similar rapid entrainment fractions across vortex rings of different fluid properties produced using the same stroke-to-diameter ratios exceeding the formation number. Although the vortex rings may have similar entrainment fractions, the vortex ring energy will vary by fluid density for a Norbury family of vortex rings (Norbury, 1973). Therefore, variation in fluid properties may result in vortex rings that do not possess the energy to adequately deform the lipid bilayer into a vesicle. Beyond the rapid fluid entrainment during formation, vortex rings continue to grow and entrain ambient fluid through slow viscous diffusion which is governed by both time and fluid viscosity. This secondary entrainment process is therefore influenced by both the fluid properties and the vortex generator parameters and must be carefully considered for each experiment.

4. Concluding Remarks

In this chapter, we have presented details of the microfluidic encapsulation technique for forming unilamellar vesicles with mixed contents, and we have demonstrated the use of the falling drop method for determining the encapsulated fraction of solutions mixed by vortex entrainment. In doing so, we have documented practical considerations that are important for forming vesicles and controlling encapsulation fraction. We find there is a marked advantage in allowing the bilayer to age prior to formation of vesicles in order to exclude oil; however, no advantage is gained by simply inverting the driving voltage signal of the inkjet. Our measurements of encapsulated fraction of $\sim$80% are significantly larger than estimates from previous vortex ring experiments, suggesting that the presence of the planar bilayer and details of the multiple inkjet pulses are significant for control of entrainment and encapsulation. We have found that ability of continuous nozzle flow to control the encapsulation fraction of two different solutions is limited, suggesting that other strategies must be used for expanding control of mixing. Combining solutions during microfluidic encapsulation is an important step toward the formation of complex unilamellar vesicles and liposomes, and continued development of the technique will be aimed at improving the ability to measure and control this property.

REFERENCES

Angelova, M., and Dimitrov, D. (1986). Liposome electroformation. *Faraday Discuss.* **81,** 303–311.

Bogy, D. B., and Talke, F. E. (1984). Experimental and theoretical-study of wave-propagation phenomena in drop-on-demand ink jet devices. *IBM J. Res. Develop.* **28,** 314–321.

Dabiri, J. O., and Gharib, M. (2004). Fluid entrainment by isolated vortex rings. *J. Fluid Mech.* **511,** 311–331.

Debbage, P. (2009). Targeted drugs and nanomedicine: Present and future. *Curr. Pharm. Des.* **15,** 153–172.

Dos Santos, N., Allen, C., Doppen, A.-M., Anantha, M., Cox, K. A. K., Gallagher, R. C., Karlsson, G., Edwards, K., Kenner, G., Samuels, L., Webb, M. S., and Bally, M. B. (2007). Influence of poly(ethylene glycol) grafting density and polymer length on liposomes: Relating plasma circulation lifetimes to protein binding. *Biochim. Biophys. Acta* **1768,** 1367–1377.

Düzgüneş, N., Flasher, D., Pretzer, E., Konopka, K., Slepushkin, V. A., Steffan, G., Salem, I. I., Reddy, M. V., and Gangadharam, P. R. J. (1995). Liposome-mediated therapy of human immunodeficiency virus type-1 and Mycobacterium infections. *J. Liposome Res.* **5,** 669–691.

Düzgüneş, N., Simões, S., Slepushkin, V., Pretzer, E., Flasher, D., Salem, I. I., Steffan, G., Konopka, K., and Pedroso de Lima, M. C. (2005). Delivery of antiviral agents in liposomes. *Methods Enzymol.* **391,** 351–373.

ElBayoumi, T. A., Pabba, S., Roby, A., and Torchilin, V. P. (2007). Antinucleosome antibody-modified liposomes and lipid-core micelles for tumor-targeted delivery of therapeutic and diagnostic agents. *J. Liposome Res.* **17,** 1–14.

Funakoshi, K., Suzuki, H., and Takeuchi, S. (2007). Formation of giant lipid vesiclelike compartments from a planar lipid membrane by a pulsed jet flow. *J. Am. Chem. Soc.* **129,** 12608–12609.

Gabizon, A., Catane, R., Uziely, B., Kaufman, B., Safra, T., Cohen, R., Martin, F., Huang, A., and Barenholz, Y. (1994). Prolonged circulation time and enhanced accumulation in malignant exudates of doxorubicin encapsulated in polyethylene-glycol coated liposomes. *Cancer Res.* **54**(4), 987–992.

Gantert, M., Lewrick, F., Adrian, J. E., Roessler, J., Steenpass, T., Schubert, R., and Peschka-Suess, R. (2009). Receptor-specific targeting with liposomes in vitro based on sterol-PEG(1300) anchors. *Pharm. Res.* **26,** 529–538.

Gharib, M., Rambod, E., and Shariff, K. (1998). A universal time scale for vortex ring formation. *J. Fluid Mech.* **360,** 121–140.

Karlsson, M., Nolkrantz, K., Davidson, M. J., Stromberg, A., Ryttsen, F., Akerman, B., and Orwar, O. (2000). Electroinjection of colloid particles and biopolymers into single unilamellar liposomes and cells for bioanalytical applications. *Anal. Chem.* **72,** 5857–5862.

Konduri, K., Nandedkar, S., Rickaby, D. A., Düzgüneş, N., and Gangadharam, P. R. J. (2005). The use of sterically stabilized liposomes to treat asthma. *Methods Enzymol.* **391,** 413–427.

Lamb, H. (1932). Hydrodynamics. The University Press, Cambridge.

Lasic, D. D., and Papahadjopoulos, D. (1995). Liposomes revisited. *Science* **267,** 1275–1276.

Martina, M. S., Fortin, J. P., Menager, C., Clement, O., Barratt, G., Grabielle-Madelmont, C., Gazeau, F., Cabuil, V., and Lesieur, S. (2005). Generation of superparamagnetic liposomes revealed as highly efficient MRI contrast agents for *in vivo* imaging. *J. Am. Chem. Soc.* **127,** 10676–10685.

Maxworthy, T. (1972). The structure and stability of vortex rings. *J. Fluid Mech.* **51,** 15–32.

Norbury, J. (1973). A family of steady vortex rings. *J. Fluid Mech.* **57,** 417–431.

Oku, N., Kendall, D. A., and Macdonald, R. C. (1982). A simple procedure for the determination of the trapped volume of liposomes. *Biochim. Biophys. Acta* **691,** 332–340.

Olson, F., Hunt, C. A., Szoka, F. C., Vail, W. J., and Papahadjopoulos, D. (1979). Preparation of liposomes of defined size distribution by extrusion through polycarbonate membranes. *Biochim. Biophys. Acta* **557,** 9–23.

Ramon, E., Alonso, C., Coderch, L., De la Maza, A., Lopez, O., Parra, J. L., and Notario, I. (2005). Liposomes as alternative vehicles for sun filter formulations. *Drug Deliv.* **12,** 83–88.

Reeves, J. P., and Dowben, R. M. (1969). Formation and Properties of thin-walled phospholipid vesicles. *J. Cell Physiol.* **1,** 49–60.

Rhodes, M. (1998). Introduction to particle technology. Wiley, New York.

Ries, R. S., Choi, H., Blunck, R., Bezanilla, F., and Heath, J. R. (2004). Black lipid membranes: Visualizing the structure, dynamics, and substrate dependence of membranes. *J. Phys. Chem. B* **108,** 16040–16049.

Safra, T., Muggia, F., Jeffers, S., Tsao-Wei, D. D., Groshen, S., Lyass, O., Henderson, R., Berry, G., and Gabizon, A. (2000). Pegylated liposomal doxorubicin (doxil): Reduced clinical cardiotoxicity in patients reaching or exceeding cumulative doses of 500 mg/m(2). *Ann. Oncol.* **11,** 1029–1033.

Salem, I. I., Flasher, D. L., and Düzgüneş, N. (2005). Liposome-encapsulated antibiotics. *Methods Enzymol.* **391,** 261–291.

Shusser, M., and Gharib, M. (2000). Energy and velocity of a forming vortex ring. *Phys. Fluids* **12,** 618–621.

Sims, R. P. A. (1954). Some observations on determining density of fluids by the falling drop method. *Can. J. Chem.* **32,** 506–511.

Stachowiak, J. C., Richmond, D. L., Li, T. H., Liu, A. P., Parekh, S. H., and Fletcher, D. A. (2008). Unilamellar vesicle formation and encapsulation by microfluidic jetting. *Proc. Natl. Acad. Sci. USA* **105,** 4697–4702.

Stachowiak, J. C., Richmond, D. L., Li, T. H., Brochard-Wyart, F., and Fletcher, D. A. (2009). Cell-like encapsulation by high-throughput inkjet printing of unilamellar lipid vesicles. *Lab Chip* doi:10.1039/B904984C.

Sugiura, S., Kuroiwa, T., Kagota, T., Nakajima, M., Sato, S., Mukataka, S., Walde, P., and Ichikawa, S. (2008). Novel method for obtaining homogeneous giant vesicles from a monodisperse water-in-oil emulsion prepared with a microfluidic device. *Langmuir* **24,** 4581–4588.

Szoka, F. Jr., and Papahadjopoulos, D. (1978). Procedure for preparation of liposomes with large internal aqueous space and high capture by reverse-phase evaporation. *Proc. Natl. Acad. Sci. USA* **75,** 4194–4198.

Weissig, V., Boddapati, S. V., Cheng, S.-M., and D'Souza, G. G. M. (2006). Liposomes and liposome-like vesicles for drug and DNA delivery to mitochondria. *J. Liposome Res.* **16,** 249–264.

Zielhuis, S. W., Seppenwoolde, J. H., Mateus, V. A. P., Bakker, C. J. G., Krijger, G. C., Storm, G., Zonnenberg, B. A., van het Schip, A. D., Koning, G. A., and Nijsen, J. F. W. (2006). Lanthanide-loaded liposomes for multimodality imaging and therapy. *Cancer Biother. Radiopharm.* **21,** 520–527.

CHAPTER FIVE

Recombinant Proteoliposomes Prepared Using Baculovirus Expression Systems

Kanta Tsumoto*,† *and* Tetsuro Yoshimura*,†,‡

Contents

Abstract

Proteoliposomes are useful for investigating the functions and properties of membrane proteins. We have established a novel method to prepare proteoliposomes using a baculovirus (*Autographa californica* nuclear polyhedrosis virus; AcNPV) gene expression system producing the recombinant membrane proteins to be reconstituted. This method consists of two key steps for production of *recombinant* proteoliposomes: (1) The cDNA of the recombinant membrane protein on a baculovirus vector is transfected into insect cells, and recombinant AcNPV budded viruses (BVs), which express the targeted proteins on their own envelopes, are collected, and (2) the BV envelopes are subjected to fusion with liposomes containing acidic phospholipids by activating the viral glycoprotein gp64 at low pH. In this chapter, we describe the reconstitution of the transmembrane proteins, thyroid-stimulating hormone receptor (TSHR, a member of G protein-coupled receptor family) and acetylcholine receptor α-subunit (AChRα), in various types of liposomes, including large unilamellar vesicles (LUVs), multilamellar vesicles (MLVs), and giant unilamellar vesicles

* Graduate School of Engineering, Mie University, Tsu, Mie, Japan
† Liposome Engineering Laboratory Inc., Tsu, Mie, Japan
‡ Nagoya Industrial Science Research Institute, Nagoya, Aichi, Japan

Methods in Enzymology, Volume 465
ISSN 0076-6879, DOI: 10.1016/S0076-6879(09)65005-9

(GUVs). Enzyme-linked immunosorbent assay (ELISA) shows that the reconstituted proteins (TSHR and AChRα) specifically bind their ligands (TSH and α-bungarotoxin), respectively, as well as antibodies against the recombinant proteoliposomes. In addition, TSHRs reconstituted on proteo-GUVs are successfully visualized using a confocal laser scanning microscope with fluorescence immunostain. These data suggest that recombinant proteoliposomes prepared using the novel method have the potential to be applied for reconstitution of complicated multicomponent membrane protein systems.

1. Introduction

Proteoliposomes are liposomes containing membrane proteins reconstituted in the lipid bilayer. They are important tools for investigation of the functions and properties of membrane proteins. Because of their amphiphilic properties, membrane proteins should be solubilized for preparation of proteoliposomes. Conventionally, proteins on cell membranes are retrieved using detergent and reconstituted in liposome membranes by removing the detergent through dialysis, gel filtration, etc. (Rigaud and Lévy, 2003). The conventional methods are, of course, very useful, but they have several disadvantages: Transbilayer orientation of membrane proteins, complete removal of detergent, and the size and lamellarity of the liposomes are not easy to control. In this chapter, we describe in detail a novel method that we have established for proteoliposome preparation (Fukushima *et al.*, 2008). In this method, a membrane protein is produced from a recombinant cDNA using a baculovirus gene expression system, and recovered by collecting amplified budded viruses (BVs) with the targeted protein in their envelopes. They are next subjected to fusion with liposomes by activating the fusogenic function of an endogenous baculovirus protein, resulting in proteoliposomes. This method is free from detergent, and may keep the orientation of membrane proteins. In addition, proteoliposomes can be prepared using a variety of liposomes, including small unilamellar vesicles (SUVs), large unilamellar vesicles (LUVs), multilamellar vesicles (MLVs), and giant unilamellar vesicles (GUVs). The proteoliposomes prepared using this method are termed *recombinant* proteoliposomes.

2. Principles of Recombinant Proteoliposome Preparation

Baculovirus expression vectors enable specific posttranslational modification and expression of large amounts of proteins (Kost and Condreay, 1999). Recently, it has been reported that some foreign membrane proteins are expressed on BV envelopes derived from baculoviruses (Hayashi *et al.*, 2004; Loisel *et al.*, 1997; Masuda *et al.*, 2003; Urano *et al.*, 2003). When the genes for G protein-coupled receptors (GPCRs), which are common drug targets, are recombined in baculoviruses, and cells are infected, GPCRs are known to be expressed on BV envelopes with normal activity levels (Loisel *et al.*, 1997). Although the normal expression of the leukotriene B_4 (LTB_4) receptor BLT1 in cell membranes is higher than that observed in BV envelopes, the binding of $[^3H]BLT_4$ to BLT1 in BV envelopes occurs at a level approximately 10 times greater than that observed in cell membranes (Masuda *et al.*, 2003). SREBP-2 and SCAP membrane proteins, which are involved in the regulation of intracellular cholesterol, are expressed on BV envelopes in their active forms (Urano *et al.*, 2003). γ-Secretase complexes, which are related to Alzheimer disease, are also expressed on BV envelopes with normal activity levels intact (Hayashi *et al.*, 2004), suggesting that it is possible to express any recombinant membrane protein in BV envelopes. Recently, thyroid-stimulating hormone receptor (TSHR) and acetylcholine receptor α-subunit (AChRα) have also been expressed in BV envelopes, while maintaining their reactivities (Fukushima *et al.*, 2008).

Enveloped viruses, such as influenza virus (IFV), feature a fusogenic protein, hemagglutinin, on their envelopes, and upon viral infection of cells, this protein induces fusion with cell membranes at low pH (Yoshimura *et al.*, 1982). IFV-infected cells fuse with liposomes at low pH in a similar manner (van Meer *et al.*, 1985), and IFV alone can also fuse with liposomes at low pH (Burger *et al.*, 1988). Similarly, BVs fuse with cell membranes at low pH in a process mediated by the envelope membrane glycoprotein gp64 (Blissard and Wenz, 1992). The fusion behavior of wild type *Autographa californica* nuclear polyhedrosis virus (AcNPV)-BVs and liposomes was examined using an R_{18}-dequenching assay, and wild type AcNPV-BVs were found to fuse with liposomes (Fukushima *et al.*, 2008). Fusion rates were higher below pH 5.0 and when liposomes contained PS rather than PG, PA, or PI, indicating that BVs specifically bind to liposomal PS (Tani *et al.*, 2001) and that fusion is induced by gp64, which is activated at low pH levels (Blissard and Wenz, 1992).

These observations demonstrate that proteoliposomes containing recombinant membrane proteins can be prepared without detergents, by fusing BVs containing target recombinant membrane proteins with liposomes, and thus, recombinant proteoliposomes can be prepared by the following two characteristic steps:

(1) A step preparing BV particles expressing a recombinant membrane protein.
(2) A step preparing recombinant proteoliposomes by fusion of the resulting BV particles with liposomes.

Concerning (1), the genes encoding targeted membrane proteins are prepared: in most cases, targeted genes are obtained by cloning the membrane protein-encoding genes from their cDNA library. In this case, for example, PCR procedures are applicable using suitable primers, and obtained genes are incorporated into transfer vectors and, together with baculovirus DNA, cotransfected into insect cultured cells (e.g., Sf9 cells). Upon incorporation of targeted genes into the transfer vector, suitable tags (e.g., His-Tag) are preferable to attach for further preparation. Recombinant viruses budded from the insect cultured cells are again infected into cultured cells. BV particles of recombinant baculovirus with targeted genes appear in the supernatant because homologous recombination between transfer vectors and baculovirus DNA occurs in the insect cells. They are prepared from the supernatant, resulting in expression of targeted membrane proteins on the prepared virus envelopes. Obtained baculoviruses are infected into insect cultured cells again if necessary, subjected to purification and amplification, and supplied for the next steps.

Concerning (2), by fusion of BVs and liposomes, recombinant proteoliposomes are prepared. Fusion of viruses with liposomes occurs at acidic pH around 4, mediated by the fusogenic protein, gp64, on the baculovirus envelope. Any one of SUV, LUV, MLV, or GUV is available, and biotinylation of any of the liposome constituents is convenient for fixation to plates. Liposomes fused with BVs can be used as they are, but it is better to prepare and recover recombinant proteoliposomes using the conventional ultracentrifugation and gel filtration.

3. Construction of Recombinant AcNPVs

There is a lot of literature on how to construct recombinant AcNPVs (e.g., see King *et al.*, 2007; Possee and King, 2007). Kits containing baculovirus transfer vectors, viral genomic DNAs, and transfection reagents are also commercially available from major suppliers. We have chosen BD BaculoGold™ (BD Biosciences), which consists of the transfer vector pVL1392/1393, linearized baculovirus DNAs, and a transfection reagent (calcium phosphate). A cDNA of the targeted protein and the pVL vector containing the foreign cDNA are prepared using PCR cloning from cDNA libraries (TSHR, AChRα) and subcloning procedures as follows (Fukushima *et al.*, 2008).

The TSHR cDNA is subcloned from a human thyroid gland cDNA library (Clontech) by PCR. The resulting DNA fragment of approximately 2.3 kb is inserted into the multiple cloning site (MCS) of pET-28a(+) (5.4 kb; Novagen) to produce pET/TSHR (7.7 kb). The DNA fragment encoding TSHR is excised from pET/TSHR by treatment with the appropriate restriction enzymes, and then inserted into the MCS of pVL1393 (9.6 kb; BD Biosciences) under the polyhedrin promoter to produce pVL1393/TSHR, which encoded a fusion protein of TSHR and six histidine (His) residues (12 kb). pVL1393/TSHR and linearized AcNPV-DNA (BD BaculoGold) are cotransfected into Sf9 cells using the calcium phosphate method to produce TSHR recombinant AcNPVs. Recombinant viruses are purified by plaque assay.

The AChRα cDNA is subcloned from a human skeletal muscle cDNA library (Clontech) by PCR. The resulting DNA fragment of approximately 1.4 kb is inserted into the MCS of pET-30a(+) (5.4 kb; Novagen) to produce pET/AChRα (6.8 kb). The DNA fragment encoding AChRα is excised from pET/AChRα by treatment with the appropriate restriction enzymes, and then inserted into the MCS of pVL1392 (9.6 kb; BD Biosciences) under the polyhedrin promoter to produce pVL1392/AChRα, which encoded a fusion protein of AChRα and six His residues (11 kb). pVL1392/AChRα and linearized AcNPV-DNA (BD BaculoGold) are cotransfected into Sf9 cells using Cellfectin reagent (Invitrogen) to produce AChRα recombinant AcNPVs, which are purified by plaque assay.

These single plaques are picked up and put into Sf9 cells cultured in a 12-well plate. After incubation for 72 h, 1 ml of the culture supernatant, which contained recombinant AcNPVs derived from a single plaque, is put into the fresh Sf9 cells in a culture flask (25 cm^2), and after 72 h incubation the supernatant is similarly put to another fresh Sf9. Three passages of the infection are conducted to amplify the virus titers, and then the supernatant is filtered with a 0.22 μm filter and stored until usage at -80 °C in separate tubes. Because recombination efficiencies are reported to be close to 100% in the case of BD BaculoGold, single plaque isolation is not necessarily required.

4. Expression of Recombinant Proteins on BV Envelopes

AcNPV-BVs are harvested as follows: Sf9 cells are cultured in 10 culture flasks (75 cm^2) containing 12 ml of Sf-900 II SFM medium. The above stock suspension of recombinant AcNPVs is added to each flask at a multiplicity of infection (MOI) of 1, incubated at 27 °C for 72 h, and then placed at 4 °C. After 120 h infection; culture supernatants are separated from cells and centrifuged at 1000$\times g$ for 5 min at 4 °C. The resulting supernatants are ultracentrifuged at 100,000$\times g$ for 60 min at 15 °C in an ultracentrifuge. Precipitates containing BV particles are resuspended in phosphate-buffered saline (PBS) (1 mM Na_2HPO_4, 10.5 mM KH_2PO_4, 140 mM NaCl, 40 mM KCl, pH 6.2) and again ultracentrifuged at 40,000$\times g$ for 30 min at 15 °C with a stepwise sucrose gradient of 10%, 15%, 20%, 25%, and 30% sucrose (w/v) in PBS (pH 6.2). Two fractions are found to contain BVs. Electron microscopic observation reveals that BVs in the upper fraction are somewhat disrupted, whereas those in the lower fraction are normal. Agarose gel electrophoresis shows that the lower fraction contained virus genomes, while the upper fraction does not. The fraction containing AcNPV-BVs without virus genomes is collected, diluted with PBS (pH 6.2), and ultracentrifuged at 100,000×*g* for 60 min at 15 °C. Precipitates are suspended in PBS (pH 6.2).

As shown in Fig. 5.1A and C, the fractions derived from TSHR recombinant AcNPV-BVs contain a protein band that is not observed in wild type AcNPV-BVs. As shown in Fig. 5.1B and D, the fractions derived

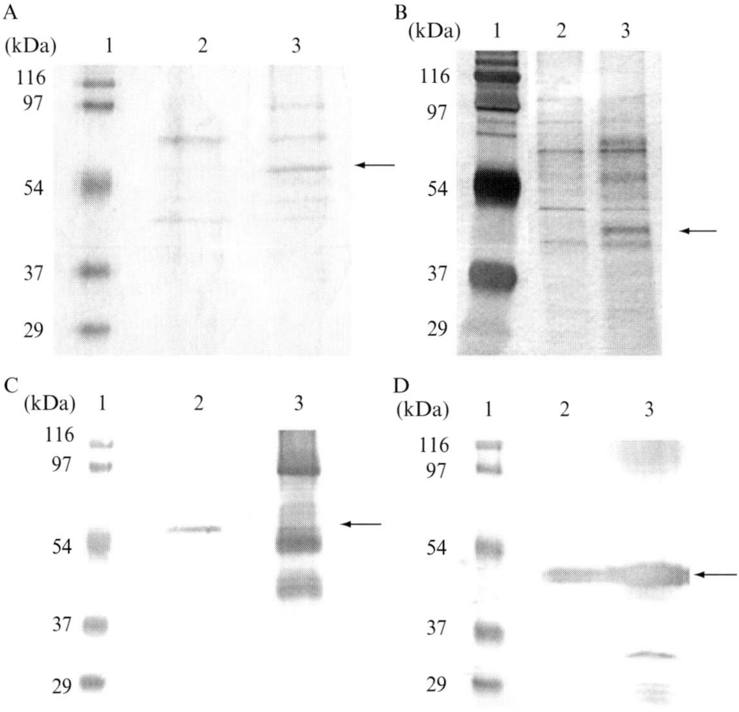

Figure 5.1 SDS–PAGE (A, B) and Western blot (C, D) analyses of recombinant AcNPV-BVs. BV fractions without DNA are obtained from TSHR and AChRα recombinant AcNPVs, and SDS–PAGE (A, B) and Western blotting is performed using anti-His-Tag antibodies (C, D). Arrows indicate the locations of TSHR (A, C) and AChRα (B, D). (A) Lane 1, protein marker; lane 2, wild-type AcNPV-BV fraction without DNA; lane 3, TSHR recombinant AcNPV-BV fraction. (B) Lane 1, protein marker; lane 2, wild type AcNPV-BV fraction without DNA; lane 3, AChRα recombinant AcNPV-BV fraction. (C) Lane 1, protein marker; lane 2, TSHR recombinant AcNPV-BV fraction; lane 3, TSHR recombinant AcNPV-infected Sf9 cell. (D) Lane 1, protein marker; lane 2, AChRα recombinant AcNPV-BV fraction; lane 3, AChRα recombinant AcNPV-infected Sf9 cell.

from AChRα recombinant AcNPV-BVs also contain a protein band that is not observed in wild type AcNPV-BVs, its molecular weight being comparable to that of the AChRα. In the Western blot analysis, multiple bands are detected in the Sf9 cell fractions (Fig. 5.1A and B), probably due to degradation of TSHR and AChRα and formation of aggregates. In contrast, only one band is detected in the virus fractions (Fig. 5.1C and D), showing that only the active form of the target recombinant membrane protein is expressed on the AcNPV-BV envelope (Hayashi *et al.*, 2004; Urano *et al.*, 2003).

5. Preparation of Proteoliposomes by Fusion of BVs with Liposomes

5.1. LUV and MLV

5.1.1. Liposome preparation

LUVs are prepared in 10 m*M* Tris–HCl/10 m*M* NaCl (pH 7.5) by the reverse-phase evaporation method and filtered through polycarbonate membranes of 0.1 μm pore size (Maezawa *et al.*, 1989). MLVs are prepared by vortex mixing of the lipid film in 10 m*M* Tris–HCl/10 m*M* NaCl (pH 7.5) at 30 °C for 30 s. The vesicles are passed through polycarbonate membranes of 0.4 μm pore size and washed 4–5 times at 6000×*g* for 20 min at 4 °C to remove small multilamellar and unilamellar vesicles. The liposome concentration is determined by measuring total lipid phosphorus by the method of Bartlett (1959) [see the detailed procedure in Düzgüneş, 2003].

5.1.2. Proteoliposome preparation

Recombinant proteoliposomes are generated by fusion of recombinant BVs with liposomes (Fukushima *et al.*, 2008). Purified recombinant BVs and LUVs or MLVs with various lipid compositions are mixed in 10 m*M* Tris–HCl/10 m*M* NaCl (pH 7.5) buffer at concentrations of 1 μg/ml and 100 μ*M*, respectively, and then, the solution is adjusted to pH 4.0 using10 m*M* CH_3COOH/10 m*M* NaCl. After incubation for 60 min at room temperature, the mixture is adjusted to pH 7 using a solution of 1 *M* Tris, and then the resulting recombinant proteoliposomes are stored on ice until use.

As mentioned above, TSHR and AChRα are expressed on BVs derived from TSHR and AChRα recombinant AcNPVs. Then, TSHR recombinant AcNPV-BVs and PC/PS(1:1)-MLVs are combined at pH 4.0 or 8.5, and incubated for 20 min. The mixture is then centrifuged at 6000×*g* for 20 min at 4 °C to yield precipitates containing MLV and supernatants containing BV, and these are subjected to SDS–PAGE and Western blot analyses using anti-His-Tag antibodies. As shown in Fig. 5.2A, protein

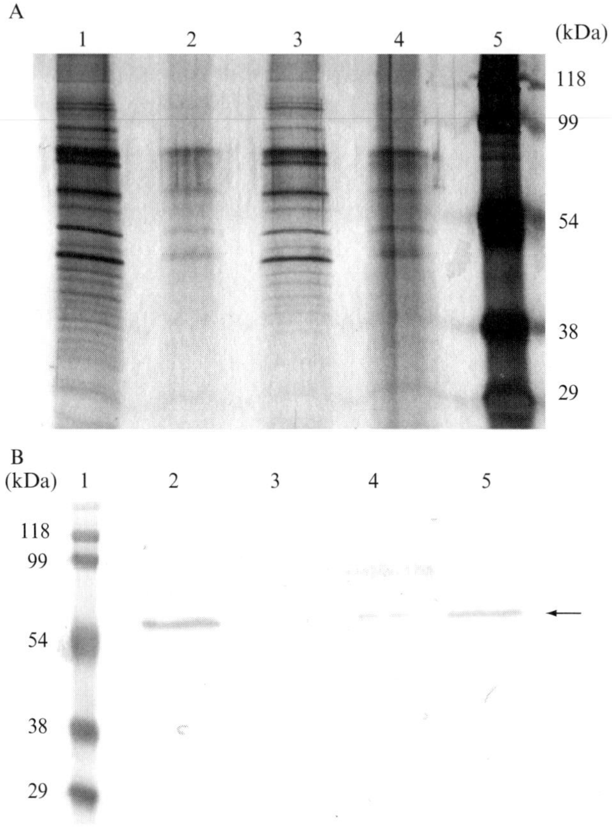

Figure 5.2 Characterization of proteoliposome samples by centrifugation. TSHR recombinant AcNPV-BV and PC:PS(1:1)-MLV are mixed at pH 4.0 or 8.5. After a 20 min incubation, the mixture is centrifuged at $6000\times g$ for 20 min at 4 °C, and the resulting precipitates and supernatants are subjected to SDS–PAGE (A) and Western blot analysis using anti-His-Tag antibodies (B). (A) Lane 1, precipitates on fusion at pH 4.0; lane 2, supernatants on fusion pH 4.0; lane 3, precipitates on fusion at pH 8.5; lane 4, supernatants on fusion at pH 8.5; lane 5, protein marker. (B) Lane 1, protein marker; lane 2, precipitates on fusion at pH 4.0; lane 3, supernatants on fusion pH 4.0; lane 4, precipitates on fusion at pH 8.5; lane 5, supernatants on fusion at pH 8.5.

bands in the supernatant fraction are detected more weakly when fusion occurred at pH 4.0, compared to pH 8.5, whereas proteins in the precipitate fraction are detected more strongly when fusion occurred at pH 4.0, compared to pH 8.5. As shown in Fig. 5.2B, a protein band corresponding to TSHR in the supernatant fraction is barely detected when fusion occurred at pH 4.0, and is detected at significant levels when fusion occurred at pH 8.5. In contrast, the TSHR band in the precipitate fraction is detected more readily when fusion occurred at pH 4.0 than at pH 8.5.

5.1.3. Reaction with ligands and autoantibodies

In addition to the methods for preparation of recombinant proteoliposomes (1) and (2) as described above, (3) a step applying the resulting proteoliposomes for enzyme-linked immunosorbent assay (ELISA) is important.

Concerning (3), recombinant proteoliposomes are coated onto the surface of well of the plate. When biotin is contained in liposome, streptavidin is fixed in advance on the surface of well; streptavidin-coated microplates on marketing are useful. For fixation of biotinylated proteoliposomes to the streptavidin-coated plates, their solutions are kept in the wells of the plate for several hours or preferably overnight. The concentration of proteoliposomes is more than 0.01 μg/ml, and preferably 0.5–20 μg/ml. Thus, ELISA systems are constructed for evaluation of ligands, antimembrane-protein antibodies, and autoantibodies using recombinant proteoliposome-fixed ELISA plate, and available for the screening test. The ELISA systems contain the primary antibody, the secondary antibody with enzyme, enzyme substrate, and the compound for color development.

Actually, recombinant proteoliposomes or liposomes with various lipid compositions are added to streptavidin-coated microplates and left at 4 °C overnight. Next, the plates are blocked at 37 °C for 1 h by addition of 1% gelatin or 3% Block Ace, and incubated at 37 °C for 1 h in the presence of ligands, antimembrane-protein antibodies, or autoantibodies. The plates are further incubated at 37 °C for 1 h with antiligand antibody and/or goat antirabbit IgG (H + L chain) or goat antihuman IgG (γ chain) peroxidase-conjugated secondary antibodies. Finally, a coloring reagent containing *o*-phenylene diamine and H_2O_2 is added to the reaction mixture, and the absorbance at 492 nm is measured.

As shown in Table 5.1, nonspecific bindings of TSH, α-bungarotoxin (α-BT, an antagonist of acetylcholine), and anti-TSHR antibody to liposomes modified with polyethylene glycol (PEG) using Block Ace as a blocking reagent were extensively lower than those to liposomes without PEG modification using gelatin as a blocking reagent, although their specific bindings were observed in any case. As shown in Fig. 5.3, under the same conditions, autoantibodies in the sera of Graves' disease patients, who are known to have hyperthyroidism, showed remarkable reactivities to TSHR expressed in recombinant proteoliposomes, and TSHR recombinant proteoliposomes were also able to detect autoantibodies in the sera of Hashimoto's disease patients, who are known to have hypothyroidism and are not detected by competition assay with TSH (Fukushima *et al.*, 2009).

5.2. GUV

5.2.1. Liposome preparation

Giant liposomes, or GUVs, have been recently used as artificial model cells to investigate cellular membrane and microcompartment functions. Although many preparation methods of GUVs have been published, the

Table 5.1 Binding of TSH, anti-TSHR antibody, and α-BT to TSHR or AChRα recombinant proteoliposomes and liposomes without TSHR or AChRα (bare liposomes)

Receptor	Ligand	Lipid composition	Blocking reagent	Liposome type	OD_{492}
TSHR	TSH	DOPC/DOPS/BtPEG(2000)-DSPE(1:1:0.06)	1% gelatin	Proteoliposome	1.53
				Bare liposome	1.26
		DOPC/DOPS/BtPEG(2000)-DSPE(1:1:0.5)	3% Block Ace	Proteoliposome	0.20
				Bare liposome	0.10
	a TSHR Ab	DOPC/DOPS/BtPEG(2000)-DSPE(1:1:0.06)	1% gelatin	Proteoliposome	1.64
				Bare liposome	1.53
		DOPC/DOPS/BtPEG(2000)-DSPE(1:1:0.5)	3% Block Ace	Proteoliposome	0.25
				Bare liposome	0.14
AChRα	α-BT	DOPC/DOPS/BtPEG(2000)-DSPE(1:1:0.06)	1% gelatin	Proteoliposome	1.89
				Bare liposome	1.47
		DOPC/DOPS/BtPEG(2000)-DSPE(1:1:0.5)	3% Block Ace	Proteoliposome	0.20
				Bare liposome	0.10

TSHR and AChRα recombinant AcNPV-BVs are mixed at pH 4.0 with DOPC/DOPS/BtPEG(2000)-DSPE (1:1:0.25)- or DOPC/DOPS/BtPEG(2000)-DSPE (1:1:0.5)-LUVs, and after incubation for 60 min at room temperature, 1 *M* Tris solution is added to their mixture to adjust their pH to pH 7. The resulting TSHR and AChRα recombinant LUVs or LUVs are added to streptavidin-coated microplates, the microplates are blocked with 1% gelatin or 3% Block Ace, and then, bindings of TSH, anti-TSHR antibody (a TSHR Ab), or α-bungarotoxin (α-BT) to the recombinant LUVs or bare LUVs are measured by ELISA.

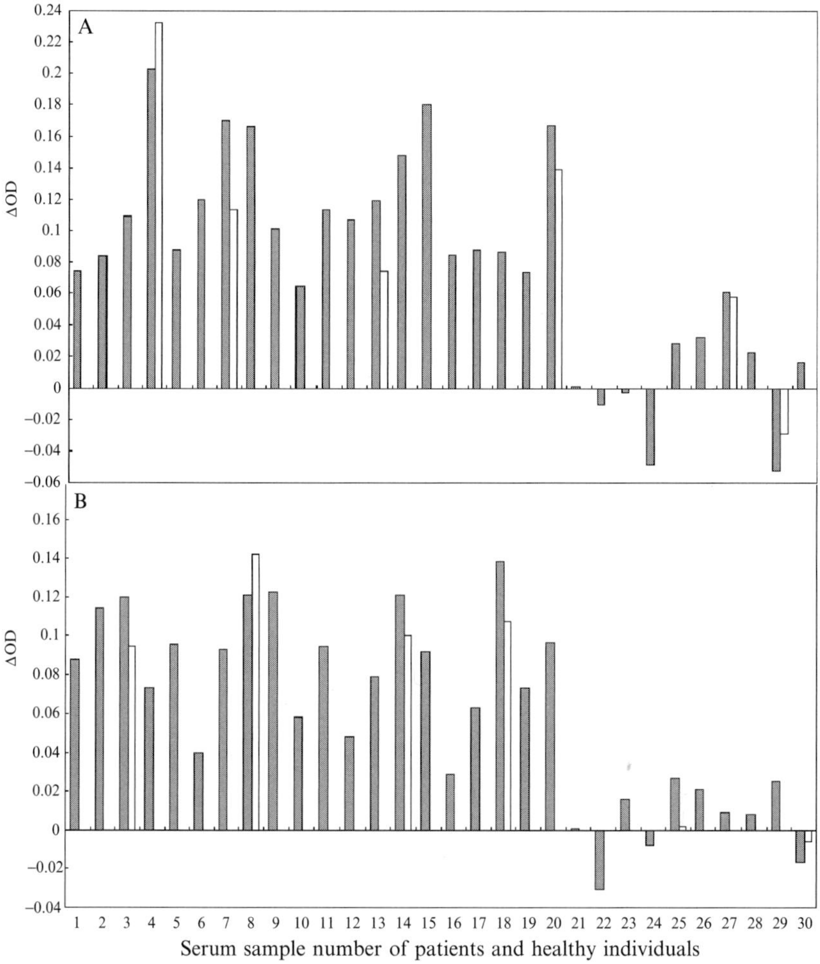

Figure 5.3 Reactivities of sera of Graves' and Hashimoto's patients toward TSHR recombinant proteoliposomes, or bare liposomes or AChR recombinant proteoliposomes as controls, at two different dilutions of antihuman IgG antibodies of 1/2000 (A) and 1/3000 (B). Reactivities are expressed as ΔOD values by subtraction of OD values of liposomes alone (gray bars) or AChR recombinant proteoliposomes (white bars) from those of TSHR recombinant proteoliposomes. Either TSHR and AChR recombinant proteoliposomes, which are obtained by fusion of TSHR and AChR recombinant AcNPV-BVs, respectively, with DOPC/DOPS/BtPEG(2000)DSPE(1:1:0.5)-LUVs, or bare DOPC/DOPS/BtPEG(2000)DSPE(1:1:0.5)-LUVs are added to streptavidin-coated microplates, the microplates are incubated with 3% Block Ace, and then their reactivities are measured by ELISA. 1–10 are sera from Graves' disease patients, 11–20 sera from Hashimoto's disease patients, and 21–30 sera from healthy individuals.

two methods—electroformation (Angelova, 2000; Fischer *et al.*, 2000; Holopainen *et al.*, 2003) and gentle hydration (Akashi *et al.*, 1998; Tsumoto *et al.*, 2009; Yamashita *et al.*, 2002)—are chosen most often. Here, we describe the use of the gentle hydration method, which is not so efficient, but easy to handle, because no special apparatus is required. First, a chloroform solution of a phospholipid mixture containing 0.05 μmol of dioleoylphosphatidylcholine (DOPC) and 0.05 μmol of dioleoylphosphatidylserine (DOPS) is placed in a glass micro test tube and evaporated under flowing inert gas (argon or nitrogen) to form dry lipid films on the bottom surface. The films are kept *in vacuo* for over a day to remove traces of chloroform. They are then gently hydrated with 0.1 ml of sodium acetate buffer (20 m*M* CH_3COOH/20 m*M* CH_3COONa, pH 4.0) or Tris buffer (10 m*M* Tris–HCl/10 m*M* NaCl) and kept overnight at room temperature, resulting in spontaneous swelling and formation of GUVs in the solution. The hydration buffer also contains 0.1 *M* sucrose. The final concentration of the GUV suspension is 1 m*M* in phospholipid. In this method, the concentration can be adjusted in the range 0.1–10 m*M*.

5.2.2. Proteoliposome preparation

GUVs are ∼5–100 μm in diameter and mechanically more fragile than the conventional liposomes mentioned above. Thus, giant proteoliposomes (proteo-GUVs) are usually difficult to prepare directly using conventional methods such as the detergent method applicable for LUVs. Instead, there are two methods to prepare proteo-GUVs: conventional proteoliposomes, prepared in advance, are (1) fused to GUVs using conjugated fusogenic peptides (Kahya *et al.*, 2001), or (2) dehydrated on an electrode to form films, which are, in turn, rehydrated using electroformation methods (Girard *et al.*, 2004). In contrast, proteo-GUVs can be produced directly, that is, not through conventional proteoliposomes, using the recombinant proteoliposome method (Tsumoto *et al.*, 2007). For reconstitution of TSHR in proteo-GUVs, 30 μl of the GUVs in the sodium acetate buffer is mixed with 30 μl of the TSHR recombinant AcNPV-BVs (1–5 μg of proteins) in PBS (pH 6.2) containing 0.1 *M* glucose, and kept for over 30 min at room temperature to induce fusion. To purify proteo-GUVs from BVs that did not fuse, the mixture is centrifuged at 1000×*g* for 5 min, and the supernatant is discarded. The precipitate containing the proteo-GUVs is resuspended with 30 μl of the Tris buffer with 0.1 *M* glucose, and centrifuged again (1000×*g*, 5 min). The supernatant is discarded and the precipitate is resuspended with 100 μl of the Tris buffer with 0.1 *M* glucose. PBS (pH 7.0) may also be used in place of the Tris buffer. As a negative control, GUVs hydrated with the Tris buffer with 0.1 *M* sucrose are subjected to the same procedure.

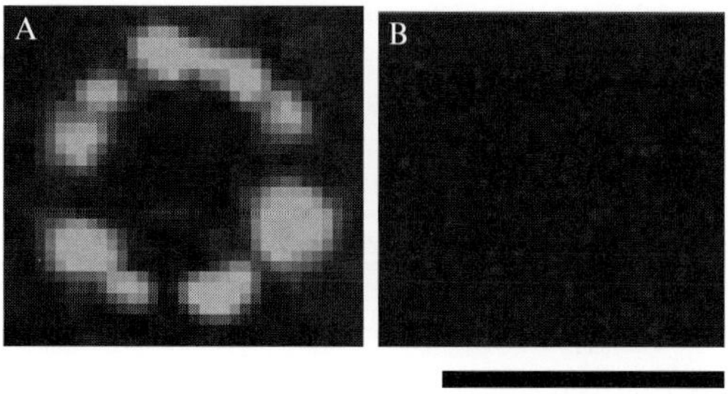

Figure 5.4 Visualization of TSHR proteins on recombinant a single proteo-GUV using fluorescence immunostain. GUVs (DOPC/DOPS = 1:1) are prepared using the gentle hydration method as mentioned in GUV section. TSHR recombinant AcNPV-BVs and the GUVs are mixed in the sodium acetate buffer (pH 4.0) (A) and in the Tris buffer (pH 7.5) (B). GUVs are treated with anti-TSHR polyclonal rabbit IgG and antirabbit IgG goat IgG that is conjugated by Alexa Fluor 488. Individual GUVs are observed by confocal laser scanning microscopy, and typical fluorescent images are shown with a bar of 10 μm.

5.2.3. Visualization of recombinant proteo-GUVs

TSHR reconstituted on proteo-GUVs is visualized using fluorescence immunostain. The anti-TSHR polyclonal rabbit IgG is added to the GUV suspension. After a 30 min incubation at room temperature, the suspension is treated with the Alexa Fluor 488-conjugated antirabbit IgG goat IgG. The aliquots are observed using a confocal laser scanning microscope (excitation 488 nm, emission 500–520 nm; Olympus FV1000). As shown in Fig. 5.4, TSHR is detected on GUVs treated with the acidic buffer, but not with the neutral buffer, indicating that TSHR recombinant AcNPV-BVs are fused with GUVs due to the function of gp64. The stimulatory G protein α subunit (GNAS), which is an acylated protein, could be incorporated into GUV membranes using the similar method, and the subunit specifically bound to GTP analogues (Tsumoto *et al.*, 2008).

6. Concluding Remarks

In this chapter, we have introduced a novel method for proteoliposome preparation, based on the baculovirus gene expression system. A recombinant membrane protein that is to be reconstituted in liposome membranes can be produced from its cDNA clone, and the expressed proteins can be recovered easily by collecting the recombinant AcNPV-BVs. Then, the BVs are mixed and fused with liposomes via activation of the viral gp64 fusion protein at acidic pH, resulting in recombinant proteoliposome formation. Generally, the use of detergent in conventional methods has disadvantages in controlling contamination of detergent, orientation of membrane proteins, and liposome types. In contrast, the novel method enables us to overcome these obstacles, because it is free from detergent and dependent on membrane fusion of liposomes and viruses. This method is applicable to a variety of liposomes, including GUVs that are difficult to convert directly into proteoliposomes using conventional methods. An additional advantage is that once expression of the targeted proteins from a recombinant cDNA on BVs is verified, not only transmembrane proteins but also anchored membrane proteins are reconstituted in liposome membranes. Thus, recombinant proteoliposomes could be a powerful tool for investigating multicomponent membrane protein systems, including those involved in signal transduction.

REFERENCES

Akashi, K., Miyata, H., Itoh, H., and Kinosita, K. Jr. (1998). Formation of giant liposomes promoted by divalent cations: Critical role of electrostatic repulsion. *Biophys. J.* **74,** 2973–2983.

Angelova, M. I. (2000). Liposome electroformation. *In* "Giant Vesicles" (P. L. Luisi and P. Walde, eds.), pp. 27–36. John Wiley & Sons Ltd, England.

Bartlett, G. R. (1959). Phosphorus assay in column chromatography. *J. Biol. Chem.* **234,** 466–468.

Blissard, G. W., and Wenz, J. R. (1992). Baculovirus gp64 envelope glycoprotein is sufficient to mediate pH-dependent membrane fusion. *J. Virol.* **66,** 6829–6835.

Burger, K. N., Knoll, G., and Verkleij, A. J. (1988). Influenza virus-model membrane interaction. A morphological approach using modern cryotechniques. *Biochim. Biophys. Acta* **939,** 89–101.

Düzgüneş, N. (2003). Preparation and quantitation of small unilamellar liposomes and large unilamellar reverse-phase evaporation liposomes. *Methods Enzymol.* **367,** 23–27.

Fischer, A., Luisi, P. L., Oberholzer, T., and Walde, P. (2000). Formation of ginat vesicles from different kinds of lipids using the electroformation method. *In* "Giant Vesicles" (P. L. Luisi and P. Walde, eds.), pp. 37–48. John Wiley & Sons Ltd, England.

Fukushima, H., Mizutani, M., Imamura, K., Morino, K., Kobayashi, J., Okumura, K., Tsumoto, K., and Yoshimura, T. (2008). Development of a novel preparation method of *recombinant* proteoliposomes using baculovirus gene expression systems. *J. Biochem.* **144,** 763–770.

Fukushima, H., Matsuo, H., Imamura, K., Morino, K., Okumura, K., Tsumoto, K., and Yoshimura, T. (2009). Diagnosis and discrimination of autoimmune Graves' disease and Hashimoto's diseases using thyroid-stimulating hormone receptor-containing recombinant proteoliposomes. *J. Biosci. Bioeng.* **108**, in press. doi:10.1016/j.jbiosc.2009.06.006.

Girard, P., Pécréaux, J., Lenoir, G., Falson, P., Rigaud, J. L., and Bassereau, P. (2004). A new method for the reconstitution of membrane proteins into giant unilamellar vesicles. *Biophys. J.* **87,** 419–429.

Hayashi, I., Urano, Y., Fukuda, R., Isoo, N., Kodama, T., Hamakubo, T., Tomita, T., and Iwatsubo, T. (2004). Selective reconstitution and recovery of functional γ-secretase complex on budded baculovirus particles. *J. Biol. Chem.* **279,** 38040–38046.

Holopainen, J. M., Angelova, M., and Kinnunen, P. K. (2003). Giant liposomes in studies on membrane domain formation. *Methods Enzymol.* **367,** 15–23.

Kahya, N., Pécheur, E. I., de Boeij, W. P., Wiersma, D. A., and Hoekstra, D. (2001). Reconstitution of membrane proteins into giant unilamellar vesicles via peptide-induced fusion. *Biophys. J.* **81,** 464–474.

King, L. A., Hitchman, R., and Possee, R. D. (2007). Recombinant baculovirus isolation. *Methods Mol. Biol.* **388,** 77–94.

Kost, T. A., and Condreay, J. P. (1999). Recombinant baculoviruses as expression vectors for insect and mammalian cells. *Curr. Opin. Biotechnol.* **10,** 428–433.

Loisel, T. P., Ansanay, H., St-onge, S., Gay, B., Boulanger, P., Strosberg, A. D., Marullo, S., and Bouvier, M. (1997). Recovery of homogeneous and functional beta 2-adrenergic receptors from extracellular baculovirus particles. *Nat. Biotechnol.* **15,** 1300–1304.

Maezawa, S., Yoshimura, T., Hong, K., Düzgüneş, N., and Papahadjopoulos, D. (1989). Mechanism of protein-induced membrane fusion: Fusion of phospholipid vesicles by clathrin associated with its membrane binding and conformational change. *Biochemistry* **28,** 1422–1428.

Masuda, K., Itoh, H., Sakihama, T., Akiyama, C., Takahashi, K., Fukuda, R., Yokomizo, T., Shimizu, T., Kodama, T., and Hamakubo, T. (2003). A combinatorial G protein-coupled receptor reconstitution system on budded baculovirus. Evidence for G_{α} and $G_{\alpha o}$ coupling to a human leukotriene B4 receptor. *J. Biol. Chem.* **278,** 24552–24562.

Possee, R. D., and King, L. A. (2007). Baculovirus transfer vectors. *Methods Mol. Biol.* **388,** 55–76.

Rigaud, J.-L., and Lévy, D. (2003). Reconstitution of membrane proteins into liposomes. *Methods Enzymol.* **372,** 65–86.

Tani, H., Nishijima, M., Ushijima, H., Miyamura, T., and Matsuura, Y. (2001). Characterization of cell-surface determinants important for baculovirus infection. *Virology* **279,** 343–353.

Tsumoto, K., Kamiya, K., and Yoshimura, T. (2007). Display of recombinant membrane receptors on giant liposomes: Attempt to construct a cell model with integrated membrane protein systems. IEEE International Symposium on Micro-NanoMechatronics and Human Science, 11–14 November 2007, Nagoya, Japan. pp. 102–107. The Institute of Electrical and Electronics Engineers, Piscataway, New Jersey.

Tsumoto, K., Yamazaki, Y., Kamiya, K., and Yoshimura, T. (2008). Reconstitution and microscopic observation of G protein subunits on giant liposomes: Attempt to construct a cell model with functional membrane protein components. IEEE International Symposium on Micro-NanoMechatronics and Human Science, 6–9 November 2008, Nagoya, Japan. pp. 145–150. The Institute of Electrical and Electronics Engineers, Piscataway, New Jersey.

Tsumoto, K., Matsuo, H., Tomita, M., and Yoshimura, T. (2009). Efficient formation of giant liposomes through the gentle hydration of phosphatidylcholine films doped with sugar. *Colloids Surf. B Biointerfaces* **68,** 98–105.

Urano, Y., Yamaguchi, M., Fukuda, R., Masuda, K., Takahashi, K., Uchiyama, Y., Iwanari, H., Jiang, S. Y., Naito, M., Kodama, T., and Hamakubo, T. (2003). A novel method for viral display of ER membrane proteins on budded baculovirus. *Biochem. Biophys. Res. Commun.* **308,** 191–196.

van Meer, G., Davoust, J., and Simons, K. (1985). Parameters affecting low-pH-mediated fusion of liposomes with the plasma membrane of cells infected with influenza virus. *Biochemistry* **24,** 3593–3602.

Yamashita, Y., Oka, M., Tanaka, T., and Yamazaki, M. (2002). A new method for the preparation of giant liposomes in high salt concentrations and growth of protein microcrystals in them. *Biochim. Biophys. Acta* **1561,** 129–134.

Yoshimura, A., Kuroda, K., Kawasaki, K., Yamashina, S., Maeda, T., and Ohnishi, S. (1982). Infectious cell entry mechanism of influenza virus. *J. Virol.* **43,** 284–293.

CHAPTER SIX

Block Liposomes: Vesicles of Charged Lipids with Distinctly Shaped Nanoscale Sphere-, Pear-, Tube-, or Rod-Segments

Alexandra Zidovska,* Kai K. Ewert,* Joel Quispe,† Bridget Carragher,† Clinton S. Potter,† *and* Cyrus R. Safinya*

Contents

Abstract

We describe the preparation and characterization of block liposomes, a new class of liquid (chain-melted) vesicles, from mixtures of the highly charged (+16 e) multivalent cationic lipid MVLBG2 and 1,2-dioleoyl-*sn*-glycero-3-phosphatidylcholine (DOPC). Block liposomes (BLs) consist of distinct spherical, tubular vesicles, and cylindrical micelles that remain connected, forming a single liposome. This is in contrast to typical liposome systems, where distinctly shaped liposomes are macroscopically separated. In a narrow composition range (8–10 mol% MVLBG2), an abundance of micrometer-scale BLs (typically sphere-tube-sphere triblocks) is observed. Cryo-TEM reveals that BLs are also present at the nanometer scale, where the blocks consist of distinctly shaped nanoscale spheres, pears, tubes, or rods. Pear-tube diblock and pear-tube-pear

* Materials, Physics, and Molecular, Cellular and Developmental Biology Departments, University of California at Santa Barbara, Santa Barbara, California, USA
† National Resource for Automated Molecular Microscopy, Department of Cell Biology, The Scripps Research Institute, La Jolla, California, USA

Methods in Enzymology, Volume 465
ISSN 0076-6879, DOI: 10.1016/S0076-6879(09)65006-0

triblock liposomes contain nanotubes with inner lumen diameter 10–50 nm. In addition, sphere-rod diblock liposomes are present, containing rigid micellar nanorods ≈4 nm in diameter and several μm in length. Block liposomes may find a range of applications in chemical and nucleic acid delivery and as building blocks in the design of templates for hierarchical structures.

1. Introduction

The landmark discovery of liposomes (self-assemblies of lipids) by Bangham *et al.* (1973) was eventually followed by enormous research efforts in both basic and applied science (Lasic, 1993). Because of their similarities to biological membranes, liposomes are used in model studies of interactions between cells and eukaryotic organelles. Liposomes can be used to encapsulate drugs and other chemicals, resulting in important applications in medicine. For example, "stealth" liposomes (Lasic and Martin, 1995) have applications in cancer therapy and cationic liposome–nucleic acid complexes (Huang *et al.*, 2005) are now used in about 10% of the ongoing human clinical trials for gene delivery (Edelstein *et al.*, 2004, 2007).

In addition to the most common spherical shape of (unilamellar) liposomes, a large variety of vesicle (bilayer-based liposome) shapes has been described, including ellipsoids and oblates, tori, and discocytes and stomatocytes (Chiruvolu *et al.*, 1994; Lipowsky and Sackmann, 1995; Seifert, 1997; Yager and Schoen, 1984). Figure 6.1 shows micrographs of complex vesicle morphologies such as a torus (Fig. 6.1A and B) (Michalet and Bensimon, 1995a), a "button" (Fig. 6.1C–E) (Michalet and Bensimon, 1995b), and a discocyte (Fig. 6.1G–I) (Kas and Sackmann, 1991). In addition, Fig. 6.2 shows vesicles shaped like tubules (Fig. 6.2A) (Chiruvolu *et al.*, 1994), a pear (Fig. 6.2B), and a dumbbell (Fig. 6.2C) (Kas and Sackmann, 1991). Current theoretical treatments, which are consistent with these shapes, include elastic free energy models of membranes described by the membrane bending and Gaussian moduli as well as the spontaneous curvature, at a constant total surface area of the vesicle (Lipowsky and Sackmann, 1995; Safran, 1994).

More recently, substantial research efforts have been directed toward producing lipid-based bio-nanotubes for a range of applications, such as storage and controlled release of chemicals and drugs, including nucleic acids, peptides, and proteins (Martin, 1994; Schnur, 1993). The approaches pursued to prepare the nanotubes include the use of templates, for example, filamentous cytoskeletal proteins such as microtubules, a tubular polymer assembled from dimers of the protein tubulin. In this way, lipid–protein nanotubes with open or closed ends were obtained (Raviv *et al.*, 2005, 2007). The quest to produce tubular structures solely from lipid components

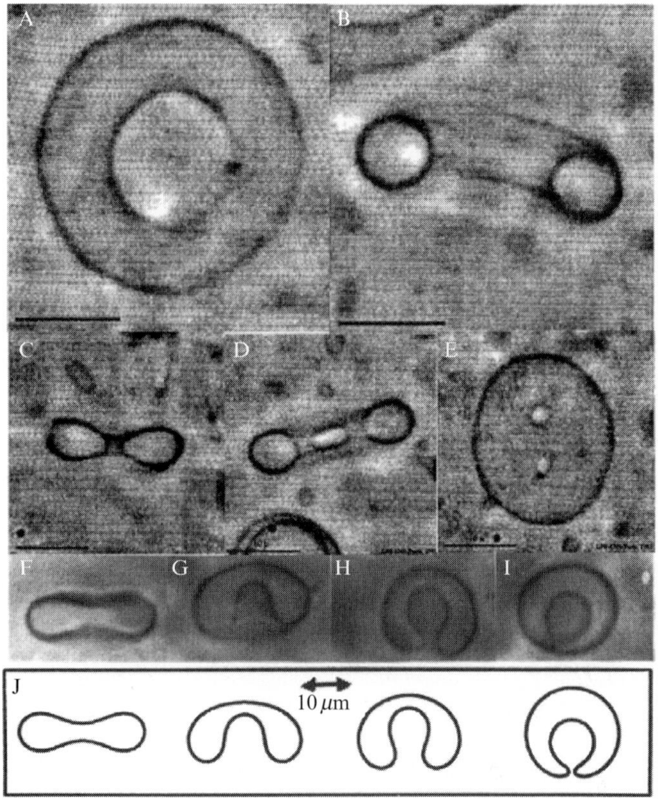

Figure 6.1 Torus, "button" and discocyte-shaped vesicles. (A, B) Micrographs of a torus vesicle shown from different angles. Reprinted in part from Michalet and Bensimon (1995a) with permission. Copyright 1995, EDP Sciences. (C–E) Different views of a button-shaped vesicle. Reprinted in part from Michalet and Bensimon (1995b) with permission. Copyright 1995, American Association for the Advancement of Science. (F–J) Different views of a discocyte vesicle (F–I) with corresponding schematics (J). Reprinted in part from Kas and Sackmann (1991) with permission. Copyright 1991, Biophysical Society.

has thus far mainly resulted in the solid-phase lipid tubules discovered many years ago (Schnur, 1993). These lipid tubules typically comprise unilamellar or multilamellar membranes in their quasi-2D solid phase, wrapped around a hollow core, with lengths on the order of several tens of micrometers, and a diameter between ≈100 nm and ≈1 μm (Singh *et al.*, 2003; Spector *et al.*, 2001). Control over the tubule dimensions and properties is important for practical application, and several procedures to generate monodisperse tubules on the micrometer scale have been developed (Shimizu *et al.*, 2005). Custom tailored lipid molecules were synthesized to systematically study the impact of the lipid architecture on the tubule formation process. This work

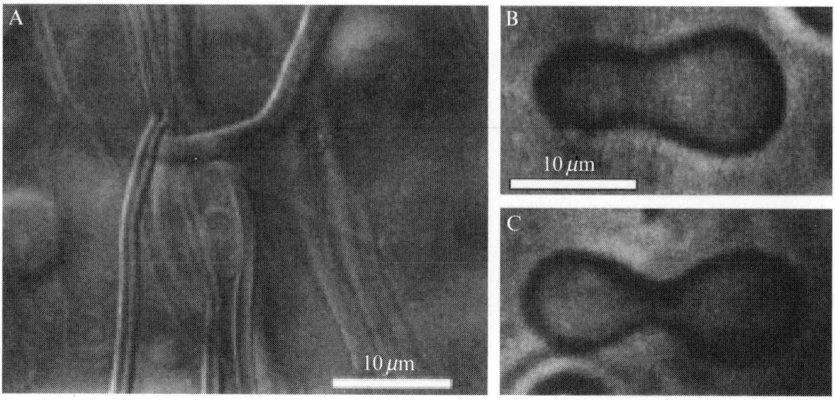

Figure 6.2 Lipid tubules, pears, and dumbbells. (A) A micrograph of micrometer-scale lipid tubules. Reprinted in part from Chiruvolu *et al.* (1994) with permission. Copyright 1994, American Association for the Advancement of Science. (B, C) Images of a pear-shaped (B) and a dumbbell-shaped (C) lipid vesicle. Reprinted in part from Kas and Sackmann (1991) with permission. Copyright 1991, Biophysical Society.

identified chirality as one of the prerequisites for forming tubules and helical ribbons from lipid molecules in their chain ordered, quasi-2D solid phase (Spector *et al.*, 2001; Thomas *et al.*, 1995).

While solid-phase lipid tubules can be obtained in systems containing only a single lipid, at least two lipid components are required to generate *liquid-phase* vesicles (consisting of chain-melted lipids above the gel-liquid crystalline phase-transition temperature, T_{m}) exhibiting the cylindrical tubular structure. One example is a mixture of two lipids, with one exhibiting a cylindrical and the other a conical molecular shape. In this system, a spontaneous breaking of symmetry in lipid composition between the outer and inner layers would generate an initial curvature of the membrane and, in principle, lead to a (energetically favored) cylindrical liquid vesicle structure (Safran, 1994; Safran *et al.*, 1990). Liquid-phase vesicles are of particularly high interest, because they allow the incorporation of functional biomolecules that typically require the lipids of the membrane to be in the chain-melted state to retain their full biological activity. Such lipid nanotubes could therefore be useful for a range of applications, including sensing and chemical delivery.

In this chapter, we describe the preparation and characterization of a new, recently discovered class of vesicles, termed block liposomes (BLs) (Zidovska *et al.*, 2009a). BLs spontaneously form in water from a lipid mixture comprising the hexadecavalent cationic lipid MVLBG2 and neutral 1,2-dioleoyl-*sn*-glycero-3-phosphatidylcholine (DOPC). They are vesicles consisting of several connected, yet distinctly shaped liposomes such as spheres, pears, tubes, and cylindrical micelles (rods). These building units are called

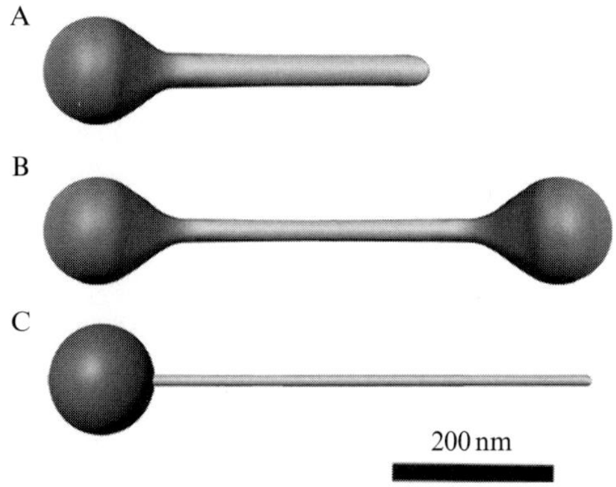

Figure 6.3 Schematic depiction of the three main types of block liposomes. Different colors represent different membrane Gaussian curvatures: positive (red), negative (blue), and zero (yellow). (A) Pear-tube diblock liposome. (B) Pear-tube-pear triblock liposome. (C) Sphere-rod diblock liposome. Reprinted in part from Zidovska *et al.* (2009a) with permission. Copyright 2008, American Chemical Society. (See Color Insert.)

blocks in analogy to block copolymers. Figure 6.3 schematically depicts a sphere-tube diblock (Fig. 6.3A), a sphere-tube-sphere triblock (Fig. 6.3B), and a sphere-rod diblock, containing a cylindrical micelle (Fig. 6.3C).

2. Liposome Preparation

DOPC is purchased from Avanti Polar Lipids (Alabaster, AL) and dissolved in chloroform. MVLBG2 trifluoroacetate is synthesized as described (Ewert *et al.*, 2006) and dissolved in chloroform/methanol (9:1, v/v). These lipid solutions (at a concentration of 10 or 30 m*M*, matching the desired concentration of the aqueous solution to be prepared) are combined in glass vials to yield the desired ratio of lipids and dried, first by a stream of nitrogen and subsequently in a vacuum for 8–12 h, to form a thin film on the vial surface. To this film, high resistivity (18.2 MΩ cm, from a Millipore system) water is added and the resulting mixture is incubated at 37 °C for at least 12 h to give a final concentration of 10 mg/ml (30 mg/ml for micellar solutions). These aqueous lipid solutions are stored at 4 °C until use.

3. Microscopy

Optical microscopy is performed using a Nikon Diaphot 300 inverted microscope equipped for epifluorescence and differential-interference-contrast (DIC) and a SensiCam^{QE} High Speed digital camera. For fluorescence microscopy, liposome stock solutions are prepared as described above, but at

1 m*M* concentration and with the inclusion of 1 mol% lipid dye Texas Red DHPE (Invitrogen/Molecular Probes).

Cryogenic TEM is used to image the liposome solutions on a nanometer scale. This part of the presented work is conducted at the National Resource for Automated Molecular Microscopy which is supported by the National Institutes of Health through the National Center for Research Resources' P41 program (RR17573). The specimens are preserved in a layer of vitreous ice suspended over a holey carbon substrate. The holey carbon films consist of a thin layer of pure carbon fenestrated by 2 μm holes spaced 4 μm apart and suspended over 400 mesh copper grids, yielding a periodic pattern of holes within a single copper grid segment (Quispe *et al.*, 2007). Carbon-coated copper grids are cleaned and activated by oxygen plasma treatment using a Solarus plasma cleaner (Gatan Inc.) and a 25% O_2, 75% Ar mixture. A total of 4 μl of the sample solution is placed on the carbon-coated side of the copper grid. The concentration of the sample solution is varied depending on the structures being imaged. A concentration of 5–10 mg/ml is used for block liposome solutions, while 30 mg/ml is used for micellar lipid solutions. The copper grid with the sample solution is mounted into a Vitrobot (FEI), which blotts the sample for 3, 4, or 5 s, thus varying the sample thickness, and subsequently vitrifies the sample by plunge-freezing into a liquid ethane bath. The vitrified sample is then imaged using a Tecnai F20 electron microscope (FEI Co.) at 120 keV. Images are collected using a 4096 $\times$ 4096 pixel CCD camera (TVPIS GmbH) at the search magnification of 5000$\times$ and final magnifications of 29,000$\times$ to 280,000$\times$ at an underfocus of $\sim$2.5 μm. The Leginon software system (Suloway *et al.*, 2005) is used for data collection, data storage, and partial data analysis.

4. Design and Synthesis of MVLBG2

The highly charged, cone-shaped lipid MVLBG2 is based on a previously described lipid building block with two oleyl chains (Schulze *et al.*, 1999). The oleyl chains are selected since they will form liquid-phase membranes at room temperature and ensure compatibility with DOPC membranes. The synthesis of the headgroup is designed in a way to allow the introduction of a high number of charges with a small number of chemical steps. To this end, ornithine is employed as a dendritic AB_2 building block. In combination with a previously prepared, Boc-protected building block for a tetravalent headgroup (Behr, 1989), this allows assembly of the headgroup using efficient peptide-coupling and -deprotection methods. The resulting, gram-scale synthesis of MVLBG2 has been described in a previous paper (Ewert *et al.*, 2006). Figure 6.4A shows the chemical structure of MVLBG2 which has a headgroup charge of +16 e at full protonation. The hydrophobic tail, spacer, ornithine branching units,

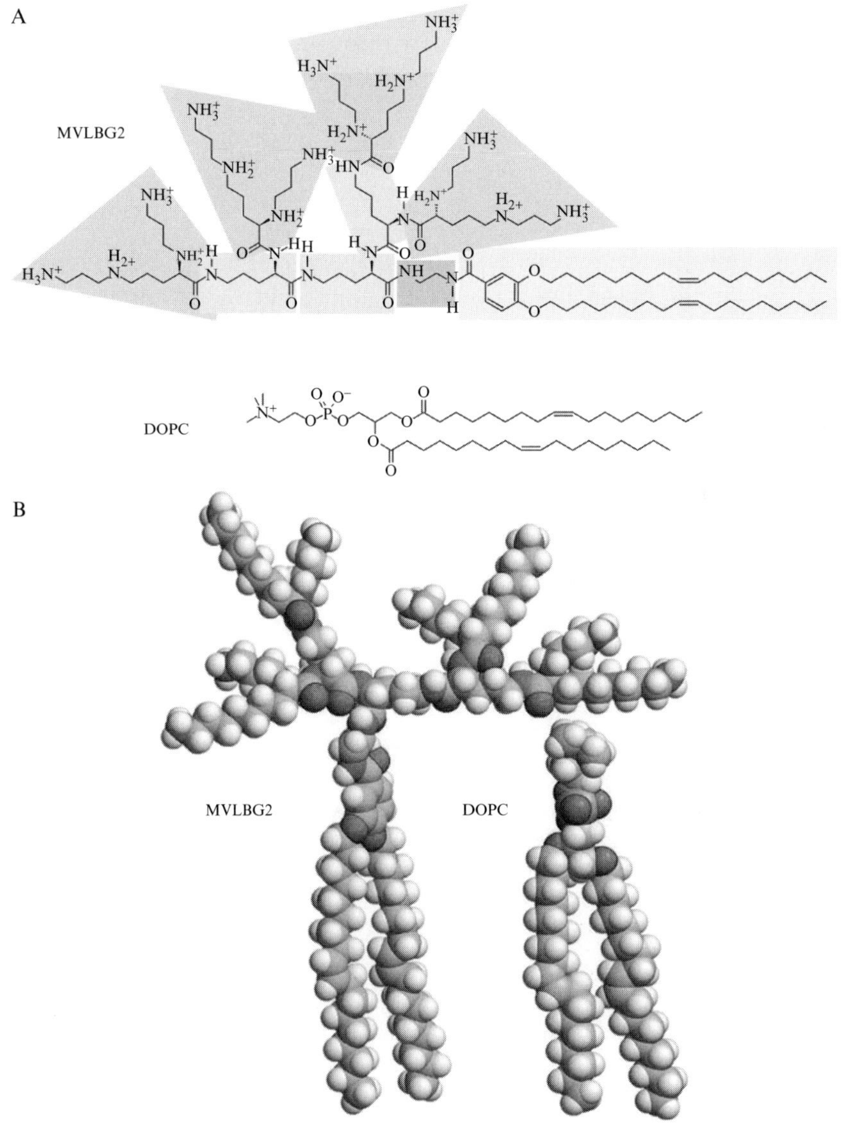

Figure 6.4 Chemical structures and molecular models of MVLBG2 and DOPC. (A) The chemical structures of MVLBG2 and DOPC. The hydrophobic tail, spacer, ornithine branching units, and charged carboxyspermine moieties of MVLBG2 are underlaid in tan, blue, green, and red, respectively. (B) Space filling molecular models of MVLBG2 and DOPC demonstrating their conical and cylindrical molecular shape, respectively. Reprinted in part from Zidovska *et al.* (2009a) with permission. Copyright 2008, American Chemical Society. (See Color Insert.)

and charged carboxyspermine moieties are underlaid in tan, blue, green, and red, respectively. The pH of the aqueous lipid solutions used in our experiments is measured using indicator paper (due to the small sample volume) as 5.5 ± 0.5, which corresponds to an average MVLBG2 charge of +14.5 e to +16 e.

5. Phase Behavior of MVLBG2/DOPC Lipid Mixtures

While the extreme mismatch in headgroup area versus tail area of MVLBG2 results in a conical molecular shape, the charge-neutral zwitterionic DOPC has a cylindrical molecular shape with its headgroup area nearly equal to its tail area. Space filling molecular models of MVLBG2 and DOPC illustrating the different molecular shapes of the two lipids are shown in Fig. 6.4B. Both lipids are in their chain-melted state at room temperature due to the presence of a *cis* double bond in their hydrophobic tails. Given this combination of molecular properties, MVLBG2/DOPC mixtures in water appeared to be a promising system for the formation of cylindrical liquid vesicles. Indeed, wide-angle X-ray scattering experiments confirmed that the lipids remained in their chain-melted liquid state in all investigated samples (data not shown).

The phase behavior of the MVLBG2/DOPC/water system is extremely rich. We initially used DIC microscopy to investigate the liposome structures formed on a micrometer scale (Zidovska *et al.*, 2009a). Figure 6.5A–G displays DIC microscopy images of vesicles undergoing shape transitions as a function of their lipid composition. Membranes with very high DOPC content exhibit predominantly polydisperse multilamellar onion-like vesicles for 0–8 mol% MVLBG2 (Fig. 6.5A). Unexpectedly, a phase consisting of a nearly uniform block liposome population was found in a very narrow composition interval of ≈8–10 mol% MVLBG2 (Fig. 6.5B). Its hallmark is the ≈5 μm long cylindrical core of diameter ≈0.5 μm, capped with quasi-spherical vesicles a few μm in diameter at both ends (Fig. 6.5C). With increasing MVLBG2 content in the sample, the system first reenters a regime of polydisperse, multilamellar onion-like vesicles (Fig. 6.5D). Then, at ≈25 mol% MVLBG2, the polydisperse vesicles are replaced by a fairly monodisperse population of spherical vesicles of diameter ≈2 μm (Fig. 6.5E). As the MVLBG2 content increases further, a regime of coexistence between micelles and vesicles is reached around 50 mol% MVLBG2 (Fig. 6.5F). In DIC microscopy, this sample appears sparsely populated with only occasional vesicles. The vesicle aggregation observed in this region results from depletion attraction forces, mediated by the micelles. At MVLBG2 contents above 75 mol%, vesicles disappear and only micelles

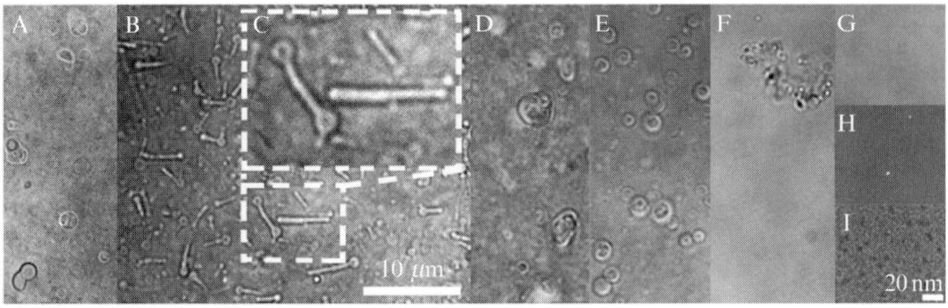

Figure 6.5 Phase behavior of the MVLBG2/DOPC/water system. (A–G) DIC images of vesicle shapes as a function of their composition: multilamellar spherical vesicles (onions) (0–8 mol% MVLBG2, A), block liposomes (8–10 mol% MVLBG2, B), reentrant onions (11–50 mol% MVLBG2, D, E), macroscopic coexistence of vesicles and micelles (≈50 mol% MVLBG2, F), and micelles (75–100 mol% MVLBG2, G). (H) Fluorescence microscopy image, demonstrating the existence of micelles. (I) Cryo-TEM image, showing micelle size and morphology. (C) An inset of (B), showing the block liposome morphology in detail: long (≈5 μm) cylindrical cores of diameter ≈0.5 μm capped at both ends with spherical vesicles of a few μm diameter. Reprinted in part from Zidovska *et al.* (2009a) with permission. Copyright 2008, American Chemical Society.

are present. Since the size of the micelles is below the optical resolution of DIC microscopy (Fig. 6.5G), we employed fluorescence imaging to prove their existence (Fig. 6.5H), and cryo-TEM to determine their size and morphology (Fig. 6.5I). As shown in Fig. 6.5I, cryo-TEM shows that samples containing 75 mol% MVLBG2 consist of a dense solution of disk-shaped micelles ≈4 nm thin and of ≈10 nm diameter. The micelle morphology changes with MVLBG2 content, and at 100 mol% MVLBG2, monodisperse spherical micelles of diameter ≈4 nm populate the sample (Zidovska *et al.*, 2009b).

5.1. Block liposomes

As mentioned earlier, DIC microscopy revealed an evolution in vesicle shape on the micrometer scale from spherical vesicles to block liposomes (BLs) in a narrow composition range (8–10 mol% MVLBG2). BLs constitute a remarkable new class of vesicles. Most importantly, BLs are unique in that well-known distinct liposome morphologies such as spheres, tubules, or cylindrical micelles, appear as connected parts within a single BL. This provided the rationale for naming these new structures, in analogy to block copolymers. As in the morphologies of block copolymers, BLs exhibit a microphase separation of well-defined shapes within a single BL, as opposed to macrophase separation, where these shapes would not be connected.

Furthermore, the Gaussian membrane curvature distinguishes BLs from previously described vesicle morphologies. Figure 6.6 shows a schematic

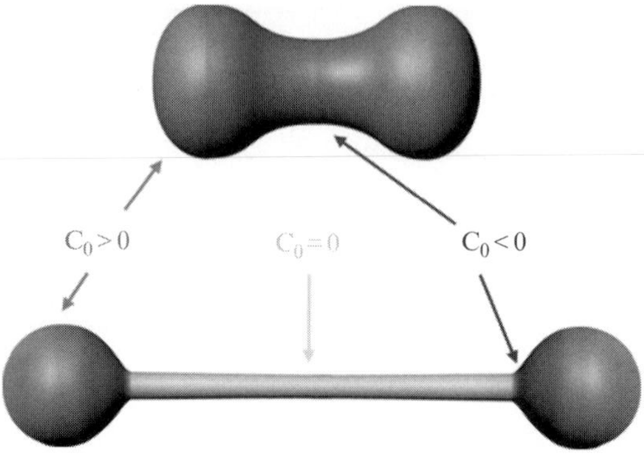

Figure 6.6 Comparison of the Gaussian curvature of a dumbbell vesicle and a triblock liposome. In the schematic, membrane regions of positive, negative, and zero Gaussian curvature are shown in red, blue, and yellow, respectively. The illustration demonstrates that regions of all three Gaussian curvatures are present in the block liposome, while the dumbbell morphology only contains regions of positive and negative Gaussian curvature. (See Color Insert.)

comparison of a block liposome and a dumbbell vesicle. This illustration visualizes membrane regions of distinct Gaussian curvature using different colors. Membrane regions exhibiting positive Gaussian curvature are depicted in red, negative Gaussian curvature is shown in blue, and zero Gaussian curvature in yellow (please see Figure 6.6 in the color plate section). The schematic demonstrates that a block liposome consists of membrane regions of three different Gaussian curvatures: positive, zero, and negative, unlike the dumbbell morphology which can be described by only two distinct Gaussian curvatures: positive and negative. Of note, the previously observed pear and dumbbell morphologies are intermediate stages in the proposed mechanisms of block liposome formation (Zidovska *et al.*, 2009a).

5.2. Nanoscale studies of block liposomes

The novel BL morphology persists from the micrometer scale (as shown by DIC microscopy) down to the nanometer scale (Zidovska *et al.*, 2009a). Cryo-TEM (Fig. 6.7) revealed diblock (sphere-tube) liposomes and triblock (sphere-tube-sphere) liposomes containing nanometer-scale tubules of diameter 10–50 nm and length >1 μm. These nanoscale block liposomes are the first examples of synthetic lipid systems forming tubular vesicles in the chain-melted liquid phase while exhibiting an inner lumen of truly nanometer-scale dimensions.

The series of cryo-TEM images shown in Fig. 6.7 reveal the detailed features of typical di- and triblock liposomes with tubular sections. Two triblock (pear-tube-pear) liposomes (with vesicles of asymmetric size

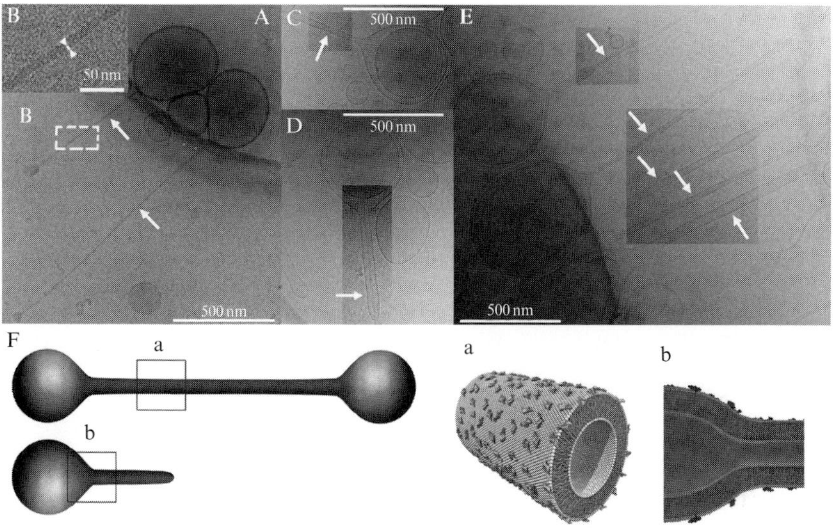

Figure 6.7 Cryo-TEM images of block liposomes containing liquid-phase lipid nanotubes. (A) Pear-tube-pear triblock liposomes. (B) An inset of (A) revealing the hollow tubular structure (white arrowheads and white bar point out the bilayer thickness of 4 nm). (C) A pear-tube diblock liposome. (D) One block liposome encapsulated within another one (also seen in E, top arrow). (E) A group of block liposomes. The block liposomes shown in (A–E) are comprised of liquid-phase lipid nanotube segments capped by spherical vesicles with diameters of a few hundred nm. The nanotubes (white arrows) are 10–50 nm in diameter and >1 μm in length. (F) Schematic depictions of the MVLBG2/DOPC tri- and diblock liposomes. Insets (A) and (B) show molecular-scale illustrations, based on the hypothesized mechanism of BL formation (Zidovska *et al.*, 2009a), manifesting the symmetry breaking between outer and inner monolayer. In (A–E), image contrast/brightness was altered in selected rectangular areas. Reprinted in part from Zidovska *et al.* (2009a) with permission. Copyright 2008, American Chemical Society.

capping the tube) can be seen in Fig. 6.7A. The high-magnification inset proves that the tubular section has an inner lumen diameter of ≈10 nm (Fig. 6.7B). A diblock (pear-tube) liposome with inner diameter of ≈50 nm is seen in Fig 6.7C. We frequently observe one block liposome encapsulated within another block liposome (Fig. 6.7D and E, top arrow). In these cases, the inner, smaller diameter tubule is observed to protrude through the encapsulating membrane, demonstrating its high-bending rigidity. This high bending rigidity and the extraordinarily large persistence length of the order of millimeter are likely a result of the high charge density of the membrane. Figure 6.7E also shows a diblock liposome (lower arrow) and parts of several block liposomes (second, third, and fourth arrows from bottom), which are either di- or triblocks. Schematics of these typical block liposomes containing tubular sections are shown in Fig. 6.7F [molecular-scale enlargements

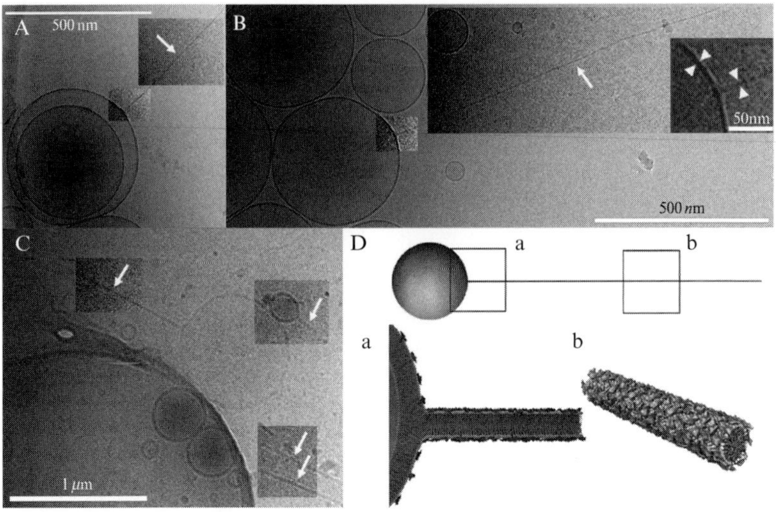

Figure 6.8 Cryo-TEM images of block liposomes containing liquid-phase lipid nanorods. (A–D) Diblock liposomes comprised of lipid nanorods (white arrows) connected to spherical vesicles. Lipid nanorods are stiff cylindrical micelles with an aspect ratio ≈1000. Their diameter equals the thickness of a lipid bilayer (≈4 nm) and their length can reach up to several μm with a persistence length of the order of mm. (C) (an inset of B) allows comparison of the thickness of the nanorod and the bilayer of the spherical vesicle: white arrow heads point out a thickness of ≈4 nm. Of note, free nanorods were not observed in any of the 426 cryo-TEM images analyzed. (E) Schematic of a MVLBG2/DOPC sphere-rod diblock liposome. Insets a and b show molecular-scale schematics based on the hypothesized mechanism of BL formation (Zidovska *et al.*, 2009a). Note the high concentration of MVLBG2 in the nanorod segment. In (A–D), image contrast/brightness was altered in selected rectangular areas (Zidovska *et al.*, 2009a). Reprinted in part from Zidovska *et al.* (2009a) with permission. Copyright 2008, American Chemical Society.

exposing the inside shown in (a) and (b)]. The length of the nanotubule sections evident in these images ranges from around 500 nm to >1 μm.

In addition to BLs with tubular sections, another type of block liposomes is found that contains highly rigid lipid nanorod sections comprising cylindrical micelles. A spontaneous topological transition from tubes (cylindrical vesicles) to rods (cylindrical micelles) leads to the formation of this population of sphere-rod diblock liposomes (Zidovska *et al.* 2009a). Figure 6.8A–D shows typical images of these remarkable diblock liposomes which demonstrate that the micellar nanorods remain attached to the spherical vesicles. The nanorod diameter equals the thickness of a lipid bilayer (≈4 nm), corresponding to the hydrophobic core of the rod which exhibits high contrast in cryo-TEM. The nanorods can reach up to several μm in length, corresponding to an aspect ratio of the order of 1000, and are very stiff, again exhibiting persistence lengths on the millimeter

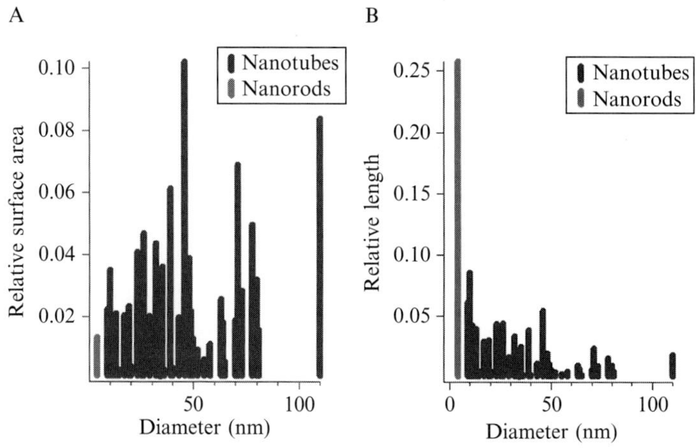

Figure 6.9 Statistical analysis of the nanotube and nanorod populations. (A) Diameter histogram weighted by the surface area, indicating the relative amount of lipid in the nanorod and nanotube state. (B) Diameter histogram weighted by the length of the structure, highlighting the striking length of nanorods compared to nanotubes. Reprinted in part from Zidovska *et al.* (2009a) with permission. Copyright 2008, American Chemical Society. (See Color Insert.)

scale. A lower magnification image (Fig. 6.8D) shows a collection of these sphere-rod BLs with a variety of spherical vesicle sizes. Schematics of this novel block liposome structure based on the hypothesized mechanism of their formation are shown in Fig. 6.8E. The region where the lipid nanorod is connected to the spherical bilayer vesicle is a true mathematical singularity, analogous to the core of a liquid crystal ($-1/2$) disclination (de Gennes and Prost, 1993). The high rigidity of the nanorods observed in cryo-TEM is consistent with the Odijk–Skolnick–Fixman (OSF) theory of polyelectrolytes (Odijk, 1977; Skolnick and Fixman, 1977), where for $b < L_B$, the electrostatic persistence length $\xi_p = \lambda_D^2/4L_B \approx 300\,\mu m$. Here, b is the distance between charges ≈ 2 Å for MVLBG2, $L_B = 7.1$ Å the Bjerrum length in water, and $\lambda_D \approx 1\ \mu m$ the Debye length in deionized water.

Figure 6.9 plots the results of a statistical analysis of the nanotube and nanorod populations, including the diameter distribution within the tube population. The diameter histograms weighted by the surface area—indicative of the amount of lipid in the nanorod and nanotube state—and by the length of the structure—highlighting the striking length of nanorods compared to nanotubes—are shown in Fig. 6.9A and B, respectively.

The evolution of shapes in the block liposome regime, from spherical vesicles to the block liposomes, is reminiscent of shape changes in biological cellular systems. The striking resemblance to the evolving shape of the plasma membrane of a cell during cytokinesis suggests that membrane-associated molecules (e.g., membrane proteins or glycophospholipids) with physical properties similar to MVLBG2 may aid the out-of-equilibrium (motor driven) process of tubule formation, during late stage cytokinesis just before fission and the splitting of the daughter cells (Reichl *et al.*, 2005; Robinson and Spudich, 2004). Similarly, the triblock liposome state with a narrow tubule connecting quasi-spherical vesicles has a striking resemblance (in shape and size) to bacterial conjugation where a tubular section connects neighboring bacteria exchanging genetic material (Brock *et al.*, 1994).

We have previously proposed possible mechanisms for the formation of BLs. Importantly, our data excludes the possibilty that nanoscale BLs are formed as a sample preparation artifact (for example, by flow effects). While the tube and rod segments seem to align to some extent—flow orientation is expected for structures with aspect ratios as high as the BLs confined in a thin layer of water (~200 nm)—their orientation is not as uniform as one would expect if they were *generated* by flow. Furthermore, Fig. 6.10A shows a rare example of a vesicle membrane shape evolving from a sphere to a sphere with four protruding symmetrical processes, resembling a clover leaf with four leaflets (instead of the more typical single process leading to the diblock (pear-tube) liposome). Such a structure could not be generated by blotting-induced flow.

Further evidence that electrostatic forces resulting from the conically shaped, charged lipid MVLBG2 are the driving force for the creation of BLs is provided by imaging of the BL-forming lipid mixture at high salt conditions (250 m*M* NaCl, Debye length ≈0.6 nm, effectively screening the

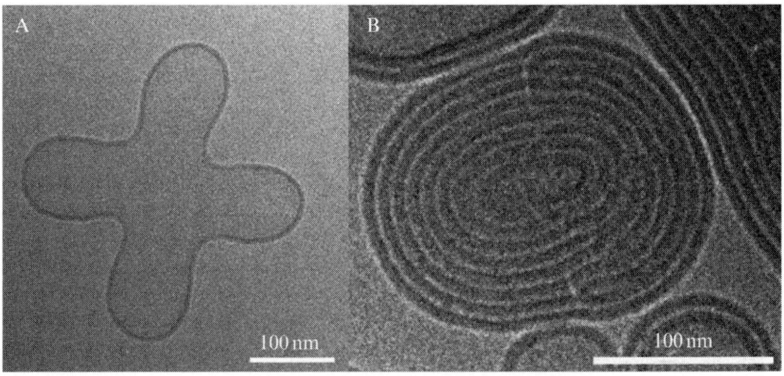

Figure 6.10 Cryo-TEM images providing insights into block liposome formation. (A) An example of four simultaneous nanotube formation processes. The lack of a preferred orientation of the protrusions proves that they were not generated by flow effects. (B) Block liposomes transform into multilamellar spherical vesicles when the electrostatic forces are screened by presence of salt (250 m*M* NaCl, Debye length ≈0.6 nm). Reprinted in part from Zidovska *et al.* (2009a) with permission. Copyright 2008, American Chemical Society.

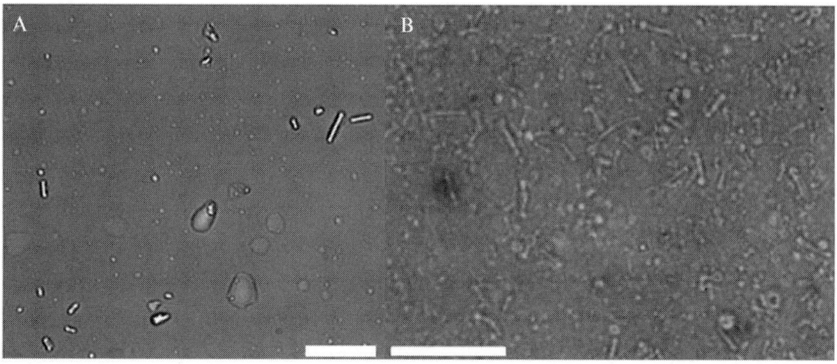

Figure 6.11 Dehydration and rehydration of BLs. (A) DIC micrograph of dehydrated block liposomes. Note that the spherical caps at the end of the cylindrical core are collapsed. (B) DIC micrograph of the sample shown in (A) after rehydration. The block liposomes have regained their original shape. Scale bars, 10 μm.

electrostatic forces). Block liposomes are not stable under these conditions and are replaced by multilamellar vesicles with spherical topology (Fig. 6.10B), independent of whether salt is added to BLs or whether the lipid mixture is hydrated with the salt solution (Zidovska *et al.*, 2009c).

While sensitive to salt, BLs are long-lived structures in water and may be dehydrated and reversibly rehydrated. When dried, the spherical caps at the end of the cylindrical core collapse (Fig. 6.11A). After rehydration, block liposome resume their original shape (Fig. 6.11B). This capacity might be of use for potential drug storage and/or drug delivery applications. DIC microscopy performed at regular intervals over the course of more than 1 year of experiments revealed no change in morphology, proving that the μm-scale multilamellar block liposomes are long-lived. Similarly, the time intervals between BL preparation and vitrification for cryo-TEM varied, reaching up to 2 months without showing an effect on the observed structures (Zidovska *et al.*, 2009a). Another piece of evidence suggesting that BLs are not only robust, long-lived structures, but equilibrium structures, is the fact that in all 426 EM images analyzed, not a single instance of macrophase separation of elongated vesicle shapes, that is an isolated tubular vesicle or cylindrical micelle which is not part of a BL, was observed. Spherical structures were the only liposomes observed in the samples alongside the BLs. While our observations suggest that BLs are equilibrium structures, we cannot rule out the possibility that they are kinetically trapped structures with barriers much larger than the thermal energy. Further investigations, for example, preparation of liposome solutions by methods other than simple film hydration may clarify this point.

6. Concluding Remarks

The discovery of block liposomes demonstrates that the addition of a single type of lipid molecule with highly conical shape—imparted to a large extent by electrostatics—to neutral vesicles can lead to a dramatic membrane shape evolution, leading to long-lived robust structures. Another recent report also described the formation of tubules upon addition of ganglioside lipids with large headgroups to DOPC-containing vesicles in the presence of salt (Akiyoshi *et al.*, 2003). These membrane shape evolutions have analogies in nature and in the flow of certain abstract geometric shapes in topology that follows the "Ricci flow equation," which is used to describe surfaces where regions of high curvature diffuse into lower curvature regions (e.g., to generate the flow of shapes from a sphere (positive curvature) to a dumbbell-shaped surface containing negative curvature regions) (Collins, 2004; Mackenzie, 2006). The nanotubes and nanorods may become desirable candidates for drug/gene delivery applications (Ewert *et al.*, 2005; Huang *et al.*, 2005; Raviv *et al.*, 2005, 2007; Schnur, 1993; Shimizu *et al.*, 2005; Singh *et al.*, 2003; Thomas *et al.*, 1995), or as a template for nanostructures such as wires or needles. Future studies involving systematic variations in the shape, size, and charge of the curvature-stabilizing lipid will be aimed at controlling the tubule diameter distribution. Of note, current state-of-the-art analytical theories (Seifert, 1997) and simulations of membrane shapes (Reynwar *et al.*, 2007) that incorporate coupling between curvature and composition are not able to predict our experimentally discovered block liposomes that have distinct shapes, but are connected to one another. This may be the result of the omission of electrostatic forces in current theories and simulations that are key to the formation of block liposomes as described earlier. New theories of charged membranes are required to describe block liposomes. Finally, it seems possible to produce analogous block polymersomes in mixtures of charged and neutral copolymers (Discher *et al.*, 1999) or peptides (Deming, 1997).

Acknowledgments

This work was supported by DOE grant DE-FG02-06ER46314, NSF grant DMR-0803103, and NIH grant GM-59288. Some of this research was conducted at the National Resource for Automated Molecular Microscopy, which is supported by the NIH National Center for Research Resources P41 program (RR17573).

REFERENCES

Akiyoshi, K., Itaya, A., Nomura, S. M., Ono, N., and Yoshikawa, K. (2003). Induction of neuron-like tubes and liposome networks by cooperative effect of gangliosides and phospholipids. *FEBS Lett.* **534,** 33–38.

Bangham, A. D., Hill, M. W., and Miller, N. G. A. (1973). Preparation and use of liposomes as models of biological membranes. *Methods Membr. Biol.* **1,** 1–68.

Behr, J. P. (1989). Photohydrolysis of DNA by polyaminobenzenediazonium salts. *J. Chem. Soc. Chem. Commun.* (2), 101–103.

Brock, T. D., Madigan, M. T., Martinko, J. M., and Parker, J. (1994). *Biology of Microorganisms* Prentice Hall, New Jersey.

Chiruvolu, S., Warriner, H. E., Naranjo, E., Idziak, S. H. J., Raedler, J. O., Plano, R. J., Zasadzinski, J. A., and Safinya, C. R. (1994). A phase of liposomes with entagled tubular vesicles. *Science* **266,** 1222–1225.

Collins, G. P. (2004). The shapes of space. *Sci. Am.* **291,** 94–103.

de Gennes, P. G., and Prost, J. (1993). *The Physics of Liquid Crystals.* Oxford University Press, Oxford.

Deming, T. J. (1997). Facile synthesis of block copolypeptides of defined architecture. *Nature* **390,** 386–389.

Discher, B. M., Won, Y. Y., Ege, D. S., Lee, J. C. M., Bates, F. S., Discher, D. E., and Hammer, D. A. (1999). Polymersomes: Tough vesicles made from diblock copolymers. *Science* **284,** 1143–1146.

Edelstein, M. L., Abedi, M. R., Wixon, J., and Edelstein, R. M. (2004). Gene therapy clinical trials worldwide 1989–2004—An overview. *J. Gen. Med.* **6,** 597–602.

Edelstein, M. L., Abedi, M. R., and Wixon, J. (2007). Gene therapy clinical trials worldwide to 2007—An update. *J. Gen. Med.* **9,** 833–842.

Ewert, K. K., Ahmad, A., Evans, H. M., and Safinya, C. R. (2005). Cationic lipid-DNA complexes for non-viral gene therapy: Relating supramolecular structures to cellular pathways. *Expert Opin. Biol. Ther.* **5,** 33–53.

Ewert, K. K., Evans, H. M., Zidovska, A., Bouxsein, N. F., Ahmad, A., and Safinya, C. R. (2006). A columnar phase of dendritic lipid-based cationic liposome-DNA complexes for gene delivery: Hexagonally ordered cylindrical micelles embedded in a DNA honeycomb lattice. *J. Am. Chem. Soc.* **128,** 3998–4006.

Huang, L., Hung, M.-C., and Wagner, E. (2005). *Non-Viral Vectors for Gene Therapy.* Elsevier, San Diego.

Kas, J., and Sackmann, E. (1991). Shape transitions and shape stability of giant phospholipid-vesicles in pure water induced by area-to-volume changes. *Biophys. J.* **60,** 825–844.

Lasic, D. D. (1993). Liposomes: From Physics to Applications. Elsevier, Amsterdam.

Lasic, D. D., and Martin, F. J. (1995). *Stealth Liposomes.* CRC Press, Boca Raton, FL.

Lipowsky, R., and Sackmann, E. (1995). *Structure and Dynamics of Membranes.* Elsevier, Amsterdam.

Mackenzie, D. (2006). Breakthrough of the year—The Poincare conjecture—Proved. *Science* **314,** 1848–1849.

Martin, C. R. (1994). Nanomaterials—A Membrane-based synthetic approach. *Science* **266,** 1961–1966.

Michalet, X., and Bensimon, D. (1995a). Vesicles of toroidal topology—Observed morphology and shape transformations. *J. Phys. II* **5,** 263–287.

Michalet, X., and Bensimon, D. (1995b). Observation of stable shapes and conformal diffusion in genus-2 vesicles. *Science* **269,** 666–668.

Odijk, T. (1977). Polyelectrolytes near rod limit. *J. Polym. Sci., Part B: Polym. Phys.* **15,** 477–483.

Quispe, J., Damiano, J., Mick, S. E., Nackashi, D. P., Fellmann, D., Ajero, T. G., Carragher, B., and Potter, C. S. (2007). An improved holey carbon film for cryo-electron microscopy. *Microsc. Microanal.* **13,** 365–371.

Raviv, U., Needleman, D. J., Li, Y. L., Miller, H. P., Wilson, L., and Safinya, C. R. (2005). Cationic liposome-microtubule complexes: Pathways to the formation of two-state lipid-protein nanotubes with open or closed ends. *Proc. Natl. Acad. Sci. USA* **102,** 11167–11172.

Raviv, U., Nguyen, T., Ghafouri, R., Needleman, D. J., Li, Y. L., Miller, H. P., Wilson, L., Bruinsma, R. F., and Safinya, C. R. (2007). Microtubule protofilament number is modulated in a stepwise fashion by the charge density of an enveloping layer. *Biophys. J.* **92,** 278–287.

Reichl, E. M., Effler, J. C., and Robinson, D. N. (2005). The stress and strain of cytokinesis. *Trends Cell Biol.* **15,** 200–206.

Reynwar, B. J., Illya, G., Harmandaris, V. A., Muller, M. M., Kremer, K., and Deserno, M. (2007). Aggregation and vesiculation of membrane proteins by curvature-mediated interactions. *Nature* **447,** 461–464.

Robinson, D. N., and Spudich, J. A. (2004). Mechanics and regulation of cytokinesis. *Curr. Opin. Cell Biol.* **16,** 182–188.

Safran, S. A. (1994). *Statistical Thermodynamics of Surfaces, Interfaces, and Membranes.* Westview Press, Boulder, CO, USA.

Safran, S. A., Pincus, P., and Andelman, D. (1990). Theory of spontaneous vesicle formation in surfactant mixtures. *Science* **248,** 354–356.

Schnur, J. M. (1993). Lipid tubules—A paradigm for molecularly engineered structures. *Science* **262,** 1669–1676.

Schulze, U., Schmidt, H. W., and Safinya, C. R. (1999). Synthesis of novel cationic poly (ethylene glycol) containing lipids. *Bioconjug. Chem.* **10,** 548–552.

Seifert, U. (1997). Configurations of fluid membranes and vesicles. *Adv. Phys.* **46,** 13–137.

Shimizu, T., Masuda, M., and Minamikawa, H. (2005). Supramolecular nanotube architectures based on amphiphilic molecules. *Chem. Rev.* **105,** 1401–1443.

Singh, A., Wong, E. M., and Schnur, J. M. (2003). Toward the rational control of nanoscale structures using chiral self-assembly: Diacetylenic phosphocholines. *Langmuir* **19,** 1888–1898.

Skolnick, J., and Fixman, M. (1977). Electrostatic persistence length of a wormlike polyelectrolyte. *Macromolecules* **10,** 944–948.

Spector, M. S., Singh, A., Messersmith, P. B., and Schnur, J. M. (2001). Chiral self-assembly of nanotubules and ribbons from phospholipid mixtures. *Nano Lett.* **1,** 375–378.

Suloway, C., Pulokas, J., Fellmann, D., Cheng, A., Guerra, F., Quispe, J., Stagg, S., Potter, C. S., and Carragher, B. (2005). Automated molecular microscopy: The new Leginon system. *J. Struct. Biol.* **151,** 41–60.

Thomas, B. N., Safinya, C. R., Plano, R. J., and Clark, N. A. (1995). Lipid tubule self-assembly—Length dependence on cooling rate through a first-order phase-transition. *Science* **267,** 1635–1638.

Yager, P., and Schoen, P. E. (1984). Formation of tubules by a polymerizable surfactant. *Mol. Cryst. Liq. Cryst.* **106,** 371–381.

Zidovska, A., Ewert, K. K., Quispe, J., Carragher, B., Potter, C. S., and Safinya, C. R. (2009a). Block liposomes from curvature-stabilizing lipids: Connected nanotubes, -rods and -spheres. *Langmuir* **25,** 2979–2985.

Zidovska, A., Evans, H. M., Ewert, K. K., Quispe, J., Carragher, B., Potter, C. S., and Safinya, C. R. (2009b). Liquid crystalline phases of dendritic lipid–DNA self-assemblies: Lamellar hexagonal and DNA bundles. *J. Phys. Chem. B* **113,** 3694–3703.

Zidovska, A., Ewert, K. K., Quispe, J., Carragher, B., Potter, C. S., and Safinya, C. R. (2009c). The effect of salt and pH on block liposomes studied by cryogenic transmission electron microscopy. *Biochim. Biophys. Acta* **1788**(9), 1869–1876.

CHAPTER SEVEN

Microfluidic Methods for Production of Liposomes

Bo Yu,[*,†] Robert J. Lee,[†,‡] *and* L. James Lee[*,†]

Contents

Abstract

Liposomes are composed of lipid bilayer membranes that encapsulate an aqueous volume. A major challenge in the development of liposomes for drug delivery is the control of size and size distribution. In conventional methods, lipids are spontaneously assembled into heterogeneous bilayers in a bulk phase. Additional processing by extrusion or sonication is required to obtain liposomes with small size and a narrow size distribution. Microfluidics is an emerging technology for liposome synthesis, because it enables precise control of the lipid hydration process. Here, we describe a number of microfluidic methods that have been reported to produce micro/nanosized liposomes with narrower size distribution in a reproducible manner, focusing on the use of continuous-flow microfluidics. The advantages of liposome formation using the microfluidic approach over traditional bulk-mixing approaches are discussed.

* Department of Chemical and Biomolecular Engineering, The Ohio State University, Columbus, Ohio, USA
† NSF Nanoscale Science and Engineering Center (NSEC), The Ohio State University, Columbus, Ohio, USA
‡ Division of Pharmaceutics, College of Pharmacy, The Ohio State University, Columbus, Ohio, USA

Methods in Enzymology, Volume 465
ISSN 0076-6879, DOI: 10.1016/S0076-6879(09)65007-2

1. Introduction

Liposomes are vesicular structures consisting of one or more lipid bilayer membranes that encapsulate an aqueous volume. Hydrophilic drugs can be loaded into the interior aqueous core of liposomes, whereas lipophilic and amphiphilic drugs can be incorporated into the lipid bilayers (Bangham *et al.*, 1965; Jesorka and Orwar, 2008; Meure *et al.*, 2008). Due to these properties, liposomes are widely used as drug and gene delivery vehicles (Torchilin, 2005). As drug delivery vehicles, liposomes can provide metabolic protection, prolong circulation time, reduce toxicity, control drug release, and enhance cell/tissue specificity of delivery. Several liposomal drugs, for example Doxil (PEGylated liposomal doxorubicin), have reached clinical use (Abraham *et al.*, 2005). In addition to conventional drugs, liposomes hold great promises as delivery vehicles for oligonucleotide-based therapeutics, including siRNA (Kawakami *et al.*, 2008; Yu *et al.*, 2009). Properties such as uniform particle size and good colloidal stability are essential for liposomes to be developed as *in vivo* drug carriers. Size characteristics of liposomes have a critical effect on the capacity of drug loading, *in vivo* biodistribution and clearance rate, etc. For instance, liposomes with a diameter of 50–200 nm have been shown to exhibit a slower clearance rate than larger ones (Li and Huang, 2008; Liu *et al.*, 1992). Therefore, it is essential for production methods of liposomes to reproducibly generate particle size distributions within a certain size range.

A number of methods have been developed to produce nano/microsized liposomes, such as thin-film hydration (Amselem *et al.*, 1990; Bhalerao and Raje Harshal, 2003), ethanol injection (Batzri and Korn, 1973; Pons *et al.*, 1993), and detergent dialysis methods (Cardoza *et al.*, 1984; Zumbuehl and Weder, 1981). However, the conventional bulk production of liposomes mainly relies on self-assembly of lipids in a bulk phase, which is heterogeneous and uncontrolled. The resultant liposomes are polydispersed in size and are multilamellar. Further post-processing by extrusion, freeze–thaw, sonication, and/or high-pressure homogenization is often required (Johnson *et al.*, 1971; Purmann *et al.*, 1993).

Microfluidics is a versatile technology to manipulate liquid flows in channels with dimensions of tens to hundreds of micrometers (Stone *et al.*, 2004; Whitesides, 2006). It can provide rapid and tunable mixing, a homogenous reaction environments and a high-throughput experimental platform. Therefore, it is an attractive technology for a variety of applications in chemical synthesis and biological analysis (Jahn *et al.*, 2008; Whitesides, 2006). The exquisite control of flow and mixing conditions in microfluidics has been applied for altering particle size and improving homogeneity of particle size distributions as well (Jahn *et al.*, 2008).

Jahn *et al.* (2004, 2007) developed a microfluidic hydrodynamic focusing (MHF) method for controlled liposome formation. Our group has successfully extended MHF technology for producing polymer-DNA (polyplex) (Koh *et al.*, 2009a) and lipid-polymer-DNA (lipopolyplex) nanoparticles (Koh *et al.*, 2009b) using three-inlet or five-inlet MHF devices. Additionally, we reported a simple and low-cost alternative method with the key feature of microfluidics for production of liposomes.

Here the use of microfluidics for production of liposomes is reviewed, emphasizing the production of nanosized liposomes based on continuous-flow microfluidics. The advantages of liposome formation using microfluidics compared to traditional bulk mixing (BM) are also discussed.

2. Conventional Technologies for Production of Liposomes

There is a wide variety of traditional methods used to prepare liposomes, including thin-film hydration, detergent dialysis, reverse-phase evaporation, and ethanol injection (Jesorka and Orwar, 2008; Meure *et al.*, 2008; Szoka and Papahadjopoulos, 1978; Düzgüneş, 2003). In these methods, lipids are generally dissolved in a transfer medium, followed by the removal of the medium. The liposomal particles are spontaneously self-assembled in the bulk phase by the hydration of a thin-film, or during the removal of the transfer medium, such as an organic solvent or a detergent solution. The self-assembly of lipid vesicles typically occurs under an environment with characteristic dimension of millimeters or centimeters, which results in local concentration fluctuations of lipids and payloads. Thus, the prepared vesicles are initially multilamellar and heterogeneous. To alter particle size and minimize the polydispersity of the size distributions, one or two postprocessing procedures, such as freeze–thaw, sonication, extrusion, and high-pressure homogenization are required (Castile and Taylor, 1999; Meure *et al.*, 2008; Plum *et al.*, 1988). For example, the extrusion method can be used to prepare liposomes with controlled sizes determined by the pore size of a track-etched polycarbonate membrane. However, this method is limited by the requirement of highly specialized equipment, and is time-consuming.

In summary, limitations of conventional preparation methods include complexity and length of procedures, low drug encapsulation efficiency and polydisperse size distributions. Recently, additional new procedures have been reported for producing liposomes to address the issues in conventional production technologies. The new procedures include freeze-drying of a monophase solution (Li and Deng, 2004), MHF method (Jahn *et al.*, 2004), and supercritical fluid method (Frederiksen *et al.*, 1997; Otake *et al.*, 2001).

3. Microfluidic Technologies for Synthesis of Nanoparticles

Microfluidics is a technology that enables precise control and manipulation of fluids and fluid interfaces at the micrometer scale. In microfluidic chips, the fluid streams can merge and form well-defined interface by laminar flow, as opposed to the typically chaotic flows in BM. According to the difference in manipulation modes of flow, microfluidics is categorized into two classes: continuous-flow microfluidics (Jahn *et al.*, 2008) and digital (droplet-based) microfluidics (Chatterjee *et al.*, 2006). In continuous-flow microfluidics, liquid flow is continuously manipulated through microfabricated channels, whereas discrete and controllable droplets are manipulated in droplet-based microfluidics. Continuous-flow microfluidic is generally more suitable for producing nanoparticles (Jahn *et al.*, 2008).

Over the past decade, microfluidics technologies have been developed as an important tool in chemical synthesis and biological analysis (Whitesides, 2006). In parallel, there has been an increasing interest in the use of microfluidics as a novel platform for the preparation of nano- and microparticles. The rapid and tunable mixing provided by microfluidic makes it suitable for the controlled synthesis of nanoparticles, including CdSe quantum dots (QDs) (Schabas *et al.*, 2008), titanium dioxide nanorods (Cottam *et al.*, 2007), and polymeric nanoparticles (Karnik *et al.*, 2008). In addition, MHF devices have been used for the self-assembly of nanoparticles, including liposomes (Jahn *et al.*, 2004, 2007). For example, microfluidics was used to synthesize complexes between calf thymus DNA and cationic lipids (Otten *et al.*, 2005) and dendrimers (Dootz *et al.*, 2006). Langer and coworkers (Karnik *et al.*, 2008) used MHF microfluidics to control nanoprecipitation of poly(lactic-*co*-glycolic acid)-*b*-poly(ethylene glycol) (PLGA-PEG) diblock copolymers as a model polymeric material for drug delivery. Polymeric nanoparticles with smaller size and narrower size distribution were obtained by varying the flow rates, polymer composition, and polymer concentration. Nanoparticles yielded by MHF synthesis had a Z-average diameter of about 24 nm at PLGA-PEG concentration of 50 mg/ml, which was smaller than the Z-average diameter of 31 nm obtained by conventional slow-BM. They also observed that with a decrease of mixing time by increasing the flow ratio (0.03–0.1), the size of nanoparticles generated by the MHF method decreased from 26 to 20 nm at 20 mg/ml PLGA-PEG, which was smaller than the 30–35 nm size prepared by BM method. Moreover, the polymeric nanoparticles prepared by MHF showed much narrower distribution, and improved drug loading and release properties (Karnik *et al.*, 2008).

4. Microfluidic Technologies for Production of Liposomes

Confinement and well-defined mixing in microfluidics makes it attractive for production of liposomes ranging from tens of nanometers to tens of micrometers in diameter. The self-assembly in microfluidics can be controlled by varying liquid flow rates, ratios of cross-flows and the composition and concentration of lipids, resulting in tunable sizes, and narrower size distributions. Recently, several studies on liposome production in microfluidics have been reported (Jahn *et al.*, 2008).

4.1. Giant liposomes produced by droplet-based microfluidics

Giant liposomes or giant unilamellar vesicles (GUVs) (over 10 μm in diameter) have a size comparable to that of cells and can encapsulate biomaterials such as DNA and proteins. Giant liposomes, therefore, are used as a model to mimic cell membranes for studing the interaction of the cell membrane with other molecules (Chatterjee *et al.*, 2006; Kuribayashi *et al.*, 2006). Several methods have been developed for production of this type of liposomes. Basically, these involve the swelling of dried lipid, with or without enhancement by an electric field.

Recently, droplet-based microfluidic systems have been used to produce giant liposomes. Tan *et al.* (2006) presented a novel method to control the formation of liposomal vesicles by a shear-focusing-based droplet microfluidic systems. Using this microfluidic system, they were able to encapsulate large biological materials ranging from cancer cells and yeast cells to nanosized proteins. The resultant vesicles showed long-term stability (>26 days) and high encapsulation efficiency. Stachowiak *et al.* (2008) also developed a microfluidics method for simultaneously creating and loading giant liposomes based on a pulsed microfluidic jet. In the microfluidic jetting, a planar lipid bilayer is deformed into a vesicle that is filled with the solution from the jet. Compared to other conventional giant liposome production methods, this method can rapidly and controllably generate multiple monodisperse and unilamellar vesicles. They demonstrated that 500 nm particles can be encapsulated into GUVs in a highly reproducible manner and functional pore proteins can also be incorporated into the vesicle membrane simultaneously.

4.2. Nanosized liposome formation by MHF

Jahn *et al.* (2004) first reported on the controlled synthesis of submicrometer-sized liposomes through MHF. As depicted in Fig. 7.1A, isopropyl alcohol (IPA) containing the dissolved lipids flows through the center inlet

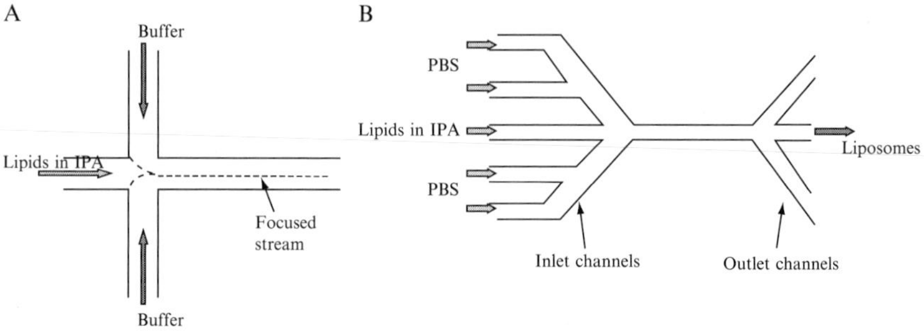

Figure 7.1 Schematic diagram of microfluidic channels for liposome formation: (A) three-inlet design; (B) five-inlet design.

channel, and an aqueous solution flows through the two side inlet channels. The stream of lipids in IPA is hydrodynamically focused by two aqueous streams at the cross junction of the microfluidic chip. The flow rates of the IPA and buffer streams are adjusted to control the width of the interface, that is the degree of hydrodynamic focusing. The liposome formation is based on a diffusion-driven process in which the dissolved lipids self-assemble into liposomes as IPA quickly diffuses and dilutes into two aqueous streams at the interfacial region. Thus, the size and size distribution of liposomes can be controlled through adjusting lipids concentrations and flow conditions. By altering the flow rate of IPA (from 2.4 to 59.8 mm/s), the size of resultant liposome can be readily controlled over the range of 100–300 nm (Jahn *et al.*, 2004). Basically, the mechanism of liposome formation in the MHF method is very similar to that in the ethanol injection method. In contrast to ethanol injection, the flow ratio between the lipid-solved stream and buffer streams becomes an important control condition in the MHF method.

Recently, Jahn *et al.* (2007) made significant modifications in their microfluidic system to greatly improve the control over size and size distribution. As shown in Fig. 7.1B, their new microfluidics device has five-inlet channels and three-outlet channels, which are fabricated in a silicon wafer. The resulting microfluidics channels have a rectangular cross section with a depth of 100 μm and a width of 46 or 64 μm. The lipid IPA solution is injected into the center channel of the microfluidics network, while phosphate-buffered saline (PBS) is injected into two side channels intersecting with the center channel. Owing to this three-outlet design, relatively high liposome concentration can be produced at the center point in the channel once the focused IPA stream is diluted to the critical concentration for formation of the more stable liposomes along the interfacial region.

Liposome formation under different shear forces was investigated by changing the buffer volumetric flow rate (VFR) of PBS from 15 to 90 μl/min and the VFR of IPA from 1 to 6 μl/min. The flow rate ratio

(FRR), defined as VFR of PBS to VFR of IPA, ranged from 10 to 60. They observed the precise control of size over a mean diameter from 50 to 150 nm by manipulating the flow conditions, especially in the FRR. As the FRR increases sixfold (from 10 to 60), the liposome size decreases in diameter from approximately 120–50 nm and the standard deviation decreases from 0.4 to 0.2. Liposome size and size distribution showed little change when VFR of IPA was varied from 30 to 180 μl/min at a constant FRR of 30. This suggests that liposome formation in MHF is more dependent on the width of the focused alcohol stream and its diffusive mixing with the aqueous stream than on the shear forces at the interfaces. The narrow liposome size and homogeneous size distribution can be facilely manipulated by reducing the width of the alcohol stream or the length scale of liquid mixing to a few micrometers.

4.3. Formation of liposomes containing oligonucleotides by MHF

Oligonucleotides (ONs), including antisense oligodeoxynucleotides (AS-ODN) and small interfering RNA (siRNA), have been emerging as therapeutic agents for treatment of numerous diseases (Dykxhoorn and Lieberman, 2006; Kurreck, 2003). Their applications in the clinic have been severely limited by the lack of safe and efficient delivery systems. To efficiently deliver ONs, liposomes are widely recognized as one of the most promising vehicles. However, it is still a challenge to produce homogeneous ON liposomes with narrow size distribution. Our group developed an MHF technology to produce DNA/polyethylenimine polyplexes for gene delivery (Koh *et al.*, 2009a). Recently, we have been extending the MHF technology to production of ON liposomes (Koh *et al.*, 2009b).

The five-inlet microfluidic device consists of three-inlet ports and one-outlet port (Fig. 7.2). The inlet ports are each connected to sterile syringes containing lipids ethanol solution, protamine or ON aqueous solutions. As shown in Fig. 7.2, a fluid stream is split into two side streams at inlet port 1 or 2, while a fluid stream directly entered the center microchannel through inlet port 3. The resulting ON liposomes solution is collected at the outlet port. The MHF device is fabricated in a poly(methyl methacrylate) (PMMA) plate using a modified microfabrication protocol described elsewhere (Koh *et al.*, 2003). The channel patterns in MHF device is first designed in AutoCAD (Autodesk, San Rafael, CA) and then is fabricated by a high-precision computer numerically controlled (CNC) machine (Aerotech, Inc.) according to the transferred g-code program from AutoCAD. The closed channels are formed by thermally laminating a 45-μm thick PMMA film onto the fabricated microfluidics chip. After lamination, fluidic connectors are bonded onto the inlet/outlet openings in PMMA plate by applying a UV curing adhesive under the exposure of UV irradiation (Novacure 2100,

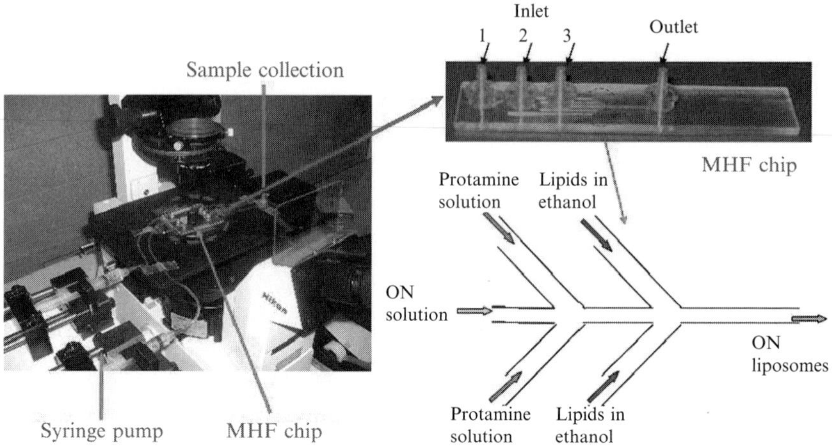

Figure 7.2 Photographs of five-inlet MHF methods to prepare liposomes containing oligonucleotides. (See Color Insert.)

EFXO Corp., Quebec, Canada) for 10 s. The degree of hydrodynamic focusing is controlled by adjusting the FRR of the side streams (lipids and protamine solutions) to the middle stream (ON solution). Two programmable syringe pumps (Pump 33, Harvard Apparatus, Holliston, MA) are used to manipulate the fluid flow rates independently.

G3139, an AS-ODN against human Bcl-2, is incorporated in liposomes to evaluate the difference between G3139 liposomes produced by the MHF method and that prepared by the BM method. For all G3139 containing liposomes, the final weight ratio of G3139:protamine:lipids (DC-Chol/egg PC/PEG-DSPE at a molar ratio of 30/68/2) is 1:0.3:12.5 and the ethanol concentration of 40% is maintained during nanoparticle synthesis. The flow rates for G3139, protamine, and lipids streams are 20, 20, and 450 μl/min, respectively. Our results demonstrate that MHF-produced G3139 liposomes have similar nanostructures but smaller size (114.8 $\pm$ 12.7 nm) and tighter size distribution comparing to BM-produced G3139 liposomes (152.7 $\pm$ 22.1 nm). Importantly, G3139 liposomes produced by the MHF method facilitated better delivery into K562 cells and also downregulated an apoptotic protein, Bcl-2 more efficiently as compared to the BM method.

To implement MHF in a laboratory setting, syringe pumps and a microfluidic chip are required. A number of commercial suppliers are now available that can provide suitable chips. For example, Translume Inc. (Ann Arbor, MI) offers all-glass microfluidic chips with T- and Y-channel designs with luer connectors with 30, 100, and 300 μm channel widths. These chips can be connected easily to a pair of syringe pumps for MHF production of liposomes.

4.4. A facile microfluidic method for production of liposomes

Although the advantages of microfluidics for liposome production have been demonstrated by Jahn *et al.* (2004, 2007) and in our group, this technology requires microfabrication expertise and facilities, which may be unavailable and unaffordable in many biomedical research laboratories. To facilely use microfluidics in the laboratory, we further developed a simple and low-cost method with the key features of microfluidics, but without the need for microfabrication (Pradhan *et al.*, 2008).

A microfluidics injection device driven by a syringe-pump is used to produce liposomes under various conditions. As demonstrated in Fig. 7.3A, the syringe carrying water is injected to an elbow connecter by plastic tubing. The other syringe containing the lipids dissolved in ethanol is connected by a needle. The internal diameters of inlet/outlet connectors are 2 mm and the internal diameter of the tubing is 1.5 mm. Needles of a series of gauges (internal diameters of 1.194, 0.838, 0.495, 0.241, and 0.191 mm) are selected for the control of shear force of lipid flow. In a typical setup, 1 ml lipid ethanol solution and 10 ml deionized water are loaded into a 1 and 10 ml syringe, respectively. Syringes are mounted on an infusion syringe pump (model 975, Harvard Apparatus, Holliston, MA, USA) and moved at the same speed (Fig. 7.3B).

Microfluidics is characterized by laminar flow, which means that the Reynolds number (*Re*) should be lower than the value of 2000. In a typical device, the calculated *Re* value is 13.5, which is much lower than 2000 and indicates strictly laminar flow. The chaotic mixing in this device is significantly different from the laminar mixing in classic microfluidics devices. As the ethanol in the lipid stream diffuses and dilutes into the water stream, the lipids tend to assemble into liposomes. Stable liposomes are eventually formed when the mixture reaches equilibrium.

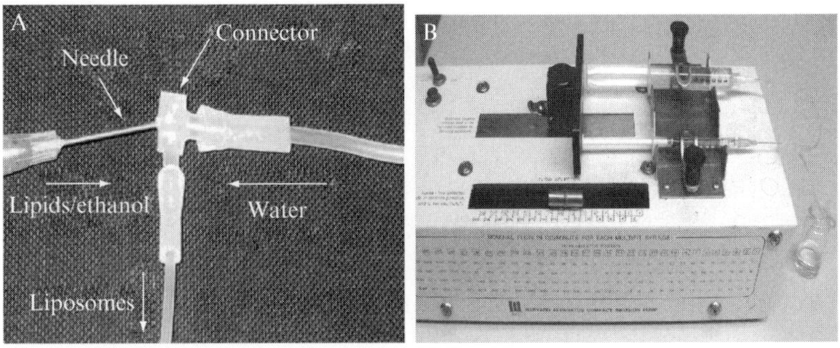

Figure 7.3 Photographs of a facile microfluidic device for production of liposomes.

This microfluidic injection method is similar to the conventional ethanol injection method, but it enables continuous manipulation of particle size by providing well-defined mixing and a constant water-to-ethanol flow ratio (10:1). More importantly, this method does not require the costly microfabrication equipment and advanced manipulation skills. The key features of a microfluidics system are kept in this nonmicrofabricated method. The average size of the liposomes produced by using this method can be conveniently and repeatedly tuned by varying the type of lipid, lipid concentration, flow rate, and temperature. Thus, the fluidity of the specific lipid composition can modulate the size of the produced liposomes. Mixtures containing cholesterol, which is known to rigidify the liquid crystalline lipid bilayers, produce larger liposomes. Lipids in the gel (solid) phase, which exhibit considerably lower fluidity than the liquid crystalline phase, generate liposomes of significantly larger size (cf. 90 nm for the liquid crystalline egg phosphatidylcholine versus 250 nm for the solid hydrogenated soy phosphatidylcholine at the same concentration and flow rate) (Pradhan *et al.*, 2008).

In contrast to other liposome manufacturing techniques, this method is simple, quick, and affordable, and it can be readily adopted in any biomedical or pharmaceutical laboratory.

5. Concluding Remarks

Microfluidics is a relatively novel technology for the production of micro- and nanosized liposomes. The characteristics of laminar flow and tunable mixing in microfluidics systems have distinctive advantages in liposome formation over traditional methods, such as thin-film hydration and reverse-phase evaporation. Reproducible control of particle size and size distribution can be implemented in continuous microfluidics flow systems. Using hydrodynamic focusing in microfluidic channels, nanosized liposomes with smaller size and narrower size distribution are easily formed by varying flow parameters. They include flow rate, flow ratio, concentration of lipids solution, and temperature as well as characteristic length in microfluidics channels. However, the problem of scaling up liposome production needs to be addressed during the implementation of microfluidics technology for practical application. Additionally, the ultrafine structure of MFH-produced liposomes needs to be further investigated. Microfluidics provides a new platform for the development and optimization of liposomes in the emerging field of nanomedicine. It can control liposome self-assembly and potentially lead to applications in instant liposome synthesis as part of point-of-care personalized therapeutics.

ACKNOWLEDGMENTS

This work was supported by grants EEC 0425626 from the National Science Foundation and R21 CA131832 and R01 CA135243 from the National Institutes of Health. The authors thank Dr. B. Tenchov and Dr. R. Koynova for their helpful suggestions.

REFERENCES

Abraham, S. A., Waterhouse, D. N., Mayer, L. D., Cullis, P. R., Madden, T. D., and Bally, M. B. (2005). The liposomal formulation of doxorubicin. *Methods Enzymol.* **391,** 71–97.

Amselem, S., Gabizon, A., and Barenholz, Y. (1990). Optimization and upscaling of doxorubicin-containing liposomes for clinical use. *J. Pharm. Sci.* **79,** 1045–1052.

Bangham, A. D., Standish, M. M., and Watkins, J. C. (1965). Diffusion of univalent ions across the lamellae of swollen phospholipids. *J. Mol. Biol.* **13,** 238–252.

Batzri, S., and Korn, E. D. (1973). Single bilayer liposomes prepared without sonication. *Biochim. Biophys. Acta* **298,** 1015–1019.

Bhalerao, S. S., and Raje Harshal, A. (2003). Preparation, optimization, characterization, and stability studies of salicylic acid liposomes. *Drug Dev. Ind. Pharm.* **29,** 451–467.

Cardoza, J. D., Kleinfeld, A. M., Stallcup, K. C., and Mescher, M. F. (1984). Hairpin configuration of H-2kk in liposomes formed by detergent dialysis. *Biochemistry* **23,** 4401–4409.

Castile, J. D., and Taylor, K. M. G. (1999). Factors affecting the size distribution of liposomes produced by freeze-thaw extrusion. *Int. J. Pharm.* **188,** 87–95.

Chatterjee, D., Hetayothin, B., Wheeler, A. R., King, D. J., and Garrell, R. L. (2006). Droplet-based microfluidics with nonaqueous solvents and solutions. *Lab. Chip.* **6,** 199–206.

Cottam, B. F., Krishnadasan, S., deMello, A. J., deMello, J. C., and Shaffer, M. S. P. (2007). Accelerated synthesis of titanium oxide nanostructures using microfluidic chips. *Lab. Chip.* **7,** 167–169.

Dootz, R., Otten, A., Koster, S., Struth, B., and Pfohl, T. (2006). Evolution of DNA compaction in microchannels. *J. Phys. Condens. Matter* **18,** S639–S652.

Düzgüneş, N. (2003). Preparation and quantitation of small unilamellar liposomes and large unilamellar reverse-phase evaporation liposomes. *Methods Enzymol.* **367,** 23–27.

Dykxhoorn, D. M., and Lieberman, J. (2006). Knocking down disease with siRNAs. *Cell* **126,** 231–235.

Frederiksen, L., Anton, K., van Hoogevest, P., Keller, H. R., and Leuenberger, H. (1997). Preparation of liposomes encapsulating water-soluble compounds using supercritical carbon dioxide. *J. Pharm. Sci.* **86,** 921–928.

Jahn, A., Vreeland, W. N., Gaitan, M., and Locascio, L. E. (2004). Controlled vesicle self-assembly in microfluidic channels with hydrodynamic focusing. *J. Am. Chem. Soc.* **126,** 2674–2675.

Jahn, A., Vreeland, W. N., DeVoe, D. L., Locascio, L. E., and Gaitan, M. (2007). Microfluidic directed formation of liposomes of controlled size. *Langmuir* **23,** 6289–6293.

Jahn, A., Reiner, J. E., Vreeland, W. N., DeVoe, D. L., Locascio, L. E., and Gaitan, M. (2008). Preparation of nanoparticles by continuous-flow microfluidics. *J. Nanopart. Res.* **10,** 925–934.

Jesorka, A., and Orwar, O. (2008). Liposomes: Technologies and analytical applications. *Annu. Rev. Anal. Chem.* **1,** 801–832.

Johnson, S. M., Bangham, A. D., Hill, M. W., and Korn, E. D. (1971). Single bilayer liposomes. *Biochim. Biophys. Acta* **233,** 820–826.

Karnik, R., Gu, F., Basto, P., Cannizzaro, C., Dean, L., Kyei-Manu, W., Langer, R., and Farokhzad, O. C. (2008). Microfluidic platform for controlled synthesis of polymeric nanoparticles. *Nano Lett.* **8,** 2906–2912.

Kawakami, S., Higuchi, Y., and Hashida, M. (2008). Nonviral approaches for targeted delivery of plasmid DNA and oligonucleotide. *J. Pharm. Sci.* **97,** 726–745.

Koh, C. G., Tan, W., Zhao, M. Q., Ricco, A. J., and Fan, Z. H. (2003). Integrating polymerase chain reaction, valving, and electrophoresis in a plastic device for bacterial detection. *Anal. Chem.* **75,** 4591–4598.

Koh, C. G., Kang, X. H., Xie, Y. B., Fei, Z. Z., Guan, J. J., Yu, B., Zhang, X. L., and Lee, L. J. (2009a). Delivery of polyethylenimine (PEI)/DNA complexes assembled in a microfluidics device. *Mol. Pharm.* (in press).

Koh, C. G., Zhang, X. L., Liu, S. J., Golan, S., Yu, B., Yang, X. J., Guan, J. J., Yan, J., Talmon, Y., Muthasamy, R., Chan, K. K., and Byrd, J. (2009b). Delivery of antisense oligodeoxyribonucleotide lipopolyplex nanoparticles assembled by microfluidic hydrodynamic focusing. *J. Control. Release* (in press).

Kuribayashi, K., Tresset, G., Coquet, P., Fujita, H., and Takeuchi, S. (2006). Electroformation of giant liposomes in microfluidic channels. *Meas. Sci. Technol.* **17,** 3121–3126.

Kurreck, J. (2003). Antisense technologies. Improvement through novel chemical modifications. *Eur. J. Biochem.* **270,** 1628–1644.

Li, C., and Deng, Y. (2004). A novel method for the preparation of liposomes: Freeze drying of monophase solutions. *J. Pharm. Sci.* **93,** 1403–1414.

Li, S. D., and Huang, L. (2008). Pharmacokinetics and biodistribution of nanoparticles. *Mol. Pharm.* **5,** 496–504.

Liu, D., Mori, A., and Huang, L. (1992). Role of liposome size and RES blockade in controlling biodistribution and tumor uptake of GM1-containing liposomes. *Biochim. Biophys. Acta* **1104,** 95–101.

Meure, L. A., Foster, N. R., and Dehghani, F. (2008). Conventional and dense gas techniques for the production of liposomes: A review. *AAPS Pharm. Sci. Technol.* **9,** 798–809.

Otake, K., Imura, T., Sakai, H., and Abe, M. (2001). Development of a new preparation method of liposomes using supercritical carbon dioxide. *Langmuir* **17,** 3898–3901.

Otten, A., Koster, S., Struth, B., Snigirev, A., and Pfohl, T. (2005). Microfluidics of soft matter investigated by small-angle X-ray scattering. *J. Synchrotron Radiat.* **12,** 745–750.

Plum, G., Korber, C., and Rau, G. (1988). Freeze/thaw response of single bilayer liposomes. *Cryobiology* **25,** 520.

Pons, M., Foradada, M., and Estelrich, J. (1993). Liposomes obtained by the ethanol injection method. *Int. J. Pharm.* **95,** 51–56.

Pradhan, P., Guan, J., Lu, D., Wang, P. G., Lee, L. J., and Lee, R. J. (2008). A facile microfluidic method for production of liposomes. *Anticancer Res.* **28,** 943–947.

Purmann, T., Mentrup, E., and Kreuter, J. (1993). Preparation of suv-liposomes by high-pressure homogenization. *Eur. J. Pharm. Biopharm.* **39,** 45–52.

Schabas, G., Yusuf, H., Moffitt, M. G., and Sinton, D. (2008). Controlled self-assembly of quantum dots and block copolymers in a microfluidic device. *Langmuir* **24,** 637–643.

Stachowiak, J. C., Richmond, D. L., Li, T. H., Liu, A. P., Parekh, S. H., and Fletcher, D. A. (2008). Unilamellar vesicle formation and encapsulation by microfluidic jetting. *Proc. Natl. Acad. Sci. USA* **105,** 4697–4702.

Stone, H. A., Stroock, A. D., and Ajdari, A. (2004). Engineering flows in small devices: Microfluidics toward a lab-on-a-chip. *Annu. Rev. Fluid Mech.* **36,** 381–411.

Szoka, F., and Papahadjopoulos, D. (1978). Procedure for preparation of liposomes with large internal aqueous space and high capture by reverse-phase evaporation. *Proc. Natl. Acad. Sci. USA* **75,** 4194–4198.

Tan, Y. C., Hettiarachchi, K., Siu, M., Pan, Y. R., and Lee, A. P. (2006). Controlled microfluidic encapsulation of cells, proteins, and microbeads in lipid vesicles. *J. Am. Chem. Soc.* **128,** 5656–5658.

Torchilin, V. P. (2005). Recent advances with liposomes as pharmaceutical carriers. *Nat. Rev. Drug Discov.* **4,** 145–160.

Whitesides, G. M. (2006). The origins and the future of microfluidics. *Nature* **442,** 368–373.

Yu, B., Zhao, X., Lee, L. J., and Lee, R. J. (2009). Targeted delivery systems for oligonucleotide therapeutics. *AAPS J.* **11,** 195–203.

Zumbuehl, O., and Weder, H. G. (1981). Liposomes of controllable size in the range of 40 to 180 nm by defined dialysis of lipid-detergent mixed micelles. *Biochim. Biophys. Acta* **640,** 252–262.

CHAPTER EIGHT

Constructing Size Distributions of Liposomes from Single-Object Fluorescence Measurements

Christina Lohr, Andreas H. Kunding, Vikram K. Bhatia, *and* Dimitrios Stamou

Contents

Abstract

We describe in detail a simple technique to construct the size distribution of liposome formulations from single-object fluorescence measurements. Liposomes that are fluorescently labeled in their membrane are first immobilized on a surface at dilute densities and then imaged individually using epi-fluorescence microscopy. The integrated intensities of several thousand single liposomes are collected and evaluated within minutes by automated image processing, using the user-friendly freeware ImageJ. The mean intensity of the liposome population is then calculated and scaled in units of length (nm) by relating the intensity data to the mean diameter obtained from a reference measurement with dynamic light scattering. We explain the process of

Bio-Nanotechnology Laboratory, Department of Neuroscience and Pharmacology and Nano-Science Center, University of Copenhagen, Copenhagen, Denmark

Methods in Enzymology, Volume 465
ISSN 0076-6879, DOI: 10.1016/S0076-6879(09)65008-4

constructing the size distributions in a step-by-step manner, starting with the preparation of liposomes through the final acquisition of size histograms. Detailed advice is given concerning critical parameters of image acquisition and processing. Size histograms constructed from single-particle measurements provide detailed information on complex distributions that may be easily averaged out in ensemble measurements (e.g., light scattering). In addition, the technique allows accurate measurements of polydisperse samples (e.g., non-extruded liposome preparations).

1. Introduction

The study of liposomes during the past 30 years has contributed tremendously to our understanding of the physical and chemical properties of biomembranes (Sessa and Weissman, 1968; Szoka and Papahadjopoulos, 1980; Torchilin, 2005; Venturoli *et al.*, 2006). Furthermore the lipid vesicle, being the ideal reconstitution platform for integral membrane proteins, has played an important role in unraveling the relationship between protein structure and function (Cooper, 2004; Palmieri *et al.*, 1993). A liposome is defined by not only a variety of parameters, mainly related to the lipid characteristics, for example charge and shape, but also the collective bilayer properties, such as elasticity and melting point (Israelachvili *et al.*, 1976, 1977). Another equally important property is the size of the vesicle.

To study the distribution of sizes for a single vesicle preparation, the two most successful techniques at present are dynamic light scattering (DLS) and electron microscopy (EM) (Coldren *et al.*, 2003; Ruf *et al.*, 1989). Transmission electron microscopy at cryogenic temperatures (Cryo-TEM) has been widely used because (i) the vesicle population is arrested and imaged in a state resembling physiological conditions and (ii) vesicle shape and structure can be resolved. Despite the wide applicability of DLS and EM to construct vesicle size distributions, both techniques are troubled by simultaneous direct measurements of size and a secondary vesicle property, for example encapsulation efficiency, density of membrane-associated molecules, ionic influx/efflux. To achieve information on the aforementioned properties, other measurements need to be performed sequentially. In this way the identity of the single vesicle is lost, and only the average properties of the ensemble are measured. Thus, to achieve more detailed knowledge, all experimental information has to be gathered at the site of the individual vesicle under native conditions.

Nanoscopic vesicles studied at the single-particle level hence constitutes an emerging field of research, providing more detailed information on the ensemble properties of vesicles and permitting to decipher population heterogeneities (Christensen properties of vesicles and Stamou, 2007).

The literature provides several examples of quantitative analyses applied on single vesicles, such as the permeation of ions across the membrane (Kuyper *et al.*, 2006), bivesicular association reactions (Chan *et al.*, 2007), vesicle-protein interactions (Hatzakis *et al.*, 2009), and membrane fusion reactions (Yoon *et al.*, 2006). Fluorescence microscopy is the most convenient technique for single-particle measurements for several reasons: (i) a wide variety of fluorescent probes have already been developed for bulk measurements, (ii) several thousands of liposomes can be individually sampled during one experiment, (iii) several fluorescence-related properties can be monitored in parallel by choosing different optical filters, and (iv) dynamic processes can be monitored. The contrast in a fluorescence micrograph is generated due to differences in the photophysical properties of the probe: hence, higher signal-to-noise ratios can be obtained as compared to TEM, where signal contributions from the background can lead to reduced contrast. Furthermore, photomultiplier tubes and charge-coupled device (CCD) cameras, both standard intensity-signal detection tools, have a broad dynamic range (16 bit), thus allowing polydisperse intensity distributions to be monitored accurately.

To achieve optimal signals it is necessary to immobilize vesicles on a microscope glass slide. Lipid vesicles generally adhere strongly to plasma-etched glass substrates, but during adsorption the lipid bilayer deforms, which leaves the membrane in a physical state differing from suspended liposomes. It is thus desirable to attach vesicles further away from the adhesive glass substrate if the native membrane properties are to be studied (Bendix *et al.*, 2009). Generally, biomolecules are applied as adhesive agents, but also because of their surface-passivating properties (Christensen and Stamou, 2007). One approach utilizes the biotin/streptavidin complex to immobilize vesicles by higher order self-assembly, (Fig. 8.1A). Here, streptavidin rests on a cushion of biotinylated bovine serum albumin (BSA) (Stamou *et al.*, 2003) or poly-L-lysine poly(ethylene-glycol) (PLL-*g*-PEG) (Yoon *et al.*, 2006), thus mediating docking of vesicles with biotinylated lipids incorporated in the bilayer. Another strategy passivates the glass substrate with a supported planar membrane, functionalized with cholesterol-anchored single-stranded DNA molecules, thus enabling the capture of vesicles with complementary DNA-sequences by the surface (Gunnarsson *et al.*, 2008; Yoshina-Ishii and Boxer, 2006). The latter immobilization scheme repels nonspecific binding efficiently, inasmuch as the adsorbent is not positively charged, but has the drawback that vesicles are free to diffuse laterally, thus not being entirely immobilized. The protein-based attachment strategy permits small amounts of nonspecific binding to take place, but ensures complete vesicle immobility. The surface density of liposomes can be modulated by varying the content of tether molecules on the surface. For fluorescence microscopy applications, an ultra-low density of vesicles bound to the functionalized glass substrate is desirable, otherwise the point spread function of adjacent particles will overlap,

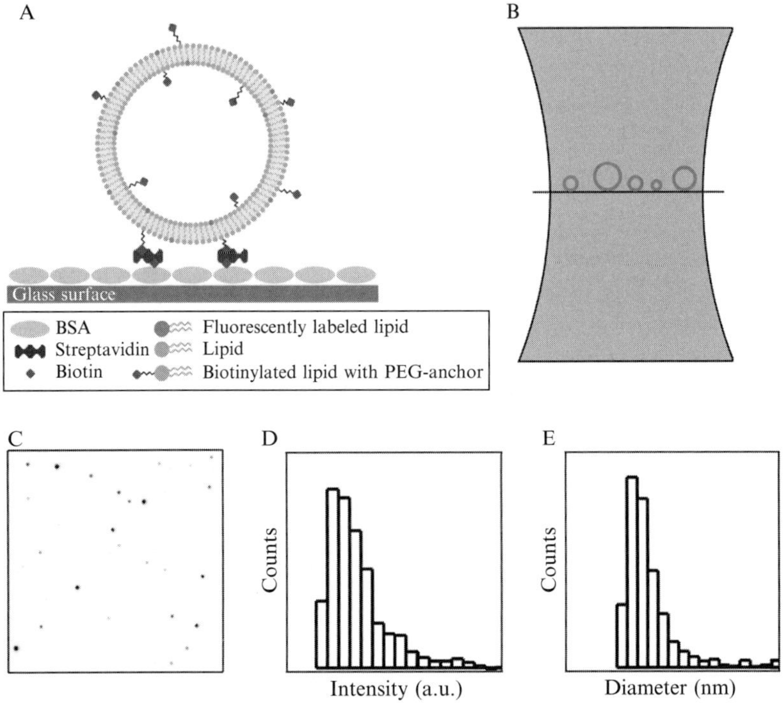

Figure 8.1 Size distribution of vesicle samples based on single-particle fluorescence measurements. (A) Single fluorescence-labeled vesicles are immobilized at low densities on a protein-functionalized glass substrate. (B) The membrane-labeled vesicles are imaged with wide-field epi-fluorescence microscopy. The large depth of field insures that all vesicles are illuminated with equal efficiency. (C) In a fluorescence micrograph vesicles of different sizes appear as diffraction-limited spots with different total intensity. An intensity histogram of all vesicles (D) can be transformed into a size distribution histogram (E) after a calibration measurement. (See Color Insert.)

thus rendering them indistinguishable from each other in a micrograph, ultimately leading to imaging artifacts.

In optical microscopy the resolving power is restrained by the diffraction limit, which is a function of objective numerical aperture and the applied wavelength. Typical values range between ~200 and 400 nm, and hence objects below this size cannot be resolved but are visualized as diffraction-limited Gaussian blurs. Thus, the size of small unilamellar vesicles (SUVs) cannot be obtained directly as is the case in various electron microscopy techniques (Cryo-TEM, freeze/fracture, negative staining). Although a single SUV will appear like a diffraction-limited spot in a micrograph (Fig. 8.1B and C), the physical shape, that is radius, remains encoded in the magnitude of the collected intensity signal. The total intensity originates from the intensity signal of individual vesicle-associated fluorophores, and is thus related to membrane area for fluorophores situated in the bilayer, and vesicle volume for soluble fluorophores in the lumen. Interpretation of the intensity signal can be challenging for advanced microscope platforms such as confocal laser scanning and total internal reflection, because of the spatially restricted illumination and detection volumes (Kunding *et al.*, 2008). However, for epi-fluorescence microscopy the total intensity is directly proportional to the amount of fluorophores situated in the vesicle, and vesicle radius values can be extracted in a straightforward manner. We will here describe in detail how to prepare and characterize the size distributions of extruded vesicle samples from single object measurements recorded with epi-fluorescence microscopy.

2. How Particle Size can be Obtained from Intensity Distributions

In epi-fluorescence microscopy, objects smaller than a few micrometers will be illuminated homogeneously throughout their entire volume. Fluorescence emission is also detected with the same efficiency within a few micrometers away from the focus plane. Thus, there exists a direct proportionality between the detected total intensity from an object and the amount of fluorophores within that object. In the case of SUVs carrying a fluorescent lipid analog in the bilayer, the intensity is directly proportional to the membrane area. Thus, *a priori* knowledge of the physical size of a vesicle obtained from another technique (DLS, electron microscopy) can be used to create a conversion table between intensities and vesicle radii (Fig. 8.1D and E). In this work we use a vesicle population, extruded several times through a filter with pore diameters of 100 nm, to create a calibration table. Below, we will describe vesicle preparation procedures, then surface-immobilization, imaging, and finally, data-processing. Finally, we describe an application of constructing vesicle size distributions by quantifying the binding of the curvature-sensing protein, Sorting Nexin-1 (SNX1).

3. Vesicle Preparation and Immobilization

3.1. Preparation of liposomes for calibration

Vesicles are prepared using the method of thin lipid film hydration (Olson *et al.*, 1979). Lipids are dissolved and stored in chloroform to assure a homogeneous mixture of lipid molecules. The different lipids are then mixed in the desired ratio in a 1.5-ml glass flask. The lipid mixture used for all preparations is mainly composed of the zwitterionic lipid 1,2-dioleoyl-*sn*-glycero-3-phosphocholine (DOPC). Additionally, the mixture contained charged lipids and poly(ethylene-glycol)$_{45}$-modified lipids in order to prevent vesicles from fusion and deformation during the process of immobilization. After the lipids have been thoroughly mixed in chloroform, the solvent is removed by evaporation under a nitrogen flow in order to create a thin-lipid film. Afterward, the flask is placed in vacuum overnight to remove residual organic solvent. Rehydration of the dry lipid film is accomplished by adding an aqueous medium (e.g., buffer solution) to the container of dry lipid and then shaking the mixture vigorously in a vortex mixer until the lipid film is completely removed from the glass walls. Extrusion can be either performed using a hand-held syringe that is connected to a filter holder or by employing pressurized gas to the sample. Because of the limited back pressure that the syringe and filter device can tolerate, we decided to work with a pressurized device to extrude the liposomes. The sample is extruded 10 times through a polycarbonate membrane with the designated pore diameter (e.g., 100 nm) at a pressure of 200 psi. This sample will be referred to as SUV-100.

3.2. Immobilization of liposomes

An essential step in working with liposomes on solid surfaces is to adsorb the vesicles from solution. To self-assemble vesicles on a glass surface, we chose the well-studied noncovalent interaction between biotin and streptavidin. The biotin–streptavidin interaction is characterized by a strong affinity constant ($K_d \approx 10^{-15}$ M) (Diamandis and Christopoulos, 1991), fast kinetics, and irreversibility with regard to experimental time scales. The exceptional stability of the biotin–streptavidin interaction allows the complex to survive harsh reaction conditions and consequently protects the system from external perturbations, such as changes in pH or rinsing procedures following the initial immobilization. The interaction is also specific enough to direct the binding only to the target of interest. In our case streptavidin serves as a linker between vesicles, functionalized with biotinylated lipid-analogs, and a surface-passivating layer of biotinylated bovine serum albumin (BSA).

BSA, though negatively charged, has positively charged residues distributed over the molecule, which enable electrostatic interactions with the negatively charged glass surface (Bajpai, 2000). We wash and store glass cover slips in methanol, before drying under nitrogen flow. To homogeneously functionalize the surface with protein, a solution of 1 g/l BSA-biotin/BSA (1:10) is added to the glass surface and incubated for 10 min at ambient temperature. Afterward, the surface is gently flushed with phosphate buffered saline (100 m*M* PBS, 100 m*M* NaCl, pH 7.4) 10 times to remove unbound proteins. Then, 0.025 g/l streptavidin solution is added to the chamber and incubated for 10 min at ambient temperature as well. After gently flushing the surface with buffer solution 10 times, the glass surface is primed for capturing biotinylated vesicles.

The functionalized surfaces are incubated with dilute bulk concentrations of vesicles. We suggest to directly add 1–2 μl of lipid solution containing approximately 1 μmol of lipid to a glass surface covered with 200 μl buffer solution, then stirring gently and incubating the sample for 5 min, to allow vesicles to bind onto the surface. Afterward, unbound vesicles can be removed by gently washing the sample 20 times with buffer solution.

4. Image Acquisition

Micrographs are recorded using a commercial wide-field epi-fluorescence microscope platform (Leica DMI6000B) equipped with a Peltier-cooled CCD camera (Roper Cascade II) and an oil-immersion objective (HCX PL APO 100×, NA 1.46). The images are 512 × 512 pixels in size, corresponding to a field of view of 81.92 μm. The bit-depth of all images is set to 16.

To obtain reliable data it is necessary to locate all particles; this can be a challenging task due to the broad range of intensity values present. Imaging of highly polydisperse populations may mean that the smallest and largest particles cannot be imaged simultaneously as the former may underexpose and the latter overexpose. However, the high dynamic range of 16-bit cameras helps avoid this problem. The optimal settings chosen for image acquisition should allow larger particles to reach maximum intensity just below saturation while exciting the micrograph background to an intensity value just above zero. Constructing an intensity line-profile through the point spread function (PSF) of the brightest particle is a convenient way to evaluate imaging settings (Fig. 8.2). The PSF, resembling a Gaussian function, should attain values just above zero far away from the center, whereas the peak value should be just below the maximum. In this way the entire dynamic range is exploited for imaging. It is worth noting that to ensure comparability between samples, identical microscope settings must be applied to all micrographs.

If the intensity signals from all vesicles within a certain population exceed the detector range, then an alternative procedure can be applied. Here, two different imaging conditions are defined by the microscope acquisition parameters; one setting with low gain enabling good imaging for the brightest particles and another setting with high gain for acquiring sufficiently exposed micrographs of small particles. In this way, each image

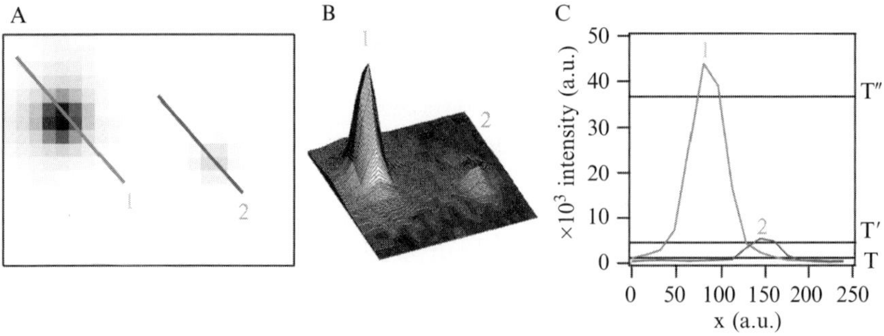

Figure 8.2 How to threshold the fluorescence image of single vesicles. (A) Two neighboring vesicles from a fluorescent micrograph exhibit different fluorescent intensities, big particle (1, red line) and small particle (2, blue line). (B) A three-dimensional view shows a strong intensity signal for particle 1 and a fluorescence signal that is similar in intensity to the background noise for particle 2. (C) The threshold value regulates the detection of particles. Fluorescence signals below the threshold value will not be detected. Different threshold values (T, T', and T'') might be applied to the image: T'' only allows the detection of particles with a strong fluorescence signal and is therefore not suitable for application. T' includes the detection of the peaks of small particles, but still excludes particles with a weaker signal that are still strong enough to be distinguished from the background. Threshold T allows the cut-off of the background noise without losing particles expressing a weak fluorescence, and represents a suitable value for application to the images. (See Color Insert.)

has to be obtained twice; once with the settings for large vesicles and once with the settings for small vesicles. The intensity signals of both images can then be mathematically related to each other. In the data presented here, it was possible to apply the same acquisition settings to all samples.

Before proceeding further we find it worthwhile to briefly discuss another vesicle labeling scheme. Here, a water-soluble fluorophore is encapsulated in the vesicle lumen; thus, the total intensity will scale as volume. We do not recommend this labeling scheme for calculating vesicle sizes because, as we recently published, the concentration of encapsulated lumen dye is highly heterogeneous (Lohse *et al.*, 2008). Vesicles with similar diameter may show variable encapsulation behavior and most importantly, a large number of vesicles may be empty. Moreover, this method requires a wider signal range. In particular, a 10-fold increase in particle diameter, for instance from 20 to 200 nm, would lead to a 1000-fold increase in the lumen signal ($r = 10$). In this case, overexposure of larger particles is unavoidable using epi-fluorescence microscopy due to the higher amount of encapsulated fluorophores. For instance, while overexposure might be observed for particles larger than 400 nm in diameter in the case of membrane labeling, lumen staining might lead to overexposure from particles larger than 200 nm and thus is limiting the range of data that can be collected.

5. Image Processing

Images might be processed manually, for example by defining and analyzing a region of interest (ROI) around a single particle or via a line scan. However, treating a series of images with approximately 300 vesicles per image would quickly transform this procedure into a very time-consuming process. Therefore, automated processing is applied to all images. The advent of digital image acquisition, as well as more powerful computer processors, has enabled fast, reproducible, and more detailed analyses of fluorescence micrographs. Specifying a few general parameters (threshold value, minimum pixel area, and circularity) can guide particle-recognition software to extract intensities, among many other properties, from several thousands of particles, thus obtaining statistically significant amounts of data. A popular, publicly available image-processing tool for scientific applications is ImageJ (Abramoff *et al.*, 2004; Collins, 2007), which we have used to quantify our presented data.

5.1. Thresholding particles

ImageJ is a freely available tool to fulfill routine image processing. The software can treat 8-bit, 16-bit, and 32-bit images. ImageJ offers a tool to separate the features of interest from the background by applying a threshold to a micrograph. The software will then analyze the brightness values of all pixels within an interval specified by a lower and upper intensity threshold (Fig. 8.2C). We suggest keeping the upper threshold at the highest possible value, thus including all particles that reached the maximum intensity signal during image acquisition.

The aim of image processing is to distinguish and locate single particles from an overall background noise. Consequently, it is of importance to choose a suitable threshold value that does not exclude any particles (Fig. 8.3). Typical difficulties that may arise during this procedure are (i) the introduction of a greater number of particles than actually exist due to a threshold value chosen too low, thus considering part of the background noise as particles (Fig. 8.3C) and (ii) the loss of small particles due to a threshold value chosen too high, thus removing background noise, but also dim particles from subsequent analysis (Fig. 8.3D). Ideally, the threshold value should be adjusted to a value slightly greater than the background intensity (Fig. 8.3E). Once established, this threshold value must be applied to all following image analyses to ensure a faithful comparison of particle intensities.

Apart from threshold, it is also possible to define a minimum pixel area (A_M), above which a collection of adjacent pixels are considered as a particle. This tool is particularly useful for the analysis of the smallest

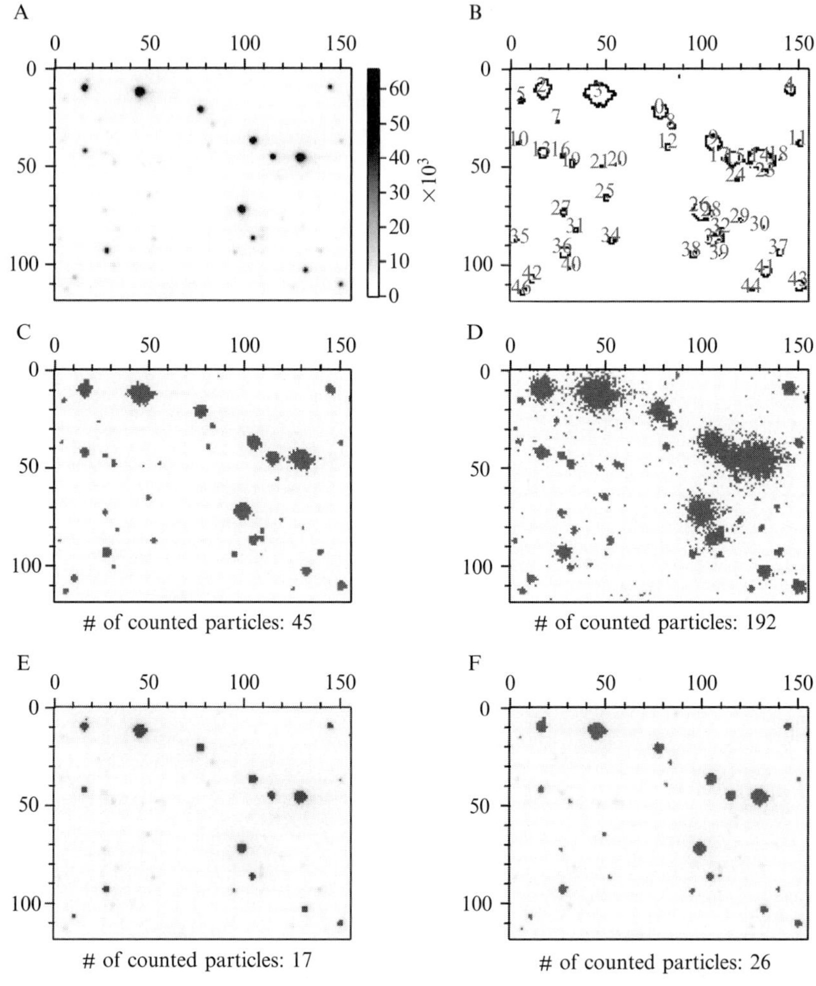

Figure 8.3 Application of different threshold values using ImageJ. (A) Original 16-bit image of immobilized vesicles. (B) Detected particles are outlined and numbered by ImageJ. The number of detected particles strongly depends on the threshold value. (C) The right threshold value allows the detection of both big and small particles and clearly separates all particles from the background. (D) When the threshold is chosen too low, part of the background noise will be considered as particles, and thus a huge error is introduced to the data. In contrast, defining a threshold that is much (E) or just a little (F) too high in relation to the background noise leads to the loss of small particles. Scale on top and left in pixels (1 pixel = 160 nm).

particles to avoid introducing artifacts, when specifying a threshold value close to the background noise. If the threshold is set too low, then random noise will appear as particles composed of single pixels. To avoid this, the A_M-value should always be set higher than 1, but in most cases visual inspection of the processed image is the most efficient method to establish suitable A_M-values.

Objects with an elongated appearance, such as aggregated particles, can be excluded from data acquisition by introducing a circularity factor, which is given as $4\pi A/C^2$, where A is area and C is circumference. For a perfect circle the circularity is 1, whereas deviations from a circle lead to lower values. For the data presented in this work, only particles with circularity values above 0.8 were accepted.

5.2. Transformation of intensity data into radius values

As previously mentioned, the fluorescence signal from the membrane (I_M) is directly proportional to the vesicle surface area (A_S): the larger a particle, the stronger the fluorescence signal will be

$$A_S \propto I_M \tag{8.1}$$

Assuming a vesicle to be a spherical shell with radius r, the surface area is

$$A_S = 4\pi r^2 \tag{8.2}$$

Combining Eqs. (8.1) and (8.2) we get an equation to calculate the radius of a single vesicle:

$$r = k\sqrt{I_M} \tag{8.3}$$

where k is a correction factor relating the intensity signal from a single vesicle to the corresponding radius in units of nm. The k-factor depends, among other things, on the density of fluorophores in the vesicle, the fluorophores' photophysical properties, and the signal detection/amplification efficiency, but can be easily obtained by experiment. To do this a collection of vesicles, from a population with a well-defined mean radius (r_{mean}), must be processed to obtain the mean value of the square root of the intensity ($(\sqrt{I_M})_{mean}$). Then k is calculated as

$$k = \frac{r_{mean}}{(\sqrt{I_M})_{mean}} \tag{8.4}$$

It is important to obtain the mean value of the distribution $(\sqrt{I_M})_{mean}$ after taking the square root of the intensity of each particle. Calculating a mean value of the intensity distribution and then taking the square root will lead to a wrong result. A comparison of the data presented in Fig. 8.4 clarifies that $\sqrt{(I_M)_{mean}} = \sqrt{299744} = 547.5 \neq (\sqrt{I_M})_{mean} = 390.1$.

Modifying Eq. (8.3) yields the final expression for the radius of a vesicle given its measured intensity:

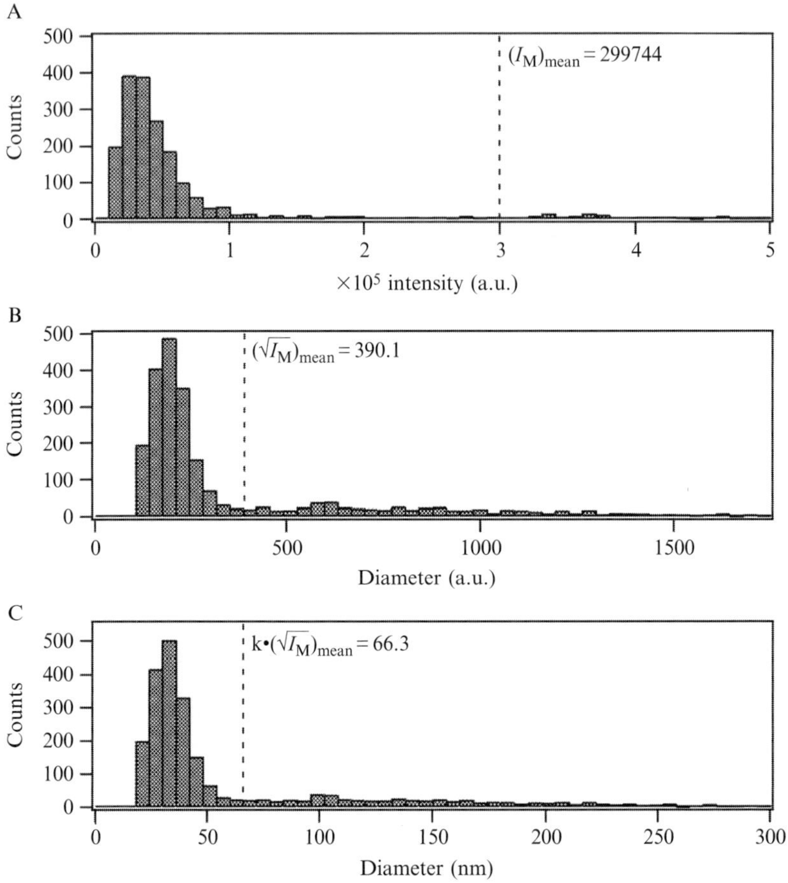

Figure 8.4 How to construct the size distribution of vesicles from the distribution of intensities. Intensity data acquired with ImageJ are plotted in histogram (A): the dotted line is the average intensity. Intensity data are transformed into an arbitrary diameter by calculating the square root of the intensity of every single vesicle (B). The arbitrary diameter is corrected to a dimension in nanometers by multiplication with a correction factor *k*, and thus leads to the final size distribution (C).

$$r = \frac{r_{mean}}{\left(\sqrt{I_M}\right)_{mean}} \cdot \sqrt{I_M} \tag{8.5}$$

To determine the r_{mean} of a calibration sample, several methods can be useful. DLS is applied to analyze the SUV-100 sample (see Section 3), but other techniques, such as Cryo-TEM, can also be used. For calibration samples, we recommend the use of vesicles extruded through pores of diameter 100 nm or less, since vesicle populations display a more narrowly size distribution (thus approaching idealized monodispersity) with decreasing pore size of the extrusion filter membrane. Sonication of a lipid samples for extended periods of time also leads to narrowly distributed vesicle populations in the range of 40–50 nm in diameter, also constituting a suitable calibration.

To perform DLS measurements on sample SUV-100, an ALV-5000 Correlator equipped with a 633-nm laserline is used. Measurements are made at 90° and the scattered intensity is recorded for 60 s. The resulting autocorrelation function is fitted with the regularized $2g(t)$ fitting routine implemented in the instrument software to yield an average diameter of 65 nm. For more detailed descriptions of DLS on liposome populations, we refer to Ruf *et al.* (1989).

6. Intensity and Size Distributions

Next, the intensity histogram of the SUV-100 population is constructed, as shown in Fig. 8.4A. The histogram is composed of data from more than 2500 individual vesicles and the mean intensity is marked with a dashed line on the graph. Because the *k*-value can only be calculated from the square root of the vesicle intensities, the intensity table is transformed into the histogram as shown in Fig. 8.4B. Marked on the graph with a dashed line is $(\sqrt{I_M})_{mean}$. Thus, with the value of r_{mean} supplied from the DLS measurements described above, the correction factors to be $k = 0.17$ and the vesicle size distribution is calculated in nanometers. To summarize, the described procedure should be sequentially executed as shown in the following list:

1. Image processing is performed on all acquired micrographs to obtain a table of vesicle intensities (I_M) (Fig. 8.4A).
2. The intensity table is transformed to the square root of intensity $(\sqrt{I_M})$ (Fig. 8.4B).
3. The mean value $(\sqrt{I_M})_{mean}$ is obtained from the table that was created during in step (2).
4. *k* is calculated from the average radius value as obtained from DLS, Cryo-TEM, etc., using Eq. (8.4).
5. The table created in step number (2) is multiplied with *k* to obtain radii-values (Fig. 8.4C).
6. The size distribution histogram is constructed.

7. Multilamellarity Assay

The lamellarity plays an important role in gaining a reliable value for the vesicle radius. A multilamellar vesicle (MLV), encapsulating one or several other vesicles, will give rise to a much stronger fluorescent signal due to the addition of fluorescent signal from the encapsulated membrane leaflets. In an intensity-based measurement, it will therefore be interpreted as a single vesicle of larger diameter. For this reason, a method to determine whether multilamellar vesicles are present in a sample, and to what extent they disturb the signal is essential. We will describe a strategy, adopted from Meers *et al.* (2000), to directly measure the amount of lipids situated in multilamellar structures.

Reductive quenching of all fluorophores in the bilayer leaflet facing the bulk medium provides an estimate of the sample multilamellarity, because the fluorescence intensity decrease will be proportional to the external membrane surface area. In order to truly reflect external surface only, several criteria have to be fulfilled: (i) the distribution of labeled lipids over all lipid layers is random, (ii) the degree of redistribution (flip-flop) of labeled lipids during the experiment is negligible, and (iii) the reagent is not permeating into the membrane during the analysis (Gruber and Schindler, 1994). The method applied in this study is based on reduction of the lipophilic fluorophore 1,1′-dioctadecyl-3,3,3′,3′-tetramethylindodicarbocyanine-5,5′-disulfonic acid (DiIC_{18}(5)-DS), by sodium dithionite, which is a membrane impermeable ion. Fluorescence changes from bulk vesicle samples is measured on a Horiba Jobin Yvon FluoroMax-4 spectrofluorometer by monitoring the temporal evolution of signal from the emission maximum (670 nm), while exciting at the peak excitation wavelength (650 nm).

Assuming the membrane dye is distributed equally in the vesicle bilayer, 50% of the dye should be located in the outer leaflet and the remaining 50% would then be located within the inner membrane leaflet. Such symmetrically labeled vesicles containing 1.0 mol% DiIC_{18}(5)-DS are prepared by adding the fluorescently labeled lipid to the lipid mixture before evaporating the chloroform from the solution. In this way, the fluorescently labeled lipid distributes between inner and outer membrane leaflets according to the distribution of total lipid. Therefore, in a sample containing only unilamellar vesicles, the expected fluorescence intensity decrease due to fluorophore quenching is 50% of the original value. In contrast, the reduction of fluorescence intensity would not be as strong for multilamellar vesicles, as more than 50% of the fluorophores will be located inside the vesicles, either as part of the inner leaflet of the outermost bilayer, or as part of the inner and outer leaflet of enclosed bilayers.

Figure 8.5A shows an example where fluorescence from liposomes containing DiIC_{18}(5)-DS is reduced by approximately 40% of the initial intensity. This means that 80% of total lipid is actually located in the outermost bilayer and is quenched by 50% ($80\% \times 0.5 = 40\%$). The remaining 20% are inner bilayers. Following this reduction, fluorescence is

A

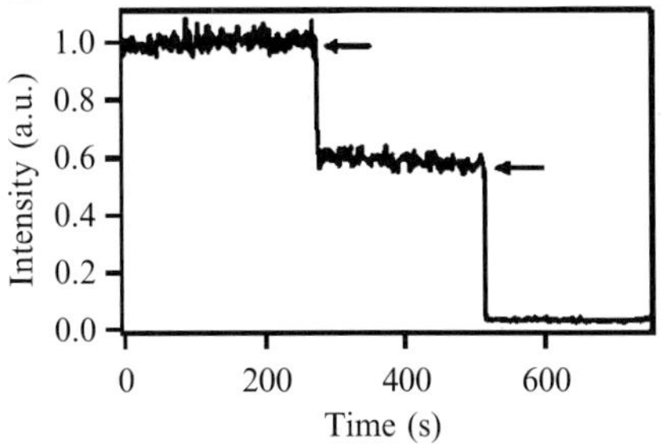

B

	Multilamellar vesicles (%)
Reference	51
100 nm	5
200 nm	10

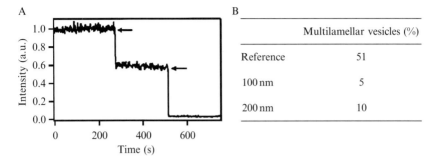

Figure 8.5 Assessing the multilamellarity of a vesicle suspension using reduction of DiIC$_{18}$(5)-DS. (A) Fluorescence is reduced after the addition of sodium dithionite at the first arrow and completely erased after treating the sample with the detergent OGP (second arrow). (B) The reduction in fluorescence intensity can be used to determine an upper limit of the amount of multilamellar vesicles within a sample. Here, samples extruded five times at 100 and 200 nm, respectively, are compared to an unextruded sample (here called reference).

completely eliminated after the addition of the detergent *n*-octyl β-D-glucopyranoside (OGP) due to exposure of the inner compartments of the vesicles to the quencher. The amount of fluorescently labeled lipid that is quenched in liposomes is calculated as follows, where RLOS is the relative loss of signal (McIntyre and Sleight, 1991):

$$\mathrm{RLOS} = \left[1 - \frac{F_\mathrm{r} - F_\mathrm{ap}}{F_0 - F_\mathrm{ap}}\right] \times 100\% \tag{8.6}$$

F_r is the fluorescence intensity after completed reduction, F_ap defines the apparent fluorescence of vesicles not containing any fluorophores, and F_0 is the fluorescence of liposomes before reaction with dithionite.

Multilamellar vesicles can exist in many different forms. Whereas MLVs are envisioned as neatly stacked, onion-like structures by most investigators, many liposome systems also posses irregularly spaced bilayers and multivesicular structures, also referred to as "liposome within liposome" structures (Perkins *et al.*, 1993). Without applying a method to visualize multilamellar structures within vesicles, we are not able to draw conclusions about the exact inner vesicular distribution of lipid mass. There are several ways to minimize the amount of multilamellar structures within a vesicle population. These include treatment with extrusion, cycles of freeze/thawing and/or sonication. All of these methods are based on the repeated fragmentation and reforming of bilayers, thus minimizing the probability of assembling into multilamellar vesicles.

8. Measuring Membrane-Curvature Selective Protein Binding

An interesting application of the technique to accurately measure the size of SUVs is to monitor the preferential binding of membrane-curvature (MC) sensitive proteins (Hatzakis *et al.*, 2009; Bhatia *et al.*, 2009). The BAR domain of SNX1 has been reported previously to show preferential binding on highly curved membranes, that is liposomes of small diameters (Carlton *et al.*, 2004). By labeling SNX1 with a second fluorophore-labeled protein, it is possible to excite it at a second wavelength and detect the total amount bound in each vesicle. Indeed, such a simple measurement (Fig. 8.6) reveals the size/curvature-selective binding properties of SNX1. The measurement of binding on single vesicles avoids ensemble averaging present in bulk ensemble assays (Carlton *et al.*, 2004) and demonstrates clearly the high specificity of SNX1. Note that each single point on the MC-sensing graph represents a *single* membrane curvature, thus allowing in a high-throughput fashion the construction of a MC-sensing graph (Fig. 8.6).

9. Concluding Remarks

We have presented a simple technique to record accurately the size of single fluorescence-labeled liposomes. The method offers a straightforward approach for the determination of size distributions, is easy to prepare and

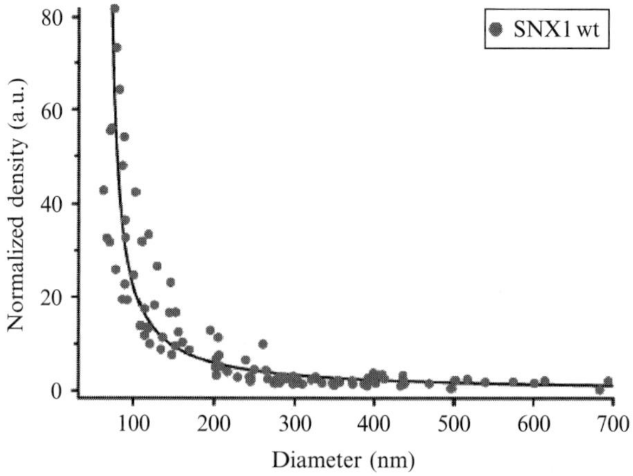

Figure 8.6 Preferential binding of SNX1 on vesicles of smaller diameter and therefore higher membrane curvature. Protein density (Ip/Iv) is plotted versus vesicle diameter. Each point is a single vesicle, the fit is for assisting in visualization.

consumes 10^{-18}–10^{-21} M of material. As the observation and analysis of each single vesicle can be regarded a single experiment, information about vesicle properties can be collected that might otherwise not be exposed by ensemble experiments. The most time-consuming part is image acquisition and processing and it has to be emphasized that a careful and critical approach is of high importance in order to define parameters such as the threshold value correctly.

The method, relying on fluorescence intensity-extraction and -conversion, can be improved even further by fitting Gaussian functions to the sampled particle intensity distributions. This, however, requires more advanced data processing and is beyond the scope of this chapter. We believe this method will prove useful for dissecting the mechanism of self-assembly at the nanoscopic scale, and furthermore, we envision that application of this technique can assay different size-dependent properties of vesicles in a high-throughput manner.

REFERENCES

Abramoff, M. D., Magelhaes, P. J., and Ram, S. J. (2004). Image processing with Image. *J. Biophotonics Intl.* **11,** 36–42.

Bajpai, A. K. (2000). Adsorption of bovine serum albumin onto glass powder surfaces coated with polyvinyl alcohol. *J. Appl. Polym. Sci.* **78,** 933–940.

Bendix, P. M., Pedersen, M. S., and Stamou, D. (2009). Quantification of nano-scale intermembrane contact areas using fluorescence resonance energy transfer. *Proc. Natl. Acad. Sci. USA.* **106**(30), 12341–12346.

Bhatia, V. K., Madsen, K. L., Bolinger, P. Y., Hedegård, P., Gether, U. and Stamou, D. (2009). Amphipathic motifs in BAR domains are essential for membrane curvature sensing. *EMBO J.* (in press).

Carlton, J., Bujny, M., Peter, B. J., Oorschot, V. M., Rutherford, A., Mellor, H., Klumperman, J., McMahon, H. T., and Cullen, P. J. (2004). Sorting Nexin-1 mediates tubular endosome-to-Tgn transport through coincidence sensing of high-curvature membranes and 3-phosphoinositides. *Curr. Biol.* **14,** 1791–1800.

Chan, Y. H., Lenz, P., and Boxer, S. G. (2007). Kinetics of DNA-mediated docking reactions between vesicles tethered to supported lipid bilayers. *Proc. Natl. Acad. Sci. USA* **104,** 18913–18918.

Christensen, S. M., and Stamou, D. (2007). Surface-based lipid vesicle reactor systems: Fabrication and applications. *Soft Matter* **3,** 828–836.

Coldren, B., van Zanten, R., Mackel, M. J., Zasadzinski, J. A., and Jung, H. T. (2003). From vesicle size distributions to bilayer elasticity via cryo-transmission and freeze-fracture electron microscopy. *Langmuir* **19,** 5632–5639.

Collins, T. J. (2007). ImageJ for microscopy. *Biotechniques* **43,** 25–30.

Cooper, M. (2004). Advances in membrane receptor screening and analysis. *J. Mol. Recognit.* **17,** 286–315.

Diamandis, E. P., and Christopoulos, T. K. (1991). The biotin-(Strept)Avidin system: Principles and applications in biotechnology. *Clin. Chem.* **37,** 625–636.

Gruber, H. J., and Schindler, H. (1994). External surface and lamellarity of lipid vesicles: A practice-oriented set of assay methods. *Biochim. Biophys. Acta* **1189,** 212–224.

Gunnarsson, A., Jonsson, P., Marie, R., Tegenfeldt, J. O., and Hook, F. (2008). Single-molecule detection and mismatch discrimination of unlabeled DNA targets. *Nano Lett.* **8,** 183–188.

Hatzakis, N. S., Bhatia, V. K., Larsen, J., Madsen, K. L., Bolinger, P. Y., Kunding, A. H., Castillo, J., Gether, U., Hedegård, P. and Stamou, D. (2009). How Curved Membranes Recognize Amphipathic Helices and Protein Anchoring Motifs. *Nat. Chem. Biol.*, advanced online publication doi:10.1038/nchembio.213.

Israelachvili, J. N., Mitchell, D. J., and Ninham, B. W. (1976). Theory of self-assembly of hydrocarbon amphiphiles into micelles and bilayers. *J. Chem. Soc., Faraday Trans. II* **72,** 1525–1568.

Israelachvili, J. N., Mitchell, D. J., and Ninham, B. W. (1977). Theory of self-assembly of lipid bilayers and vesicles. *Biochim. Biophys. Acta* **470,** 185–201.

Kunding, A. H., Mortensen, M. W., Christensen, S. M., and Stamou, D. (2008). A fluorescence-based technique to construct size distributions from single-object measurements: Application to the extrusion of lipid vesicles. *Biophys. J.* **95,** 1176–1188.

Kuyper, C. L., Kuo, J. S., Mutch, S. A., and Chiu, D. T. (2006). Proton permeation into single vesicles occurs via a sequential two-step mechanism and is heterogeneous. *J. Am. Chem. Soc.* **128,** 3233–3240.

Lohse, B., Bolinger, P. Y., and Stamou, D. (2008). Encapsulation efficiency measured on single small unilamellar vesicles. *J. Am. Chem. Soc.* **130,** 14372–14373.

McIntyre, J. C., and Sleight, R. G. (1991). Fluorescence assay for phospholipid membrane asymmetry. *Biochemistry* **30,** 11819–11827.

Meers, P., Ali, S., Erukulla, R., and Janoff, A. S. (2000). Novel inner monolayer fusion assays reveal differential monolayer mixing associated with cation-dependent membrane fusion. *Biochim. Biophys. Acta* **1467,** 227–243.

Olson, F., Hunt, C. A., Szoka, F. C., Vail, W. J., and Papahadjopoulos, D. (1979). Preparation of liposomes of defined size distribution by extrusion through polycarbonate membranes. *Biochim. Biophys. Acta* **557,** 9–23.

Palmieri, F., Indiveri, C., Bisaccia, F., and Kramer, R. (1993). Functional properties of purified and reconstituted mitochondrial metabolite carriers. *J. Bioenerg. Biomembr.* **25,** 525–535.

Perkins, W. R., Minchey, S. R., Ahl, P. L., and Janoff, A. S. (1993). The determination of liposome captured volume. *Chem. Phys. Lipids* **64,** 197–217.

Ruf, H., Georgalis, Y., and Grell, E. (1989). Dynamic laser light scattering to determine size distributions of vesicles. *Methods Enzymol.* **172,** 364–390.

Sessa, G., and Weissman, G. (1968). Phospholipid spherules (Liposomes) as a model for biological membranes. *J. Lipid Res.* **9,** 310–318.

Stamou, D., Duschl, C., Delamarche, E., and Vogel, H. (2003). Self-assembled microarrays of attoliter molecular vessels. *Angew. Chem. Int. Ed. Engl.* **42,** 5580–5583.

Szoka, F. Jr., and Papahadjopoulos, D. (1980). Comparative properties and methods of preparation of lipid vesicles (liposomes). *Annu. Rev. Biophys. Bioeng.* **9,** 467–508.

Torchilin, V. P. (2005). Recent advances with liposomes as pharmaceutical carriers. *Nat. Rev. Drug. Discov.* **4,** 145–160.

Venturoli, M., Sperotto, M., Kranenburg, M., and Smit, B. (2006). Mesoscopic models of biological membranes. *Phys. Rep.* **437,** 1–54.

Yoon, T. Y., Okumus, B., Zhang, F., Shin, Y. K., and Ha, T. (2006). Multiple intermediates in snare-induced membrane fusion. *Proc. Natl. Acad. Sci. USA* **103,** 19731–19736.

Yoshina-Ishii, C., and Boxer, S. G. (2006). Controlling two-dimensional tethered vesicle motion using an electric field: Interplay of electrophoresis and electro-osmosis. *Langmuir* **22,** 2384–2391.

CHAPTER NINE

Giant Unilamellar Vesicle Electroformation: From Lipid Mixtures to Native Membranes Under Physiological Conditions

Philippe Méléard,* Luis A. Bagatolli,† *and* Tanja Pott*

Contents

Abstract

Giant unilamellar vesicles (GUVs) are well-known model systems, especially because they are easily observable using optical microscopy. In this chapter, we revisit in detail the versatile GUV electroformation protocol. We demonstrate how GUV electroformation can be adapted to various membrane systems including synthetic lipid mixtures, natural lipid extracts, and bilayers containing membrane proteins. Further, we show how to adjust this protocol to a given aqueous environment and prove that GUVs can be obtained under physiologically relevant conditions, that is, in the presence of electrolytes. Finally, we provide firm evidence that electroformation is a method of choice to produce giant vesicles from native cell membranes. This is illustrated with the example of GUV electroformation from red blood cell ghosts in a physiologically

* CNRS UMR 6226, ENSCR, Avenue du Général Leclerc, Cedex, France
† Membrane Biophysics and Biophotonics group/MEMPHYS Center of Biomembrane Physics, Department of Biochemistry and Molecular Biology, University of Southern Denmark, Odense M, Denmark

Methods in Enzymology, Volume 465
ISSN 0076-6879, DOI: 10.1016/S0076-6879(09)65009-6

pertinent buffer. GUVs obtained in this manner maintain the native membrane asymmetry, thereby validating the physiological relevance of GUV electroformation.

1. Introduction

Giant unilamellar vesicles (GUVs), a popular membrane model, were described first about 40 years ago (Reeves and Dowben, 1969). Contrariwise to other unilamellar vesicles (small unilamellar vesicles, SUV; large unilamellar vesicles, LUV), GUVs are cell-sized or larger, that is, large enough to be observable by optical microscopy. The GUV membranes can be easily observed using either natural contrast enhancement techniques (Zernicke phase contrast, differential interference contrast or Hoffman modulation contrast) or fluorescence microscopy (confocal/two-photon excitation). In the latter case, specifically designed fluorescent probes have to be inserted into the phospholipid bilayers.

Giant vesicles became popular when it became apparent that the approaches of physics could be applied to the study of membrane mechanical properties (Bivas *et al.*, 1987; Engelhardt *et al.*, 1985; Faucon *et al.*, 1989; Schneider *et al.*, 1984; Servuss *et al.*, 1976), inspired by the interpretation of the erythrocyte flicker phenomenon by Brochard and Lennon (1975). In those days, the lipid composition of GUVs was restricted to either pure synthetic phospholipids or phosphatidylcholine extracts from egg yolk (Evans and Needham, 1986; Needham *et al.*, 1988) and the most popular preparation method was the swelling method (Faucon *et al.*, 1989), that is, an adaptation of the initial Reeves and Dowben protocol (1969). The swelling method is an "easy to do" procedure that can, in principle, be adapted to any environment, including physiologically relevant conditions. To achieve GUVs in the presence of physiological concentrations of salts, high proportions of negatively charged lipids ($\sim$15 mol %) in the lipid mixture is necessary (Akashi *et al.*, 1996), impeding free choice of the desired lipid composition. Additionally, using this swelling method, a few days may be necessary to obtain GUVs. This slowness precludes intrinsically its application to lipid mixtures that show rapid degradation, for example, unsaturated lipid species (Angelova *et al.*, 1992).

GUV electroformation, an alternative method to spontaneous swelling initially proposed in 1986 (Angelova and Dimitrov, 1986), gives the opportunity to produce GUVs rapidly (typically, in 1–3 h; Angelova *et al.*, 1992), thereby enabling work with fragile membrane compositions. GUVs obtained by this method are immobilized on the electrodes, but can be detached to observe them freely moving in the aqueous environment. In this case, GUVs have a fixed area and a volume controlled mainly by water permeability. Their center-of-mass movement is Brownian and the analysis of their membrane fluctuations allows measurement of the bending elasticity (Faucon *et al.*, 1989). Freely moving GUVs can be trapped further by a micropipette to determine their membrane stretching elasticity (Evans and Needham, 1986; Evans and Rawicz, 1990). Additionally, immobilization on the electrodes may be required for some techniques, such as in confocal microscopy (Fidorra *et al.*, 2006) or in studies related to membrane budding (Roux *et al.*, 2005).

The electroformation method was also applied to generate GUVs containing cholesterol molar fractions similar to that of mammalian cells (Méléard *et al.*, 1997), and original studies involving direct visualization of membrane domains using fluorescent dyes appeared in the literature (Bagatolli, 2006; Bagatolli and Gratton, 1999, 2000; Korlach *et al.*, 1999). Since then, GUVs with more and more complex membrane compositions were made and binary or ternary lipid mixtures, as well as natural lipid extracts were studied (Bagatolli, 2006; Bernardino de la Serna *et al.*, 2004; Dietrich *et al.*, 2001; Méléard *et al.*, 1997; Montes *et al.*, 2007; Plasencia *et al.*, 2007; Veatch and Keller, 2005). Prior to the electroformation process, however, the lipid has to be left on the electrodes. These deposits were generally made from phospholipid solutions in organic solvents (Dimova *et al.*, 2006; Evans and Needham, 1986; Méléard *et al.*, 1997). Although GUVs containing bacteriorhodopsin, that is, a special membrane protein, were prepared in this way (Manneville *et al.*, 1999), this method prohibits the incorporation of most membrane proteins. Yet, it is well known that small-scale vesicles (LUV, SUVs, etc.) can be made with highly complex compositions including fragile membrane proteins. Our group therefore attempted proteo-GUV electroformation from deposits made from proteoliposomes (Valverde, 2000). The method being successful and leading to high GUV formation rates, it was applied by Girard *et al.* (2004) following our suggestion, and by others (Bacia *et al.*, 2004). Thereafter, we generalized the exploration of other vesicle or liposome preparations classically used in a variety of biochemical or biophysical studies (Pott *et al.*, 2008).

However, there were still some shortcomings concerning GUV formation. First, the general belief was that GUVs electroformation would work only for low salt concentrations, that is ≤ 10 mM NaCl (Bagatolli *et al.*, 2000; Dimova *et al.*, 2006), so that studies involving physiological electrolyte concentrations were largely prohibited. Likewise, neither the slow and gentle swelling procedure (Faucon *et al.*, 1989; Reeves and Dowben, 1969), nor the rapid electroformation method (Angelova *et al.*, 1992) gives rise to asymmetric membranes. However, GUVs with asymmetric membranes can be obtained using a recently reported protocol that makes use of an inverse emulsion (Pautot *et al.*, 2003). Briefly, the inner monolayer is obtained by forming a water-in-oil emulsion stabilized by one lipid. The outer monolayer is then applied by transfer of these droplets into the aqueous phase through the interface loaded with a second type of lipid. Albeit promising, it is clear that such a protocol precludes the incorporation of membrane proteins.

This drawback has recently been overcome by Baumgart *et al.* (2007) who generated asymmetric plasma membrane GUVs using chemically induced vesiculation or "blebbing." However, it is unclear how the polar organic solvent used in their protocol affects the overall behavior of these giant vesicles. As pointed out by the authors, individual vesicles obtained by "blebbing" show large variations in transition temperatures (Baumgart *et al.*, 2007). This indicates that giant plasma membrane vesicles obtained by organic solvent-induced vesiculation, may have varying composition.

Recently, we optimized the GUV electroformation protocol, focusing specifically on high salinity conditions (up to 250 m*M* NaCl) and showed that, despite common wisdom, it is possible to attain high GUV electroformation rates in electrolyte containing buffers at physiologically relevant concentrations (Pott *et al.*, 2008). This protocol was the starting point for Bagatolli and colleagues to be promptly successful in the GUV electroformation from native membranes using red blood cell ghosts as deposits (Montes *et al.*, 2007). The obtained giant ghosts were further shown to maintain the initial asymmetry of the plasma membrane (Montes *et al.*, 2007). This excellent illustration reveals the general character of GUV electroformation and demonstrates that a big hurdle in GUV production has been positively overcome, which should obviously widen the use of GUV for biochemical, biological, or biophysical studies.

In this chapter, we present the detailed procedures for successful GUV electroformation from quite simple lipid systems up to native membrane GUVs under physiologically relevant electrolyte concentrations. The general GUV electroformation procedure and its basic setup are presented first, followed by methods for GUV electroformation from lipid mixtures, liposomes, and proteoliposomes, as well as using different aqueous environments (low and high salt concentrations, etc.). Ultimately, we show that electroformation can be used to prepare GUVs from erythrocyte ghost membranes in physiologically relevant buffers and prove that so-obtained GUVs maintain red blood cell asymmetry.

2. General GUV Electroformation Protocol

GUV electroformation should be considered as a two-step procedure. The first step is the deposition of the lipid matrix on the electrode(s), the second step, the electroformation as such. Although the electroformation process itself is poorly understood from a theoretical point of view, it is intuitive that the first step, that is, the formation of the lipid deposit on the electrodes and its quality, is important for the subsequent GUV production and is addressed in the following sections.

For GUV electroformation, a function generator (e.g., the 3324A generator; Hewlett-Packard, Germany) or a simpler generator able to produce square or sinusoidal voltages at different frequencies (from 1 Hz to 1 kHz), is needed, as well as a homemade electroformation cell. Two inexpensive types can be worked out, depending of the working distance of the microscope objective (Méléard *et al.*, 1997). When using a high magnification objective and a working distance smaller than about 0.3 mm, the electroformation cell should be rather thin (typically 1.5 mm thick, including inner volume, cover glass, and microscope slide thicknesses) and can readily be made from indium tin oxide (ITO) coated microscope slides (Méléard *et al.*, 1997). The main disadvantage of this cell type is the stability of the ITO coating when applying electric fields in aqueous solutions for a rather long time. This problem can be circumvented using cylindrical platinum electrodes (0.5 mm $\leq \phi \leq$ 0.8 mm; Goodfellow, UK) inserted into a standard optical glass cuvette with 1 mm inner thickness (Hellma-France) (Fig. 9.1). The two platinum electrodes are held at about 3 mm axis-to-axis, using a homemade Plexiglas cap. Two more holes are also made inside the cap. The first one is close to the chamber side and is used to fill the cell with water or any other convenient buffer, while the second (e.g., drilled between the electrodes) lets the air flowing out the chamber during the filling process (Fig. 9.1). A minor drawback of this practical electroformation cell is its thickness, necessitating a microscope objective with a rather long working distance (>1.5 mm).

Once several phospholipid deposits have been made on the inner sides of platinum electrodes, the electrodes are connected to the function generator and an alternating potential difference ΔV (peak-to-peak) is applied immediately. ΔV corresponds to an electric field of 50–100 mV/mm, at a given electroformation frequency (see next sections), for example 10 Hz for a

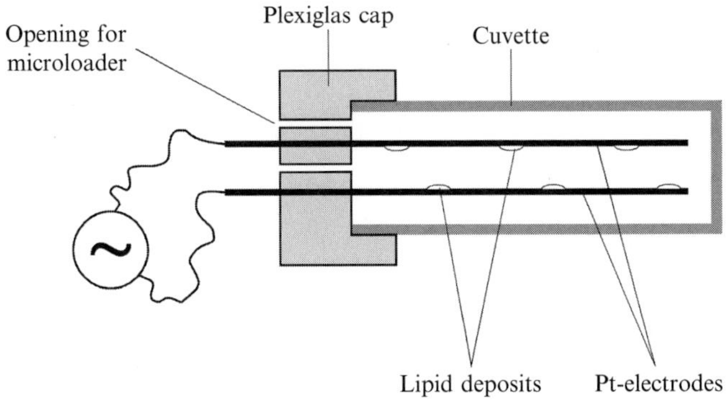

Figure 9.1 The chamber used for GUV electroformation.

buffer containing a low salt concentration. Then, the chamber can be filled with the chosen buffer. It should be noted that the filling of the electroformation cell should be carried out very carefully to keep to lipid films on the electrodes. A microloader tip (Eppendorf) mounted onto a micropipette tip, already filled with the desired aqueous solution, works well. The microloader tip is inserted into the appropriate opening (see Fig. 9.1). Filling of the cell is done slowly starting at the bottom of the chamber. To limit buffer turbulence, the microloader tip has to be maintained close to the air/buffer interface during the chamber filling. The electroformation cell is sealed and positioned on a microscope stage to have a visual control of the electroformation process. As soon as the electroformation protocol is finely tuned to any peculiar conditions, there is no further need to visually control the electroswelling. Also, it is quite easy to increase the GUV amount by connecting several electroformation cells in parallel to the same function generator, without the need to visually observe the formation process (Fidorra *et al.*, 2006).

2.1. Basic electroformation protocol

The electroformation is started as soon as the electroformation cell is filled with the desired medium to avoid any spontaneous swelling. Generally speaking, we identify three different steps in the electroformation procedure. The first step, lasting a time t, will correspond to an increase of the field amplitude E at fixed frequency f up to its maximum value E_{max}, the second one to the swelling period when electric field parameters ($E = E_{max}$ and f) are constant, and the third one to the "rebounding" period. During the first step, GUVs gradually start to form from the deposits on the electrodes. Typically within 20–30 min, grape-like GUVs increase progressively in size (Fig. 9.2). To control this step, the electric field has to be strong

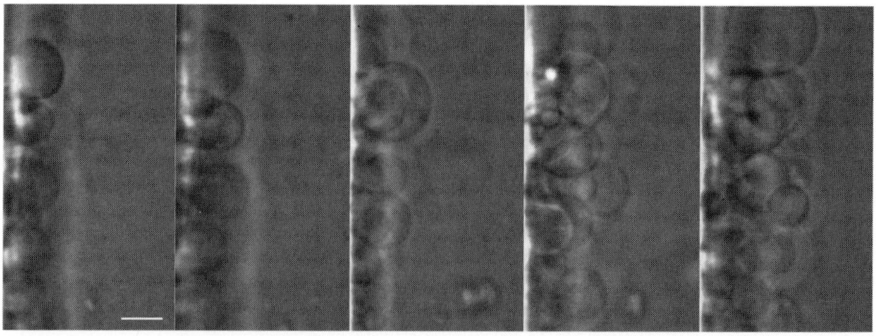

Figure 9.2 GUV production by electroformation at different time points (bar = 10 μm). Typically, 20–30 min are necessary to obtain such grape-like GUVs.

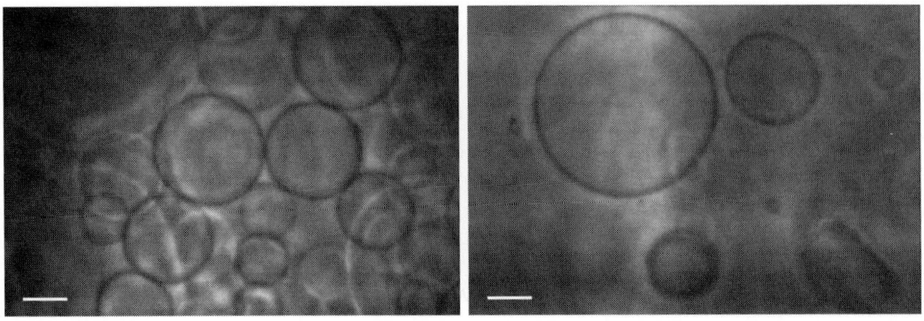

Figure 9.3 GUV grapes obtained at the end of the electroformation protocol (left) and isolated GUVs after detachment from their electrodes (right). The bar length corresponds to 10 μm.

enough to maintain a spherical-like shape for the growing vesicles. The second step, the swelling period, is optional and may be omitted at low salinity when the electric field amplitude increases slowly to its maximum value, E_{max}. Otherwise, this swelling period is essential to control the final size of the GUVs. For instance, if E_{max} is reached very fast (in about 20 min), the GUVs might be quite small (from 5 to 10 μm) and it may be beneficial to include this swelling period to adapt the vesicle size to the desired purpose (e.g., objective magnification, micromanipulation setup) (Fig. 9.3). During the "rebounding" period and final step of the electroformation, the frequency is reduced to promote vesicle closure and future detachment of the GUVs from their breeding electrodes. This period takes its name from the bouncing periodic movement of the vesicle accompanying the low frequency electric field oscillations like a rebounding ball. Here again, one can note that this last "rebounding step" might be omitted if the electroformed GUVs are spherical at the end of the second step. Finally, the GUVs can be left on the electrodes or liberated from the electrodes by gently patting the electroformation cell (Fig. 9.3).

3. Methods for GUV Electroformation from Lipid Mixtures, Liposomes, and Native Membranes

3.1. GUV formation from lipid mixture or liposomes in an aqueous medium at low electrolyte concentration

Zwitterionic phosphatidylcholines (PCs), as well as cholesterol, are readily dissolved in a chloroform/methanol mixture (9:1, v/v). For GUV electroformation, this desired lipid mixture is adjusted to concentrations of about

0.25 mg/ml. Deposits, at least 3 per electrode, can be made directly from such an organic solution using not more than 2 μl for each deposit. After evaporation of the solvent, remaining traces are removed under reduced pressure. As aqueous medium, either pure water or a low electrolyte buffer, for example, 10 m*M* Tris, 2 m*M* EDTA, and pH 7.4, can be used. Electroformation is done using the low electrolyte parameters (Table 9.1).

In the case of mixed dimyristoyl-PC (DMPC)-cholesterol GUVs obtained by electroformation, it was found that bending elasticity increases significantly when increasing the cholesterol content from 10 to 50 mol% (Méléard *et al.*, 1997). At a given cholesterol concentration, individual DMPC-cholesterol GUVs were shown to have bending elasticities with standard deviations typical of monocomponent GUVs, showing that GUVs from such mixtures are homogeneous in composition.

In the case of binary mixtures showing bilayer phase separation into gel and fluid domains, GUVs are formed from organic solvent deposits (Bagatolli and Gratton, 2000). In this experiment, one assumes that the solubility of both lipids is about the same, since the observed liquid-disordered to gel/liquid-disordered phase transition temperature obtained at the level of single vesicles is in line with that predicted from the particular lipid phase diagram. Recently, a more rigorous method was presented where the gel and the fluid domain areas were calculated from reconstructed fluorescence microscopy 3D images of GUVs composed of dilauroyl-PC/dipalmitoyl-PC. The analysis with groups of over 10 GUVs selected randomly from different preparations agreed well with the known phase diagram (Fidorra *et al.*, 2008). Indeed, these experiments show that even for GUVs composed of binary phospholipid mixtures, their phospholipid composition is identical to that of the initial organic solution.

For some type of lipid mixtures, however, it can be difficult to identify a common organic solvent, that is, a solvent in which all lipid compounds have about the same solubility. Whenever these solubilities are significantly different, solvent evaporation from the lipid deposits will lead to the precipitation of the less soluble ingredient, leaving large doubts about the exact composition of the final GUV membrane. This problem can be

Table 9.1 Electroformation parameters in water, in "low electrolyte" aqueous medium

Step	E (V/m)	f (Hz)	t (min)
1	50 → 700	10	60–90
2	700	10	0–60
3	700	10 → 4	30–60

E is the alternating electric field amplitude (peak-to-peak), *f* is its frequency, and *t* is the application time.

circumvented using the protocol developed for mixtures of soy PC with a hydrophobic two-photon fluorescence dye, NH2-PV2-FL (Pott *et al.*, 2008). Due to the overall low solubility of the dye, formation of a homogeneous deposit on the electrodes from an organic solvent is not possible. The dye can nevertheless be incorporated into small-scale liposomes by colyophilization with soy PC from cyclohexane (lipid to dye molar ratio of 1000). The dry and homogeneously mixed powder is then hydrated and subsequently subjected to violent mechanical agitation in a grinder mill (Retsch, Inc., Newton, PA) leading to polydisperse liposomes. Liposomes are diluted to a concentration of about 0.25 mg/ml. About 2 μl of the aqueous liposome dispersion are applied per deposit. To speed the drying, the electrodes with their lipid deposits are placed in a desiccator at reduced pressure. Alternatively, deposits can be partially dehydrated against a saturated salt solution, thereby fixing the lipid water content. Deposition of such liposomes on electrodes following drying and electroformation (low electrolyte parameters) results easily in many quasi-spherical vesicles, and fluorescence microscopy showed indeed the homogeneous incorporation of the dye into the soy PC GUVs.

The use of liposome dispersions as deposits, now generalized by our laboratory, has important advantages. First, GUV electroformation seems to work more readily, probably due to the better quality of the deposits (Pott *et al.*, 2008). Second, the amount of phospholipid set down on the electrodes is better defined, a point that is important when working with aqueous media containing solutes that partition into the membrane (see later).

Third, membrane proteins can be reconstituted into liposomes, making possible the preparation of proteo-GUVs (Girard *et al.*, 2004). Bacteriorhodopsin, (BR from *Halobacterium salinarum* purple membranes, kindly provided by Drs. Rigaud and Lévy) is incorporated into stearoyloleoyl-PC (SOPC) or egg PC LUVs by reverse-phase evaporation (Düzgüneş, 2003; Rigaud *et al.*, 1983). The liposomes are checked by UV/VIS spectroscopy to ensure that the BR is incorporated completely into the membranes (BR specific adsorption of the LUVs is in agreement with the initially provided ratio). The functionality of the reconstituted membranes is monitored by the spectral changes induced by the light- and dark-adapted forms of BR (Manneville *et al.*, 1999). GUV electroformation in diluted buffer (1 m*M* Tris, 0.2 m*M* EDTA adjusted to pH 7.4) using low electrolyte parameters (Table 9.1) is carried out without any significant influence of the protein content (phospholipid to protein molecular ratio $R_i = 400$, 300, and 200). Once detached from the electrodes, GUVs at $R_i = 400$ fluctuate, that is, their membrane undulates due to Brownian motion, allowing bending elasticity measurements as explained in details by Méléard *et al.* (1997). The bending elasticity is measured to be equal to $1.2 \pm 0.17 \times 10^{-19}$ J, that is, a decrease of about 35% compared to the system in the absence of BR ($1.81 \pm 0.08 \times 10^{-19}$ J) (Méléard *et al.*, 1997). The situation is dramatically changed at $R_i = 200$, where the GUV membranes are inhomogeneous, displaying fluctuating domains as well as huge rigid zones of high contrast. This observation is a further indication that GUVs maintain the initial composition of the liposome deposits.

3.2. GUV electroformation in different aqueous media, including high electrolyte buffer

When using an aqueous medium containing solutes that partition into the membrane, such as amphipathic peptides, GUV electroformation from pure lipid deposits becomes sometimes difficult above a given solute concentration. In the case of the amphipathic peptides melittin or alamethicin, this happens at about 1 μM (Gerbeaud, 1998; Vitkova *et al.*, 2006). In this case, we found that GUV electroformation in low electrolyte buffer (Table 9.1) works fine, provided that the deposit is made of a lipid/peptide mixture. For example, defining the partition coefficient K_p to be equal to $C_\mathrm{p}^\mathrm{b}/C_\mathrm{p}^\mathrm{f}C_\mathrm{l}$ where C_p^b is the bound peptide concentration, C_p^f is the free peptide concentration and C_l is the lipid concentration, we assume a total peptide concentration $C_\mathrm{p}^\mathrm{t} = C_\mathrm{p}^\mathrm{b} + C_\mathrm{p}^\mathrm{f} = 1\,\mu M$. One finds $C_\mathrm{p}^\mathrm{f} = C_\mathrm{p}^\mathrm{t}/(K_\mathrm{p}C_\mathrm{l} + 1)$ and $C_\mathrm{p}^\mathrm{b} = C_\mathrm{p}^\mathrm{t}[K_\mathrm{p}C_\mathrm{l}/(K_\mathrm{p}C_\mathrm{l} + 1)]$. In the case of alamethicin, $K_\mathrm{p} = 1300$ l/mol (Rizzo *et al.*, 1987). Working with six deposits of 2 μl at a concentration of 0.25 mg/ml, one obtains $C_\mathrm{l} \sim 13\ \mu M$, assuming a lipid molecular weight equal to 780 g/mol and a buffer volume equal to 300 μl. This leads to a peptide membrane concentration $C_\mathrm{p}^\mathrm{b} \sim 17\,\mathrm{n}M$ and $C_\mathrm{p}^\mathrm{f} \sim 0.98\,\mu M$, that is very similar to C_p^t.

Whenever more relevant physiological environments are considered for GUV electroformation, osmolarity and electrolyte content have to be distinguished. Osmotic pressure can simply be adjusted to that of biological fluids using uncharged species such as glucose or sucrose. Deposits are made from SOPC liposomes in pure water. After drying, the electroformation cell is filled with glucose or sucrose solution, and electroformation is started using the low electrolyte parameters (Table 9.1). GUVs form readily with high production rates in glucose as well as sucrose solutions up to 200 mM. After detachment of the GUVs from the electrodes, bending elasticity is determined from thermal fluctuations and found to be $1.28 \pm 0.11 \times 10^{-19}$ J for 200 mM glucose and $1.17 \pm 0.05 \times 10^{-19}$ J for 200 mM sucrose. For comparison, the bending elasticity of SOPC in pure water and in buffer (1 mM Tris, 0.2 mM EDTA adjusted to pH 7.4) is respectively $1.27 \pm 0.07 \times 10^{-19}$ J and $1.81 \pm 0.08 \times 10^{-19}$ J (Méléard *et al.*, 1998). These results show clearly that the membrane mechanical properties depend at least partly on the solutes of the dispersion medium. In addition, the small but significant difference in k_c obtained in glucose and sucrose shows that care has to be taken before using these molecules at physiologically relevant concentration. For example, it is common to produce GUVs with sucrose inside and glucose outside as these two solutions have different refraction indexes, thereby enhancing the optical contrast of GUVs (Needham and Nunn, 1990).

Nevertheless, sugar solutions alone are of limited use as a physiological medium since electrolytes are crucial for biological fluids. We therefore explored the influence of the electric field parameters on GUV electroformation as a function of salt concentration in the aqueous medium (Pott *et al.*, 2008). For these experiments, SOPC or POPC SUVs in water are used as lipid deposits. Setting the frequency to 500 Hz during the first step in a 100 m*M* NaCl buffer (also containing 10 m*M* Tris, 2 m*M* EDTA, pH 7.4) enhance substantially the formation rate of SOPC or POPC giant vesicles when E_{max} is set to 1300 V/m (Table 9.2). The GUV nests are very large, highly productive and with typical vesicle piles with thickness up to 200–300 μm at every spot where SUVs are deposited. SOPC or POPC GUVs can also be produced by electroformation in our usual buffer (10 m*M* Tris, 2 m*M* EDTA, pH 7.4) containing 250 m*M* NaCl. The electroformation parameters are left unchanged (Table 9.2); but a reduction in GUV size is observed. Noteworthy, these new electroformation parameters are specific to high electrolyte content. To succeed readily in GUV formation in physiologically relevant buffers, the dried or partly dehydrated lipid deposits should not contain high amounts of salt. High salt content in the deposits may lead to osmotic pressure-triggered spontaneous swelling, which prevents well-controlled GUV electroformation.

The formation of GUVs using lipid binary and ternary mixtures under physiological conditions is tested using the above-presented electroformation protocol. By applying the sequential steps described in Table 9.2, formation of GUVs is successful for binary phospholipid mixtures (DOPC:DPPC), and cholesterol-containing ternary mixtures (DOPC: DPPC:cholesterol) (Fig. 9.4A) (Montes *et al.*, 2007). The same protocol

Table 9.2 Electroformation parameters in "high electrolyte" aqueous medium

Step	E (V/m)	f (Hz)	t (min)
1	50 → 1300	500	30
2	1300	500	90
3	1300	500 → 50	30–60

E is the alternating electric field amplitude (peak-to-peak), *f* is its frequency, and *t* is the application time.

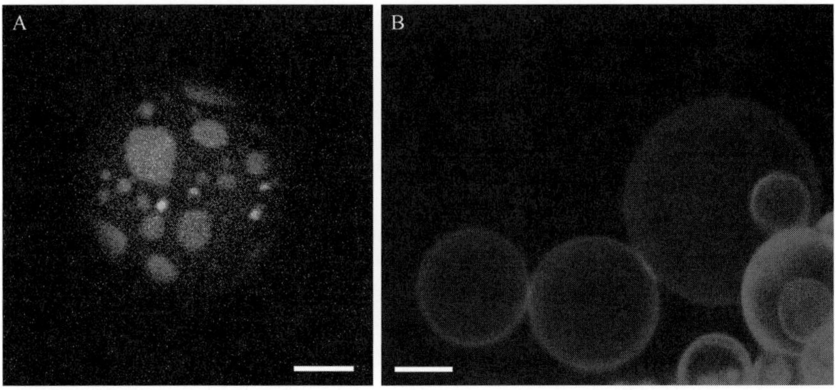

Figure 9.4 Fluorescent images of giant unilamellar vesicles composed of (A) DOPC/DPPC (1:1, mol/mol) plus 20 mol% cholesterol, showing coexistence of liquid ordered (red, fluorescent probe naphthopyrene) and liquid disordered (green, fluorescent probe rhodamine-DOPE), and (B) erythrocyte ghosts labeled with the fluorescent probe DiIC18. The GUVs were prepared using the electroformation protocol under physiological conditions (Table 9.2). The bar corresponds to 10 μm. (See Color Insert.)

has been applied successfully to ternary mixtures containing sphingomyelin/POPC/cholesterol (Bernardino de la Serna *et al.*, 2009).

3.3. GUVs from native membrane extracts and ghosts

Ghosts from red blood cells are prepared according to well-known procedures, and their corresponding lipid extracts can be obtained following the protocols detailed in (Méléard *et al.*, 1997; Montes *et al.*, 2007). However, different parameters for GUV electroformation are used, depending on the salt concentration in the buffer.

Total lipid extract GUVs are electroformed in deionized water after first dissolving the lipids in chloroform/methanol (9:1, v/v), at a concentration of 0.2 mg/ml. Several 2 μl droplets are put down on the electrodes, and after solvent removal, the electric field parameters commonly used at low salt concentration are applied (Table 9.1). Electroformation works smoothly in this case also. GUV bending elasticity, k_c; measured both at ambient and at 37 °C, was found to be temperature dependent with a typical dispersion (Méléard *et al.*, 1997), indicating that the composition of the bilayer made with the RBC total lipid extract is reproducible. Natural lipid extracts obtained from brush border membranes from rat kidney cells (Dietrich *et al.*, 2001), or pulmonary surfactant lipid extracts (Bernardino de la Serna *et al.*, 2004; Nag *et al.*, 2002) can also be used to generate GUVs, using the standard electroformation parameters corresponding to low concentration buffers (Table 9.1). In the first and second cases, the preparation starts with the deposition of organic solutions of lipid extracts (0.2 mg/ml total phospholipids in chloroform) followed with hydration in water (Dietrich *et al.*, 2001). In the third case, the obtained pulmonary surfactant membrane dispersed in buffer are deposited on the Pt electrodes, dried and rehydrated in sucrose solutions (Bernardino de la Serna *et al.*, 2004). Although GUVs are obtained in both cases, the procedure is not successful when using buffers at physiological salt concentrations.

On the contrary, the formation of GUVs using erythrocyte ghost membranes (Fig. 9.4B) or their lipid extracts is possible under physiological conditions using the protocol presented in Table 9.2 (Montes *et al.*, 2007). GUVs can be electroformed in the same typical environment using native pulmonary surfactant membranes as well as their corresponding lipid extracts (Bernardino de la Serna *et al.*, 2009).

Particularly for the GUVs originating from red blood cell ghosts, the membrane phase state (presence of a liquid-ordered phase), and the membrane asymmetry are conserved after the preparation of GUVs (Montes *et al.*, 2007). These conclusions are based on (1) a comparative LAURDAN GP image analysis among red blood cells, ghosts, and GUVs composed of ghost membranes, and (2) the analysis of the orientation of a cytoplasmic domain of band III, an extracellular domain of glycophorin A and globosides in GUVs prepared from right side out and inside out ghosts, using immunofluorescence approaches (Montes *et al.*, 2007), respectively. In this respect, it is important to point out that all GUVs obtained from organic solvents will show symmetry in the membrane (because of lipid miscibility). However, the results obtained from the GUVs composed of ghosts strongly point out that the obtained giant structures retain the asymmetric nature of the natural membranes used to originate the GUVs.

4. Concluding Remarks

Electroformation is a rapid and easy method to obtain GUVs from various lipid mixtures, liposomes, and proteosomes. The "high-salt" electroformation protocol makes it further possible to generate GUV under physiologically pertinent electrolyte concentrations. One of the most exciting potentialities of this new method is the possibility to generate GUVs composed of a variety of native membranes under physiologically relevant conditions. The alternate existing method to generate GUV from native membranes, that is cell blebbing (Baumgart *et al.*, 2007), is only suitable to form GUVs from plasma membranes avoiding the possibility to prepare GUVs from internal organelles.

REFERENCES

Akashi, K. I., Miyata, H., Itoh, H., and Kinosita, K. (1996). Preparation of giant liposomes in physiological conditions and their characterization under an optical microscope. *Biophys. J.* **71,** 3242–3250.

Angelova, M. I., and Dimitrov, D. (1986). Liposome electroformation. *Faraday Discuss. Chem. Soc.* **81,** 303–311.

Angelova, M. I., Soléau, S., Méléard, P., Faucon, J.-F., and Bothorel, P. (1992). Preparation of giant vesicles by external AC fields. Kinetics and applications. *Prog. Colloid Polym. Sci.* **89,** 127–131.

Bacia, K., Schuette, C. G., Kahya, N., Jahn, R., and Schwille, P. (2004). SNAREs prefer liquid-disordered over "raft" (liquid-ordered) domains when reconstituted into giant unilamellar vesicles. *J. Biol. Chem.* **279,** 37951–37955.

Bagatolli, L. A. (2006). To see or not to see: Lateral organization of biological membranes and fluorescence microscopy. *Biochim. Biophys. Acta* **1758,** 1541–1556.

Bagatolli, L. A., and Gratton, E. (1999). Two-photon fluorescence microscopy observation of shape changes at the phase transition in phospholipid giant unilamellar vesicles. *Biophys. J.* **77,** 2090–2101.

Bagatolli, L. A., and Gratton, E. (2000). Two photon fluorescence microscopy of coexisting lipid domains in giant unilamellar vesicles of binary phospholipid mixtures. *Biophys. J.* **78,** 290–305.

Bagatolli, L. A., Parasassi, T., and Gratton, E. (2000). Giant phospholipid vesicles: comparison among the whole lipid sample characteristics using different preparation methods: A two photon fluorescence microscopy study. *Chem. Phys. Lipids* **105,** 135–147.

Baumgart, T., Hammond, A. T., Sengupta, P., Hess, S. T., Holowka, D. A., Baird, B. A., and Webb, W. W. (2007). Large-scale fluid/fluid phase separation of proteins and lipids in giant plasma membrane vesicles. *Proc. Natl. Acad. Sci. USA* **104,** 3165–3170.

Bernardino de la Serna, J., Perez-Gil, J., Simonsen, A. C., and Bagatolli, L. A. (2004). Cholesterol rules: Direct observation of the coexistence of two fluid phases in native pulmonary surfactant membranes at physiological temperatures. *J. Biol. Chem.* **279,** 40715–40722.

Bernardino de la Serna, J., Hansen, S., Hannibal-Bach, H. K., Simonsen, A. C., Knudsen, J., and Bagatolli, L. A. (2009). Biophysical, structural and compositional characterization at the molecular level of native pulmonary surfactant membranes directly isolated from mice wild-type and knocked-out protein D bronco-alveolar lavage fluid. Biophysical Society Meeting Abstracts. *Biophys. J.* **96,** 451a.

Bivas, I., Hanusse, P., Bothorel, P., Lalanne, J., and Aguerre-Chariol, O. (1987). An application of the optical microscopy to the determination of the curvature elastic modulus of biological and model membranes. *J. Phys.* **48,** 855–867.

Brochard, F., and Lennon, J.-F. (1975). Frequency spectrum of the flicker phenomenom in erythrocytes. *J. Phys.* **36,** 1035–1047.

Dietrich, C., Bagatolli, L. A., Volovyk, Z. N., Thompson, N. L., Levi, M., Jacobson, K., and Gratton, E. (2001). Lipid rafts reconstituted in model membranes. *Biophys. J.* **80,** 1417–1428.

Dimova, R., Aranda, S., Bezlyepkina, N., Nikolov, V., Riske, K. A., and Lipowsky, R. (2006). A practical guide to giant vesicles, probing the membrane nanoregime via optical microscopy. *J. Phys.: Condens. Matter* **18,** S1151–S1176.

Düzgüneş, N. (2003). Preparation and quantification of small unilamellar liposomes and large unilamellar reverse-phase evaporation liposomes. *Methods Enzymol.* **367,** 23–27.

Engelhardt, H., Duwe, H. P., and Sackmann, E. (1985). Bilayer bending elasticity measured by Fourier analysis of thermally excited surface undulations of flaccid vesicles. *J. Phys. Lett.* **46,** L395–L400.

Evans, E., and Needham, D. (1986). Giant vesicle bilayers composed of mixtures of lipids, cholesterol and polypeptides. *Faraday Discuss. Chem. Soc.* **81,** 267–280.

Evans, E., and Rawicz, W. (1990). Entropy-driven tension and bending elasticity in condensed-fluid membranes. *Phys. Rev. Lett.* **64,** 2094–2097.

Faucon, J. F., Mitov, M. D., Méléard, P., Bivas, I., and Bothorel, P. (1989). Bending elasticity and thermal fluctuations of lipid membranes. Theoretical and experimental requirements. *J. Phys.* **50,** 2389–2414.

Fidorra, M., Duelund, L., Leidy, C., Simonsen, A. C., and Bagatolli, L. A. (2006). Absence of fluid-ordered/fluid-disordered phase coexistence in ceramide/POPC mixtures containing cholesterol. *Biophys. J.* **90,** 4437–4451.

Fidorra, M., Hartel, S., Ipsen, J. H., and Bagatolli, L. A. (2008). Do GUVs composed of binary lipid mixtures obey the lever rule? A quantitative microscopy imaging approach *2008 Biophysical Society Meeting Abstracts. Biophys. J. Suppl.* **94**(2S), 1198–Pos.

Gerbeaud, C. (1998). Effets de l'insertion de protéines et de peptides membranaires sur les propriétés mécaniques et les changements morphologiques de vésicules géantes. Ph.D. thesis, Université Bordeaux I.

Girard, P., Pécréaux, J., Lenoir, G., Falson, P., Rigaud, J.-L., and Bassereau, P. (2004). A new method for the reconstitution of membrane proteins into giant unilamellar vesicles. *Biophys. J.* **87,** 419–429.

Korlach, J., Schwille, P., Webb, W. W., and Feigenson, G. W. (1999). Characterization of lipid bilayer phases by confocal microscopy and fluorescence correlation spectroscopy. *Proc. Natl. Acad. Sci. USA* **96,** 8461–8466.

Manneville, J. B., Bassereau, P., Lévy, D., and Prost, J. (1999). Activity of transmembrane proteins induces magnification of shape fluctuations of lipid membranes. *Phys. Rev. Lett.* **82,** 4356–4359.

Méléard, P., Gerbeaud, C., Pott, T., Fernandez-Puente, L., Bivas, I., Mitov, M. D., Dufourcq, J., and Bothorel, P. (1997). Bending elasticities of model membranes. Influences of temperature and cholesterol content. *Biophys. J.* **72,** 2616–2629.

Méléard, P., Gerbeaud, C., Bardusco, P., Jeandaine, N., Mitov, M. D., and Fernandez-Puente, L. (1998). Mechanical properties of model membranes studied from shape transformations of giant vesicles. *Biochimie* **80,** 401–413.

Montes, L. R., Alonso, A., Goni, F. M., and Bagatolli, L. A. (2007). Giant unilamellar vesicles electroformed from native membranes and organic lipid mixtures under physiological conditions. *Biophys. J.* **93,** 3548–3554.

Nag, K., Pao, J. S., Harbottle, R. R., Possmayer, F., Petersen, N. O., and Bagatolli, L. A. (2002). Segragation of saturated chain lipids in pulmonary surfactant films and bilayers. *Biophys. J.* **82,** 2041–2051.

Needham, D., and Nunn, R. S. (1990). Elastic deformation and failure of lipid bilayer membranes containing cholesterol. *Biophys. J.* **58,** 997–1009.

Needham, D., McIntosh, T. J., and Evans, E. (1988). Thermomechanical and transition properties of dimyristoylphosphatidylcholine/cholesterol bilayers. *Biochemistry* **27,** 4668–4673.

Pautot, S., Frisken, B. J., and Weitz, D. A. (2003). Engineering asymmetric vesicles. *Proc. Natl. Acad. Sci. USA* **100,** 10718–10721.

Plasencia, I., Norlén, L., and Bagatolli, L. A. (2007). Direct visualization of lipid domains in human skin stratum corneum's lipid membranes: Effect of pH and temperature. *Biophys. J.* **93,** 3142–3155.

Pott, T., Bouvrais, H., and Méléard, P. (2008). Giant unilamellar vesicle formation under physiologically relevant conditions. *Chem. Phys. Lipids* **154,** 115–119.

Reeves, J. P., and Dowben, R. M. (1969). Formation and properties of thin walled phospholipid vesicles. *J. Cell Physiol.* **73,** 49–60.

Rigaud, J. L., Bluzat, A., and Buschlen, S. (1983). Incorporation of bacteriorhodopsin into large unilamellar liposomes by reverse phase evaporation. *Biochem. Biophys. Res. Commun.* **111,** 373–382.

Rizzo, V., Stankowski, S., and Schwarz, G. (1987). Alamethicin incorporation in lipid bilayers: A thermodynamic study. *Biochemistry* **26,** 2751–2759.

Roux, A., Cuvellier, D., Nassoy, P., Prost, J., Bassereau, P., and Goud, B. (2005). Role of curvature and phase transition in lipid sorting and fission of membrane tubules. *EMBO J.* **24,** 1537–1545.

Schneider, M. B., Jenkins, J. T., and Webb, W. W. (1984). Thermal fluctuations of large quasi spherical bimolecular phospholipid vesicles. *J. Phys.* **45,** 1457–1472.

Servuss, R. M., Harbich, W., and Helfrich, W. (1976). Measurement of the curvature-elastic modulus of egg lecithin bilayers. *Biochim. Biophys. Acta* **436,** 900–903.

Valverde, A. (2000). Corrélation entre les fluctuations thermiques des membranes de vésicules géantes et la présence de bactériorhodopsine au sein de ces bicouches. Master thesis, Université Bordeaux I, Talence.

Veatch, S. L., and Keller, S. L. (2005). Seeing spots: Complex phase behavior in simple membranes. *Biochim. Biophys. Acta* **1746,** 172–185.

Vitkova, V., Méléard, P., Pott, T., and Bivas, I. (2006). Alamethicin influence on the membrane bending elasticity. *Eur. Biophys. J.* **35,** 281–286.

SECTION TWO

LIPOSOMES IN THERAPEUTICS

CHAPTER TEN

Liposomal Boron Delivery for Neutron Capture Therapy

Hiroyuki Nakamura

Contents

Abstract

Tumor cell destruction in boron neutron capture therapy (BNCT) is due to the nuclear reaction between ^{10}B and thermal neutrons. The thermal neutrons have an energy of 0.025 eV, clearly below the threshold energy required to ionize

Department of Chemistry, Faculty of Science, Gakushuin University, Toshima-ku, Tokyo, Japan

Methods in Enzymology, Volume 465
ISSN 0076-6879, DOI: 10.1016/S0076-6879(09)65010-2

tissue components. However, neutron capture by ^{10}B produces lithium ion and helium (α-particles), which are high linear energy transfer (LET) particles, and dissipate their kinetic energy before traveling one cell diameter (5–9 μm) in biological tissues, ensuring their potential for precise cell killing. BNCT has been applied clinically for the treatment of malignant brain tumors, malignant melanoma, head and neck cancer, and hepatoma using two boron compounds: sodium borocaptate ($Na_2{}^{10}B_{12}H_{11}SH$; $Na_2{}^{10}BSH$) and L-*p*-boronophenylalanine (L-^{10}BPA). These low molecular weight compounds are cleared easily from the cancer cells and blood. Therefore, high accumulation and selective delivery of boron compounds into tumor tissues are most important to achieve effective BNCT and to avoid damage of adjacent healthy cells. Much attention has been focused on the liposomal drug delivery system (DDS) as an attractive, intelligent technology of targeting and controlled release of ^{10}B compounds. Two approaches have been investigated for incorporation of ^{10}B into liposomes: (1) encapsulation of ^{10}B compounds into liposomes and (2) incorporation of ^{10}B-conjugated lipids into the liposomal bilayer. Our laboratory has developed boron ion cluster lipids for application of the latter approach. In this chapter, our boron lipid liposome approaches as well as recent developments of the liposomal boron delivery system are summarized.

1. Introduction

Boron neutron capture therapy (BNCT) was first proposed as a binary approach to cancer treatment (Locher, 1936). The cell-killing effect of BNCT is due to a nuclear reaction of two essentially nontoxic species, boron-10 (^{10}B) and thermal neutrons (Eq. (10.1)).

$$^{10}B + {}^{1}n \rightarrow {}^{4}He(\alpha) + {}^{7}Li + 2.4\,MeV \qquad (10.1)$$

The resulting α-particles and Li nuclei are high linear energy transfer (LET) particles and dissipate their kinetic energy before traveling one cell diameter (5–9 μm) in biological tissues ensuring their potential for precise cell killing. Their destructive effect is highly observed in boron-loaded tissues. Therefore, high accumulation and selective delivery of boron-10 into the tumor tissue are the most important requirements to achieve efficient neutron capture therapy of cancer (Barth *et al.*, 1990; Hawthorne, 1993; Soloway *et al.*, 1998). The amounts of boron-10 required to obtain fatal tumor cell damage has been calculated to be more than 20–30 μg/g of tumor tissue (Barth *et al.*, 1992). At the same time, the boron concentration in the surrounding normal tissues and blood should be kept low to minimize damage to the normal tissues. BNCT has been applied clinically for the treatment of patients with malignant brain tumors and malignant melanoma, using sodium mercaptoundecahydrododecaborate ($Na_2{}^{10}B_{12}H_{11}SH$; $Na_2{}^{10}BSH$) (Nakagawa and Hatanaka, 1997;

Soloway *et al.*, 1967) and L-*p*-boronophenylalanine (L-^{10}BPA) (Mishima *et al.*, 1989; Synder *et al.*, 1958), respectively. Furthermore, positron emission tomography (PET) using ^{18}F-BPA has been developed (Imahori *et al.*, 1998). The structures of boron compounds which have been utilized for clinical treatment of BNCT are shown in Fig. 10.1. Since the achievement of ^{18}F-BPA PET imaging, we have been able to predict tumor/blood and tumor/normal tissue ratios of L-^{10}BPA before neutron irradiation. This PET technology also displayed selective accumulation of ^{18}F-BPA in various tumors. Thus, BNCT has been applied for various cancers including head and neck cancer, lung cancer, hepatoma, chest wall cancer, and mesothelioma (Aihata *et al.*, 2006; Kato *et al.*, 2004; Suzuki *et al.*, 2007). Number of cases for treatment of cancers with BNCT at KUR (Kyoto University Reactor) is summarized in Fig. 10.2. Although the number of

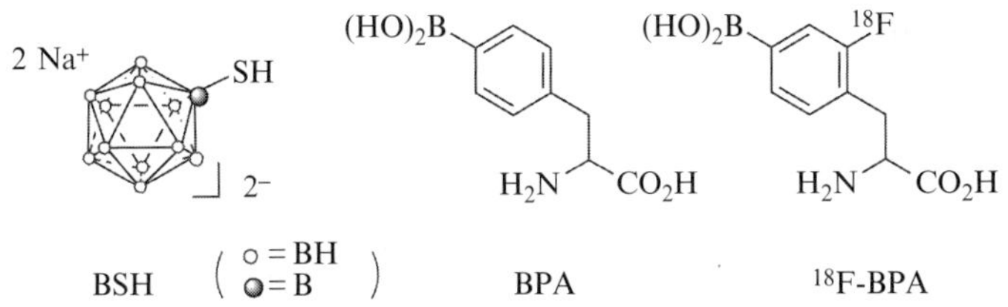

Figure 10.1 Structures of boron compounds utilized for clinical treatment of BNCT.

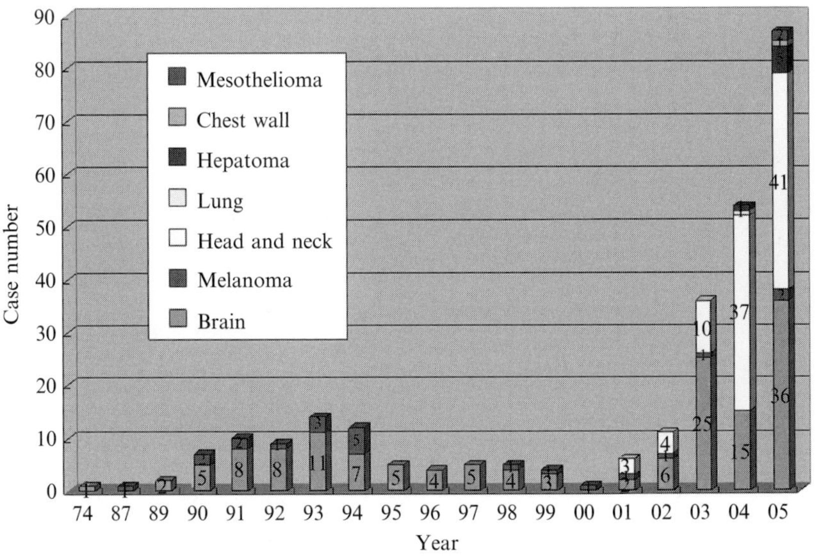

Figure 10.2 Number of cases for treatment of cancers with BNCT at KUR. (See Color Insert.)

cases is increasing, development of new ^{10}B-carriers that deliver an adequate concentration of ^{10}B atoms to tumors is still an important requirement for effective and extensive cancer therapy in BNCT.

Recent promising approaches that meet the requirement entail the use of small boron molecules (Cai *et al.*, 1997; Gedda *et al.*, 1997; Kelly *et al.*, 1994; Nakamura *et al.*, 1997; Yamamoto and Nakamura, 1993; Yamamoto *et al.*, 1995), such as porphyrins (Alam *et al.*, 1989; Kahl and Li, 1996; Kahl *et al.*, 1990; Miura *et al.*, 1996; Murakami *et al.*, 1993; Woodburn *et al.*, 1993), nucleosides (Rong and Soloway, 1994; Schinazi and Prusoff, 1985; Sood *et al.*, 1989; Yamamoto *et al.*, 1992), and amino acids (Kahl and Kasar, 1996; Kirihata *et al.*, 1995; Malan and Morin, 1996; Nakamura *et al.*, 1998, 2000; Srivastava *et al.*, 1997; Takagaki *et al.*, 1996), and boron-conjugated biological complexes, such as monoclonal antibodies (Alam *et al.*, 1985; Goldenberg *et al.*, 1984; Pak *et al.*, 1995), epidermal growth factors (Capala *et al.*, 1996; Gedda *et al.*, 1996; Yang *et al.*, 1997), carborane oligomers (Cai *et al.*, 1997; Fulcrand-El Kattan *et al.*, 1994; Nakanishi *et al.*, 1999; Sood *et al.*, 1990), micells (Wei *et al.*, 2003), and dendrimers (Backer *et al.*, 2005; Shukla *et al.*, 2003; Wu *et al.*, 2004).

Liposomes are efficient drug delivery vehicles, because encapsulated drugs can be delivered selectively to tumors. Therefore, liposomal boron delivery system, in this context, is also considered to be potent for BNCT due to the possibility to carry a large amount of ^{10}B compound. Two approaches have been investigated for liposomes as boron delivery vehicles: (1) encapsulation of boron compounds into liposomes and (2) incorporation of boron-conjugated lipids into the liposomal bilayer, as shown in Fig. 10.3. Our laboratories first synthesized a boron ion cluster lipid which have double alkyl chains in the molecule and investigated the vesicle formation of the boron ion cluster lipids (Nakamura *et al.*, 2004). The boron lipid liposomes are highly potent because drugs, including boron compounds, can be encapsulated into the vacant inner cell of a liposome. Furthermore,

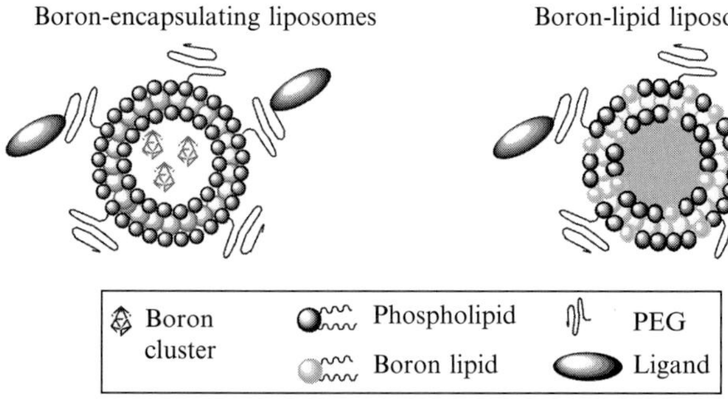

Figure 10.3 Boron-encapsulating liposomes and boron-lipid liposomes. (See Color Insert.)

functionalization of liposomes is possible by combination of lipid contents. Therefore, ^{10}B and drugs may be simultaneously delivered to tumor and this will lead to the combination therapy of both BNCT and chemotherapy for cancers. In this chapter, liposomal boron delivery systems of both boron-encapsulation and boron-lipid liposome approaches are described.

2. Boron-Encapsulation Approach

Liposomal boron delivery system was first reported by Yanagië *et al.* (1989, 1991). They investigated a BSH-encapsulated liposome which was conjugated with a monoclonal antibody specific for carcinoembryonic antigen (CEA). A new murine monoclonal antibody (2C-8) was prepared by immunizing mice *i.p.* with CEA producing human pancreatic cancer cell line, AsPC-1. SDS–PAGE and Western blot analysis showed that 2C-8 monoclonal antibody recognized CEA. This anti-CEA monoclonal antibody was conjugated with large multilamellar liposomes incorporated ^{10}B compound ($Cs_2{}^{10}BSH$). The liposome was prepared from egg yolk phosphatidylcholine, cholesterol, and dipalmitoylphosphatidylethanolamine (1/1/0.05), and $Cs_2{}^{10}BSH$ was encapsulated. The liposomes were treated with dithiothreitol and suspended in the *N*-hydroxysuccinimidyl-3-(2-pyridyldithio)propionate-treated antibody solution for conjugation. This immunoliposome was shown to bind selectively to human pancreatic carcinoma cells (AsPC-1) bearing CEA on their surface and inhibit tumor cell growth on thermal neutron irradiation (5×10^{12} neutrons/cm^2) *in vitro*. Furthermore, the theraputic effects of locally injected BSH-encapsulated immunoliposome on AsPC-1 xenografts in nude mice were evaluated. After intratumoral injection of the immunoliposomes, boron concentrations in tumor tissue and blood were 49.6 ± 6.6 and 0.30 ± 0.08 ppm, respectively. Tumor growth of mice with intratumoral injection of BSH-encapsulated immunoliposomes was suppressed with thermal neutron irradiation (2×10^{12} neutrons/cm^2) *in vivo*. Histopathologically, hyalinization and necrosis were found in the immunoliposome-treated tumors (Yanagië *et al.*, 1997).

Hawthorne and coworkers succeeded in the preparation of boron-encapsulating liposomes with mean diameters of 70 nm or less from distearoylphosphatidylcholine (DSPC) and cholesterol in 1992. The hydrolytically stable borane anions $B_{10}H_{10}{}^{2-}$, $B_{12}H_{11}SH^{2-}$, $B_{20}H_{17}OH^{4-}$, $B_{20}H_{19}{}^{3-}$, and the normal form and photoisomer of $B_{20}H_{18}{}^{2-}$ were encapsulated in liposomes as their soluble sodium salts. Although the boron compounds used do not exhibit an affinity for tumors and are normally rapidly cleared from the body, liposomes were observed to selectively deliver the borane anions to tumors. High tumor concentrations were achieved in the therapeutic range (>15 μg of boron per g of tumor) while maintaining high tumor-boron/blood-boron ratios (>3). The most favorable results were obtained with the two isomers of $B_{20}H_{18}^{2-}$. These boron compounds have the capability to react with intracellular components after they have been deposited within tumor cells by the liposome, thereby preventing the borane ion from being released into blood (Shelly *et al.*, 1992). Furthermore, an apical–equatorial (*ae*) isomer of the $B_{20}H_{17}NH_3^{3-}$ ion, $[1\text{-}(2'\text{-}B_{10}H_9)\text{-}2\text{-}NH_3B_{10}H_8]^{3-}$, which was produced from the reaction of the polyhedral borane ion $B_{20}H_{18}^{2-}$ with liquid ammonia, was encapsulated into liposomes prepared with 5% PEG-200-distearoyl phosphatidylethanolamine. The PEGylated liposomes exhibited a long circulation lifetime due to escape from reticuloendothelial system (RES), resulting in the continued accumulation of boron in the tumor over the entire 48-h experiment and reaching a maximum of 47 μg of boron per g of tumor (Feakes *et al.*, 1994).

Boron-containing folate receptor-targeted liposomes have been developed by Lee and coworkers (Pan *et al.*, 2002). Expression of the folate receptor (FR) is frequently is amplified among human tumors. Two highly ionized boron compounds, $Na_2[B_{12}H_{11}SH]$ and $Na_3(B_{20}H_{17}NH_3)$, were incorporated into liposomes by passive loading with encapsulation efficiencies of 6% and 15%, respectively. In addition, five weakly basic boronated polyamines investigated were incorporated into liposomes by a pH-gradient-driven remote-loading method with varying loading efficiencies. Greater loading efficiencies were obtained with lower molecular weight boron derivatives, using ammonium sulfate as the trapping agent, compared to those obtained with sodium citrate. The *in vitro* uptake of folate-conjugated boron-encapsulating liposomes was investigated using human KB squamous epithelial cancer cells, which have amplified FR expression. Higher cellular boron uptake (up to 1584 μg per 10^9 cells) was observed with FR-targeted liposomes than with nontargeted control liposomes (up to 154 μg per 10^9 cells), irrespective of the chemical form of the boron and the method used for liposomal preparation.

Kullberg and coworkers investigated EGF-conjugated PEGylated liposome delivery vehicle, containing water-soluble boronated phenanthridine, WSP1, or water-soluble boronated acridine, WSA1, for EGFR targeting. In the case of WSA1, a ligand-dependent uptake was obtained and the boron uptake was as good as if free WSA1 was given. No ligand-dependent boron uptake was seen for WSP1-containing liposomes. Thus, WSA1 is a candidate for further studies. Approximately 10^5 boron atoms were in each liposome. A critical assessment indicates that after optimization up to 10^6 boron atoms can be loaded. *In vitro* boron uptake by glioma cells (6.29 ± 1.07 μg/g cells) was observed with WSA1-encapsulated EGF-conjugated PEGylated liposomes (Kullberg *et al.*, 2003).

Cetuximab-conjugated liposome was also investigated as an althernative immunoliposome for targeting of EGFR(+) glioma cells. Lee and coworkers developed cetuximab-immunoliposomes via a cholesterol-based membrane anchor, maleimido-PEG-cholesterol (Mal-PEG-Chol), to incorporate cetuximab into liposomes. BSH-encapsulated cetuximab-immunoliposomes were evaluated for targeted delivery to human EGFR gene transfected $F98_{EGFR}$ glioma cells. Much greater (approximately eightfold) cellular uptake of boron was obtained using cetuximab-immunoliposomes in EGFR(+) $F98_{EGFR}$ compared with nontargeted human IgG-immunoliposomes (Pan *et al.*, 2007).

Maruyama and coworkers developed a new type of target-sensitive liposomes, in which transferrin-coupling pendant-type PEG liposomes were extravasated effectively into solid tumor tissue in colon 26 tumor-bearing mice, and internalized into tumor cells (Ishida *et al.*, 2001). Transferrin (TF) receptor-mediated endocytosis is a normal physiological process by which TF delivers iron to the cells and higher concentration of TF receptor has been observed on most tumor cells in comparison with normal cells. TF-PEG liposomes showed a prolonged residence time in the circulation and low RES uptake in tumor-bearing mice, resulting in enhanced extravasation of the liposomes into the solid tumor tissue. Once at the tumor site, TF-PEG liposomes were internalized into tumor cells by receptor-mediated endocytosis. TF-PEG liposomes were taken up into endosome-like intracellular vesicles. Therefore, the clearance of TF-PEG liposomes from tumor tissue is so impaired that they remain in the tumor interstitium for a long time. Thus, the potential of liposomes for selective delivery of therapeutic quantities of ^{10}B to tumors has been studied (Maruyama *et al.*, 2004). TF-PEG liposomes and PEG liposomes encapsulating $Na_2{}^{10}BSH$ were prepared and their tissue distributions in colon 26 tumor-bearing mice after *i.v.* injection were compared with those of bare liposomes and free $Na_2{}^{10}BSH$. When TF-PEG liposomes were injected at a dose of 35 mg ^{10}B/kg, a prolonged residence time in the circulation and low uptake by the RES were observed in colon 26 tumor-bearing mice, resulting in enhanced accumulation of ^{10}B into the solid tumor tissue (e.g., 35.5 μg of boron per g of tumor). TF-PEG liposomes maintained a high ^{10}B level in the tumor, with concentrations over 30 μg of boron per g of tumor for at least 72 h after injection. On the other hand, the plasma level of ^{10}B decreased, resulting in a tumor/plasma ratio of 6.0 at 72 h after injection. Administration of $Na_2{}^{10}BSH$ encapsulated in TF-PEG liposomes at a dose of 5 or 20 mg ^{10}B/kg and irradiation with 2×10^{12} neutrons/cm^2 for 37 min produced tumor growth suppression and improved long-term survival compared with PEG liposomes, bare liposomes, and free $Na_2{}^{10}BSH$. Masunaga and coworkers evaluated biodistribution of $Na_2{}^{10}BSH$- and $Na_2{}^{10}B_{10}H_{10}$-encapsulated TF-PEG liposomes in SCC VII tumor-bearing mice (Masunaga *et al.*, 2006). The time course of the change in the ^{10}B concentration in tumors loaded with both liposomes were similar except that ^{10}B concentrations were greater 24 h after the loading of $Na_2{}^{10}B_{10}H_{10}$ than $Na_2{}^{10}BSH$ in TF-PEG liposomes and ^{10}B concentration in tumors was 35.6 μg of boron per g of tumor with injection of $Na_2{}^{10}B_{10}H_{10}$-encapsulated TF-PEG liposomes (35 mg ^{10}B/kg).

3. Boron Lipid-Liposome Approach

Development of lipophilic boron compounds, embedded within the liposome bilayer, provides an attractive method to increase the overall efficiency of incorporation of boron-containing species, as well as raise the gross boron content of the liposomes in the formulation. Hawthorne and coworkers first introduced *nido*-carborane as hydrophilic moiety into the amphiphile and this single-tailed *nido*-carborane amphiphile was utilized for liposomal boron delivery using tumor-bearing mice (Feakes *et al.*, 1995; Watson-Clark *et al.*, 1998). They synthesized the *nido*-carborane amphiphile **1** (Fig. 10.4) and prepared boronated liposomes composed of DSPC, cholesterol, and **1** in the bilayer. After the injection of liposomal suspensions in BALB/c mice bearing EMT6 mammary adenocarcinomas, the time-course biodistribution of boron was examined. At the low injected doses normally used (5–10 mg ^{10}B/kg), peak tumor boron concentrations of

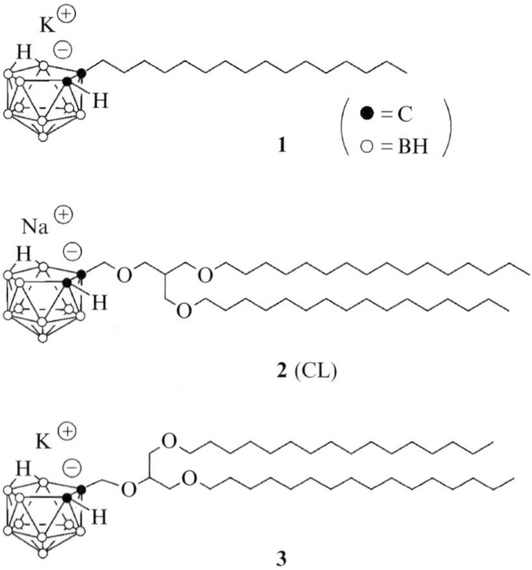

1

2 (CL)

3

Figure 10.4 Structures of *nido*-carborane lipids.

35 μg of boron per g of tumor and tumor/blood boron ratios of ~8 were achieved. These values are sufficiently high for the successful application of BNCT. The incorporation of both **1** and the hydrophilic species, $Na_3[1\text{-}(2'\text{-}B_{10}H_9)\text{-}2\text{-}NH_3B_{10}H_8]$, within the same liposomes demonstrated significantly enhanced biodistribution characteristics, exemplified by maximum tumor boron concentration of 50 μg of boron per g of tumor and tumor/blood boron ratio of 6.

Our laboratories designed the *nido*-carborane lipid **2** (CL), which has a double-tailed moiety conjugated with *nido*-carborane as a hydrophilic moiety (Nakamura *et al.*, 2004). This lipid has a symmetric carbon in the lipophilic alkyl chain. Analysis in a transmission electron microscope by negative staining with uranyl acetate showed a stable vesicle formation of CL. Furthermore, we focused on overexpression of TF receptor on most cell surfaces. Thus we investigated active targeting of the boron liposomes to solid tumor by functionalization of TF on the surface of their liposomes and achieved a boron concentration of 22 μg ^{10}B per g of tumor by the injection of the liposomes at 7.2 mg ^{10}B/kg body weight with longer survival rates of tumor-bearing mice after BNCT (Miyajima *et al.*, 2006). The detailed protocols for synthesis of the CL and *in vivo* BNCT effects of the CL-liposomes are described.

4. *NIDO*-CARBORANE LIPID LIPOSOMES

4.1. Synthesis of the *nido*-carborane lipid (CL)

Chemical synthesis of the *nido*-carborane lipid (CL) is shown in Scheme 10.1. Reaction of two equivalents of heptadecanol with 3-chloro-2-chloromethyl-1-propene using NaH as base gives the diether **4** in 93% yield and the hydroboration of **4** gives the corresponding alcohol **5** in 71% yield. The alcohol **5** is converted into the propargyl ether **6** in 48% yield by the treatment with propargyl bromide, and the decaborane coupling of **6** is carried out in the presence of acetonitrile in toluene under reflux condition to give the corresponding *ortho*-carborane **7** in 80% yield. The degradation of the carborane cage by the treatment with sodium methoxide in methanol affords the *nido*-carborane lipid **2** (CL) in 57% yield.

4.2. Stability of the *nido*-carborane lipid vesicles

The stability of the boron cluster vesicles in fetal bovine serum (FBS), which can be considered as a model of blood, is examined using the vesicle solution. A boron cluster vesicle fraction is added to FBS (the volume ratio FBS:vesicle solution = 9:1) and the mixture is incubated at 37 °C with stirring. The fluorescence of the FBS solution is measured at 0–18 h.

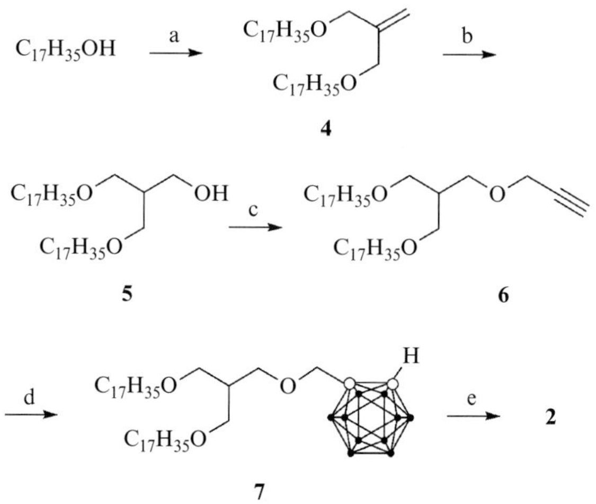

Scheme 10.1 Synthesis of the boron lipid 2. *Reagents*: (a) 1—NaH, THF, 2—$CH_2NC(CH_2Cl)_2$, 93%; (b) 1—$BH_3 \cdot Me_2S$, 2—H_2O_2, NaOH, 71%; (c) 1—NaH, THF, 2—propargyl bromide, 58%; (d) $B_{10}H_{14}$, CH_3CN, toluene, 80%; (e) NaOMe, MeOH, 57%.

No increase of the fluorescence intensity of the FBS solutions is observed during 18 h. Therefore, the calcein encapsulated into the boron cluster vesicle is not released. This result indicates that the boron cluster vesicle prepared from the *nido*-carborane lipid is quite stable in the FBS solution at 37 °C.

4.3. Incorporation of the *nido*-carborane lipid into liposomal membranes

The effect of the accumulation ratio of DSPC and CL on the liposome formation is examined under various mixing ratios. PEG liposomes are prepared from DSPC, CH, CL, and PEG-distearoylphosphatidylethanolamine (DSPE) (1:1:x:0.11, x = 0–1). The lipid concentration is estimated by phosphorus assay (Fiske and Subbarow, 1925). Boron content is determined by inductively coupled plasma atomic emission spectroscopy (ICP-AES). Very interestingly, the ratio of DSPC and CL in the liposomes increased in proportion to the increase of the mixing ratio of CL to DSPC in the solution. Furthermore, it was observed that CL was incorporated into the liposome membranes with five times higher concentration than DSPC (Nakamura *et al.*, 2004) (Fig. 10.5).

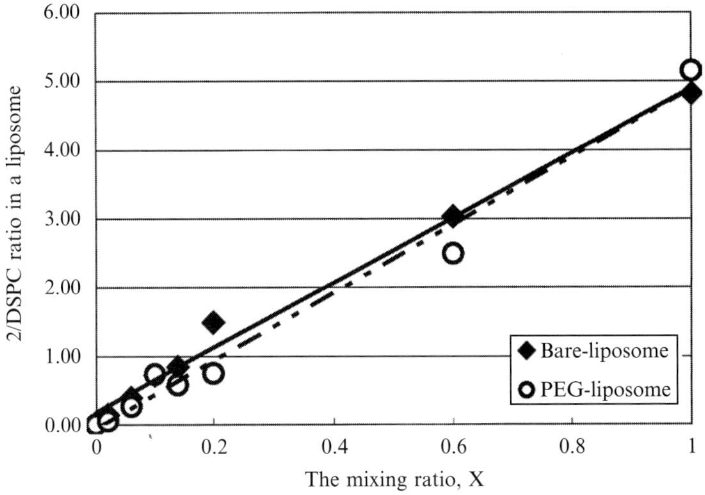

Figure 10.5 Incorporation of the *nido*-carborane lipid (CL) into liposomal membranes. The bare-liposome was prepared from DSPC, CH, and CL (the mixing ratio of 1:1:x; $x =$ 0–1), and the PEG-liposome was prepared from DSPC, CH, CL, and PEG-DSPE (the mixing ratio of 1:1:x:0.11, $x =$ 0–1).

5. Transferrin-Conjugated *Nido*-Carborane Lipid Liposomes

5.1. Preparation of liposomes

TF(−)-PEG-CL liposomes are prepared from DSPC, CH, DSPE-PEG-OMe, DSPE-PEG-O-$(CH_2)_5CO_2H$, and CL (molar ratio 1:1:0.11:0.021:0.25) by the reverse-phase evaporation (REV) method. A mixture of DSPC, CH, DSPE-PEG-OMe, and CL are dissolved in chloroform/diisipropylether mixture (1:1, v/v) in a round-bottomed flask. The volume ratio of the aqueous phase to the organic phase is maintained at 1:2. The emulsion is sonicated for 1 min, and then the organic solvent is removed under vacuum in a rotary evaporator at 37 °C for 1 h to form a lipid gel. The gel obtained is subjected to extrusion through a polycarbonate membrane of 100 nm pore size, using an extruder device (Lipex Biomembrane, Canada) thermostated at 60 °C. Purification is accomplished by ultracentrifuging at 200,000×g for 20 min at 4 °C, and the pellets obtained are resuspended in PBS buffer. Liposome size is measured with an electrophoretic light scattering spectrophotometer (ELS-700, Otsuka Electronics, Tokyo).

TF(+)-PEG-CL liposomes are prepared by the coupling of TF to the PEG-CO_2H moieties of PEG-CL liposomes (Ishida *et al.*, 2001). To 1 ml of PEG-CL liposomes (5 μmol lipids) in Mes buffer (10 m*M* Mes/150 m*M* NaCl, pH 5.5), 21 μmol of EDC and 28 μmol of S-NHS are added, and the mixture is incubated for 15 min at room temperature. The mixture is loaded into a Sephadex G25 column equilibrated with Mes buffer and liposome fractions are collected. The desired amount of TF, and if necessary, a trace amount of ^{125}I-thyraminyl inulin, is then added to the liposome solution, and the mixture is incubated for 3 h at room temperature with gentle stirring. Purification is accomplished by ultracentrifuging at 200,000×*g* for 20 min at 4 °C, and the pellets obtained are resuspended in PBS. TF-PEG-CL liposomes obtained are converted into the diferric form by treatment with $FeCl_3$–nitriloacetic acid solution. After the reaction, the suspension is purified by ultracentrifugation and resuspended in PBS.

5.2. Biodistribution of TF-PEG-CL and PEG-CL liposomes

Biodistribution studies are performed using male BALB/c mice (6 weeks old, 16–18 g, Nihon SLC). Tumor-bearing mice are prepared by inoculating subcutaneous (s.c.) injection of a suspension (5×10^6 cells) of colon 26 cells directly into their back. The mice are kept on regular mouse diet and water, and maintained under a standard light/dark cycle in an ambient atmosphere. These experiments are performed when the tumor is 7–9 mm in diameter. ^{125}I-Thylaminyl inuline solution is encapsulated in the liposomes, and 100–400 μl of liposomes is injected into the mice (three per group) via the tail vein at a selected dose of ^{10}B. At selected time intervals after administration, the mice are lightly anesthetized, bled via the retro-orbital sinus, sacrificed by cervical dislocation and dissected. Their organs are excised and their ^{125}I content is estimated by a liquid scintillation counter.

The time-dependent distributions of TF(−)-PEG-CL and TF(+)-PEG-CL liposomes in various tissues are shown in Fig. 10.6. The rapid clearance of TF(−)-PEG-CL and TF(+)-PEG-CL liposomes was observed in the blood, lung, and kidney after 3 h of injection. In general, PEGylated liposomes possess a longer circulation time compared to nonstealth liposomes (Mumtaz *et al.*, 1991; Vaage *et al.*, 1992). We examined the effect of CL on the circulation time of PEGylated liposomes and found that CL influenced a stealth property of the PEG-CL liposome. The enhanced accumulation of TF(+)-PEG-CL liposomes in comparison with TF(−)-PEG-CL liposome accumulation was observed in the spleen within 72 h. TF(+)-PEG-CL liposomes accumulated in the liver and tumor gradually, whereas TF(−)-PEG-CL liposomes were released from those organs 72 h after injection, although percentage doses of these liposomes were similar 24 h after injection. Surprisingly, more than 1.5% of the total dose injected

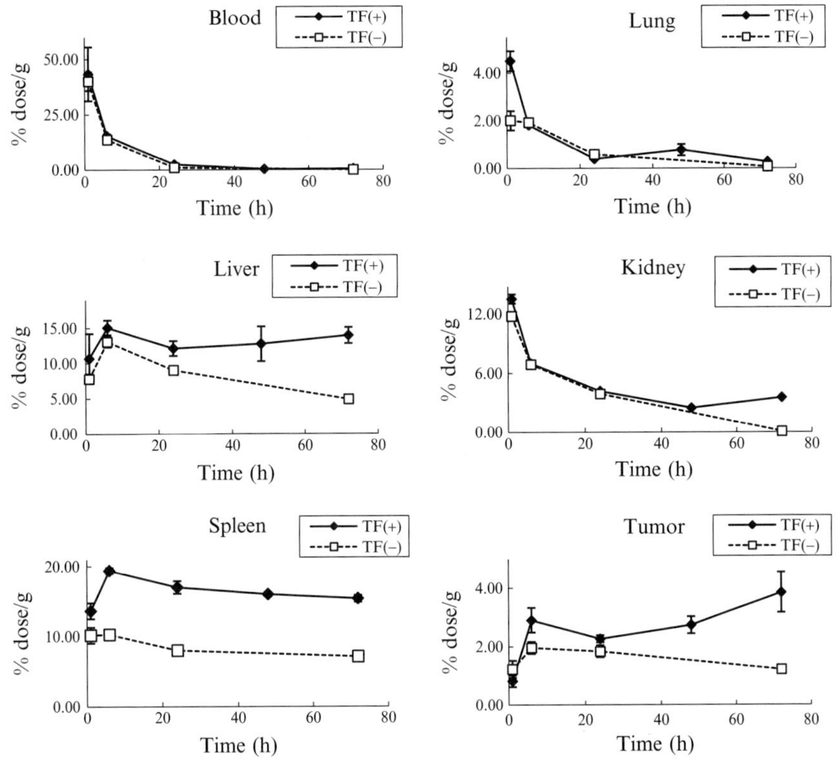

Figure 10.6 Time course of biodistribution of TF(−)-PEG-CL liposome (TF(−)) and the TF(+)-PEG-CL liposome (TF(+)). Liposomes encapsulating ^{125}I-tyraminyl inulin (500 g lipid/200 μl) were injected into male BALB/c mice (7 weeks old, weighing 20–25 g) via the tail vein. The distribution of liposomes was measured by determining the radioactivity of each tissue. The % dose/g in each tissue is plotted on the vertical axis, and the time (h) after administration is plotted on the horizontal axis.

accumulated in tumor tissues. This enhanced accumulation of TF(+)-PEG-CL liposomes may reflect marked receptor-mediated endocytosis after binding to tumor cells (Iinuma *et al.*, 2002).

^{10}B-enriched TF(+)-PEG-CL liposomes are injected into tumor-bearing mice, in which colon 26 cells were transplanted into their left thigh, via the tail vein at a dose of 7.2 mg ^{10}B/kg (200 μl of a liposome solution). Seventy-two hours after administration, ^{10}B concentration in each organ is measured by prompt γ-ray spectroscopy. The results are shown in Fig. 10.7. No boron accumulation was observed in the muscle, heart, and brain; however, the boron concentrations in the lung, kidney, and blood were approximately 10 ppm. Since no accumulation of TF(+)-PEG-CL liposomes labeled with ^{125}I-tyraminyl inulin was observed, as shown in

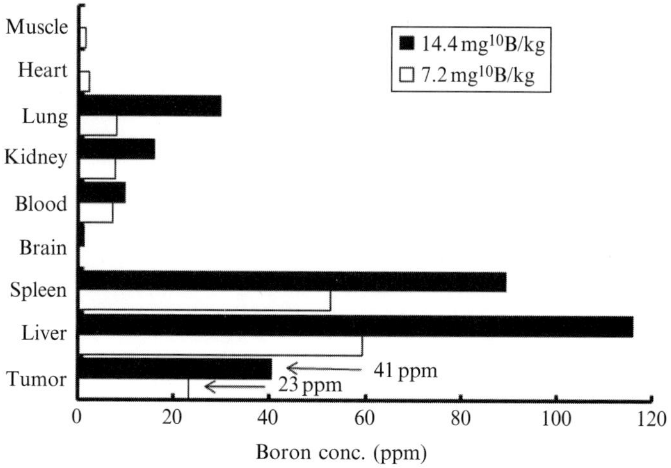

Figure 10.7 ^{10}B concentration in various tissues 72 h after injection of TF(+)-PEG-CL liposomes into tumor-bearing mice. ^{10}B-enriched TF(+)-PEG-CL liposomes were injected into tumor-bearing mice, in which colon 26 cells were transplanted into the left thigh, via the tail vein with 200 μl of liposome solutions (7.2 and 14.4 mg^{10}B/kg).

Fig. 10.6, it is considered that the ^{10}B concentrations detected in such organs may be due to the accumulation of the *nido*-carborane lipid, which was caused by the degradation of the parent liposomes. Enhanced accumulation of ^{10}B was observed in the spleen and liver, and this does not conflict with the result of the biodistribution of TF(+)-PEG-CL liposomes as shown in Fig. 10.6. A high level of ^{10}B concentration (22 ppm) in the tumor was observed in tumor tissues 72 h after the administration of TF(+)-PEG-CL liposomes. Furthermore, almost twice ^{10}B concentrations in each organ were observed in the mice injected with double dose of TF(+)-PEG-CL liposomes (14 mg ^{10}B/kg body weight).

5.3. Survival of tumor-bearing mice after BNCT

Besides the determination of ^{10}B concentration in various organs, the mice are anesthetized with sodium pentobarbital solution 72 h after the administration of TF(+)-PEG-CL liposomes and placed in an acrylic mouse holder, where their whole bodies, except their tumor-implanted leg, are shielded with acrylic resin. Neutron irradiation is carried out for 37 min at a rate of 2×10^{12} neutrons/cm^2 in the KUR atomic reactor. The antitumor effect of BNCT is evaluated on the basis of the survival of the mice, as shown in Fig. 10.8. The untreated mice did not survive after 32 days of neutron irradiation, and their average survival rate was 21 days. Long survival rates were observed in the mice treated with TF(+)-PEG-CL liposomes; one of

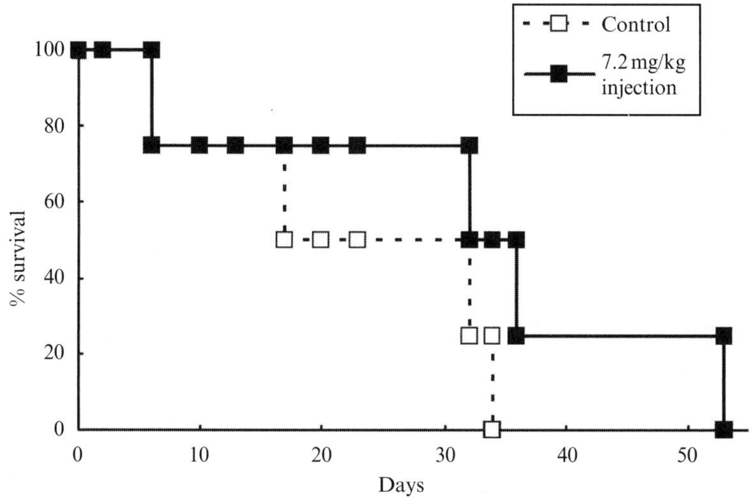

Figure 10.8 Survival curve of tumor-bearing mice after neutron irradiation for 37 min in KUR atomic reactor. The mice were injected with 7.2 mg ^{10}B/kg of the TF(+)-PEG-CL liposome and incubated for 72 h before irradiation. Control indicates survival rates of tumor-bearing mice after neutron irradiation without administration of TF(+)-PEGCL liposomes.

them even survived for 52 days after neutron irradiation. The average survival rate of the treated mice was 31 days.

However, the acute toxicity has been observed in the mice with a double dose injection (14 mg ^{10}B/kg body weight) within 1 day. Similar toxicity has been observed in the liposome prepared from the *nido*-carborane lipid **3** (Li *et al.*, 2006). Therefore, we modified the boron lipids based on biomimetic composition of phosphatidylcholines combined with the *closo*-type boron anion cluster to meet a sufficiently low toxic requirement. We introduced BSH, as an alternative hydrophilic function, to the boron lipids and examined the BNCT effects of *closo*-dodecaborate lipid liposomes on tumor-bearing mice (Lee *et al.*, 2007; Nakamura *et al.*, 2007).

6. *closo*-Dodecaborate Lipid Liposomes

BSH is known as a water-soluble divalent "*closo*-type" anion cluster and significantly lowered toxicity, and thus has been utilized for clinical treatment of BNCT. We have succeeded in the synthesis of the *closo*-dodecaborate lipids (**8** and **9**) (Lee *et al.*, 2007), which have a $B_{12}H_{11}S$-moiety as a hydrophilic function with a chirality similar to that of natural phospholipids, such as DSPC, in their lipophilic tails (Fig. 10.9).

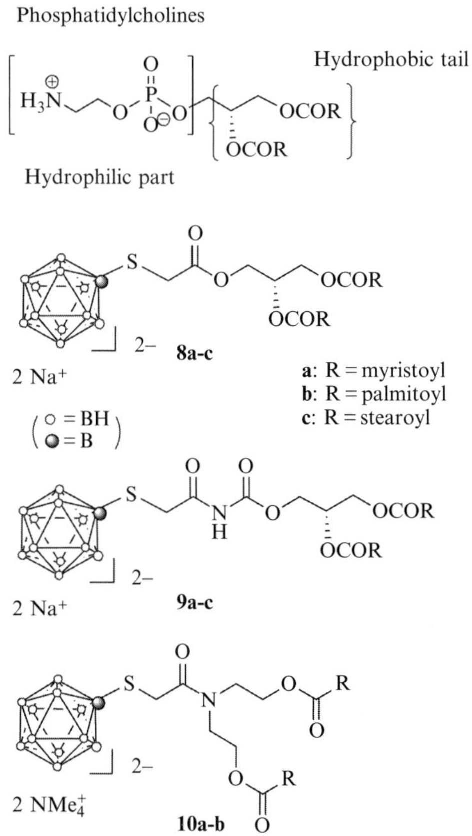

Figure 10.9 Design of *closo*-dodecaborate lipids based on biomimetic composition of phosphatidylcolines.

Recently, the symmetric *closo*-dodecaborate lipids **10** were reported by Gabel and coworkers (Justus *et al.*, 2007). The detailed protocols for synthesis of the the *closo*-dodecaborate lipids (**8** and **9**) and *in vivo* BNCT effects of the *closo*-dodecaborate lipid-liposomes are described.

6.1. Synthesis of *closo*-dodecaborate lipids

Synthesis of the hydrophobic tail functions of **8** and **9** is shown in Schemes 10.2 and 10.3. The chiral alcohol **11** is protected with benzylbromide using NaH and the resulting dioxolane **12** is converted into the diol **13** using aqueous AcOH in 83% yield. The ester formation of the diol **13** with various carboxylic acids is promoted by dicyclohexylcarbodiimide in the presence of catalytic amounts of *N*,*N*-dimethylaminopyridine in CH_2Cl_2 to

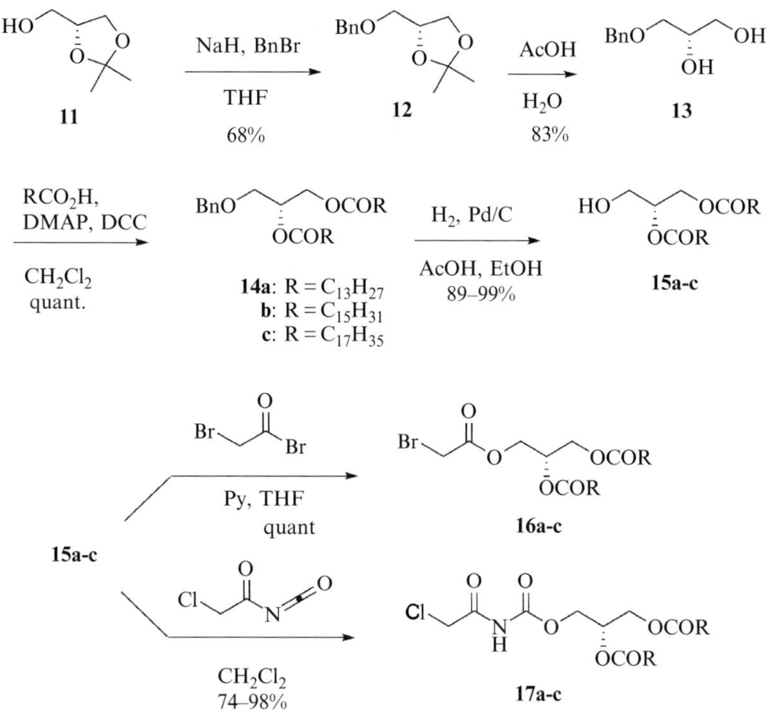

Scheme 10.2 Synthesis of the hydrophobic tail functions.

afford the precursors **14a–c**, quantitatively. Deprotection of the benzyl group of **14a–c** by hydrogenation gives the corresponding alcohols **15a–c** in 89–99% yields. The ester formation of **15a–c** with bromoacetyl bromide in pyridine gave **16a–c**, quantitatively, and the carbamate formation with chloroacetyl isocyanate in CH_2Cl_2 gave **17a–c** in 74–98% yields.

Introduction of BSH into the hydrophobic tail functions **16** and **17** is examined using the "protected BSH (**20**)," which is prepared according to Gabel's protocol, as shown in Scheme 10.3. Briefly, Na_2BSH is treated with 2 equiv. of 2-iodopropionitrile in acetonitrile and the resulting dicyanoethylated BSH **19** is precipitated as tetramethyl ammonium salts. Dealkylation of **19** proceeds in the presence of 1 equiv. of tetramethylammonium hydroxide in acetone to afford the protected BSH **20**. The *S*-alkylation of **20** with **16a–c** proceeds in acetonitrile at 70 °C for 12–24 h, giving the corresponding *S*-dialkylated products (**21**), which are immediately treated with tetramethylammonium hydroxide (1 equiv.) in acetone followed by ion exchange with Dowex X-100 to give **8a–c** as sodium salts. In a similar manner, **9a–c** are obtained from **17a–c**.

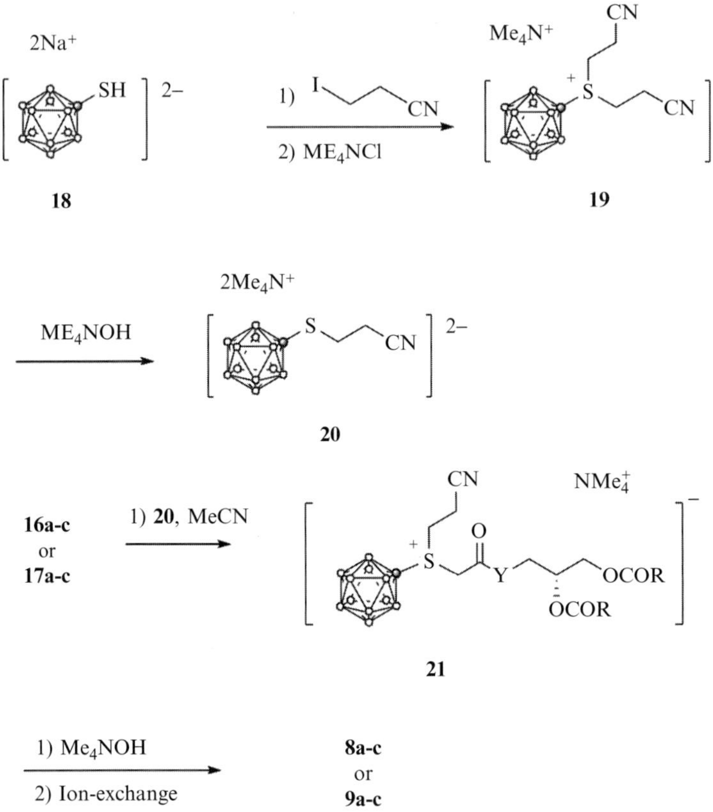

Scheme 10.3 Synthesis of *closo*-dodecaborate lipids **8** and **9**.

6.2. Synthesis of *closo*-dodecaborate cholesterols

Carborane-conjugated cholesterols (Feakes *et al.*, 1999) and carborane-containing cholesterol mimic (Thirumamagal *et al.*, 2006) have been developed as an alternative content of liposomal membranes. Our laboratory has focused on the structure of BSH, as an alternative water-soluble boron cluster and succeeded in the synthesis of dodecaborate-conjugated cholesterols **23a–b** (Scheme 10.4) for liposomal boron delivery systems. Cholesterol is treated with bromoacetyl bromide in the presence of pyridine to give the corresponding ester **22a** in 83% yield. The chloroacetylcarbamate **22b** is also synthesized from chlesterol treating with chloroacetyl isocyanate in the presence of trimethylamine. The protected BSH **20** with **22a–b** proceeds in acetonitrile and the resulting sulfoniums are treated with 1 equiv. of

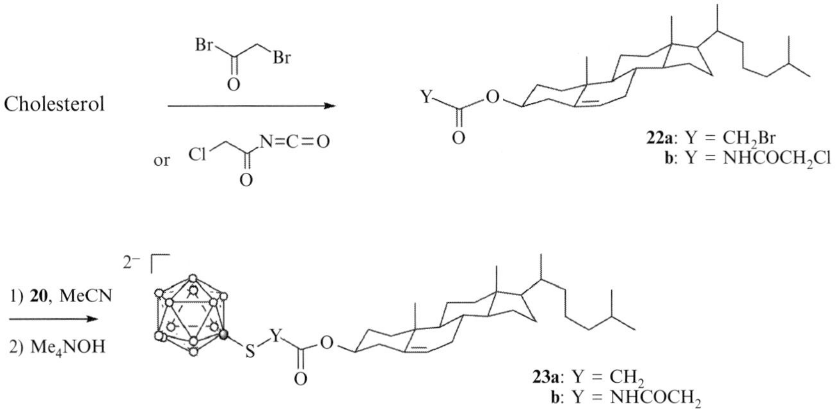

Scheme 10.4 Synthesis of *closo*-dodecaborate cholesterols **23**.

tetramethylammonium hydroxide in acetone to give the corresponding thioesters **23a–b** in 73 and 47% yields, respectively, in two steps.

6.3. Preparation of *closo*-dodecaborate lipid-liposomes

Bare boron liposomes and PEG boron liposomes are prepared from DSPC, boron lipids, Chol (X:1 − X:1, molar ratio, $0 < X < 1$) and DSPC, boron lipids, Chol, DSPE-PEG (X:1 − X:1:0.11, molar ratio, $0 < X < 1$), respectively. These boron liposomes are prepared according to the REV method. Total lipids of 200 mg are dissolved in 6 ml of chloroform/diisopropyl ether mixture (1:1, v/v) and 3 ml of distillated water is added to the mixture to form a w/o emulsion. The emulsion is sonicated for 3 min, and then, the organic solvents are removed under the reduced pressure in a rotary evaporator at 60 °C for 30 min to obtain a suspension of liposomes. The liposomes obtained are subjected to extrusion 10 times through a polycarbonate membrane of 100 nm pore size, using an extruder device thermostated at 60 °C. Purification is accomplished by ultracentrifugation at 200,000×g for 60 min at 4 °C, and the pellets obtained are resuspended in 0.9% NaCl solution or PBS buffer. The size distribution of the boron liposomes are measured by an electrophoretic light scattering spectrophotometer. The composition of boron lipids and DSPC in liposome is calculated from simultaneous measurement of boron and phosphine concentrations by ICP-AES. Transmission electron micrograph (TEM) is carried out by using negative attaining method with uranyl acetate. Figure 10.10 shows the TEM image of 25% DSBL liposomes after sizing with 100 nm pore-diameter polycarbonate membranes.

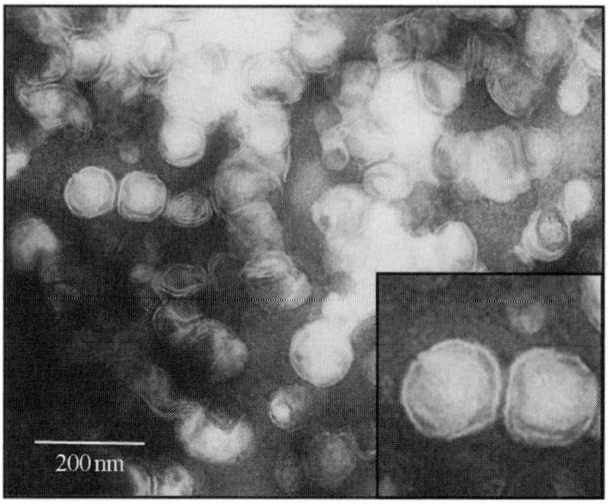

Figure 10.10 Electron micrographs of the boron liposomes composed of 25% DSBL after extrusion using 100 nm filter.

6.4. Fluorescent-labeled *closo*-dodecaborate lipid-liposomes

PKH67-labeled boron liposomes are prepared according to the conventional cell membrane labeling method. Briefly pellets of boron liposomes are dissolved in 250 μl of Diluent C, and then the boron liposomes solution are dropped into 1 μl of PKH67 dye stock solution. The mixture is maintained at 20 °C for 5 min, and then free PKH67 is removed by ultracentrifugation at 200,000×*g* for 60 min at 4 °C, and the obtained PKH67-labeled boron liposomes are resuspended in PBS.

Human epithelial carcinoma cell line HeLa cells are maintained at 37 °C under 5% CO_2 atmosphere in RPMI 1640 medium supplemented with 10% FBS, 100 U/ml of penicillin, and 100 μg/ml of streptomycin. For the subsequent experiments, the cells are seeded at a density of 5×10^4 cells in a φ-35 mm diameter dish (Greiner) and incubated at 37 °C for 20 h. The cells are incubated for further 3 h in the presence of PKH67-labeled boron liposomes in medium. After incubation, the cells are fixed with 4% paraformaldehyde in PBS for 10 min, and then treated with 0.1% Triton X-100 in PBS for 10 min. The cells are mounted on a slide after incubation for 1 h in PBS containing Hoechst 33342 nuclear stain at room temperature and analyzed by fluorescent microscope (IX71, OLYMPUS, Japan). Intracellular localization of the DSBL liposomes is visualized in the cytoplasm of the cells as shown in Fig. 10.11. Suspension of PKH67-labeled DSBL liposomes in PBS is detected by fluorescent microscopy (Fig. 10.11A). After HeLa cells are incubated with this liposome suspension for 3 h, the liposomes are internalized into the cell cytoplasm, but not the cell nucleus (Fig. 10.11(B–D)), without any ligands conjugated on the surface of the liposomes.

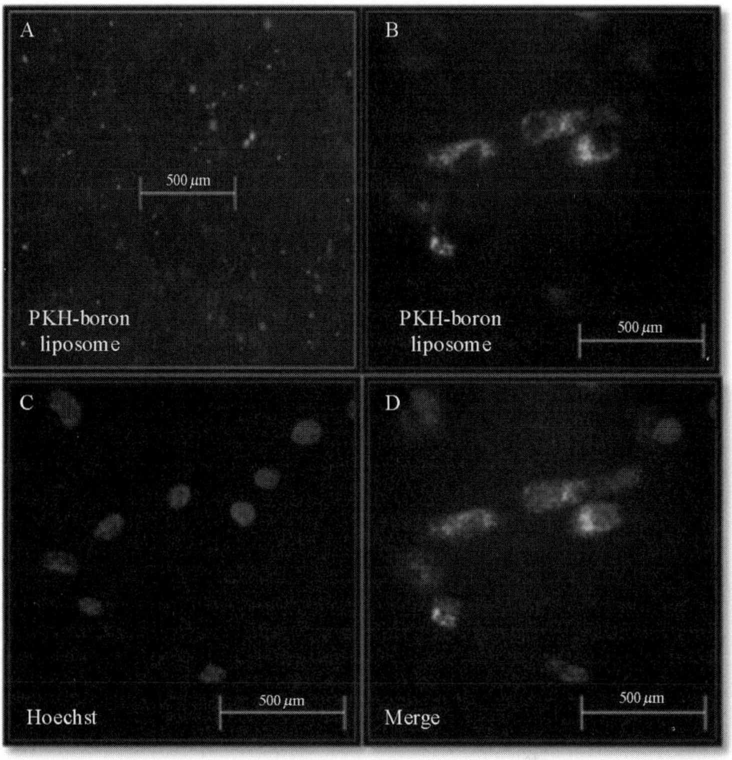

Figure 10.11 PKH67-labeled boron liposomes and intracellular location of PKH-labeled boron liposomes. (A) PKH67-labeled 25% DSBL liposomes in PBS were visualized in fluorescent microscope. (B) Intracellular (HeLa) location of PKH67-labeled 25% DSBL liposomes were visualized in fluorescent microscope. (C) Hoechst-labeled nucleuses in HeLa were visualized in fluorescent microscope. (D) The merge image of PKH67-labeled 25% DSBL liposomes (B) and Hoechst-labeled nucleus (C) is shown in (D). (See Color Insert.)

6.5. Acute toxicity and accumulation of liposomes prepared from *closo*-dodecaborate lipids and cholesterols in healthy mice

Acute toxicity and distribution of boron liposomes in mice are examined. The results are summarized in Table 10.1. Although the acute toxicity was observed in mice injected with the 50% DSBL and BCC liposomes at a dose of 30 mg ^{10}B/kg body weight, the lethal toxicity was not observed at lower boron concentrations. The boron concentration in various organs including liver and spleen is measured 3 weeks after injection of various boron liposomes. Boron accumulation was not detected in liver and spleen of the healthy mice injected with boron liposomes composed of double-tailed

Table 10.1 Acute toxicity and accumulation of boron liposomes in mice

Boron lipids		Content (%)	Injection dose (mg B/kg)	Liver (ppm B)	Spleen (ppm B)	Kidney (ppm B)	Toxicity in 72 h (%)
DSBL	**8c**	25	15	0.71 ± 0.05	1.01 ± 0.07	0.18 ± 0.01	0
		50	30	0.59 ± 0.18	0.96 ± 0.29	0.20 ± 0.00	33
			20	0.22 ± 0.00	0.17 ± 0.00	0.07 ± 0.00	0
DPBL	**8b**	25	15	0.77 ± 0.00	1.28 ± 0.00	0.19 ± 0.00	0
			10	0.27 ± 0.06	0.34 ± 0.10	0.24 ± 0.03	0
		50	20	0.39 ± 0.10	0.36 ± 0.08	0.56 ± 0.11	0
			15	0.26 ± 0.00	0.61 ± 0.00	0.50 ± 0.00	0
DSCBL	**9c**	25	15	0.79 ± 0.01	1.59 ± 1.00	0.19 ± 0.00	0
			10	0.44 ± 0.08	0.88 ± 0.57	0.41 ± 0.08	0
		50	20	0.78 ± 0.04	1.38 ± 0.32	0.26 ± 0.03	0
DPCBL	**9b**	25	10	1.03 ± 0.03	1.59 ± 0.26	0.24 ± 0.01	0
			5	0.83 ± 0.27	0.64 ± 0.21	0.30 ± 0.02	0
		50	20	0.68 ± 0.21	0.82 ± 0.36	0.24 ± 0.02	0
			15	0.73 ± 0.21	0.96 ± 0.38	0.19 ± 0.03	0
			10	0.61 ± 0.12	0.99 ± 0.03	0.18 ± 0.03	0
BC	**23a**	25	10	34.3 ± 2.65	34.3 ± 1.21	1.94 ± 0.20	0
			5	22.4 ± 0.10	19.5 ± 0.34	1.36 ± 0.11	0
		50	10	42.9 ± 0.00	37.3 ± 0.00	1.10 ± 0.00	0
			5	21.2 ± 1.66	21.3 ± 2.25	0.74 ± 0.07	0
BCC	**23b**	25	15	60.8 ± 8.84	38.0 ± 2.95	4.66 ± 0.22	0
			10	39.1 ± 5.14	15.9 ± 0.33	1.76 ± 0.26	0
		50	30	78.5 ± 12.5	60.7 ± 3.62	7.11 ± 0.00	33
			20	79.1 ± 0.00	54.9 ± 0.00	3.98 ± 0.00	0

Data are expressed as mean ± S.E.M.

boron lipids, **8a–c** and **9a–c**. However, high boron concentrations in those organs were observed in the mice injected with the liposomes composed of the boron cholesterols **23a–b**.

6.6. Biodistribution of *closo*-dodecaborate lipid-liposomes in mice

Tumor-bearing mice (female, 5–6 weeks old, 16–20 g) are prepared by inoculating subcutaneously a suspension (2.5×10^6 cell/mouse) of colon 26 cells directly into the right thigh. The mice are kept on a regular mouse diet and water, and maintained under a standard light/dark cycle at an ambient atmosphere. Biodistribution experiments are performed when the tumor size is in the range from 7 to 9 mm in diameter. The tumor-bearing mice are injected via the tail vein with 200 μl of Na_2BSH (6000 ppm B) in 0.9% NaCl solution or 200 μl solutions of 25% DSBL (**8c**) PEG-liposomes. The mice are lightly anesthetized and blood samples are collected from the retro-orbital sinus 24 h after injection. The mice are then sacrificed by cervical dislocation and dissected. The various organs, including liver spleen, kidney, heart, brain, lung, muscle, and tumor, are excised, washed with 0.9% NaCl solution, and weighed. The excised organs are digested with 2 ml of conc. HNO_3 (ultratrace analysis grade; Wako, Tokyo, Japan) at 90 °C for 1–3 h, and then the digested samples are diluted with distillated water. After filtration with hydrophobic filter (13JP050AN, ADVANTEC), the boron concentration is measured by ICP-AES. Figure 10.12 shows the boron concentrations in various organs of tumor-bearing mice 24 h after injection of the boron liposomes composed of DSBL (25% and 50%) at doses of 20 mg B/kg, and 1 h after injection of

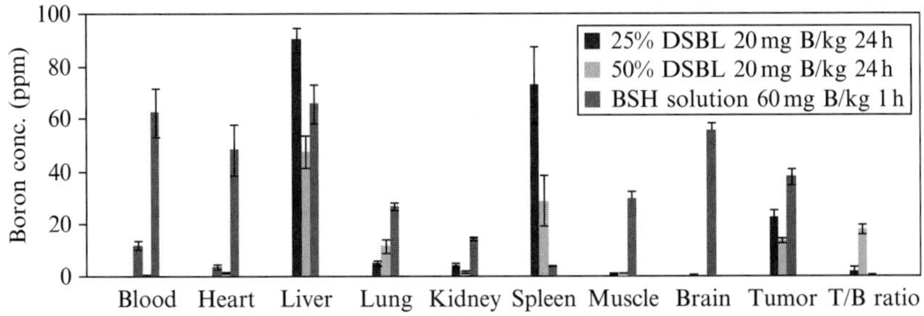

Figure 10.12 Biodistributions of mice (Balb/c, female, 6 weeks old, 14–20 g) bearing colon 26 solid tumors 24 h after i.v. injection of 25% and 50% DSBL PEG-liposomes and BSH solution. ■, 25% DSBL PEG-liposomes (20 mg B/kg); ■, 50% DSBL PEG-liposomes (20 mg B/kg); ■, BSH solution (60 mg B/kg). Data are expressed as mean ± S.E.M. (n = 4–5).

Na_2BSH solution at a dose of 60 mg B/kg. The boron concentrations in tumor were 22.7 and 13.4 ppm in the mice injected with the 25% and 50% DSBL liposomes (20 mg B/kg), respectively. The 25% DSBL liposomes showed higher boron concentrations in blood compared to the 50% DSBL liposomes. The tumor/blood ratio of the 50% DSBL liposomes increased to 7.22, which was higher than those of 25% DSBL liposomes (1.94) and BSH solution (0.48). These boron liposomes also accumulated in the liver and spleen. Na_2BSH accumulated in various tissues including the liver, spleen, heart, muscle, blood, and tumor, nonselectively and a higher boron concentration was also obtained in the tumor in comparison with DSBL liposomes. However, the injected boron dose of BSH was much higher than that of DSBL liposomes; therefore the percent injected boron dose per tumor (%ID) accumulation of BSH and DSBL liposomes is calculated. The %ID value of 25% DSBL liposomes was 5.68, where as those of 50% DSBL liposomes and Na_2BSH were 3.36 and 3.15, respectively. These results indicate that the 25% DSBL liposomes accumulate in tumors more selectively than BSH and 50% DSBL liposomes.

6.7. Tumor growth in mice administer DSBL-liposomes after neutron irradiation

DSBL-25% liposomes, which are prepared from the ^{10}B-enriched DSBL (**8c**), DSPC, Chol, and DSPE-PEG (0.25:0.75:1:0.11, molar ratio), are injected into colon 26 tumor-bearing mice (female, 6–7 weeks old, 16–20 g) via the tail at a dose of 20 mg ^{10}B/kg (2000 ppm of ^{10}B concentration; 200 μl of boron liposome solution). The mice are anesthetized with isoflurane (Forane, Abbott, Japan) and placed in an acrylic mouse holder 24 h after *i.v.* injection. The mice are irradiated in the JRR4 for 30 min at a rate of $0.9–1.4 \times 10^{12}$ neutrons/cm^2. The antitumor effects of BNCT are evaluated on the basis of the changes in tumor volume of the mice. The mortality is monitored daily and tumor volume is measured at intervals of a few days. For determining the tumor volume, two perpendicular diameters of the tumor are measured with a slide caliper, and calculation is carried out using the formula $0.5\ (A \times B^2)$, where A and B are the longest and shortest dimensions of the tumor in millimeters, respectively. All protocols are approved by the Institutional Animal Care and Use Committee in Gakushuin University. As shown in Fig. 10.13, the tumor growth rate in mice treated with the boron liposomes was significantly inhibited after thermal neutron irradiation. Suppression of tumor growth was observed during 2 weeks after neutron irradiation, although rapid tumor growth was observed in the mice without injection of the boron liposomes after neutron irradiation.

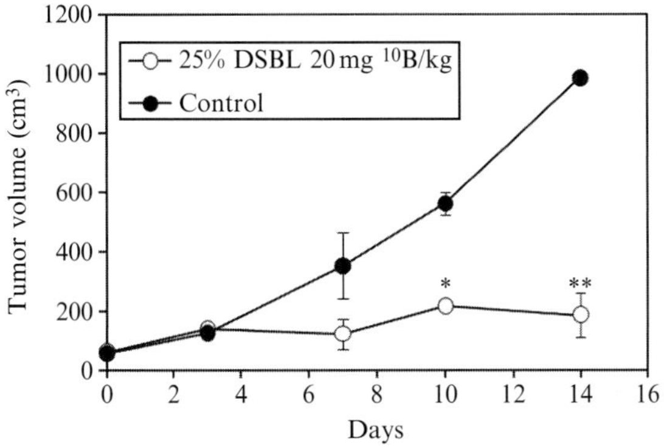

Figure 10.13 Tumor growth of mice (Balb/c, female, 6 weeks old, 14–20 g) bearing colon 26 solid tumor after thermal neutron irradiation for 30 min. The mice were injected with 25% DSBL PEG-liposomes (20 mg B/kg) incubated for 24 h before irradiation. Control indicated tumor growth of mice after neutron irradiation without administration of 25% DSBL PEG-liposomes. Data are expressed as mean ± S.E. ($n = 4$).

7. Concluding Remarks

BNCT is a binary system of thermal neutrons and neutron absorbers for the treatment of cancer. The clinical treatment with BNCT has been limited to the location and number of patients, because thermal neutrons are available only from atomic reactors. Recent development of accelerator technologies displays a possibility of accelerator-based BNCT in the near future, and in fact several accelerators are now under development for this purpose. The second component of this binary system involves the boron delivery system. Sufficient boron accumulation in the tumor tissues is the most important requirement for efficient BNCT. Liposomal boron delivery technologies described in this chapter may become one of the effective tools for boron delivery to tumor tissues. Accompanied with the establishment of hospital-based accelerators and development of new boron delivery technologies, BNCT will become the major modality for the next generation of cell-selective radiation therapy of cancers.

REFERENCES

Aihata, T., Hiratsuka, J., Morita, N., Uno, M., Sakurai, Y., Maruhashi, A., Ono, K., and Harada, T. (2006). First clinical case of boron neutron capture therapy for head and neck malignancies using ^{18}F-BPA PET. *Head Neck* **28,** 850–855.

Alam, F., Soloway, A. H., McGuire, J. E., Barth, R. F., Carey, W. E., and Adams, D. (1985). Dicesium *N*-succinimidyl-3-(undecahydro-closo-dodecaboranyldithio)propionate, a novel heterobifunctional boronating agent. *J. Med. Chem.* **28,** 522–525.

Alam, F., Soloway, A. H., Bapat, B. V., Barth, R. F., and Adams, D. M. (1989). Boron compounds for neutron capture therapy. *Basic Life Sci.* **50,** 107–111.

Backer, M. V., Gaynutdinov, T. I., Patel, V., Bandyopadhyaya, A. K., Thirumamagal, B. T., Tjarks, W., Barth, R. F., Claffey, K., and Backer, J. M. (2005). Vascular endothelial growth factor selectively targets boronated dendrimers to tumor vasculature. *Mol. Cancer Ther.* **4,** 1423–1429.

Barth, R. F., Soloway, A. H., and Fairchild, R. G. (1990). Boron neutron capture therapy of cancer. *Cancer Res.* **50,** 1061–1070.

Barth, R. F., Soloway, A. H., Fairchild, R. G., and Brugger, R. M. (1992). Boron neutron capture therapy for cancer, realities and prospects. *Cancer* **70,** 2995–3007.

Cai, J., Soloway, A. H., Barth, R. F., Adams, D. M., Hariharan, J. R., Wyzlic, I. M., and Radcliffe, K. (1997). Boron-containing polyamines as DNA targeting agents for neutron capture therapy of brain tumors: Synthesis and biological evaluation. *J. Med. Chem.* **40,** 3887–3896.

Capala, J., Barth, R. F., Bendayan, M., Lauzon, M., Adams, D. M., Soloway, A. H., Fenstermaker, R. A., and Carlsson, J. (1996). Boronated epidermal growth factor as a potential targeting agent for boron neutron capture therapy of brain tumors. *Bioconjug. Chem.* **7,** 7–15.

Feakes, D. A., Shelly, K., Knobler, C. B., and Hawthorne, M. F. (1994). $Na_3[B_{20}H_{17}NH_3]$: Synthesis and liposomal delivery to murine tumors. *Proc. Natl. Acad. Sci. USA* **91,** 3029–3033.

Feakes, D. A., Shelly, K., and Hawthorne, M. F. (1995). Selective boron delivery to murine tumors by lipophilic species incorporated in the membranes of unilamellar liposomes. *Proc. Natl. Acad. Sci. USA* **92,** 1367–1370.

Feakes, D. A., Spinler, J. K., and Harris, F. R. (1999). Synthesis of boron-containing cholesterol derivatives for incorporation into unilamellar liposomes and evaluation as potential agents for BNCT. *Tetrahedron* **55,** 11177–11186.

Fiske, C. H., and Subbarow, Y. (1925). The colorimetric determination of phosphorus. *J. Biol. Chem.* **66,** 375–400.

Fulcrand-El Kattan, G., Lesnikowski, Z. J., Yao, S., Tanious, F., Wilson, W. D., and Schinazi, R. F. (1994). Carboranyl oligonucleotides. Synthesis and physicochemical properties of dodecathymidylate containing 5-(*o*-carboran-1-yl)-2′-deoxyuridine. *J. Am. Chem. Soc.* **116,** 7494–7501.

Gedda, L., Olsson, P., and Carlsson, J. (1996). Development and *in vitro* studies of epidermal growth factor-dextran conjugates for boron neutron capture therapy. *Bioconjug. Chem.* **7,** 584–591.

Gedda, L., Silvander, M., Sjoberg, S., Tjarks, W., and Carlsson, J. (1997). Cytotoxicity and subcellular localization of boronated phenanthridinium analogues. *Anticancer Drug Des.* **12,** 671–685.

Goldenberg, D. M., Sharkey, R. M., Primus, F. J., Mizusawa, E., and Hawthorne, M. F. (1984). Neutron-capture therapy of human cancer: *In vivo* results on tumor localization of boron-10-labeled antibodies to carcinoembryonic antigen in the GW-39 tumor model system. *Proc. Natl. Acad. Sci. USA* **81,** 560–563.

Hawthorne, M. F. (1993). The role of chemistry in the development of boron neutron capture therapy of cancer. *Angew. Chem. Int. Ed. Engl.* **32,** 950–984.

Iinuma, H., Maruyama, K., Okinaga, K., Sasaki, K., Sekine, T., Ishida, O., Ogiwara, N., Johkura, K., and Yonemura, Y. (2002). Intracellular targeting therapy of cisplatin-encapsulated transferrin-polyethylene glycol liposome of peritoneal dissemination of gastric cancer. *Int. J. Cancer* **99,** 130–137.

Imahori, Y., Ueda, S., Ohmori, Y., Sakae, K., Kusuki, T., Kobayashi, T., Takagaki, M., Ono, K., Ido, T., and Fujii, R. (1998). Positron emission tomography-based boron neutron capture therapy using boronophenylalanine for high-grade gliomas: Part I. *Clin. Cancer Res.* **4,** 1825–1832.

Ishida, O., Maruyama, K., Tanahashi, H., Iwatsuru, M., Sasaki, K., Eriguchi, M., and Yanagie, H. (2001). Liposomes bearing polyethyleneglycol coupled transferrin with intracellular targeting property to the solid tumors *in vivo*. *Pharm. Res.* **18,** 1042–1048.

Justus, E., Awad, D., Hohnholt, M., Schaffran, T., Edwards, K., Karlsson, G., Damian, L., and Gabel, D. (2007). Synthesis, liposomal preparation, and *in vitro* toxicity of two novel dodecaborate cluster lipids for boron neutron capture therapy. *Bioconjug. Chem.* **18,** 1287–1293.

Kahl, S. B., and Kasar, R. A. (1996). Simple, high-yield synthesis of polyhedral carborane amino acids. *J. Am. Chem. Soc.* **118,** 1223–1234.

Kahl, S. B., and Li, J. (1996). Synthesis and characterization of a boronated metallophthalocyanine for boron neutron capture therapy. *Inorg. Chem.* **35,** 3878–3880.

Kahl, S. B., Joel, D. D., Nawrocky, M. M., Micca, P. L., Tran, K. P., Finkel, G. C., and Slatkin, D. N. (1990). Uptake of a *nido*-carboranylporphyrin by human glioma xenografts in athymic nude mice and by syngeneic ovarian carcinomas in immunocompetent mice. *Proc. Natl. Acad. Sci. USA* **87,** 7265–7269.

Kato, I., Ono, K., Sakurai, Y., Ohmae, M., Maruhashi, A., Imahori, Y., Kirihata, M., Nakazawa, M., and Yura, Y. (2004). Effectiveness of BNCT for recurrent head and neck malignancies. *Appl. Radiat. Isot.* **61,** 1069–1073.

Kelly, D. P., Bateman, S. A., Martin, R. F., Reum, M. E., Rose, M., and Whittaker, A. D. (1994). Synthesis and characterization of boron-containing bibenzimidazoles related to the minor groove binder, Hoechst 33258. *Aust. J. Chem.* **47,** 247–262.

Kirihata, M., Morimoto, T., Mizuta, T., and Ichimoto, I. (1995). Synthesis of *p*-boronophenylserine, a new boron-containing amino acid for boron neutron capture therapy. *Biosci. Biotechnol. Biochem.* **59,** 2317–2318.

Kullberg, E. B., Carlsson, J., Edwards, K., Capala, J., Sjöberg, S., and Gedda, L. (2003). Introductory experiments on ligand liposomes as delivery agents for boron neutron capture therapy. *Int. J. Oncol.* **23,** 461–467.

Lee, J.-D., Ueno, M., Miyajima, Y., and Nakamura, H. (2007). Synthesis of boron cluster lipids: *closo*-Dodecaborate as an alternative hydrophilic function of boronated liposomes for neutron capture therapy. *Org. Lett.* **9,** 323–326.

Li, T., Hamdi, J., and Hawthorne, M. F. (2006). Unilamellar liposomes with enhanced boron content. *Bioconjug. Chem.* **17,** 15–20.

Locher, G. L. (1936). Biological effects and the therapeutic possibilities of neutrons. *Am. J. Roentgenol.* **36,** 1–13.

Malan, C., and Morin, C. (1996). Synthesis of 4-borono-L-phenylalanine. *Synletters* **2,** 167–168.

Maruyama, K., Ishida, O., Kasaoka, S., Takizawa, T., Utoguchi, N., Shinohara, A., Chiba, M., Kobayashi, H., Eriguchi, M., and Yanagie, H. (2004). Intracellular targeting of sodium mercaptoundecahydrododecaborate (BSH) to solid tumors by transferrin-PEG liposomes, for boron neutron-capture therapy (BNCT). *J. Control. Release* **98,** 195–207.

Masunaga, S., Kasaoka, S., Maruyama, K., Nigg, D., Sakurai, Y., Nagata, K., Suzuki, M., Kinashi, Y., Maruhashi, A., and Ono, K. (2006). The potential of transferrin-pendant-type polyethyleneglycol liposomes encapsulating decahydrodecaborate-10B (GB-10) as 10B-carriers for boron neutron capture therapy. *Int. J. Radiat. Oncol. Biol. Phys.* **66,** 1515–1522.

Mishima, Y., Ichihashi, M., Htta, S., Honda, C., Yamamura, K., and Nakagawa, T. (1989). New thermal neutron capture therapy for malignant melanoma: Melanogenesis-seeking

^{10}B molecule-melanoma cell interaction from *in vitro* to first clinical trial, pigment. *Cell Res.* **2,** 226–234.

Miura, M., Micca, P. L., Fisher, C. D., Heinrichs, J. C., Donaldson, J. A., Finkel, G. C., and Slatkin, D. N. (1996). Synthesis of a nickel tetracarboranylphenylporphyrin for boron neutron-capture therapy: Biodistribution and toxicity in tumor-bearing mice. *Int. J. Cancer* **68,** 114–119.

Miyajima, Y., Nakamura, H., Kuwata, Y., Lee, J.-D., Masunaga, S., Ono, K., and Maruyama, K. (2006). Transferrin-loaded *nido*-carborane liposome: Synthesis and intracellular targeting to solid tumors for boron neutron capture therapy. *Bioconjug. Chem.* **17,** 1314–1320.

Mumtaz, S., Ghosh, P. C., and Bachhawat, B. K. (1991). Design of liposomes for circumventing the reticuloendothelial cells. *Glycobiology* **1,** 505–510.

Murakami, H., Nagasaki, T., Hamachi, I., and Shinkai, S. (1993). Sugar sensing utilizing aggregation properties of boronic-acid-appended porphyrins. *Tetrahedron Lett.* **34,** 6273–6276.

Nakagawa, Y., and Hatanaka, H. (1997). Boron neutron capture therapy. Clinical brain tumor studies. *J. Neurooncol.* **33,** 105–115.

Nakamura, H., Sekido, M., and Yamamoto, Y. (1997). Synthesis of carboranes containing an azulene framework and *in vitro* evaluation as boron carriers. *J. Med. Chem.* **40,** 2825–2830.

Nakamura, H., Fujiwara, M., and Yamamoto, Y. (1998). A concise synthesis of enantiomerically pure l-(4-boronophenyl)alanine from l-tyrosine. *J. Org. Chem.* **63,** 7529–7530.

Nakamura, H., Fujiwara, M., and Yamamoto, Y. (2000). A practical method for the synthesis of enantiomerically pure 4-borono-L-phenylalanine. *Bull. Chem. Soc. Jpn.* **73,** 231–235.

Nakamura, H., Miyajima, Y., Takei, T., Kasaoka, T., and Maruyama, K. (2004). Synthesis and vesicle formation of a *nido*-carborane cluster lipid for boron neutron capture therapy. *Chem. Commun.* **17,** 1910–1911.

Nakamura, H., Ueno, M., Lee, J.-D., Ban, H. S., Justus, E., Fan, P., and Gabel, D. (2007). Synthesis of dodecaborate-conjugated cholesterols for efficient boron delivery in neutron capture therapy. *Tetrahedron. Lett.* **48,** 3151–3154.

Nakanishi, A., Guan, L., Kane, R. R., Kasamatsu, H., and Hawthorne, M. F. (1999). Toward a cancer therapy with boron-rich oligomeric phosphate diesters that target the cell nucleus. *Proc. Natl. Acad. Sci. USA* **96,** 238–241.

Pak, R. H., Primus, F. J., Rickard-Dickson, K. J., Ng, L. L., Kane, R. R., and Hawthorne, M. F. (1995). Preparation and properties of *nido*-carborane-specific monoclonal antibodies for potential use in boron neutron capture therapy for cancer. *Proc. Natl. Acad. Sci. USA* **92,** 6986–6990.

Pan, X. Q., Wang, H., Shukla, S., Sekido, M., Adams, D. M., Tjarks, W., Barth, R. F., and Lee, R. J. (2002). Boron-Containing folate receptor-targeted liposomes as potential delivery agents for neutron capture therapy. *Bioconjug. Chem.* **13,** 435–442.

Pan, X., Wu, G., Yang, W., Earth, R. F., Tjarks, W., and Lee, R. J. (2007). Synthesis of cetuximab-immunoliposomes via a cholesterol-based membrane anchor for targeting of EGFR. *Bioconjug. Chem.* **18,** 101–108.

Rong, F.-G., and Soloway, A. H. (1994). Synthesis of 5-tethered carborane-containing pyrimidine nucleosides as potential agents for DNA incorporation. *Nucleosides Nucleotides* **13,** 2021–2034.

Schinazi, R. F., and Prusoff, W. H. (1985). Synthesis of 5-(dihydroxyboryl)-2′-deoxyuridine and related boron-containing pyrimidines. *J. Org. Chem.* **50,** 841–847.

Shelly, K., Feakes, D. A., Hawthorne, M. F., Schmidt, P. G., Krisch, T. A., and Bauer, W. F. (1992). Model studies directed toward the boron neutron-capture therapy of cancer: Boron delivery to murine tumors with liposomes. *Proc. Natl. Acad. Sci. USA* **89,** 9039–9043.

Shukla, S., Wu, G., Chatterjee, M., Yang, W., Sekido, M., Diop, L. A., Müller, R., Sudimack, J. J., Lee, R. J., Barth, R. F., and Tjarks, W. (2003). Synthesis and biological evaluation of folate receptor-targeted boronated PAMAM dendrimers as potential agents for neutron capture therapy. *Bioconjug. Chem.* **14,** 158–167.

Soloway, A. H., Hatanaka, H., and Davis, M. A. (1967). Penetration of brain and brain tumor. VII. Tumor-binding sulfhydryl boron compounds. *J. Med. Chem.* **10,** 714–717.

Soloway, A. H., Tjarks, W., Barnum, B. A., Rong, F. G., Barth, R. F., Codogni, I. M., and Wilson, J. G. (1998). The chemistry of neutron capture therapy. *Chem. Rev.* **98,** 1515–1562.

Sood, A., Shaw, B. R., and Spielvogel, B. F. (1989). Boron-containing nucleic acids. Synthesis of cyanoborane adducts of 2′-deoxynucleosides. *J. Am. Chem. Soc.* **111,** 9234–9235.

Sood, A., Shaw, B. R., and Spielvogel, B. F. (1990). Boron-containing nucleic acids. Synthesis of oligodeoxynucleoside boranophosphates. *J. Am. Chem. Soc.* **112,** 9000–9001.

Srivastava, R. R., Singhaus, R. R., and Kabalka, G. W. (1997). Synthesis of 1-amino-3-[2-(1,7-dicarba-closo-dodecaboran(12)-1-yl)ethyl]cyclobutanecarboxylic acid: A potential BNCT agent. *J. Org. Chem.* **62,** 4476–4478.

Suzuki, M., Sakurai, Y., Hagiwara, S., Masunaga, S., Kinashi, Y., Nagata, K., Maruhashi, A., Kudo, M., and Ono, K. (2007). First attempt of boron neutron capture therapy (BNCT) for hepatocellular carcinoma. *Jpn. J. Oncol.* **37,** 376–381.

Synder, H. R., Reedy, A. J., and Lennarz, W. J. (1958). Synthesis of aromatic boronic acids. Aldehydo boronic acids and a boronic acid analog of tyrosine. *J. Am. Chem. Soc.* **80,** 835–838.

Takagaki, M., Ono, K., Oda, Y., Kikuchi, H., Nemoto, H., Iwamoto, S., Cai, J., and Yamamoto, Y. (1996). Hydroxylforms of *p*-boronophenylalanine as potential boron carriers on boron neutron capture therapy for malignant brain tumors. *Cancer Res.* **56,** 2017–2020.

Thirumamagal, B. T. S., Zhao, X. B., Bandyopadhyaya, A. K., Narayanasamy, S., Johnsamuel, J., Tiwari, R., Golightly, D. W., Patel, V., Jehning, B. T., Backer, M. V., Barth, R. F., Lee, R. J., *et al.* (2006). Receptor-targeted liposomal delivery of boron-containing cholesterol mimics for boron neutron capture therapy (BNCT). *Bioconjug. Chem.* **17,** 1141–1150.

Vaage, J., Mayhew, E., Lasic, D., and Martin, F. (1992). Therapy of primary and metastatic mouse mammary carcinomas with doxorubicin encapsulated in long circulating liposomes. *Int. J. Cancer.* **51,** 942–948.

Watson-Clark, R. A., Banquerigo, M. L., Shelly, K., Hawthorn, M. F., and Brahn, E. (1998). Model studies directed toward the application of boron neutron capture therapy to rheumatoid arthritis: Boron delivery by liposomes in rat collagen-induced arthritis. *Proc. Natl. Acad. Sci. USA* **95,** 2531–2534.

Wei, Q., Kullberg, E. B., and Gedda, L. (2003). Trastuzumab-conjugated boron-containing liposomes for tumor-cell targeting; development and cellular studies. *Int. J. Oncol.* **23,** 1159–1165.

Woodburn, K., Phadke, A. S., and Morgan, A. R. (1993). An *in vitro* study of boronated porphyrins for potential use in boron neutron capture therapy. *Bioorg. Med. Chem.* **3,** 2017–2022.

Wu, G., Barth, R. F., Yang, W., Chatterjee, M., Tjarks, W., Ciesielski, M. J., and Fenstermaker, R. A. (2004). Site-specific conjugation of boron-containing dendrimers to anti-EGF receptor monoclonal antibody cetuximab (IMC-C225) and its evaluation as a potential delivery agent for neutron capture therapy. *Bioconjug. Chem.* **15,** 185–194.

Yamamoto, Y., and Nakamura, H. (1993). 1-Carboranyl-3-(2-methylaziridino)-2-propanol. Synthesis, selective uptake by B-16 melanoma, and selective cytotoxicity toward cancer cells. *J. Med. Chem.* **36,** 2232–2234.

Yamamoto, Y., Seko, T., Nakamura, H., Nemoto, H., Hojo, H., Mukai, N., and Hashimoto, Y. (1992). Synthesis of carboranes containing nucleoside bases. Unexpectedly high cytostatic and cytocidal toxicity toward cancer cells. *J. Chem. Soc., Chem. Commun.* 157–158.

Yamamoto, Y., Cai, J., Nakamura, H., Sadayori, N., Asao, N., and Nemoto, H. (1995). Synthesis of netropsin and distamycin analogs bearing *o*-carborane and their DNA recognition. *J. Org. Chem.* **60,** 3352–3357.

Yanagiё, H., Fujii, Y., Takahashi, T., Tomita, T., Fukano, Y., Hasumi, K., Nariuchi, H., Yasuda, T., Sekiguchi, M., and Uchida, H. (1989). Boron neutron capture therapy using ^{10}B entrapped anti-CEA immunoliposome. *Hum. Cell* **2,** 290–296.

Yanagiё, H., Tomita, T., Kobayashi, H., Fujii, Y., Takahashi, T., Hasumi, K., Nariuchi, H., and Sekiguchi, M. (1991). Application of boronated anti-cea immunoliposome to tumour cell growth inhibition in in vitro boron neutron capture therapy model. *Br. J. Cancer* **63,** 522–526.

Yanagiё, H., Tomita, T., Kobayashi, H., Fujii, Y., Nonaka, Y., Saegusa, Y., Hasumi, K., Eriguchi, M., Kobayashi, T., and Ono, K. (1997). Inhibition of human pancreatic cancer growth in nude mice by boron neutron capture therapy. *Br. J. Cancer* **75,** 660–665.

Yang, W., Barth, R. F., Adams, D. M., and Soloway, A. H. (1997). Intratumoral delivery of boronated epidermal growth factor for neutron capture therapy of brain tumors. *Cancer Res.* **57,** 4333–4339.

CHAPTER ELEVEN

Production of Recombinant Proteoliposomes for Therapeutic Uses

Lavinia Liguori *and* Jean Luc Lenormand

Contents

Abstract

One of the major challenges in human therapy is to develop delivery systems that are convenient and effective for tackling problems in disease treatments. In the past 20 years, liposomes have represented promising pharmaceutical carriers for drug delivery. Due to their biophysical properties, liposomes can deliver and specifically target a large set of bioactive molecules, they can protect molecules from degradation, and their composition is easily modifiable.

The use of recombinant proteoliposomes containing therapeutic membrane proteins is a recently developed technology that allows biologically active proteins to penetrate across the plasma membrane of eukaryotic cells. One of the bottlenecks in this powerful delivery system lies in the production of functional therapeutic membrane proteins mainly due to their biophysical characteristics. Membrane proteins represent about 30% of the total proteins

HumProTher Laboratory, TIMC-ThereX, UMR 5525 CNRS-UJF, Université Joseph Fourier, UFR de Médecine, Domaine de la Merci, La Tronche, France

Methods in Enzymology, Volume 465
ISSN 0076-6879, DOI: 10.1016/S0076-6879(09)65011-4

from an organism, and play a central role in drug discovery as potential pharmaceutical targets. This chapter describes the methodology for the production of bioactive proteoliposomes containing therapeutic, proapoptotic membrane proteins synthesized with an optimized cell-free expression system. We will examine (1) the design of the expression vectors and the liposome compositions compatible with the cell-free expression system; (2) the production of membrane proteins using a cell-free expression system in combination with liposomes, to obtain in a one-step reaction functional therapeutic proteoliposomes; (3) proteoliposome purification for further use in the treatment of cancer cells; and (4) the methodology for detecting apoptosis in cells after treatment. Furthermore, this system can be easily adapted for producing "difficult to express proteins" compared with the classical overexpression (bacterial or eukaryotic) systems.

1. Introduction

One of the actual challenges in human therapy today is to develop new strategies to efficiently deliver active macromolecules specifically in diseased cells and not in normal cells. Over the past two decades, important progress has been made in the optimization of drug delivery technologies for human therapy (Torchilin, 2006). However, the efficacy of biologically active drugs depends on the characteristics of the drug delivery system. Such delivery systems result in the release of drugs with increased efficiency and a decrease of side effects. The development of a perfect controlled drug delivery technology is still in progress, and the recent cutting-edge technologies, including polymeric micelles and liposomes, are tackling this challenge.

Among the delivery systems, liposomes have represented good candidates for the delivery of macromolecules into cells (Düzgüneş *et al.*, 2001). Liposomes are pharmaceutical carriers of choice for various applications, because they are noncytotoxic, they can deliver and specifically target a large set of bioactive molecules (such as proteins, DNA, ribozymes, etc.), they can protect molecules from degradation, and their composition is easily modifiable (Düzgüneş *et al.*, 2005, 2007; Maeda *et al.*, 2009; Salem *et al.*, 2005; Torchilin, 2006, 2007). For instance, hydrophilic drugs can be encapsulated into the aqueous solution of the liposomes, whereas hydrophobic chemicals can be entrapped in the membrane.

Liposomes have been one of the first nanotechnologies to benefit patients with cancer or infectious disease. Recently, several liposomal anthracycline formulations have demonstrated to have a great impact in oncology. For instance, one of the most impressive effects of liposome-based drug is the delivery of the highly toxic anticancer agent doxorubicin [Doxil®/Caelyx®] in patients with breast-carcinoma metastasis, squamous cell cancer of the head and neck, ovarian cancer, or unresectable hepatocellular carcinoma resulting in a subsequent improvement of patient survival.

These liposome carriers can be surface modified to achieve long circulation of liposomes *in vivo*, and to increase the targeting of the carriers to tissues and organs (Torchilin, 2006). For instance, coating the liposome surface with inert, biocompatible polymers such as polyethylene glycol (PEG) resulted in a decrease of the clearance of liposomes and an increase in bioavailability. These long-circulating liposomes (LCL; also called Stealth liposomes) can access the angiogenic vessels in tumors by circulating for days as stable constructs, and accumulate in the tumors (Williams *et al.*, 1993). The LCL accumulate in the tumors, allow a better targeting and interaction with the target and result in delivery of 3–10 times more drug to the solid tumors as compared to free drug (Torchilin, 2005). Attaching tumor-specific targeting ligands such as monoclonal antibodies to the liposome surface (immunoliposomes) also results in an accumulation of the drug at the tumor site. These immunoliposomes are designed for cellular internalization by targeting cell surface receptors in the tumor cells (Kamps *et al.*, 2000; Lukyanov *et al.*, 2004). For example, immunoliposomes have been engineered by linking an antibody fragment against the human transferrin receptor, which is overexpressed on the surface of many tumor cells, and used *in vivo* to deliver tumor-suppressing genes into tumors. Different tissue-specific gene delivery immunoliposomes have also been achieved in brain and breast cancer tissue (Torchilin, 2005).

One of the major problems in the treatment of cancers by chemotherapy is the potential toxic side effects of the drugs on healthy cells. Liposomes used as drug delivery vehicles can overcome this problem as a result of their biochemical properties. For instance, an important property of liposomes is their ability to spontaneously penetrate into the interstitium in the body compartment and accumulate in various pathological sites such as solid tumors via the so-called enhanced permeability and retention (EPR) effect (Maeda *et al.*, 2000, 2009). The EPR effect is based on a passive targeting dependent of the cutoff size of the tumor blood vessel wall which ranges from 200 to 600 nm. Liposomes tend to accumulate in tumor tissue much more than in normal tissues due to the fact that blood vessels around the solid tumor mass usually present abnormalities in form and architecture. Consequently, this results in an accumulation of the loaded drug at the tumor site. The EPR concept has already been validated in clinical settings with different solid tumors.

In comparison to small-molecule drugs, protein therapeutics present some advantages including a large and complete set of cellular functions without side effects and immune responses. Therefore, most protein therapeutics are well tolerated after administration, and for this reason the clinical development and the approval by the national regulatory agencies are faster than that of small-molecule drugs (Leader *et al.*, 2008; Pavlou and Reichert, 2004; Tauzin, 2008). However, delivery of protein therapeutics in the human body presents a significant problem arising from the biochemical properties of the proteins, such their large sizes, their optimal activities changing as a function of the environmental pH or temperature, and their susceptibility to degradation. Different strategies are used to increase the effectiveness of therapeutic proteins. These include PEGylation, liposome formulations, protein glycoengineering, and conjugation or fusion of therapeutic proteins to proteins or transport peptides (Torchilin, 2007). All these methods have provided exceptional potency for the use of protein therapeutics in the treatment of some unmanageable diseases. Nevertheless, the hydrophobic nature of the plasma membrane impedes the uptake of large molecules such as peptides and proteins. The bioavailability constraints of the plasma membrane may represent a bottleneck in the delivery of therapeutic proteins. This situation is even worse for therapeutic membrane proteins due to their high content of hydrophobic amino acids.

Membrane proteins represent about 30% of the total proteins from an organism and are involved in essential biological processes. Dysregulation of their biological activity is one of the primary responses of the cell to bacterial or viral infections, as well as in cancer or genetic diseases. Nevertheless, one of the major difficulties in the study of membrane proteins is to recover sufficient amounts of recombinant proteins from the classical overexpression techniques (in mammalian cells, yeast or bacteria). These overexpression methods for protein production still display several kinds of limiting features. It is still difficult in the classical expression systems to make functional membrane proteins in a reasonable amount for structural studies, or functional applications (to date, only 182 membrane protein structures have been solved; see http://blanco.biomol.uci.edu/Membrane_Proteins_xtal.html). By using these protein expression systems, production of membrane proteins very often results in low yields, or in the production of aggregated or misfolded proteins.

A very interesting and attractive alternative for producing such proteins is the use of cell-free transcription/translation systems. These cell-free protein synthesis systems are essentially derived from rabbit reticulocytes, *Escherichia coli* lysates, or wheat germ (Spirin, 2004; Swartz *et al.*, 2004). Beside the production of recombinant proteins, one of the advantages of these *in vitro* systems is their ability to synthesize cytotoxic membrane proteins, and regulatory or unstable proteins that cannot be expressed in living organisms.

Apoptosis defects are recognized as an important complement to protooncongene activation in cancer progression and pathological cell expansion (Fesik, 2005). It has been reported that overexpression of one or more antiapoptotic members of the Bcl-2 family occurs in a variety of cancers, rendering tumor cells resistant to myriad apoptotic stimuli, including most cytotoxic anticancer drugs. Thus, the restoration of the apoptotic program in tumor cells by specifically targeting protein–protein interactions represents a promising and innovative therapeutic strategy to promote tumor cell death (Fantin and Leder, 2006).

Proapoptotic Bak protein belongs to the multidomain proapoptotic members of the Bcl-2 family. Bak is a membrane protein containing a transmembrane domain at the C-terminal part, and resides in complexes in the outer membrane of mitochondria (Reed, 2006). Upon cytotoxic signals, Bak forms homooligomers leading to membrane permeabilization and release of the proapoptogenic cytochrome *c* and Smac/DIABLO proteins from the intermembrane space of mitochondria to the cytoplasm. The release of proapoptotic mitochondrial proteins leads to caspase activation (e.g., caspases 9 and 3) and to promote apoptosis (Reed, 2006). It has been demonstrated that the restoration of apoptosis in cancer cells by activation of Bak using peptidomimetics or drugs, represents a promising therapeutic strategy (Walensky, 2006). However, the use of full-length Bak protein as a protein therapeutic to mimic the full activation of apoptosis was not possible due to the difficulty to produce and to deliver functional Bak protein.

In this chapter, we describe a new overexpression system that allows producing in one-step, high yields of recombinant proapoptotic Bak protein coupled to liposomes (proteoliposomes). In addition, we provide protocols for directly using the purified Bak proteoliposomes for treating cancer cells. This technology has been successfully applied to produce proteoliposomes containing different membrane proteins from mammalian sources, viruses, and porins from plants or Gram-negative bacteria (Liguori *et al.*, 2008a,b).

2. Expression of Bak Protein Using a Bacterial Cell-Free Expression System

The rapid translation system (RTS; Roche Applied Science) is a high-yield cell-free protein expression system based on T7 RNA polymerase and an optimized *E. coli* lysate. A coupled transcription/translation reaction is employed for *in vitro* protein synthesis. To optimize and enhance protein synthesis and the correct folding of the recombinant membrane proteins, different substances can be added to the reaction mix, including detergents, chaperones, and lipids (Klammt *et al.*, 2004, 2005). The proapoptotic Bak protein has been selected as a potential therapeutic candidate to restore apoptosis in cancer cells. Bak belongs to the Bcl-2 family and is located in the outer mitochondrial membrane. Bak is involved in the activation of the intrinsic apoptotic pathway (Walensky, 2006). The cDNA coding for human Bak is first cloned into the pIVEX expression vectors (see Table 11.1 for vectors compatibles with the RTS system), and then preliminary expression tests are performed in reaction volumes of 25–100 μl (Liguori *et al.*, 2008a; Marques *et al.*, 2007). The expression vectors contain a cDNA coding for a truncated form of the human Bak. The resulting protein is a 141 amino acid protein lacking the first 70 residues at the N-terminal, but

Table 11.1 Overview—RTS and different expression vector systems

	Vector compatibility with RTS	Yield of plasmid preparation
pIVEX	Yes	+++
pET family (Novagen)	Yes	+
pET Blue-1 AccepTor™ (TA cloning)	Yes	+++
pCRT7/CT-TOPO® (TA/ Topo cloning)	Weak (sequence between somal-binding site and codon not optimized)	+++
pDEST17 (Gateway™)	Yes	+
pEXP1-Dest (Gateway™)	Yes	+++

www.roche-applied-science.com.

containing the BH1, BH2, BH3 motifs and the complete carboxyl-terminal hydrophobic domain (Fig. 11.1A).

1. The PCR amplification is performed by designing primers to directly subclone Bak cDNA into a pIVEX 2.4dNde1 or a pIVEX 2.3 vector containing a histidine tag (6xHis) at the N terminus or at the C terminus, respectively. Primer sequences and cloning protocols are described in Liguori *et al.* (2008b).
2. After purification of the PCR products from agarose gel, ligation into the expression vector and bacterial transformations are performed as previously described (Liguori *et al.*, 2008a). Protein synthesis is carried out in a batch format (1.5 ml Eppendorf tube) using the RTS 100 *E. coli* HY system (from 12 to 100 μl final volume reactions) following the manufacturer's recommendations (Liguori *et al.*, 2008a). The GFP containing vector is used as a positive control for protein expression.
3. After an overnight incubation at 30 °C, the reaction mixture is centrifuged at 13,000×*g* for 10 min at 4 °C. The pellet is resuspended in Tris–Cl 50 m*M*, pH 7.5, and 1/10 of the final volume reaction are loaded onto a 15% SDS–PAGE. Proteins are transferred onto a nitrocellulose membrane (0.22 μm, Biorad) 1 h at 0.8 mA/cm^2 gel (~50 mA/gel) and the membrane stained with Ponceau Red to verify the transfer efficiency. After 1 h in a blocking solution (4% fat-free milk in TBS-T: Tris-HCl 0.05 *M*, pH 8.3, 0.2 *M* NaCl, 0.1% Tween 20), the membrane is incubated for 1 h at room temperature (RT) with a monoclonal anti-His HRP-conjugated antibody (1:5000 in blocking solution, Sigma-Aldrich). The membrane is washed three times (10 min) in TBS-T and positive bands are detected using an ECL detection kit (Amersham, GE Healthcare).

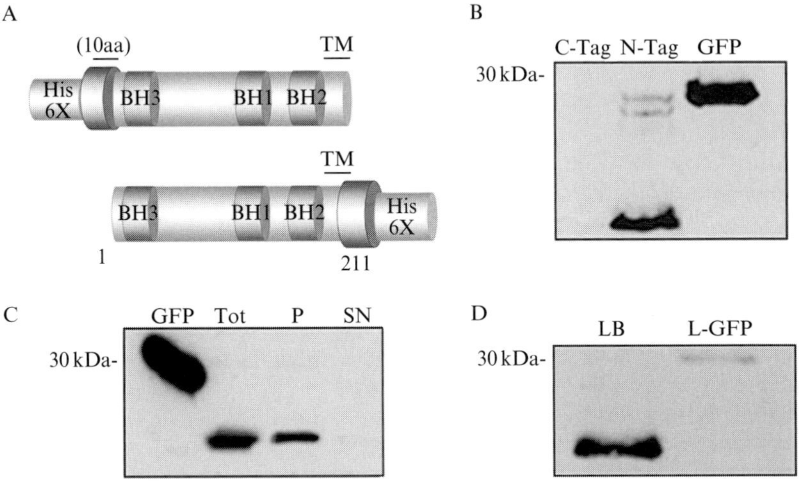

Figure 11.1 (A) Scheme of the constructs used to express human recombinant Bak. (B) Expression tests for Bak cloned in pIVEX 2.3MCS (C-Tag) and in pIVEX 2.4bNdeI (N-Tag). (C) Solubility tests for Bak, both N-tagged: P, pellet of the expression mix; SN, supernatant; Tot, total reaction mix. (D) Expression tests for Bak and GFP in the presence of liposomes.

The position of the histidine tag can interfere on the expression level of the protein as shown in Fig. 11.1B. The solubility tests confirm that the hydrophobic nature of Bak protein can only be detected in the insoluble fraction (Fig. 11.1C).

Note and reagents: Conditions that can be tested to increase expression of Bak.

- *GSH-GSSG* 0.1 m*M*:1 m*M* (final ratio, Sigma-Aldrich).
- *GroE* used according to the instruction manual (Roche Applied Science).
- *GroE* + GSH-GSSG (0.1 m*M*:1 m*M* final ratio) mix.
- *DnaK* used according to the instruction manual (Roche Applied Science).
- *Protein inhibitors:* complete protease inhibitor cocktail tablets, EDTA-free, (Roche Applied Science) Leupeptin, suggested final concentration: 1–10 μM (SigmaAldrich), Pepstatin (15 μM), specific for aspartic proteinases (Sigma-Aldrich).

3. Scale-Up Production of Bak Proteoliposomes

After an optimization of the reaction conditions, a scale-up reaction is performed in order to obtain large yields of Bak in the presence of liposomes. Reactions are carried out using the RTS 500 ProteoMaster *E. coli* HY Kit (Roche Applied Science). The reaction device is composed by two compartments: a 1-ml reaction chamber and a 10-ml feeding chamber separated by a dialysis semipermeable membrane (continuous-exchange cell-free (CECF) system). The feeding mixture (amino acids, nucleotides, and energy substrates) is supporting the reaction by diffusion exchange across the membrane. The recombinant protein is accumulated into the reaction chamber (Liguori *et al.*, 2008a). To produce Bak proteoliposomes, empty liposomes are directly added to the reaction mixture at a final concentration of 5 mg/ml and the reaction is performed either in the RTS ProteoMaster Instrument or in the Eppendorf Thermomixer. With this optimized protocol, it is possible to obtain recombinant Bak proteoliposomes in a one-step reaction. The liposomes concentration may be a critical parameter, because high concentrations of liposomes may result in an inhibition of the transcription/translation process and low liposomes concentration often result in protein precipitation.

4. Bak Proteoliposome Production

1. According to the instruction manual (Roche Applied Science), the 1-ml reaction mixture is prepared and incubated for 72 h in the RTS ProteoMaster Instrument or in the Eppendorf Thermomixer at 26 °C. Liposomes (40 mg/ml stock solution) are added to the reaction mixture for reaching a final concentration of 5 mg/ml.
2. The mixture is centrifuged for 20 min at 13,000×*g*, at 4 °C.
3. The supernatant, containing proteins from the *E. coli* lysate and other contaminants, is discarded and the pellet is resuspended in 1 ml of 50 m*M* Tris-HCl 7.5.

The final yield of Bak integrated into liposomes is between 0.5 and 1 mg (synthesized in a one-step reaction).

5. Liposome Preparation

Liposomes are prepared by mixing commercially available lipids (lipids are from Avanti Polar or Lipoid). The general protocol to prepare liposomes is described in Liguori *et al.* (2008a). Lipids of different compositions such as 1,2-dioleoyl-*sn*-glycero-3-phosphocholine (DOPC), 1,2-dilinoleoyl-*sn*-glycero-3-phosphoethanolamine (DOPE), cholesterol, di-palmitoyl-phosphatidylcholine (DPPC), and ratio can be used in the cell-free expression system. A general protocol is given which can be used with different liposome formulations.

1. Lipids (powder) are dissolved in methanol:chloroform ($CHCl_3$/MeOH) (1:1) at a final concentration of 10 mg/ml, and stored at −20 °C until use.

2. Liposomes are obtained by evaporation of $CHCl_3$/MeOH by placing the mix on ice 10 min and evaporate to dryness with SpeedVac System or nitrogen vacuum.
3. To obtain a concentrated stock solution (40 mg/ml), the lipid film is resuspended in 250 μl of Hepes buffer, pH 7.4 or DEPC water.
4. After sonication (three times for 1 min on ice) (Branson Sonic Power, Smithkline Company Brentford, Middlesex, UK), lipids are filtered through a 0.22 μm filter. The vesicles (final concentration 40 mg/ml) present a high homogeneity and their size ranges between 90 and 120 nm.

The composition in lipids of liposomes can be modified according to the protein expression levels. Furthermore, it has been shown that addition of PEG can protect liposomes from rapid clearance by the reticuloendothelial system (RES) and can be included in the formulation to increase the particles stability (Vail *et al.*, 2004). By using this expression strategy, Bak is directly expressed and integrated into liposomes in a one-step reaction.

6. Proteoliposome Purification

To purify the recombinant Bak proteoliposomes, the resuspended pellet (step 3 in Bak proteoliposomes production) is loaded onto a discontinuous sucrose gradient (Fig. 11.2). After centrifugation, Bak proteoliposomes appear as a white ring at the interface between the 25% and 60% sucrose layers according to the particles density. The detailed protocol is described in Liguori *et al.* (2008a). Briefly:

1. 60%, 25%, and 10% sucrose solutions are prepared in 50 m*M* Tris-HCl 7.5.
2. The gradient is loaded in specific ultracentrifuge tubes: 3.5 ml of 60% sucrose, 3.5 ml of 25% sucrose.
3. The 1-ml Bak proteoliposomes sample (resuspended in Tris–HCl 50 m*M*, pH 7.5) is loaded onto the top of the 25% sucrose layer.
4. Slowly, the gradient is filled up with the 10% sucrose solution (3.5 ml).
5. After ultracentrifugation for 1 h at 200,000$\times g$ at 4 °C (Beckman L2 65B ultracentrifuge, rotor SW2), 1-ml fractions are collected from the top to the bottom of the gradient.

7. Analysis of the Purified Bak Proteoliposomes

After purification, an SDS–PAGE is performed to analyze the purity of each collected fractions. A silver staining or a Coomassie blue staining can be performed to determine the degree of purity of the sample.

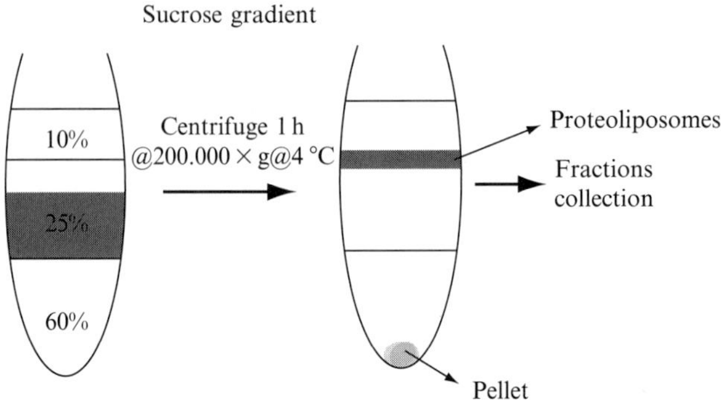

Figure 11.2 Sucrose gradient for proteoliposomes purification. A discontinuous three layers sucrose gradient is made to purify by density, liposomes containing inserted membrane proteins into the lipid bilayer. After ultracentrifugation, gradient fractions are collected and proteins not integrated into the vesicles are recovered in the pellet.

A 15% SDS–PAGE is loaded with fractions containing Bak proteoliposomes and Coomassie blue staining allows to determine the degree of purification. Coomassie blue staining solution is purchased from Fermentas (Liguori *et al.*, 2008a).

1. The SDS–PAGE gel is first washed in dH_2O for 30 min at RT.
2. The gel is then treated with Coomassie blue solution for 30 min at RT under gentle agitation.
3. To clearly visualize the protein bands, an extensive wash with dH_2O is performed until the gel background becomes clear.

Bak proteoliposome purification always results in two bands: the first at 19 kDa, corresponding to the full-length recombinant Bak, and the second at 17 kDa which corresponds to a shorter translation product. Interestingly, only fractions containing highly concentrated Bak revealed the presence of the protein organized in homodimers and trimers (Fig. 11.3). Furthermore, it has been shown previously that Bak oligomerization is involved in exerting its apoptotic effect (Letai *et al.*, 2002).

8. Transmission Electron Microscopy

The size and homogeneity of purified Bak proteoliposomes are analyzed by transmission electron microscopy (TEM) (Fig. 11.4). To heighten the contrast of biological materials, an electron-dense stain is used.

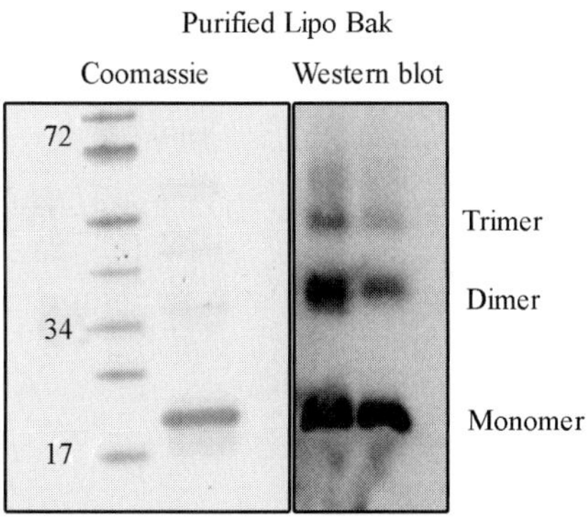

Figure 11.3 Coomassie blue staining and Western blotting of purified Bak proteoliposomes. After purification onto a sucrose gradient, Bak proteoliposomes were loaded on a SDS gel and blotted with an anti-His antibody. To check the purity of the preparation, a Coomassie blue staining is performed. Bak proteins in the liposomes are detected as monomeric or oligomeric forms.

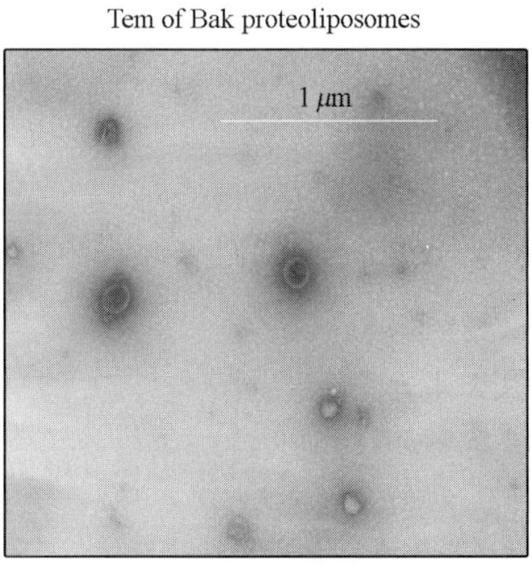

Figure 11.4 Transmission electron microscopy of Bak proteoliposomes. After purification, Bak proteoliposomes are washed in PBS and negatively stained with 1% (w/v) uranyl acetate. Uranyl acetate is added to increase the contrast to the liposomes vesicle preparation. The bar represents 1 μm.

1. Proteoliposomes are first washed (15 min at 15,000×g at 4 °C) in PBS (1× PBS Sigma-Aldrich) and resuspended in PBS at the final concentration of 0.5 mg/ml.
2. Liposomes are then negatively stained with 1% (w/v) uranyl acetate (20 min at RT). Uranyl acetate is added to increase the contrast of the vesicles preparation and the samples are analyzed using a Philips CM12 electron microscope equipped with a LaB6 filament operating at 100 kV. The average size for recombinant purified Bak proteoliposomes ranged from 100 to120 nm.

9. Apoptosis Induction in Cancer Cell Lines: Caspase 9 Activation

In determining the critical steps of the apoptotic process, mitochondria represent one of the crucial players. Several membrane proteins located at the outer or at the inner mitochondrial membranes are involved in the activation of the intrinsic apoptotic pathway. Different strategies to induce apoptosis are thus employed to target therapeutic agents to the mitochondria such as positively charged α-helical peptides (Yin *et al.*, 2005), BH3 peptidomimetics for blocking antiapoptotic Bcl-2-like proteins (Letai *et al.*, 2002; Walensky *et al.*, 2006), ampholytic cations, metals, and steroid-like compounds (Galluzzi *et al.*, 2006). In this section, we describe a protocol to analyze proapoptotic effects of Bak proteoliposomes in the cell death process (Fantin and Leder, 2006; Liguori *et al.*, 2008b). Upon appropriate stimuli, Bax and Bak homo- and heterooligomers form a pore at the mitochondrial membrane, leading to the release of cytochrome *c* and caspase 9 activation (Hemmati *et al.*, 2006). Cytochrome *c* is often released from mitochondria during the early stages of apoptosis. Once in the cytosol and in the presence of ATP, caspase-9 is directly activated by Apaf-1 and cytochrome *c*. Active caspase-9 initiates the caspases cascade that leads to DNA fragmentation and cell death (Gupta *et al.*, 2009). It has been previously demonstrated that purified proteoliposomes are capable to deliver therapeutic Bak into different cancer cell lines (human and mouse) and that a large fraction of recombinant Bak is colocalized with mitochondria, resulting in specific caspase 9 activation (Liguori *et al.*, 2008b). To check for this apoptotic marker, murine cell lines are transduced with purified Bak proteoliposomes:

1. Murine glioma cells (GL26) are cultured in 6-well plates in DMEM medium (GIBCO, Invitrogen Corporation) modified with fetal bovine serum, 10% and incubated for 12 h in the presence of different concentrations of Bak proteoliposomes (2.5, 5, 10, 20 μg/million of cells).
2. After incubation, total cell lysates are prepared by washing GL26 twice in PBS and by addition of 100 μl of 1× FastBreak Cell Lysis reagent (Promega, Madison, WI, USA) with protease inhibitors (Complete Mini Protease Cocktail, Roche, IN, USA) for 30 min on ice.
3. The samples are centrifuged at 12,000×*g* for 15 min and the pellet discarded. Protein concentration is determined with a colorimetric dye-binding assay (BCA Assay, Pierce, Rockford, IL, USA).
4. For Western blot analysis, 50–75 μg of proteins are loaded, separated by SDS–PAGE and electrotransferred to a nitrocellulose membrane (Bio-Rad, Hercules, CA, USA).
5. Membranes are blocked with 4% milk in TBS-T, 3 h at RT and incubated overnight at 4 °C with an antibody recognizing the full length

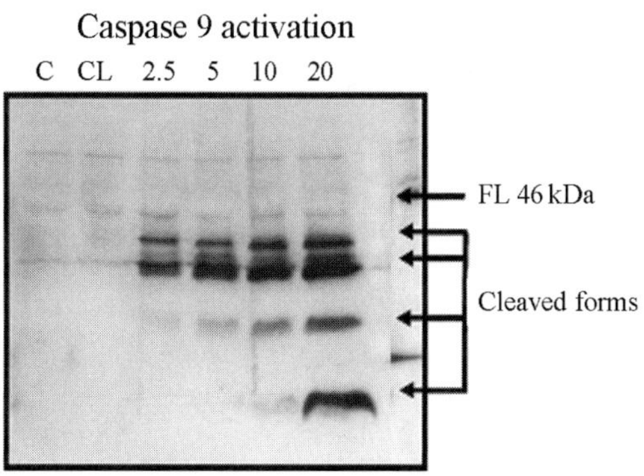

Figure 11.5 Dose-dependent caspase 9 activation after 12 h treatment with Bak proteoliposomes. Seventy-five micrograms of proteins from total lysate (GL26 cells) are loaded: C, no treated cells; CL, cells treated with empty liposomes; 2.5, 5, 10, 20 μg of Bak proteoliposomes (μg/million of cells). No activation was detected in the controls and a dose-dependent cleavage of caspase 9 is present when Bak proteoliposomes are internalized into the cells.

and cleaved caspase 9 forms (1:500 in TBS-T, BSA 5 mg/ml; Calbiochem, San Diego, CA, USA). After three washes in TBS-T (10 min, RT), the membrane is incubated with an antirabbit peroxydase-labeled secondary antibody (Amersham Corp., Arlington Heights, IL, USA) 1:10000 in TBS-T, 1 h at RT.

6. Positive bands are detected using the Lumi-Light chemiluminescence kit (ECL, Amersham, Buckinghamshire, UK).

In untreated cells and cells incubated with empty liposomes, the only band detected is the caspase 9 full length (46 kDa) whereas in Bak proteoliposomes treated cells display the cleaved forms at 37/35, 18, and 10 kDa (Fig. 11.5). The cleavage is also dose dependent as shown in Fig. 11.5.

10. Concluding Remarks

This chapter presents the "proof of concept" that the use of cell-free expression technology is a powerful method for producing proteoliposomes containing functional membrane proteins. These recombinant proteoliposomes are capable to deliver therapeutic Bak into cancer cells and to specifically induce apoptosis. The development of a Bak therapeutic proteoliposome containing a tumor-targeting peptide is currently under investigation.

ACKNOWLEDGMENTS

This work was supported by a grant from the European Commission: Marie Curie Excellence Grant #014320.

REFERENCES

Düzgüneş, N., Simões, S., Konopka, K., Rossi, J. J., and Pedroso de Lima, M. C. (2001). Delivery of novel macromolecular drugs against HIV-1. *Expert Opin. Biol. Ther.* **1,** 949–970.

Düzgüneş, N., Simões, S., Slepushkin, V., Pretzer, E., Flasher, D., Salem, I. I., Steffan, G., Konopka, K., and Pedroso de Lima, M. C. (2005). Delivery of antiviral agents in liposomes. *Methods Enzymol.* **391,** 351–373.

Düzgüneş, N., Simões, S., Lopez-Mesas, M., and Pedroso de Lima, M. C. (2007). Intracellular delivery of therapeutic oligonucleotides in pH-sensitive and cationic liposomes. *In* "Liposome Technology. Vol. III" (G. Gregoriadis, ed.), pp. 253–275. Informa Healthcare USA, New York.

Fantin, V. R., and Leder, P. (2006). Mitochondriotoxic compounds for cancer therapy. *Oncogene* **25,** 4787–4797.

Fesik, S. W. (2005). Promoting apoptosis as a strategy for cancer drug discovery. *Nat. Rev. Cancer* **5,** 876–885.

Galluzzi, L., Larochette, N., Zamzami, N., and Kroemer, G. (2006). Mitochondria as therapeutic targets for cancer chemotherapy. *Oncogene* **25,** 4812–4830.

Gupta, S., Kass, G. E., Szegezdi, E., and Joseph, B. (2009). The mitochondrial death pathway: A promising therapeutic target in diseases. *J. Cell. Mol. Med.* **13**(6), 1004–1033.

Hemmati, P. G., Guner, D., Gillissen, B., Wendt, J., von Haefen, C., Chinnadurai, G., Dorken, B., *et al.* (2006). Bak functionally complements for loss of Bax during p14ARF-induced mitochondrial apoptosis in human cancer cells. *Oncogene* **25,** 6582–6594.

Kamps, J. A., Koning, G. A., Velinova, M. J., Morselt, H. W., Wilkens, M., Gorter, A., Donga, J., and Scherphof, G. L. (2000). Uptake of long-circulating immunoliposomes, directed against colon adenocarcinoma cells, by liver metastases of colon cancer. *J. Drug Target.* **8,** 235–245.

Klammt, C., Lohr, F., Schafer, B., Haase, W., Dotsch, V., Ruterjans, H., Glaubitz, C., *et al.* (2004). High level cell-free expression and specific labeling of integral membrane proteins. *Eur. J. Biochem.* **271,** 568–580.

Klammt, C., Schwarz, D., Fendler, K., Haase, W., Dotsch, V., and Bernhard, F. (2005). Evaluation of detergents for the soluble expression of alpha-helical and beta-barrel-type integral membrane proteins by a preparative scale individual cell-free expression system. *FEBS J.* **272,** 6024–6038.

Leader, B., Baca, Q. J., and Golan, D. E. (2008). Protein therapeutics: A summary and pharmacological classification. *Nat. Rev. Drug Discov.* **7,** 21–40.

Letai, A., Bassik, M. C., Walensky, L. D., Sorcinelli, M. D., Weiler, S., and Korsmeyer, S. J. (2002). Distinct BH3 domains either sensitize or activate mitochondrial apoptosis, serving as prototype cancer therapeutics. *Cancer Cell* **2,** 183–192.

Liguori, L., Marques, B., and Lenormand, J. L. (2008a). A bacterial cell-free expression system to produce membrane proteins and proteoliposomes: From cDNA to functional assay. *Curr. Protoc. Protein Sci.* Chapter 5: Unit 5.22.

Liguori, L., Marques, B., Villegas-Mendez, A., Rothe, R., and Lenormand, J. L. (2008b). Liposomes-mediated delivery of pro-apoptotic therapeutic membrane proteins. *J. Control. Release* **126,** 217–227.

Lukyanov, A. N., Elbayoumi, T. A., Chakilam, A. R., and Torchilin, V. P. (2004). Tumor-targeted liposomes: Doxorubicin-loaded long-circulating liposomes modified with anti-cancer antibody. *J. Control. Release* **100,** 135–144.

Maeda, H., Wu, J., Sawa, T., Matsumura, Y., and Hori, K. (2000). Tumor vascular permeability and the EPR effect in macromolecular therapeutics: A review. *J. Control. Release* **65,** 271–284.

Maeda, H., Bharate, G. Y., and Daruwalla, J. (2009). Polymeric drugs for efficient tumor-targeted drug delivery based on EPR-effect. *Eur. J. Pharm. Biopharm.* **71,** 409–419.

Marques, B., Liguori, L., Paclet, M. H., Villegas-Mendez, A., Rothe, R., Morel, F., and Lenormand, J. L. (2007). Liposome-mediated cellular delivery of active gp91. *PLoS ONE* **2,** e856.

Pavlou, A. K., and Reichert, J. M. (2004). Recombinant protein therapeutics—Success rates, market trends and values to 2010. *Nat. Biotechnol.* **22,** 1513–1519.

Reed, J. C. (2006). Proapoptotic multidomain Bcl-2/Bax-family proteins: Mechanisms, physiological roles, and therapeutic opportunities. *Cell Death Differ.* **13,** 1378–1386.

Salem, I. I., Flasher, D. L., and Düzgüneş, N. (2005). Liposome-encapsulated antibiotics. *Methods Enzymol.* **391,** 261–291.

Spirin, A. S. (2004). High-throughput cell-free systems for synthesis of functionally active proteins. *Trends Biotechnol.* **22,** 538–545.

Swartz, J. R., Jewett, M. C., and Woodrow, K. A. (2004). Cell-free protein synthesis with prokaryotic combined transcription-translation. *Methods Mol. Biol.* **267,** 169–182.

Tauzin, B (2008). Biotechnology research continues to bolster arsenal against disease with 633 medicines in development. *Biotechnology* 1–60.

Torchilin, V. P. (2005). Recent advances with liposomes as pharmaceutical carriers. *Nat. Rev. Drug Discov.* **4,** 145–160.

Torchilin, V. P. (2006). Recent approaches to intracellular delivery of drugs and DNA and organelle targeting. *Annu. Rev. Biomed. Eng.* **8,** 343–375.

Torchilin, V. P. (2007). Targeted pharmaceutical nanocarriers for cancer therapy and imaging. *AAPS J.* **9,** 128–147.

Vail, D. M., Amantea, M. A., Colbern, G. T., Martin, F. J., Hilger, R. A., and Working, P. K. (2004). Pegylated liposomal doxorubicin: Proof of principle using preclinical animal models and pharmacokinetic studies. *Semin. Oncol.* **31,** 16–35.

Walensky, L. D. (2006). BCL-2 in the crosshairs: Tipping the balance of life and death. *Cell Death Differ.* **13,** 1339–1350.

Walensky, L. D., Pitter, K., Morash, J., Oh, K. J., Barbuto, S., Fisher, J., Smith, E., *et al.* (2006). A stapled BID BH3 helix directly binds and activates BAX. *Mol. Cell* **24,** 199–210.

Williams, S. S., Alosco, T. R., Mayhew, E., Lasic, D. D., Martin, F. J., and Bankert, R. B. (1993). Arrest of human lung tumor xenograft growth in severe combined immunodeficient mice using doxorubicin encapsulated in sterically stabilized liposomes. *Cancer Res.* **53,** 3964–3967.

Yin, H., Lee, G. I., Sedey, K. A., Kutzki, O., Park, H. S., Orner, B. P., Ernst, J. T., *et al.* (2005). Terphenyl-based Bak BH3 alpha-helical proteomimetics as low-molecular-weight antagonists of Bcl-xL. *J. Am. Chem. Soc.* **127,** 10191–10196.

CHAPTER TWELVE

Liposome-Mediated Therapy of Neuroblastoma

Daniela Di Paolo,*[,1] Monica Loi,*[,1] Fabio Pastorino,* Chiara Brignole,* Danilo Marimpietri,* Pamela Becherini,* Irene Caffa,* Alessia Zorzoli,* Renato Longhi,† Cristina Gagliani,‡,§,¶ Carlo Tacchetti,‡,§,¶ Angelo Corti,‖ Theresa M. Allen,** Mirco Ponzoni,* *and* Gabriella Pagnan*

Contents

* Experimental Therapy Unit, Laboratory of Oncology, G. Gaslini Children's Hospital, Genoa, Italy
† Istituto di Chimica del Riconoscimento Molecolare, Consiglio Nazionale delle Ricerche, Milan, Italy
‡ MicroSCBio Research Center, University of Genoa, Genoa, Italy
§ IFOM (FIRC Institute of Molecular Oncology), Milan, Italy
¶ Department of Experimental Medicine, Human Anatomy Section, University of Genoa, Genoa, Italy
‖ Division of Molecular Oncology, IIT Network Research Unit of Molecular Neuroscience, San Raffaele Scientific Institute, Milan, Italy
** Department of Pharmacology, University of Alberta, Edmonton, Alberta, Canada
[1] Contributed equally to the work and should be considered joint first authors

Methods in Enzymology, Volume 465
ISSN 0076-6879, DOI: 10.1016/S0076-6879(09)65012-6

Abstract

Neuroblastoma (NB) is the most common extracranial solid tumor in childhood and the most frequently diagnosed neoplasm during infancy. Despite of aggressive treatment strategies, the 5-year survival rate for metastatic disease is still less than 60% and, consequently, novel therapeutic approaches are needed. For increasing the therapeutic index of anticancer drugs, while reducing side effects, one of the most promising strategies in modern chemotherapy is based on the development of innovative drug delivery systems, such as liposomes. "Anticancer drug"-loaded liposomes have demonstrated enhanced ability to target to the affected area, as well as increased antitumor efficacy compared to conventional drugs. Liposomes tend to extravasate preferentially and to accumulate into tumor interstitial fluids, due to the defective structure of the new angiogenic vessels within the tumor masses. This inherent tumor selectivity can be increased further by coupling tumor-specific antibodies or other targeting moieties to the surface of the lipid envelope. Here, we describe the methodology used in these studies, as well as the antitumor results obtained by the use of several "anticancer drugs," encapsulated into antibody- and peptide-targeted liposomal formulations, against NB.

1. Introduction

Neuroblastoma (NB) is the most common extracranial solid tumor in childhood (Conte *et al.*, 2006; Maris and Matthay, 1999) and the most frequently diagnosed neoplasm during infancy (Maris *et al.*, 2007). NB originates from the neural crest and, therefore, tumors can develop anywhere in the sympathetic nervous system (Maris and Matthay, 1999). Typically, 65% of primary tumors are localized in the abdomen, with at least half of these arising in the adrenal medulla, while neck, chest, and pelvis are other common sites of disease (Brodeur and Maris, 2006). Presenting signs and symptoms are highly variable and dependent on the site of primary tumor as well as on the presence or the absence of metastatic disease of paraneoplastic syndromes (Maris *et al.*, 2007). Approximately half of all patients with NB present with widely disseminated disease that is often refractory to intensive chemo- radiotherapy. Despite aggressive treatment strategies, such as high-dose chemotherapy and autologous hematopoietic stem cell transplantation, the overall long-term disease-free survival rate is marginally prolonged (De Bernardi *et al.*, 2003; Miano *et al.*, 2001; von Allmen *et al.*, 2005).

The main limitation of current chemotherapeutic drug regimens in cancer therapy is the lack of specificity of the drugs, that can lead to severe systemic and organ toxicity. In this scenario, the need of developing new therapeutic approaches is an emerging query.

One of the most promising strategies in modern chemotherapy is the development of tools that enable the modification of the kinetic features of chemotherapeutic drugs, thus increasing their efficacy and decreasing their toxicity. This goal may be achieved both by specifically targeting the drug to the tumor and by using innovative drug delivery systems, such as liposome formulations (Allen *et al.*, 1992; Gabizon *et al.*, 2002; Pastorino *et al.*, 2008). Because of their physicochemical features, liposomes can be manipulated with relative easiness, allowing to design more sophisticated cancer treatment strategies.

In this chapter, we describe our liposomal preparation methods and the antitumor effects obtained in the treatment of NB. In particular, we present data based on the "antitumor compound" encapsulated into liposomes.

2. Materials

Liposomes are synthesized by mixing the following lipids: hydrogenated soy phosphatidylcholine (HSPC), cholesterol (CHOL), 1,2-distearoyl-*sn*-glycero-3-phosphoethanolamine-*N*-[methoxy(polyethylene glycol)-2000] (DSPE-PEG_{2000}), 1,2-distearoyl-*sn*-glycero-3-phosphoethanolamine-*N*-[methoxy(polyethylene glycol)-2000] modified with a maleimido group at the distal terminus chain (1,2-distearoyl-*sn*-glycero-3-phosphoethanolamine-*N*-[maleimide(polyethylene glycol)-2000], DSPE-PEG_{2000}-MAL), 1,2-distearoylglycero-3-phosphoethanolamine-*N*-[carboxy(polyethylene glycol) 2000] (DSPE-PEG_{2000}-COOH), 1,2-distearoylglycero-3-phosphoethanolamine-*N*-[amino(polyethylene glycol)2000] (DSPE-PEG_{2000}-NH_2), and 1,2-dioleoy-1-3-trimethylammonium-propane (DOTAP).

In some preparations, cholesteryl-[1,2-^{3}H-(*N*)]-hexadecyl ether ([^{3}H]-CHOL) is added as a nonexchangeable, nonmetabolizable lipid tracer. In some experiments, 1,2-dipalmitoyl-*sn*-glycero-3-phosphoethanolamine-*N*-(Lissamine Rhodamine B Sulfonyl) (Rhoda-PE) is used at 1 mol% of total phospholipids (PLs) as a fluorimetric lipid tracer.

3. Untargeted Liposomes Entrapping Doxorubicin

Doxorubicin (DXR) is a potent antineoplastic agent active against a wide range of human neoplasms (Injac and Strukelj, 2008). However, administration of this drug is associated with severe acute toxicities (including myeloid suppression and gastrointestinal toxicity) as well as a cumulative dose-limiting cardiotoxicity (Broder *et al.*, 2008). Many reports reveal that the liposomal encapsulation of DXR decreased these side effects in various animal models and now used in clinical trials. (Allen and Cullis, 2004; Torchilin, 2006)

3.1. Liposomes preparation

DXR-entrapped stealth liposomes (SL[DXR]) are synthesized from HSPC: CHOL:DSPE-PEG_{2000}, 2:1:0.1 molar ratio (Allen *et al.*, 1991). The lipids are first dissolved in chloroform and then combined in appropriate ratios. A thin lipid film is formed under nitrogen by rotator evaporation under controlled vacuum and temperature. DXR is then loaded into the liposomes via an ammonium sulfate gradient, using a method adapted from Bolotin *et al.* (1994) and reported previously (Moase *et al.*, 2000). Specifically, after evaporation under nitrogen, the lipid film is hydrated at 10 m*M* PLs in about 1 ml of 155 m*M* ammonium sulfate, pH 5.5 and then the hydrated liposomes are sequentially extruded through a series of polycarbonate filters of pore sizes ranging from 0.2 to 0.1 μm to produce primarily unilamellar vesicles. The liposomal mixture is then eluted through a Sephadex G-50 column with sodium acetate buffer (100 m*M* $NaCH_3COO$, 140 m*M* NaCl) pH 5.5, then DXR is added at a ratio of 1 mg PL:0.2 mg DXR, and incubated for 1 h at 65 °C. The mixture is then applied to a Sephadex G-50 column eluted with Hepes buffer (25 m*M* Hepes, 140 m*M* NaCl), pH 7.4, to remove the unloaded DXR. The concentration of DXR encapsulated is quantified by measuring absorbance at 480 nm after dissolving liposomes in 1 ml MeOH. The loading efficiency ofDXR is around 95% and liposomes routinely contained DXR at a concentration of 150–180 μg DXR/μmol PLs (Pastorino *et al.*, 2003b).

Leakage of DXR from liposomes is measured by dialysis in complete medium against a large volume of 25% human plasma, at 37 °C, sampling the contents of the dialysis bag at increasing time intervals and determining the absorbance. The same procedure is also performed at 4 °C, in Hepes buffer, pH 7.4.

4. Tumor-Targeted Liposomal Chemotherapy

In the last few years, antibody-based therapeutics have emerged as important tools in anticancer therapies against an increasing number of human malignancies (Adams and Weiner, 2005) and it is expected that several immunoliposomes will be in clinical trials in the near future. DXR-loaded, long-circulating liposomes, modified by coupling a monoclonal antibody are referred as "stealth immunoliposomes" (SILs) (Allen *et al.*, 1995).

So far, the only immunoliposome formulation, used for a phase I study in patients with metastatic stomach cancer, consists of the DXR-incorporated PEG liposomes, targeted via the $F(ab')_2$ fragments of the human GAH mAb, which are also able to recognize gastric, colon, and breast cancer cells (Hamaguchi *et al.*, 2004; Matsumura *et al.*, 2004).

The mechanism by which SILs appear to act is related to the specific binding and, subsequently, the internalization of the liposomal drug into the cell (Lopes de Menezes *et al.*, 1998). Interestingly, it has been shown that approximately 400-fold more monoclonal antibody is required to achieve similar results with antibody–drug conjugates (Ding *et al.*, 1990). Hence, high drug:antibody ratios can be achieved with SILs, thus decreasing the need for expensive and potentially immunogenic antibodies. Therefore, immunoliposomes have been recognized as a promising tool for the site-specific delivery of drugs and diagnostic agents (Allen *et al.*, 1995; Park *et al.*, 1995). Moreover, the coupling of antibody Fab′ fragments instead of whole immunoglobulin molecules to SL have increased their circulation times in blood (Allen, 2002) and improved their *in vivo* antitumor efficacy (Brignole *et al.*, 2003a; Moase *et al.*, 2000; Sugano *et al.*, 2000).

Among the antigens found on malignant cells, the disialoganglioside GD_2 is an attractive target for immunoliposomal therapy of neuroectoderma-derived tumors, especially of NB (Frost *et al.*, 1997; Uttenreuther-Fischer *et al.*, 1995), because it is extensively expressed in this cancer, but it is not expressed in non malignant tissues (Mujoo *et al.*, 1987; Schulz *et al.*, 1984). Monoclonal antibodies recognizing this antigenic determinant have been developed and shown to bind to NB cells with high affinity and specificity (Kowalczyk *et al.*, 2009).

4.1. Preparation of anti-GD_2 immunoliposomes

SILs, coupled with the anti-GD_2 mAb or its Fab′ fragments, are synthesized from HSPC:CHOL:DSPE-PEG_{2000}:DSPE-PEG_{2000}-MAL, 2:1:0.08:0.02 and 2:1:0.06:0.04 molar ratio, respectively (Pagnan *et al.*, 1999; Pastorino *et al.*, 2003b). Liposomes are prepared as described above. For preparation of $F(ab')_2$ fragments, purified antibody is digested with lysyl endopeptidase in 50 m*M* Tris–HCl (pH 8.5), at 37 °C for 3 h and the reaction mix is passed through a protein-A/G column equilibrated with 50 m*M* Tris–HCl (pH 7.5). $F(ab')_2$ fragments are then reduced to Fab′ with 7.5 m*M* β-mercaptoethylamine for 90 min at 37 °C in acetate buffer (pH 6.5), after which the reducing agent is removed by passing the sample over a Sephadex G-25 column pre-equilibrated with the same buffer (Pastorino *et al.*, 2003b). The concentration of mAbs is determined by spectrophotometry ($\lambda = 280$ nm) and their purity assessed by SDS-PAGE. The reactivity of anti-GD_2 mAbs and their fragments is checked on GD_2-positive NB cell lines, by FACS analysis. To activate the GD_2 mAb or its fragments for reactivity toward the maleimide (MAL) 2-iminothiolane is used to convert exposed amino groups on the Ab surface into free sulfhydryl groups. Thiolation of mAb/Fab′ by this reagent does not interfere with their target recognition, as already reported (Huwyler *et al.*, 1996). A 2-iminothiolane/IgG molar ratio of 20:1 and 1 h of incubation at room temperature (RT), with occasional mixing, appears to be optimal for their activation. After separation of the activated IgGs from unreacted iminothiolane over a Sephadex G-25 column, the activated Ab is slowly added to a 5-ml test tube containing the liposomes and a small stirring bar. Since the MAL group slowly hydrolyzes when in contact with water, it is essential to proceed to the synthesis of immunoliposomes without unnecessary delay. Oxygen is displaced by running a slow stream of N_2 over the reaction mixture. The tube is capped and sealed with Teflon tape and the mixture incubated overnight at RT with continuous slow stirring. Immunoliposomes are then removed from coupled IgG by chromatography over a Sepharose CL-4B column. Liposomes are sterilized by filtration through 0.22 μm pore cellulose membranes and stored at 4 °C until use. (Scheme 12.1A)

The efficiency of coupling is not affected by using different molar ratios (1000:1–2000:1) of PL/mAb. An average of 60% (±15%) of the Ab is coupled to the liposomes. The antibody density (evaluated by BioRad protein assay) is in the range of 40–60 μg Ab/μmol PL, which means, approximately, 30 ± 5 IgG per liposome. Coupling expressed as nmol mAb/μmol PL is in the range of 0.33–0.53 nmol of anti-GD_2 mAb/μmol PL (0.66–1.06 nmol binding sites, because each mAb has two binding sites) and 0.54–0.91 nmol of anti-GD_2 Fab′ fragments/μmol PL (Pastorino *et al.*, 2003b). The conversion from μg of IgG/μmol PL to the number of molecules conjugated/liposomes is based on the assumption that a 100 nm liposome contains about 100,000 molecules of PL (Hansen *et al.*, 1995). It has already been reported that SILs, with the above Abs density, have good binding capability *in vitro* and retain long circulation times sufficient to bind to target cells *in vivo* (Lopes de Menezes *et al.*, 1998).

Anti-GD_2 whole antibody or their Fab′ fragments are then used as a ligand for targeting liposomal formulation to human NB animal models (Brignole *et al.*, 2003a; Pastorino *et al.*, 2006, 2003b). The efficacy of Fab′ fragments of anti-GD_2 immunoliposomes containing DXR is evaluated in terms of metastasis growth inhibition and increased life span, which represent the principal end points for establishing the antitumor activity of a new drug formulation. Pharmacokinetics (PKs) studies shows that Fab′–SILs have more long-circulating profiles in blood compared with anti-GD_2-SILs, with the PKs profile for Fab′–SILs being almost identical to that obtained with untargeted stealth liposomes (SLs) (Pastorino *et al.*, 2003b).

In vivo, long-term survivors are obtained in mice treated with Fab′-SILs containing DXR, but not in untreated animals, or those treated with free anti-GD_2 Fab′, Fab′-SILs (no drug), free-DXR or untargeted, DXR-loaded, SLs.

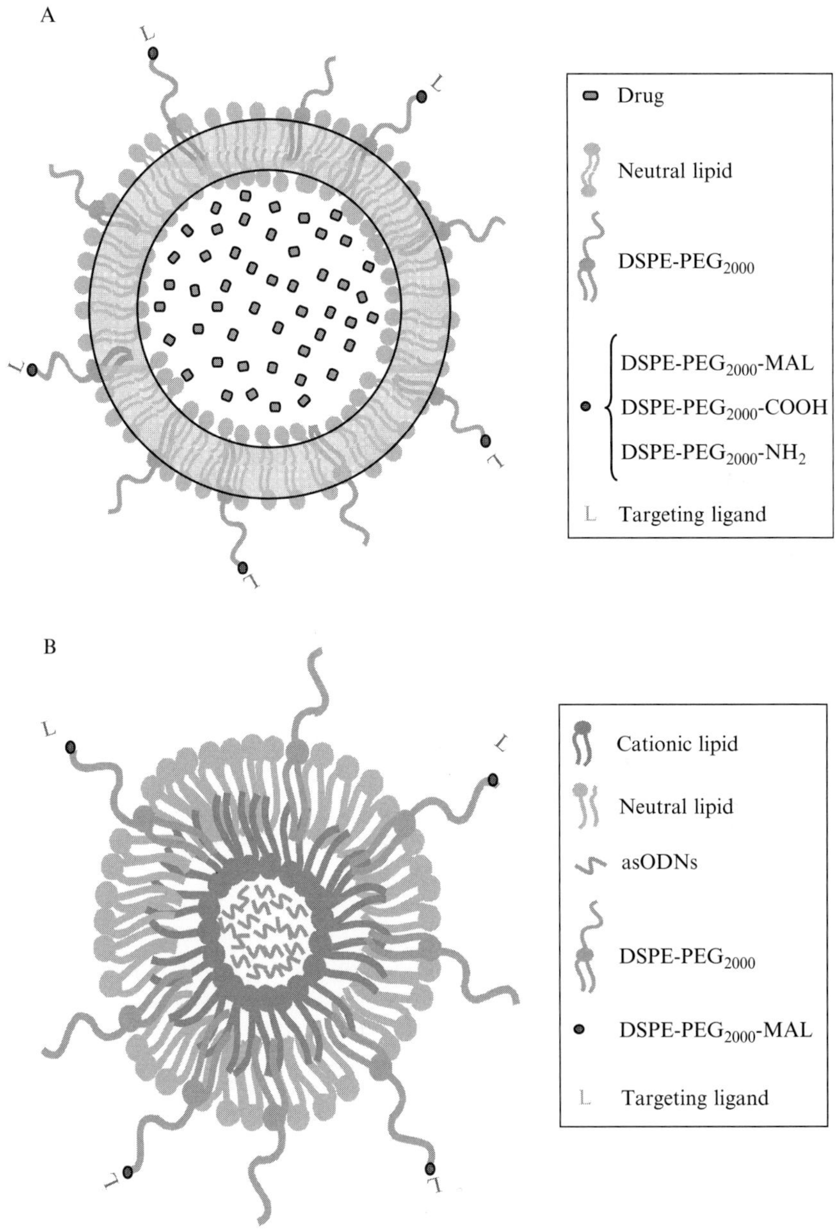

Scheme 12.1 Representative images of targeted liposomes (A) and targeted asODN-entrapping coated cationic liposomes (B). (See Color Insert.)

Immunoliposomes containing DXR prevented the establishment and growth of the tumor in all of the organs examined. In conclusion, our formulation led to the total inhibition of metastatic growth of human NB in a nude mouse metastatic model (Pastorino *et al.*, 2003b).

5. Vascular-Targeted Liposomal Chemotherapy

Growing tumors rely on recruitment of new blood vessels (angiogenesis) to support tumor expansion (Hillen and Griffioen, 2007). If the angiogenic blood vessels are selectively destroyed, tumor cells will not receive the oxygen and nutrients they need for their survival, resulting in tumor cell killing and necrosis (Fidler and Ellis, 1994). Owing to the "angiogenesis dependence" of solid tumors predicted by Folkman nearly 30 years ago, selective inhibition or destruction of the tumor vasculature (using antiangiogenic or antivascular treatments, respectively) could trigger tumor growth inhibition, regression, and/or a state of dormancy and thereby offer a novel approach to cancer treatment (Folkman, 1971).

Currently, several innovative and combined methods are being evaluated to overcome the limits of classical and antiangiogenic and anticancer therapies that include the specific targeting of the drugs to endothelial cells and the use of drug delivery systems. Vascular targeting offers therapeutic promise for the delivery of drugs (Arap *et al.*, 1998; Colombo *et al.*, 2002), radionuclides (Sipkins *et al.*, 1998), and genes (Hood *et al.*, 2002; Niethammer *et al.*, 2002). This approach has the advantage that the delivered vehicle, once in the bloodstream, should have direct access to the target tumor endothelial cells.

Vascular targeting exploits molecular differences that exist in blood vessels of different organs and tissues as well as differences between normal and angiogenic or remodeled blood vessels. Differences in plasma-membrane proteins (the "vascular zip code") can be used to target therapeutic or imaging agents directly to a particular organ or tumor (Arap *et al.*, 2002; Trepel *et al.*, 2002). Endothelial cells in the angiogenic vessels within solid tumors express several proteins that are absent or barely detectable in established blood vessels, including α_v integrins (Jin and Varner, 2004), receptors for angiogenic growth factors (Mahabeleshwar *et al.*, 2007), and other types of membrane-spanning molecules, such as aminopeptidase N (CD13) (Curnis *et al.*, 2000) and aminopeptidase A (APA) (Marchio *et al.*, 2004).

Because liposomes represent an effective drug delivery system, tumor vessel-specific liposomes would be a useful strategy for delivering already known as well as novel cytotoxic drugs. The vascular network is a highly accessible target for tumor therapy but, to be successful, an antivascular strategy should target as many tumor vessels as possible. We reasoned that the multivalent display of specific peptides at the surface of liposomes could be greatly effective in active targeting of the tumor vasculature.

In vivo panning of phage libraries in tumor-bearing mice have proven useful for selecting peptides that interact with proteins expressed within tumor-associated vessels and home to neoplastic tissues (Arap *et al.*, 1998; Hood *et al.*, 2002; Trepel *et al.*, 2002; Zurita *et al.*, 2003). Among the various tumor-targeting ligands identified so far, the NGR (asparagine-glycine-arginine) peptide, which is the ligand for aminopeptidase N (APN), was able to deliver various "antitumor compounds" to tumor vessels (Arap *et al.*, 1998; Curnis *et al.*, 2000, 2002b; Ellerby *et al.*, 1999). Indeed, NGR peptide binds to an isoform of aminopeptidase N specifically expressed on endothelial cells of tumor-associated vessels (Curnis *et al.*, 2002a; Pasqualini *et al.*, 2000). In this context, by coupling NGR peptide to the surface of sterically stabilized liposomes containing DXR, we show tumor regression, pronounced destruction of the tumor vasculature, and prolonged survival of orthotopic human NB-, lung adenocarcinoma-, and ovarian carcinoma-bearing mice (Pastorino *et al.*, 2003a, 2006, 2008). Moreover, this NGR-targeted approach has the advantage to extravasate from the tumor vasculature and to deliver DXR directly into the tumor interstitial space (Pastorino *et al.*, 2003b, 2006).

While endothelial cells are increasingly accepted as a valid target for cancer therapies, another vascular cell type, the pericytes, has been recently recognized as a potentially important, complementary, target for cancer treatment (Bagley *et al.*, 2006; Chantrain *et al.*, 2006). The membrane-associated protease, APA, is upregulated and enzymatically active in perivascular cells (i.e., pericytes) of tumor blood vessels. Moreover, the expression of APA has been correlated to neoplastic progression (Hillen and Griffioen, 2007) and, recently, it has been demonstrated the ability of an APA-recognizing peptide to home to tumor vasculature and to inhibit tumor growth *in vivo* (Marchio *et al.*, 2004).

In conclusion, the availability of ligands binding to additional tumor-associated antigens and to additional targets on both endothelial and perivascular tumor cells would allow one to design more sophisticated cancer treatment strategies that exhibit high levels of selective toxicity for the neoplastic cells. This would improve therapeutic outcome, decrease dose-limiting side effects and improve patient quality of life.

5.1. Preparation of APN peptide-targeted liposomes

APN-targeted liposomes are synthesized from HSPC:CHOL:DSPE-PEG_{2000}:DSPE-PEG_{2000}-MAL, 2:1:0.08:0.02 molar ratio (Pastorino *et al.*, 2003a) and DXR is loaded into the liposomes as described in Section 3.1.

The NGR peptide consists of a linear pentapeptide, GNGRG, fused to the N-terminal sequence of h-TNF followed by a Tyr to enable peptide spectrophotometric detection (Pastorino *et al.*, 2003a). To enhance the accessibility of APN-recognizing peptides (NGR) when bound to liposomes and also to permit coupling via a thiol to a maleimido moiety on the liposomes, additional residues are added to the peptide carboxyl terminus to provide the peptides with a C-terminal Cys (Colombo *et al.*, 2002). The peptides have been synthesized by maintaining the NGR frame responsible for the binding to APN and the C-terminal Cys, for the coupling to liposomes. Specifically, peptides are conjugated to the liposomes by mixing freshly prepared liposomes with an equimolar (with respect to DSPE-PEG_{2000}-MAL) quantity of peptides at 4 °C, for 16 h under argon, followed by a 10-fold excess of 2-mercapethanol for 1 h at RT to derivatize remaining free maleimido groups. Uncoupled peptides are separated from the liposomes by passing the mixture through a Sepharose CL-4B column in Hepes buffer (pH 7.4). Liposomes are sterilized by filtration through 0.22 μm pore cellulose membranes and stored at 4 °C until use. (Scheme 12.1A)

The efficiency of coupling is determined by estimating the amount of liposome-associated peptides using the CBQCA Protein Quantification Kit, as reported (Pastorino *et al.*, 2003a).

NGR-targeted liposomes entrapping DXR are used to treat orthotopically implanted, human NB xenografts in immunodeficient mice. Specifically, mice are anesthetized, subjected to laparatomy, and orthotopically injected with NB cells in the capsule of the left adrenal gland, as reported previously (Pastorino *et al.*, 2003a, 2007). PK profiles indicate that systemic administration of NGR-targeted liposomes have long-circulating profiles in blood (Pastorino *et al.*, 2006). They are removed only slightly more rapidly than the untargeted formulations, with approximately 20% of both drug and carrier remaining in the blood 24 h after liposomes inoculation. Their uptake into NB tumors is at least three times higher than that of untargeted liposomes after 24 h, with DXR spreading outside the blood vessels into the tumors. Five percent of both the liposomes and the drug has localized to tumor by 24 h postinjection and this increases to about 13% by 48 h postinjection. No tumor uptake is observed with liposomes coupled with a control peptide having the ARA motif in place of NGR. Finally, tumor uptake of DXR is completely blocked when mice are coinjected with an excess of the soluble NGR peptide (Pastorino *et al.*, 2006). In therapeutic experiments, tumor-bearing mice treated with NGR-targeted liposomal DXR partly outlive the control mice, and histopathological analysis of tumors reveal pronounced destruction of the tumor vasculature with a marked decrease in vessel density (Pastorino *et al.*, 2006). Moreover, mice injected with higher doses of NGR-targeted liposomal DXR display rapid tumor regression, as well as inhibition of metastatic growth. One day after the end of the treatment, 70% of mice show no evidence of tumors, and the others >80% reduction in tumor mass, and >90% suppression of blood vessel density (Pastorino *et al.*, 2006). In conclusion, this strategy markedly enhances the therapeutic index of DXR. A dual mechanism of action is proposed: indirect tumor cell killing as a consequence of the tumor endothelium destruction by NGR-targeted liposomes, and direct tumor cell killing as a consequence of the increased uptake of liposomal DXR into the tumor interstitial space (Pastorino *et al.*, 2006).

A bioluminescence imaging system has been recently used to test the *in vivo* expression of transfected NB cells (Pastorino *et al.*, 2007). We used a reporter gene that code for a bioluminescent marker to stably label different NB cell lines and we monitored disease growth and metastasis following orthotopic implantation into recipients. The initial trafficking of the malignant cells through the body, the organ-specific homing, and orthotopic expansion over time, as well as the minimal residual disease after surgical removal of the primary mass and the response to NGR-targeted liposomal DXR therapy, is readily visualized, and quantified by a highly sensitive, cooled CCD camera mounted in a light-tight specimen box. BLI intensity over time clearly confirms that DXR encapsulated, APN-targeted liposomal formulations leads to partial primary tumor growth arrest in NB-bearing mice, with respect to untreated mice (Pastorino *et al.*, 2007).

5.2. Preparation of APA peptide-targeted liposomes

For peptides not provided with extra C-terminal Cys, due to possible incorrect folding with other Cys present in the peptide sequence, we assessed novel coupling technologies. For building up APA-targeted liposomes, coupling is allowed between either the COOH (DSPE-PEG$_{2000}$-COOH) or NH_2 (DSPE-PEG$_{2000}$-NH_2) of the derivatized terminal end of DSPE-PEG$_{2000}$ of liposomes, with the amino group of the C-terminal lysine of peptides, as reported (Curnis *et al.*, 2008; Ishida *et al.*, 2001). The APA-recognizing peptides consist of the CPRECES sequence (Marchio *et al.*, 2004) fused to the NH_2-terminal sequence of human TNF and followed by a Tyr.

5.2.1. Preparation of DSPE-PEG$_{2000}$-COOH peptide-targeted liposomes

Accordingly to Ishida *et al.* (2001) with slight modifications, DSPE-PEG$_{2000}$-COOH liposomes are prepared from HSPC, CHOL, and DSPE-PEG$_{2000}$-COOH (1:0.5:0.26 molar ratio). Liposome preparation and DXR loading are performed as described in Section 3.1.

1-Ethyl-3-(3-dimethylaminopropyl)carbodiimide (EDC) and *N*-hydroxysulfosuccinimide (S-NHS) are used for the activation of DSPE-PEG$_{2000}$-COOH on liposomes before the peptides coupling. Briefly, 21 μM of EDC and 28 μM of S-NHS are added to 5 μM lipids in MES buffer (100 mM MES, 150 mM NaCl), pH 5.5 and the mixture is incubated for 15 min at RT with gentle stirring. 2-Mercaptoethanol is added to the mixture at the final concentration of 20 m*M*, then immediately applied to a Sephadex G-50 column, equilibrated with MES buffer and the liposomes fractions collected. Then, APA-recognizing peptide (molar ratio DSPE-PEG_{2000}-COOH:peptide 1:1) is added to liposomes solution and incubated for 2 h at RT, with gentle stirring. Uncoupled peptides are separated from the liposomes by passing the coupling mixture through a Sepharose CL-4B column in Hepes buffer. The efficiency of coupling is determined as reported in Section 5.1. (Scheme 12.1A)

5.2.2. Preparation of DSPE-PEG_{2000}-NH_2 peptide-targeted liposomes

Liposomes are prepared, DXR encapsulation is carried out as described in Section 3.1. DSPE-PEG_{2000}-NH_2 liposomes are prepared from HSPC, CHOL, and DSPE-PEG_{2000}-NH_2 (1:0.5:0.1 molar ratio). In this method, bis[sulfosuccinimiddyl] suberate (BS3) is used for the activation of DSPE-PEG_{2000}-NH_2 on liposomes, according to Curnis *et al.* (2008) with slight modifications. Briefly, the homobifunctional cross-linker BS3 is added at the final concentration 1 m*M*. The *N*-hydroxysuccinimide-liposomes are purified from unreacted cross-linker by a Sepharose CL-4B column equilibrated with PBS. Liposomes are then collected, and peptide is added at the ratio of 8:1 (v/v). After incubation (2 h at RT with gentle stirring), glycine is added at a final concentration of 50 m*M* at RT for 15 min to neutralize the unreacted groups. The conjugates are then separated from free peptide by a Sepharose CL-4B column equilibrated with PBS. The coupling efficiency of peptide is estimated as reported in Section 5.1. (Scheme 12.1A)

Since the coupling efficiency obtained by the use of DSPE-PEG_{2000}-NH_2 technology is higher than that obtained with the DSPE-PEG_{2000}-COOH one, in subsequent experiments, the "NH_2 method" is used also to prepare a new formulation of APN-targeted liposomes by coupling to the liposome surface the cyclic APN-specific peptide (CNGRC) synthesized and purified as described by Di Matteo *et al.* (2006).

6. Liposomes Entrapping Fenretinide (HPR)

Fenretinide (*N*-4 hydroxyphenyl retinamide, HPR) is a synthetic retinoid, structurally related to vitamin A, with biological effects in the control of cell proliferation, differentiation, and fetal development. It has been reported to be cytotoxic to, or inhibit the growth of, primary tumor cells, cell lines, and xenografts of various cancers, including NB (Brewer *et al.*, 2006; Kouhara *et al.*, 2007; Kuefer *et al.*, 2007; Ponzoni *et al.*, 1995; Wei *et al.*, 2005). Moreover, we also reported that HPR exerted an antiangiogenic activity through both a direct inhibitory effect on endothelial cell proliferation and on the response of the endothelial cells to the proliferative stimuli mediated by angiogenic growth factors (Ribatti *et al.*, 2001).

6.1. Preparation of anti-GD_2-targeted, HPR-entrapped, liposomes

Briefly, untargeted liposomes and anti-GD_2-targeted immunoliposomes are prepared as described in Sections 3.1 and 4.1, respectively. The external buffer is then exchanged by passing the liposomes over a Sephadex G-50 column eluted with 123 m*M* sodium citrate, pH 5.5. HPR is then loaded actively into the liposome water phase by 1 h incubation at 62.5 °C (about 1 mol HPR/μmol PL) (Montaldo *et al.*, 1999; Pagnan *et al.*, 1999). Free HPR is separated over a Sephadex G-50 column in Hepes buffer. HPR concentration is evaluated by absorbance at 345 nm. Approximately 75–90% of original HPR is incorporated into the liposomes, as previously shown (Pagnan *et al.*, 1999). The HPR-liposome preparations are sterilized by filtration through 0.22 μm pore cellulose membranes and stored at 4 °C until use. Leakage of HPR from liposomes is measured by dialysis, as described in Section 3.1.

To increase the specific accumulation of HPR in tumor and, consequently, the antitumor effects of this drug, we incorporated HPR into anti-GD_2-targeted liposomes (Raffaghello *et al.*, 2003a) (Scheme 12.1A). Using an NB-metastatic mouse model, we demonstrated that this formulation is able to inhibit the development of macroscopic and microscopic metastasis at a higher extent compared to free HPR and untargeted HPR-loaded formulation (Raffaghello *et al.*, 2003a,b).

7. Antisense Oligonucleotide-Entrapped Liposomes

Inhibition of gene expression by antisense oligonucleotides has become a viable possibility for gene-targeted therapy. The asODNs are single-stranded DNA molecules (traditionally 15–25 bases long) that work by hybridization to corresponding RNA by Watson–Crick binding, resulting in a specific down modulation of gene expression at the transcriptional stage (Stein and Cheng, 1993). asODN therapy became useful in the treatment of cardiovascular diseases and even in oncology, a number of clinical trials have been developed with asODNs directed against molecular targets including p53, bcl-2, raf kinase, protein kinase C-alpha, c-myb, and c-myc (Iversen *et al.*, 2003; Shishodia *et al.*, 2008; Tamm, 2005; Tamm *et al.*, 2001). However, endogenous uptake pathways have insufficient capacity to deliver the quantities of asODNs required to suppress gene expression (Beltinger *et al.*, 1995; Corrias *et al.*, 1997). Moreover, therapeutic application of asODNs may be hampered by their weak physiologic instability, unfavorable PK and lack of tissue specificity (Wagner, 1995). This problem can be circumvented by the synthesis of chemically modified asODNs (Crooke, 1995). However, structural modifications of the asODNs molecules to increase their resistance to nucleases degradation may compromise their biological activities or lead to nonspecific binding (Agrawal and Iyer, 1997; Galderisi *et al.*, 1999; Henry *et al.*, 1999). An alternative tool to increase blood stability, cellular uptake, and specificity of asODNs is represented by the use of lipid-based delivery systems. Cationic liposomal carriers have achieved widespread use for their ability to increase the intracellular delivery of asODNs (Brignole *et al.*, 2004; Felgner and Rhodes, 1991; Pagnan *et al.*, 2000). Nevertheless, the clinical use of cationic liposomes is limited, if repeatedly administered, by their toxicity with induction of immunostimulation and complement activation (Gutierrez-Puente *et al.*, 1999; Mui *et al.*, 2001; Tamm *et al.*, 2001).

For these reasons, we design a different liposome formulation termed as "coated cationic liposomes" (CCLs). We prepare charge-neutral inverted micelles containing cationic lipids, (i.e., DOTAP), in association with negatively charged asODNs. Then, using a reverse phase evaporation procedure, a layer of neutral lipids consisting of HSPC, CHOL, and DSPE-PEG$_{2000}$ are added to form an outer monolayer, coating the cationic lipid-asODN complex.

With this method, CCLs are characterized by a high asODN loading efficiency (90–99%), an average diameter of 130–170 nm, a good stability in terms of both size and retention of contents in the presence of plasma. These lipid-based delivery system have shown improved *in vivo* delivery of asODNs through increase of blood stability, cellular uptake, and specificity (Pagnan *et al.*, 2000; Pastorino *et al.*, 2001; Stuart and Allen, 2000). Particularly, we focused our attention to the use of asODNs against c-myb since the proliferation and differentiation of different types of cancers, including NB, has been associated with the expression of this protooncogene (Alitalo *et al.*, 1984; Raschella *et al.*, 1992). Downregulation of the expression of this protein in tumor cells is associated with an inhibition of cell proliferation (Ekwobi, 2008; Mannava *et al.*, 2008; Pagnan *et al.*, 2000).

7.1. Preparation of anti-GD$_2$-targeted, ODN-encapsulated, liposomes

A phosphorothioate 24-mer oligonucleotide, which is complementary to codons 2–9 of human c-myb messenger RNA (mRNA) (Brignole *et al.*, 2004; Pagnan *et al.*, 2000), is used. Briefly, a 700 μg of a given asODNs with a trace of ^{125}I-labeled asODNs (for allow estimation of the percent of asODNs at various stages of the procedure) is dissolved in 0.25 ml of distilled deionized water. A 0.25 ml $CHCl_3$ containing 2.17 μmol of DOTAP is added, and the mixture is gently mixed. The addition of 510 μl of methanol forms a Bligh–Dyer monophase (Bligh and Dyer, 1959), allowing electrostatic interactions between the positively charged cationic lipids and the negatively charged asODNs: in this way, micelle-like structures are formed, with the hydrophobic PL tails facing the organic phase and the asODNs sequestered inside (Stuart and Allen, 2000). After 30 min incubation at RT, 0.25 ml of distilled deionized water and 0.25 ml of $CHCl_3$ are added. Then mixing and centrifugation (800*g* for 7 min at RT), the aqueous methanol layer is removed. In fact, the system reverts back into a biphase, following the addiction of $CHCl_3$ and water, with the hydrophobic asODNs-cationic lipid particles now concentrated in the organic phase. Under these conditions, 90–95% of the asODNs are recovered in the organic phase. The resulting asODNs-to-lipid mole ratio is 1:24 and a positive-to-negative charge ratio of 1:1 is then obtained (Pagnan *et al.*, 2000). Following extraction, neutral "coating lipids", composed of 4.3 μmol of CHOL and appropriate amounts of HSPC, DSPE-PEG_{2000}, or DSPE-PEG_{2000}-MAL (to give CHOL:PL mole ratio of 1:2), are added to the organic phase. Subsequently, 0.2 ml distilled deionized water is added, and the mixture is mixed and then emulsified by sonication for 2 min: a "water-in-oil emulsion" containing inverse micelles of neutral lipids is established (Pagnan *et al.*, 2000).

Liposomes, which form as the organic phase evaporates, are reduced in size to approximately 100 nm by extrusion through 0.4 μm, five times through 0.2 μm, and 0.1 μm polycarbonate membranes, under nitrogen pressure in a high-pressure LiposoFast-100 extruder, as reported (Pagnan *et al.*, 2000). Thus, as previously described in Section 4.1 either anti-GD_2 mAb or its Fab′ fragments are covalently linked to the MAL terminus of DSPE-PEG_{2000}-MAL and the reaction is incubated overnight, at RT, with continuous slow stirring. The anti-GD_2-targeted asODNs-CCL are then separated from uncoupled IgG by chromatography over a Sepharose CL-4B column. The resulting anti-GD_2-CCL are sterilized by filtration through 0.22 μm pore cellulose membranes and stored at 4 °C until use. Coupling efficiency is quantified as described in Section 4.1. Optimal coupling is obtained with the use of a PL:anti-GD_2 mole ratio of 1500–2000:1, leading to a final antibody density in a range of 40–70 μg Ab/μmol PL. The efficiency of asODNs entrapped in liposomes is approximately 90% (Pagnan *et al.*, 2000). Leakage of aODNs from liposomes is measured by dialysis using the same procedure described in Section 4.1. (Scheme 12.1B)

During 4 h of dialysis at 37 °C in 25% human plasma, 10–20% of asODNs are released; continued dialysis for up to 1 week produced no further free asODNs. Finally, our dialyzed liposomes retain their ability to bind to GD_2-expressing cells over these same time period.

We reported that these anti-GD_2-CCLs, containing c-myb-asODNs, resulted in a selective inhibition of the proliferation of GD_2-positive NB cells *in vitro* (Pagnan *et al.*, 2000). Compared to free asODNs, PK studies show that these targeted liposomal formulations are long circulating in blood. In biodistribution studies, while free c-myb-asODNs are widely distributed (mainly in liver, kidney, and spleen) even after 30 min postinjection, myb-asODNs, entrapped into CCLs or anti-GD_2-CCL, present only an accumulation in the spleen after 24 h (Brignole *et al.*, 2003b).

Recently, it has been reported that CpG motif-containing asODNs, used in clinical therapy to inhibit the oncogene transcription process, can activate various immune cell subsets, inducing the production of numerous cytokines in several syngenic experimental tumor animal models (Hartmann and Krieg, 2000; Klinman, 2004; Krieg, 2006).

Furthermore, the encapsulation and delivery of CpG motif-containing asODNs delivered by lipid vesicles can enhance the process of immune stimulation. In agreement with that, we demonstrated that the encapsulation of CpG-enriched, c-myb-asODNs, into liposomes, targeted to NB cells by GD_2 mAb, led to a great antitumor effect in NB-bearing mice (Brignole *et al.*, 2004). The effectiveness of our liposomal formulations came from two mechanisms of action, the direct and selective downmodulation of the c-myb oncogene expression, mediated by asODNs and the stimulation of the immune system, due to the CpG islands (Brignole *et al.*, 2004). Moreover, we found that the systemic administration of anti-GD_2-targeted CCL containing CpG-c-myb-asODNs has potent anticancer effects, leading to long-term survival in NB-bearing mice (Brignole *et al.*, 2004). It is important to underline that an antitumor activity, mediated by the immune stimulation, was also achieved in NB-bearing mice treated with both targeted and untargeted CCL formulations of scrambled ODNs (which maintain the CpG islands), in agreement with previous findings (Brignole *et al.*, 2004; Carpentier *et al.*, 1999).

The achievement of long-term survival in NB-bearing mice treated with anti-GD_2-targeted, CpG-enriched c-myb asODN, CCLs is related to several factors working in synergism: the selective targeting of the CCLs to NB tumors via GD_2 (required to reduce nonspecific distribution of the CCLs), with the specific antisense activity of c-myb-asODNs (required for downmodulation of c-myb expression) and the CpG-mediated immune system stimulation (Brignole *et al.*, 2004, 2005). Secondly, based on our *in vivo* results, we hypothesize that macrophages are the first activators of a cascade that ends with activated NK cells killing NB cells. This is confirmed by the observations that abrogation of macrophages by Clodronate-entrapped liposomes abolishes the antitumor activity of CpG-enriched myb-as-liposomes. The critical involvement of NK cells is demonstrated by the total loss of antitumor activity in experiments in which targeted formulations containing a scrambled CpG-containing sequence are administered to NB-bearing, NK-depleted, SCID-beige mice. However, in the same model, the treatment with the anti-GD_2-targeted, c-myb-asODNs-

CCLs still results in 20% long-term survivors. These experiments allow us to dissect out antitumor activity related to the delivery of the CpG sequence to macrophages from that obtained by the specific down-modulation of c-myb protein, which is a result of localization of the targeted CCL to NB tumors, followed by the GD_2-mediated internalization of the CCL-asODN package. In conclusion, the encapsulation of asODNs in targeted CCLs significantly enhances their therapeutic activity against NB tumor by a direct mechanism involving the downmodulation of the expression of the proliferative c-myb protooncogene. Antitumor activity is synergistically enhanced via an indirect mechanism, when the asODNs contain immune-stimulatory CpG sequences.

8. Gold-Containing Liposomes

Understanding and controlling the interactions between nanoscale objects and living cells is of great importance for arising diagnostic and therapeutic applications of nanoparticles and for nanotoxicology studies. Colloidal gold nanoparticles have shown great promise for many biomedical applications as therapeutics, therapeutic delivery vectors, intracellular imaging agents, and histological marker, owing to their high electron density, uniform size, shape, surface chemistry, versatility, and optical–electronic properties (Handley *et al.*, 1981).

The first method to prepare liposomes containing gold as an electron-dense marker to trace liposomes–cell interactions was described in 1983 by Hong *et al.* (1983). They ascertained that the optimal conditions for encapsulating the gold particles into liposomes were achieved by the method of reverse-phase evaporation. Gold particles formed rapidly at ambient temperature and without organic solvent producing homogeneous populations of gold granules inside liposomes. These gold-containing liposomes allowed to determine unambiguously the intracellular fate of liposomes and of their contents (Hong *et al.*, 1983).

Recently, Kirpotin *et al.* performed experiments using colloidal gold-labeled liposomes targeted to HER2 (ErbB2, Neu) by the conjugation of anti-HER2 mAb fragments to the external surface of liposomes. They evaluated the distribution of these particles within tumor tissue by light microscopy, revealing differences in microdistribution and cellular localization for gold-containing anti-HER2 immunoliposomes in HER2 overexpressing breast tumor model in comparison with nontargeted formulations (Kirpotin *et al.*, 2006).

Here, in order to obtain the same active targeting either to NB tumor cell or to tumor endothelial cells within the tumor, we couple cell-targeting, antibodies and peptides onto colloidal gold-entrapped liposomes.

8.1. Preparation of anti-GD_2-targeted, gold-encapsulated, liposomes

According to Hong *et al.* (1983), gold chloride/citrate solution is prepared adding chloroauric acid (12.72 m*M* $HAuCl_4$ in distilled water) to basic citrate solution (13.6 m*M* $Na_3C_6H_5O_7$/3.33 m*M* K_2CO_3), immediately before the preparation of liposomes. The final concentration in gold chloride/citrate solution is 3.18 m*M* $HAuCl_4$/2.5 m*M* K_2CO_3/ 10.2 m*M* trisodium citrate, pH 6.0–6.2 to avoid the premature nucleation of gold sols during liposomes preparation. Gold chloride/citrate solution is added to HSPC, CHOL, and DSPE-PEG_{2000} or HSPC, CHOL, DSPE-PEG_{2000}, and DSPE-PEG_{2000}-MAL (2:2:1:0.1 or 2:2:1:0.08:0.02 molar ratio), and the mixture is emulsified by sonication at 4 °C. Gold unilamellar vesicles are then prepared by reverse-phase evaporation in aqueous phase after removing chloroform under reduced pressure by a rotary evaporator at 30 °C. Gold-lipid film is subsequently rehydrated with distilled water. To ensure a narrow distribution of size, liposomes are extruded through polycarbonate membranes of 0.2 μm of pore diameter to reach a size of about 170 nm. Finally, the liposome suspension is incubated at 37 °C for 1 h to obtain gold sols formation. Gold-entrapped liposomes are applied to a Sephadex G-50 column preequilibrated with Hepes buffer, pH 7.4 to remove gold granules either free or associated with liposomes surface. Gold encapsulation is quantified by measuring absorbance at 525 nm after gold liposomes dissolving in 1 ml MeOH. For preparing gold-entrapped anti-GD_2-targeted immunoliposomes, anti-GD_2 mAb is coupled to the MAL terminus of DSPE-PEG_{2000}-MAL, as previously described in Section 4.1.

8.2. Characterization of anti-GD_2-targeted, gold-encapsulated, liposomes by TEM

Anti-GD_2-immunoliposomes are prepared as describe in Section 4.1 and observed by transmission electron microscopy (TEM) after staining with 2% uranyl acetate in solution and deposition on a formvar resin-coated electron microscopy copper grid. Liposomes ranged from 100 to 200 nm in diameter, and displayed an irregular roundish or elliptical shape. The surrounding membrane is often multilayered (Fig. 12.1).

Electron-dense gold precipitates are observed in a fraction of the liposomes in the form of clusters or single particles. A minor portion of the gold particles are observed outside the boundary of liposomes, likely due to mechanical rupture of some liposomes during the drying step required for TEM observation.

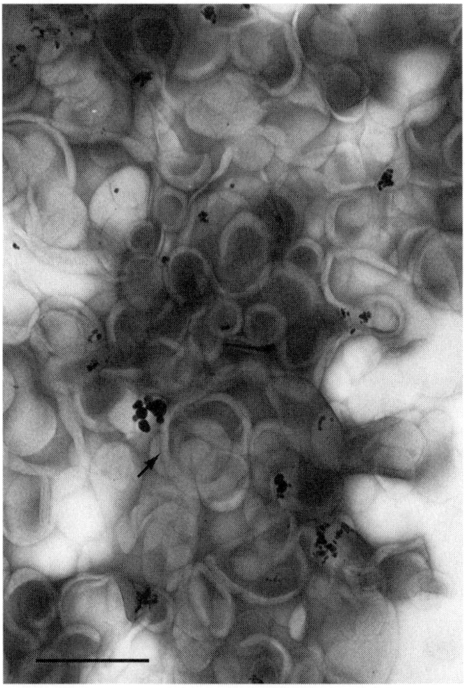

Figure 12.1 Anti-GD_2-targeted, gold-entrapping immunoliposomes are observed by electron microscopy. Liposomes are irregularly round or elliptical in shape. Dark electron-dense dots indicate the presence of gold precipitates. The arrow points to a multilayered membrane bounding one liposome. Bar: 250 nm.

ACKNOWLEDGMENTS

This work is supported by Fondazione Italiana per la lotta al Neuroblastoma, Associazione Italiana Ricerca Cancro (AIRC, MFAG to Pastorino and IG to Ponzoni) and Italian Ministry of Health. D. Di Paolo is a recipient of a Fondazione Italiana Ricerca Cancro (FIRC) fellowship. F. Pastorino and G. Pagnan are recipients of a Fondazione Italiana per la Lotta al Neuroblastoma fellowship.

REFERENCES

Adams, G. P., and Weiner, L. M. (2005). Monoclonal antibody therapy of cancer. *Nat. Biotechnol.* **23,** 1147–1157.

Agrawal, S., and Iyer, R. P. (1997). Perspectives in antisense therapeutics. *Pharmacol. Ther.* **76,** 151–160.

Alitalo, K., Winqvist, R., Lin, C. C., de la Chapelle, A., Schwab, M., and Bishop, J. M. (1984). Aberrant expression of an amplified c-myb oncogene in two cell lines from a colon carcinoma. *Proc. Natl Acad. Sci. USA* **81,** 4534–4538.

Allen, T. M. (2002). Ligand-targeted therapeutics in anticancer therapy. *Nat. Rev. Cancer* **2,** 750–763.

Allen, T. M., and Cullis, P. R. (2004). Drug delivery systems: Entering the mainstream. *Science* **303,** 1818–1822.

Allen, T. M., Hansen, C., Martin, F., Redemann, C., and Yau-Young, A. (1991). Liposomes containing synthetic lipid derivatives of poly(ethylene glycol) show prolonged circulation half-lives *in vivo. Biochim. Biophys. Acta* **1066,** 29–36.

Allen, T. M., Mehra, T., Hansen, C., and Chin, Y. C. (1992). Stealth liposomes: An improved sustained release system for 1-beta-D-arabinofuranosylcytosine. *Cancer Res.* **52,** 2431–2439.

Allen, T. M., Brandeis, E., Hansen, C. B., Kao, G. Y., and Zalipsky, S. (1995). A new strategy for attachment of antibodies to sterically stabilized liposomes resulting in efficient targeting to cancer cells. *Biochim. Biophys. Acta* **1237,** 99–108.

Arap, W., Pasqualini, R., and Ruoslahti, E. (1998). Cancer treatment by targeted drug delivery to tumor vasculature in a mouse model. *Science* **279,** 377–380.

Arap, W., Kolonin, M. G., Trepel, M., Lahdenranta, J., Cardo-Vila, M., Giordano, R. J., Mintz, P. J., Ardelt, P. U., Yao, V. J., Vidal, C. I., Chen, L., Flamm, A., *et al.* (2002). Steps toward mapping the human vasculature by phage display. *Nat. Med.* **8,** 121–127.

Bagley, R. G., Rouleau, C., Morgenbesser, S. D., Weber, W., Cook, B. P., Shankara, S., Madden, S. L., and Teicher, B. A. (2006). Pericytes from human non-small cell lung carcinomas: An attractive target for anti-angiogenic therapy. *Microvasc. Res.* **71,** 163–174.

Beltinger, C., Saragovi, H. U., Smith, R. M., LeSauteur, L., Shah, N., DeDionisio, L., Christensen, L., Raible, A., Jarett, L., and Gewirtz, A. M. (1995). Binding, uptake, and intracellular trafficking of phosphorothioate-modified oligodeoxynucleotides. *J. Clin. Invest.* **95,** 1814–1823.

Bligh, E. G., and Dyer, W. J. (1959). A rapid method of total lipid extraction and purification. *Can. J. Biochem. Physiol.* **37,** 911–917.

Bolotin, E. M., Cohen, R., Bar, L. K., Emanuel, N., Ninio, S., Lasic, D. D., and Barenholz, Y. (1994). Ammonium sulphate gradients for efficient and stable remote loading of amphipathic weak bases into liposomes and ligandoliposomes. *J. Liposmes Res.* **4,** 455–479.

Brewer, M., Kirkpatrick, N. D., Wharton, J. T., Wang, J., Hatch, K., Auersperg, N., Utzinger, U., Gershenson, D., Bast, R., and Zou, C. (2006). 4-HPR modulates gene expression in ovarian cells. *Int. J. Cancer* **119,** 1005–1013.

Brignole, C., Marimpietri, D., Gambini, C., Allen, T. M., Ponzoni, M., and Pastorino, F. (2003a). Development of Fab′ fragments of anti-GD(2) immunoliposomes entrapping doxorubicin for experimental therapy of human neuroblastoma. *Cancer Lett.* **197,** 199–204.

Brignole, C., Pagnan, G., Marimpietri, D., Cosimo, E., Allen, T. M., Ponzoni, M., and Pastorino, F. (2003b). Targeted delivery system for antisense oligonucleotides: A novel experimental strategy for neuroblastoma treatment. *Cancer Lett.* **197,** 231–235.

Brignole, C., Pastorino, F., Marimpietri, D., Pagnan, G., Pistorio, A., Allen, T. M., Pistoia, V., and Ponzoni, M. (2004). Immune cell-mediated antitumor activities of GD2-targeted liposomal c-myb antisense oligonucleotides containing CpG motifs. *J. Natl. Cancer Inst.* **96,** 1171–1180.

Brignole, C., Marimpietri, D., Pagnan, G., Di Paolo, D., Zancolli, M., Pistoia, V., Ponzoni, M., and Pastorino, F. (2005). Neuroblastoma targeting by c-myb-selective antisense oligonucleotides entrapped in anti-GD2 immunoliposome: Immune cell-mediated anti-tumor activities. *Cancer Lett.* **228,** 181–186.

Broder, H., Gottlieb, R. A., and Lepor, N. E. (2008). Chemotherapy and cardiotoxicity. *Rev. Cardiovasc. Med.* **9,** 75–83.

Brodeur, G. M., and Maris, J. M. (2006). *Neuroblastoma.* J. B. Lippincott Company, Philadelphia.

Carpentier, A. F., Chen, L., Maltonti, F., and Delattre, J. Y. (1999). Oligodeoxynucleotides containing CpG motifs can induce rejection of a neuroblastoma in mice. *Cancer Res.* **59,** 5429–5432.

Chantrain, C. F., Henriet, P., Jodele, S., Emonard, H., Feron, O., Courtoy, P. J., DeClerck, Y. A., and Marbaix, E. (2006). Mechanisms of pericyte recruitment in tumour angiogenesis: A new role for metalloproteinases. *Eur. J. Cancer* **42,** 310–318.

Colombo, G., Curnis, F., De Mori, G. M., Gasparri, A., Longoni, C., Sacchi, A., Longhi, R., and Corti, A. (2002). Structure-activity relationships of linear and cyclic peptides containing the NGR tumor-homing motif. *J. Biol. Chem.* **277,** 47891–47897.

Conte, M., Parodi, S., De Bernardi, B., Milanaccio, C., Mazzocco, K., Angelini, P., Viscardi, E., Di Cataldo, A., Luksch, R., and Haupt, R. (2006). Neuroblastoma in adolescents: The Italian experience. *Cancer* **106,** 1409–1417.

Corrias, M. V., Guarnaccia, F., and Ponzoni, M. (1997). Bioavailability of antisense oligonucleotides in neuroblastoma cells: Comparison of efficacy among different types of molecules. *J. Neurooncol.* **31,** 171–180.

Crooke, S. T. (1995). Progress in antisense therapeutics. *Hematol. Pathol.* **9,** 59–72.

Curnis, F., Sacchi, A., Borgna, L., Magni, F., Gasparri, A., and Corti, A. (2000). Enhancement of tumor necrosis factor alpha antitumor immunotherapeutic properties by targeted delivery to aminopeptidase N (CD13). *Nat. Biotechnol.* **18,** 1185–1190.

Curnis, F., Arrigoni, G., Sacchi, A., Fischetti, L., Arap, W., Pasqualini, R., and Corti, A. (2002a). Differential binding of drugs containing the NGR motif to CD13 isoforms in tumor vessels, epithelia, and myeloid cells. *Cancer Res.* **62,** 867–874.

Curnis, F., Sacchi, A., and Corti, A. (2002b). Improving chemotherapeutic drug penetration in tumors by vascular targeting and barrier alteration. *J. Clin. Invest.* **110,** 475–482.

Curnis, F., Sacchi, A., Gasparri, A., Longhi, R., Bachi, A., Doglioni, C., Bordignon, C., Traversari, C., Rizzardi, G. P., and Corti, A. (2008). Isoaspartate–glycine–arginine: A new tumor vasculature-targeting motif. *Cancer Res.* **68,** 7073–7082.

De Bernardi, B., Nicolas, B., Boni, L., Indolfi, P., Carli, M., Cordero Di Montezemolo, L., Donfrancesco, A., Pession, A., Provenzi, M., di Cataldo, A., Rizzo, A., Tonini, G. P., *et al.* (2003). Disseminated neuroblastoma in children older than one year at diagnosis: Comparable results with three consecutive high-dose protocols adopted by the Italian Co-Operative Group for Neuroblastoma. *J. Clin. Oncol.* **21,** 1592–1601.

Di Matteo, P., Curnis, F., Longhi, R., Colombo, G., Sacchi, A., Crippa, L., Protti, M. P., Ponzoni, M., Toma, S., and Corti, A. (2006). Immunogenic and structural properties of the Asn–Gly–Arg (NGR) tumor neovasculature-homing motif. *Mol. Immunol.* **43,** 1509–1518.

Ding, L., Samuel, J., MacLean, G. D., Noujaim, A. A., Diener, E., and Longenecker, B. M. (1990). Effective drug-antibody targeting using a novel monoclonal antibody against the proliferative compartment of mammalian squamous carcinomas. *Cancer Immunol. Immunother.* **32,** 105–109.

Ekwobi, C. C. (2008). The prognostic value of c-myc oncoprotein levels in malignant melanoma. *J. Plast. Reconstr. Aesthet. Surg.* **61,** 464.

Ellerby, H. M., Arap, W., Ellerby, L. M., Kain, R., Andrusiak, R., Rio, G. D., Krajewski, S., Lombardo, C. R., Rao, R., Ruoslahti, E., Bredesen, D. E., and Pasqualini, R. (1999). Anti-cancer activity of targeted pro-apoptotic peptides. *Nat. Med.* **5,** 1032–1038.

Felgner, P. L., and Rhodes, G. (1991). Gene therapeutics. *Nature* **349,** 351–352.

Fidler, I. J., and Ellis, L. M. (1994). The implications of angiogenesis for the biology and therapy of cancer metastasis. *Cell* **79,** 185–188.

Folkman, J. (1971). Tumor angiogenesis: Therapeutic implications. *N. Engl. J. Med.* **285,** 1182–1186.

Frost, J. D., Hank, J. A., Reaman, G. H., Frierdich, S., Seeger, R. C., Gan, J., Anderson, P. M., Ettinger, L. J., Cairo, M. S., Blazar, B. R., Krailo, M. D., Matthay, K. K., *et al.* (1997). A phase I/IB trial of murine monoclonal anti-GD2 antibody

14.G2a plus interleukin-2 in children with refractory neuroblastoma: A report of the Children's Cancer Group. *Cancer* **80,** 317–333.
Gabizon, A., Tzemach, D., Mak, L., Bronstein, M., and Horowitz, A. T. (2002). Dose dependency of pharmacokinetics and therapeutic efficacy of pegylated liposomal doxorubicin (DOXIL) in murine models. *J. Drug Target.* **10,** 539–548.
Galderisi, U., Cascino, A., and Giordano, A. (1999). Antisense oligonucleotides as therapeutic agents. *J. Cell. Physiol.* **181,** 251–257.
Gutierrez-Puente, Y., Tari, A. M., Stephens, C., Rosenblum, M., Guerra, R. T., and Lopez-Berestein, G. (1999). Safety, pharmacokinetics, and tissue distribution of liposomal P-ethoxy antisense oligonucleotides targeted to Bcl-2. *J. Pharmacol. Exp. Ther.* **291,** 865–869.
Hamaguchi, T., Matsumura, Y., Nakanishi, Y., Muro, K., Yamada, Y., Shimada, Y., Shirao, K., Niki, H., Hosokawa, S., Tagawa, T., and Kakizoe, T. (2004). Antitumor effect of MCC-465, pegylated liposomal doxorubicin tagged with newly developed monoclonal antibody GAH, in colorectal cancer xenografts. *Cancer Sci.* **95,** 608–613.
Handley, D. A., Arbeeny, C. M., Witte, L. D., and Chien, S. (1981). Colloidal gold—Low density lipoprotein conjugates as membrane receptor probes. *Proc. Natl. Acad. Sci. USA* **78,** 368–371.
Hansen, C. B., Kao, G. Y., Moase, E. H., Zalipsky, S., and Allen, T. M. (1995). Attachment of antibodies to sterically stabilized liposomes: Evaluation, comparison and optimization of coupling procedures. *Biochim. Biophys. Acta* **1239,** 133–144.
Hartmann, G., and Krieg, A. M. (2000). Mechanism and function of a newly identified CpG DNA motif in human primary B cells. *J. Immunol.* **164,** 944–953.
Henry, S. P., Templin, M. V., Gillett, N., Rojko, J., and Levin, A. A. (1999). Correlation of toxicity and pharmacokinetic properties of a phosphorothioate oligonucleotide designed to inhibit ICAM-1. *Toxicol. Pathol.* **27,** 95–100.
Hillen, F., and Griffioen, A. W. (2007). Tumour vascularization: Sprouting angiogenesis and beyond. *Cancer Metastasis Rev.* **26**(3-4)**,** 489–502.
Hong, K., Friend, D. S., Glabe, C. G., and Papahadjopoulos, D. (1983). Liposomes containing colloidal gold are a useful probe of liposome-cell interactions. *Biochim. Biophys. Acta* **732,** 320–323.
Hood, J. D., Bednarski, M., Frausto, R., Guccione, S., Reisfeld, R. A., Xiang, R., and Cheresh, D. A. (2002). Tumor regression by targeted gene delivery to the neovasculature. *Science* **296,** 2404–2407.
Huwyler, J., Wu, D., and Pardridge, W. M. (1996). Brain drug delivery of small molecules using immunoliposomes. *Proc. Natl. Acad. Sci. USA* **93,** 14164–14169.
Injac, R., and Strukelj, B. (2008). Recent advances in protection against doxorubicin-induced toxicity. *Technol. Cancer Res. Treat.* **7,** 497–516.
Ishida, O., Maruyama, K., Tanahashi, H., Iwatsuru, M., Sasaki, K., Eriguchi, M., and Yanagie, H. (2001). Liposomes bearing polyethyleneglycol-coupled transferrin with intracellular targeting property to the solid tumors *in vivo*. *Pharm. Res.* **18,** 1042–1048.
Iversen, P. L., Arora, V., Acker, A. J., Mason, D. H., and Devi, G. R. (2003). Efficacy of antisense morpholino oligomer targeted to c-myc in prostate cancer xenograft murine model and a Phase I safety study in humans. *Clin. Cancer Res.* **9,** 2510–2519.
Jin, H., and Varner, J. (2004). Integrins: Roles in cancer development and as treatment targets. *Br. J. Cancer* **90,** 561–565.
Kirpotin, D. B., Drummond, D. C., Shao, Y., Shalaby, M. R., Hong, K., Nielsen, U. B., Marks, J. D., Benz, C. C., and Park, J. W. (2006). Antibody targeting of long-circulating lipidic nanoparticles does not increase tumor localization but does increase internalization in animal models. *Cancer Res.* **66,** 6732–6740.
Klinman, D. M. (2004). Immunotherapeutic uses of CpG oligodeoxynucleotides. *Nat. Rev. Immunol.* **4,** 249–258.

Kouhara, J., Yoshida, T., Nakata, S., Horinaka, M., Wakada, M., Ueda, Y., Yamagishi, H., and Sakai, T. (2007). Fenretinide up-regulates DR5/TRAIL-R2 expression via the induction of the transcription factor CHOP and combined treatment with fenretinide and TRAIL induces synergistic apoptosis in colon cancer cell lines. *Int. J. Oncol.* **30,** 679–687.

Kowalczyk, A., Gil, M., Horwacik, I., Odrowaz, Z., Kozbor, D., and Rokita, H. (2009). The GD2-specific 14G2a monoclonal antibody induces apoptosis and enhances cytotoxicity of chemotherapeutic drugs in IMR-32 human neuroblastoma cells. *Cancer Lett.* **281**(2)**,** 171–182.

Krieg, A. M. (2006). Therapeutic potential of Toll-like receptor 9 activation. *Nat. Rev. Drug Discov.* **5,** 471–484.

Kuefer, R., Genze, F., Zugmaier, W., Hautmann, R. E., Rinnab, L., Gschwend, J. E., Angelmeier, M., Estrada, A., and Buechele, B. (2007). Antagonistic effects of sodium butyrate and N-(4-hydroxyphenyl)-retinamide on prostate cancer. *Neoplasia* **9,** 246–253.

Lopes de Menezes, D. E., Pilarski, L. M., and Allen, T. M. (1998). *In vitro and in vivo* targeting of immunoliposomal doxorubicin to human B-cell lymphoma. *Cancer Res.* **58,** 3320–3330.

Mahabeleshwar, G. H., Feng, W., Reddy, K., Plow, E. F., and Byzova, T. V. (2007). Mechanisms of integrin-vascular endothelial growth factor receptor cross-activation in angiogenesis. *Circ. Res.* **101,** 570–580.

Mannava, S., Grachtchouk, V., Wheeler, L. J., Im, M., Zhuang, D., Slavina, E. G., Mathews, C. K., Shewach, D. S., and Nikiforov, M. A. (2008). Direct role of nucleotide metabolism in C-MYC-dependent proliferation of melanoma cells. *Cell Cycle* **7,** 2392–2400.

Marchio, S., Lahdenranta, J., Schlingemann, R. O., Valdembri, D., Wesseling, P., Arap, M. A., Hajitou, A., Ozawa, M. G., Trepel, M., Giordano, R. J., Nanus, D. M., Dijkman, H. B., *et al.* (2004). Aminopeptidase A is a functional target in angiogenic blood vessels. *Cancer Cell* **5,** 151–162.

Maris, J. M., and Matthay, K. K. (1999). Molecular biology of neuroblastoma. *J. Clin. Oncol.* **17,** 2264–2279.

Maris, J. M., Hogarty, M. D., Bagatell, R., and Cohn, S. L. (2007). Neuroblastoma. *Lancet* **369,** 2106–2120.

Matsumura, Y., Gotoh, M., Muro, K., Yamada, Y., Shirao, K., Shimada, Y., Okuwa, M., Matsumoto, S., Miyata, Y., Ohkura, H., Chin, K., Baba, S., *et al.* (2004). Phase I and pharmacokinetic study of MCC-465, a doxorubicin (DXR) encapsulated in PEG immunoliposome, in patients with metastatic stomach cancer. *Ann. Oncol.* **15,** 517–525.

Miano, M., Garaventa, A., Pizzitola, M. R., Piccolo, M. S., Dallorso, S., Villavecchia, G. P., Bertolazzi, C., Cabria, M., and De Bernardi, B. (2001). Megatherapy combining I(131) metaiodobenzylguanidine and high-dose chemotherapy with haematopoietic progenitor cell rescue for neuroblastoma. *Bone Marrow Transplant.* **27,** 571–574.

Moase, E. H., Qi, W., Ishida, T., Gabos, Z., Longenecker, B. M., Zimmermann, G. L., Ding, L., Krantz, M., and Allem, T. M. (2000). Anti-MUC-1 immunoliposomal doxorubicin in the treatment of murine models of metastatic breast cancer. *Biochim. Biophys. Acta* **77996,** 1–13.

Montaldo, P. G., Pagnan, G., Pastorino, F., Chiesa, V., Raffaghello, L., Kirchmeier, M., Allen, T. M., and Ponzoni, M. (1999). N-(4-hydroxyphenyl) retinamide is cytotoxic to melanoma cells in vitro through induction of programmed cell death. *Int. J. Cancer* **81,** 262–267.

Mui, B., Raney, S. G., Semple, S. C., and Hope, M. J. (2001). Immune stimulation by a CpG-containing oligodeoxynucleotide is enhanced when encapsulated and delivered in lipid particles. *J. Pharmacol. Exp. Ther.* **298,** 1185–1192.

Mujoo, K., Cheresh, D. A., Yang, H. M., and Reisfeld, R. A. (1987). Disialoganglioside GD2 on human neuroblastoma cells: Target antigen for monoclonal antibody-mediated cytolysis and suppression of tumor growth. *Cancer Res.* **47,** 1098–1104.

Niethammer, A. G., Xiang, R., Becker, J. C., Wodrich, H., Pertl, U., Karsten, G., Eliceiri, B. P., and Reisfeld, R. A. (2002). A DNA vaccine against VEGF receptor 2 prevents effective angiogenesis and inhibits tumor growth. *Nat. Med.* **8,** 1369–1375.

Pagnan, G., Montaldo, P. G., Pastorino, F., Raffaghello, L., Kirchmeier, M., Allen, T. M., and Ponzoni, M. (1999). GD2-mediated melanoma cell targeting and cytotoxicity of liposome-entrapped fenretinide. *Int. J. Cancer* **81,** 268–274.

Pagnan, G., Stuart, D. D., Pastorino, F., Raffaghello, L., Montaldo, P. G., Allen, T. M., Calabretta, B., and Ponzoni, M. (2000). Delivery of c-myb antisense oligodeoxynucleotides to human neuroblastoma cells via disialoganglioside GD(2)-targeted immunoliposomes: Antitumor effects. *J. Natl. Cancer Inst.* **92,** 253–261.

Park, J. W., Hong, K., Carter, P., Asgari, H., Guo, L. Y., Keller, G. A., Wirth, C., Shalaby, R., Kotts, C., Wood, W. I., *et al.* (1995). Development of anti-p185HER2 immunoliposomes for cancer therapy. *Proc. Natl. Acad. Sci. USA* **92,** 1327–1331.

Pasqualini, R., Koivunen, E., Kain, R., Lahdenranta, J., Sakamoto, M., Stryhn, A., Ashmun, R. A., Shapiro, L. H., Arap, W., and Ruoslahti, E. (2000). Aminopeptidase N is a receptor for tumor-homing peptides and a target for inhibiting angiogenesis. *Cancer Res.* **60,** 722–727.

Pastorino, F., Stuart, D., Ponzoni, M., and Allen, T. M. (2001). Targeted delivery of antisense oligonucleotides in cancer. *J. Control Release* **74,** 69–75.

Pastorino, F., Brignole, C., Marimpietri, D., Cilli, M., Gambini, C., Ribatti, D., Longhi, R., Allen, T. M., Corti, A., and Ponzoni, M. (2003a). Vascular damage and anti-angiogenic effects of tumor vessel-targeted liposomal chemotherapy. *Cancer Res.* **63,** 7400–7409.

Pastorino, F., Brignole, C., Marimpietri, D., Sapra, P., Moase, E. H., Allen, T. M., and Ponzoni, M. (2003b). Doxorubicin-loaded Fab' fragments of anti-disialoganglioside immunoliposomes selectively inhibit the growth and dissemination of human neuroblastoma in nude mice. *Cancer Res.* **63,** 86–92.

Pastorino, F., Brignole, C., Di Paolo, D., Nico, B., Pezzolo, A., Marimpietri, D., Pagnan, G., Piccardi, F., Cilli, M., Longhi, R., Ribatti, D., Corti, A., *et al.* (2006). Targeting liposomal chemotherapy via both tumor cell-specific and tumor vasculature-specific ligands potentiates therapeutic efficacy. *Cancer Res.* **66,** 10073–10082.

Pastorino, F., Marimpietri, D., Brignole, C., Di Paolo, D., Pagnan, G., Daga, A., Piccardi, F., Cilli, M., Allen, T. M., and Ponzoni, M. (2007). Ligand-targeted liposomal therapies of Neuroblastoma. *Curr. Med. Chem.* **14**(29), 3070–3078.

Pastorino, F., Di Paolo, D., Piccardi, F., Nico, B., Ribatti, D., Daga, A., Baio, G., Neumaier, C. E., Brignole, C., Loi, M., Marimpietri, D., Pagnan, G., *et al.* (2008). Enhanced antitumor efficacy of clinical-grade vasculature-targeted liposomal doxorubicin. *Clin. Cancer Res.* **14,** 7320–7329.

Ponzoni, M., Bocca, P., Chiesa, V., Decensi, A., Pistoia, V., Raffaghello, L., Rozzo, C., and Montaldo, P. G. (1995). Differential effects of N-(4-hydroxyphenyl)retinamide and retinoic acid on neuroblastoma cells: Apoptosis versus differentiation. *Cancer Res.* **55,** 853–861.

Raffaghello, L., Pagnan, G., Pastorino, F., Cosimo, E., Brignole, C., Marimpietri, D., Bogenmann, E., Ponzoni, M., and Montaldo, P. G. (2003a). Immunoliposomal fenretinide: A novel antitumoral drug for human neuroblastoma. *Cancer Lett.* **197,** 151–155.

Raffaghello, L., Pagnan, G., Pastorino, F., Cosimo, E., Brignole, C., Marimpietri, D., Montaldo, P. G., Gambini, C., Allen, T. M., Bogenmann, E., and Ponzoni, M. (2003b). *In vitro and in vivo* antitumor activity of liposomal fenretinide targeted to human neuroblastoma. *Int. J. Cancer* **104,** 559–567.

Raschella, G., Negroni, A., Skorski, T., Pucci, S., Nieborowska-Skorska, M., Romeo, A., and Calabretta, B. (1992). Inhibition of proliferation by c-myb antisense RNA and oligodeoxynucleotides in transformed neuroectodermal cell lines. *Cancer Res.* **52,** 4221–4226.

Ribatti, D., Surico, G., Vacca, A., De Leonardis, F., Lastilla, G., Montaldo, P. G., Rigillo, N., and Ponzoni, M. (2001). Angiogenesis extent and expression of matrix metalloproteinase-2 and -9 correlate with progression in human neuroblastoma. *Life Sci.* **68,** 1161–1168.

Schulz, G., Cheresh, D. A., Varki, N. M., Yu, A., Staffileno, L. K., and Reisfeld, R. A. (1984). Detection of ganglioside GD2 in tumor tissues and sera of neuroblastoma patients. *Cancer Res.* **44,** 5914–5920.

Shishodia, S., Harikumar, K. B., Dass, S., Ramawat, K. G., and Aggarwal, B. B. (2008). The guggul for chronic diseases: Ancient medicine, modern targets. *Anticancer Res.* **28,** 3647–3664.

Sipkins, D. A., Cheresh, D. A., Kazemi, M. R., Nevin, L. M., Bednarski, M. D., and Li, K. C. (1998). Detection of tumor angiogenesis *in vivo* by alphaVbeta3-targeted magnetic resonance imaging. *Nat. Med.* **4,** 623–626.

Stein, C. A., and Cheng, Y. C. (1993). Antisense oligonucleotides as therapeutic agents—Is the bullet really magical? *Science* **261,** 1004–1012.

Stuart, D. D., and Allen, T. M. (2000). A new liposomal formulation for antisense oligodeoxynucleotides with small size, high incorporation efficiency and good stability. *Biochim. Biophys. Acta* **1463,** 219–229.

Sugano, M., Egilmez, N. K., Yokota, S. J., Chen, F. A., Harding, J., Huang, S. K., and Bankert, R. B. (2000). Antibody targeting of doxorubicin-loaded liposomes suppresses the growth and metastatic spread of established human lung tumor xenografts in severe combined immunodeficient mice. *Cancer Res.* **60,** 6942–6949.

Tamm, I. (2005). Antisense therapy in clinical oncology: Preclinical and clinical experiences. *Methods Mol. Med.* **106,** 113–134.

Tamm, I., Dorken, B., and Hartmann, G. (2001). Antisense therapy in oncology: New hope for an old idea? *Lancet* **358,** 489–497.

Torchilin, V. P. (2006). Application of nanomedical approaches in experimental and clinical oncology. *Anticancer Agents Med. Chem.* **6,** 501.

Trepel, M., Arap, W., and Pasqualini, R. (2002). *In vivo* phage display and vascular heterogeneity: Implications for targeted medicine. *Curr. Opin. Chem. Biol.* **6,** 399–404.

Uttenreuther-Fischer, M. M., Huang, C. S., Reisfeld, R. A., and Yu, A. L. (1995). Pharmacokinetics of anti-ganglioside GD2 mAb 14G2a in a phase I trial in pediatric cancer patients. *Cancer Immunol. Immunother.* **41,** 29–36.

von Allmen, D., Grupp, S., Diller, L., Marcus, K., Ecklund, K., Meyer, J., and Shamberger, R. C. (2005). Aggressive surgical therapy and radiotherapy for patients with high-risk neuroblastoma treated with rapid sequence tandem transplant. *J. Pediatr. Surg.* **40,** 936–941discussion: 941.

Wagner, R. W. (1995). The state of the art in antisense research. *Nat. Med.* **1,** 1116–1118.

Wei, J. S., Whiteford, C. C., Cenacchi, N., Son, C. G., and Khan, J. (2005). BBC3 mediates fenretinide-induced cell death in neuroblastoma. *Oncogene* **24,** 7976–7983.

Zurita, A. J., Arap, W., and Pasqualini, R. (2003). Mapping tumor vascular diversity by screening phage display libraries. *J. Control Release* **91,** 183–186.

CHAPTER THIRTEEN

Tumor-Specific Liposomal Drug Release Mediated by Liposomase

Ian Cheong *and* Shibin Zhou

Contents

Abstract

Despite the large arsenal of anticancer agents currently available and recent efforts in developing molecularly targeted therapies, the prognosis for many solid cancers remains dismal. A major obstacle to successful cancer therapy is the limited specificity of most anticancer agents. One approach to this problem is to construct drug carriers that preferentially accumulate at the cancer site, thus targeting otherwise nonselective cytotoxic chemotherapeutic agents to cancer cells. Liposomes stand out in this regard as the most successful drug carrier that has been approved for clinical use. Currently, most clinical liposomal formulations involve the use of PEGylated phospholipids that help prolong their residence time in the systemic circulation. Paradoxically, the robustness of these long-circulating formulations also obstructs the release of their payloads at the cancer site. This chapter describes a recently discovered bacterial protein capable of targeted liposome disruption within tumors.

The Ludwig Center for Cancer Genetics and Therapeutics, Johns Hopkins Kimmel Comprehensive Cancer Center, Baltimore, Maryland, USA

Methods in Enzymology, Volume 465
ISSN 0076-6879, DOI: 10.1016/S0076-6879(09)65013-8

1. Introduction

The major challenge facing cancer therapy is the development of therapeutic agents or approaches that are lethal to cancer tissues, but not to normal ones. Over the past three decades, powerful molecular biology techniques have allowed cancer researchers to identify a number of signal transduction pathways critical to tumorigenesis (Vogelstein and Kinzler, 2004). Consequently, both academic and pharmaceutical communities have been consolidating their resources in a major effort to develop drugs selectively targeting those pathways, ushering in a new era of molecularly targeted therapy (Allgayer and Fulda, 2008). Other approaches exploiting cancer-associated pathological features for targeting otherwise nondiscriminating cytotoxic agents to cancer cells have also been developed. One such approach exploits the well-known "enhanced permeability and retention (EPR)" effect that allows particulates within an appropriate size range to preferentially accumulate in tumors (Maeda *et al.*, 2003). Drug carriers in this size range have thus been fabricated for the tumor-selective delivery of cytotoxic drugs. These drug formulations have the advantage of targeting cancer cells regardless of their genetic makeup. The key to success, however, lies in whether a high level of tumor specificity can be achieved. Among the most successful tumor-selective drug formulations are liposome-encapsulated drugs (Drummond *et al.*, 1999; Torchilin, 2005). The best example is Doxil which has been approved by the US FDA for treating ovarian cancer, multiple myeloma, and AIDS-related Kaposi's sarcoma. Earlier versions of liposomal formulations suffered from rapid elimination by the mononuclear phagocyte system, resulting in an insufficient accumulation at the cancer site. To overcome this shortcoming, sterically stabilized liposomes (SSLs) were developed (Papahadjopoulos *et al.*, 1991). SSLs were characterized by the use of abundant cholesterol, hydrogenated phospholipids, and surface PEGylation (Allen and Cleland, 1980; Horowitz *et al.*, 1992; Yuan *et al.*, 1994). These modifications led to a high bilayer phase-transition temperature and general physical robustness, thus preserving the physical integrity of SSLs in circulation. Furthermore, PEGylation markedly reduced SSL uptake by the mononuclear phagocyte system, leading to further improvement in the circulation half-life. However, this physical robustness also made it more difficult for SSLs to degrade and release their therapeutic cargos, resulting in suboptimal local drug concentrations (Horowitz *et al.*, 1992). This may explain why the inherent low toxicity of SSLs has yet to translate into dramatically improved efficacy, even though greater therapeutic doses are now possible (White *et al.*, 2006). It is clear, therefore, that tumor-specific destabilization is key for SSLs to realize their full therapeutic potential.

One of the hallmarks of rapidly growing solid tumors is hypoxia, a consequence of tumor cells outgrowing their blood supply (Dang *et al.*, 2001; Vaupel and Harrison, 2004). The hypoxic tumor compartment has been known to pose challenges for both chemotherapy and radiation therapy (Brown, 2007). Conversely, this tumor compartment, hypoxic and immune-privileged, provides a unique niche for anaerobic bacteria to grow. Thus, a variety of obligate and facultative anaerobes have been used as tumor-selective therapeutic agents in both animal and human studies (Wei *et al.*, 2007). The idea of using bacteria for cancer therapy is not novel. More than a century ago, Coley (1991) treated inoperable human tumors with live bacteria, and later with bacterial extracts. Coley's pioneering work laid the foundation for the field of cancer immunotherapy. In the 1960s, Mose *et al.* treated patients suffering from a variety of cancer types with spores of *Clostridium sporogenes* (Mose *et al.*, 1967a,b). More recently, an attenuated strain of *Salmonella typhimurium* (VNP20009) has been used in a phase I clinical trial to treat metastatic melanomas and renal cell carcinomas (Toso *et al.*, 2002). These studies were based on the premise that anaerobic bacteria could selectively target the hypoxic tumor compartment.

Our group screened a panel of 22 anaerobic species for their ability to germinate and spread in murine tumor models (Dang *et al.*, 2001). The obligate anaerobe *Clostridium novyi* was chosen for further study because it exhibited extensive colonization throughout the necrotic/hypoxic tumor regions. Although wild-type *C. novyi* was extremely toxic to the animals, this problem was solved by creating a nonpathogenic strain in which the residential phage carrying the lethal *α-toxin* gene had been removed. As expected, this new strain of *C. novyi* was considerably less toxic: no adverse effect was observed following injection of up to 10^8 spores into nude mice with tumors. As a result, this strain was dubbed *C. novyi-NT* because of its nontoxic status. In contrast, all tumor-bearing mice died after the injection of 5×10^7 wild-type *C. novyi* spores. Despite its attenuated character, *C. novyi-NT* spores retained the capacity to germinate and spread within tumors, resulting in extensive tumor necrosis. Interestingly, *C. novyi-NT* spores were unable to colonize poorly vascularized non-neoplastic lesions such as myocardial infarcts (Diaz *et al.*, 2005). In general, infection by *C. novyi-NT* alone is insufficient for tumor eradication because the well-oxygenated tumor rim eventually grows back and causes a relapse. Cures could be achieved through combination therapy involving both *C. novyi-NT* and chemotherapy or radiation therapy (Bettegowda *et al.*, 2003; Dang *et al.*, 2004). Such combination therapies, however, showed excessive toxicity, largely resulting from the nondiscriminating cytotoxic effects of these conventional therapeutic agents.

We recently made the observation that *C. novyi-NT* possesses the ability to perturb lipid membranes and theorized that this ability could be used to target the release of liposomal formulations within tumors (Cheong *et al.*,

2006). This would hypothetically achieve high drug concentrations in the tumor while simultaneously reducing systemic toxicity. This theory turned out to be correct. Combination therapy employing Doxil, a liposomal formulation of doxorubicin, or liposomal CPT-11 (irinotecan) led to cures in ~70% of the treated mice in both syngeneic and xenograft tumor models. Importantly, no clinically significant toxicity was observed, underscoring the improved safety profile of liposomal drugs. In accordance with this observation, administration of Doxil plus *C. novyi-NT* spores resulted in a sixfold increase in intratumoral drug concentration compared to Doxil alone, without increasing drug concentrations in normal tissues. This enhanced local drug release within the tumor, but not otherwise, was fully consistent with a drug-release mechanism mediated by *C. novyi-NT*. We used biochemical and proteomic methods to identify the bacterial factor responsible for this phenomenon, naming it liposomase. Liposomase is a novel secreted bacterial lipase highly expressed in *C. novyi-NT*-infected tumors. Interestingly, its liposome-disrupting activity is not mediated by the classic catalytic domain. Instead, liposome destabilization is likely to result from the decreased lipid order and enhanced membrane polarity in the liposome lipid bilayer after incubation with liposomase.

In the following sections, we describe detailed methods for preparing drug-loaded liposomes and studying their targeted release in tumors mediated by *C. novyi-NT*. We also describe our methods for purifying secreted liposomase from media conditioned by the growth of *C. novyi-NT*.

2. Tumor Models

Two tumor models are employed to study the *in vivo* efficacy of liposomal drugs enhanced by liposomase. For a syngeneic tumor model, BALB/c mice (Harlan Breeders, Indianapolis, IN) bearing subcutaneously implanted CT26 tumors are used. CT26 (CRL-2638, ATCC) is an *N*-nitroso-*N*-methylurethane (NNMU)-induced, undifferentiated murine colon carcinoma cell line representing a minimally to moderately immunogenic tumor cell. This explains why CT26 tumors can develop when a relatively small number of cells are injected into syngeneic mice but cannot form in allogeneic mice even when a large number of cells are injected (Fearon *et al.*, 1990). For a human xenograft model, athymic nu/nu mice (Harlan Breeders) bearing HCT116 tumors are used. HCT116 (CCL-247, ATCC) is an established cell line derived from a human colorectal cancer. Generally, CT26 and HCT116 cells are grown as monolayers in McCoy's 5A medium (modified, plus L-glutamine; Invitrogen) supplemented with 10% FBS (HyClone, Logan, UT) at 37 °C in the presence of 5% carbon dioxide until ~80% confluency is reached. Cells are then harvested by trypsinization and kept on ice. Five million cells are injected subcutaneously into the right flank of each mouse at 6–8 weeks of age. The animals are housed within positive-pressure, individually ventilated caging systems in a microisolation facility, and observed for 10–14 days until tumor volumes reach 200–500 mm^3 (tumor volume is calculated as length $\times$ width^2 $\times$ 0.5). Mice are then randomized into various treatment groups of at least five animals per group. Mice with tumors that grow too slowly or develop ulcerations are excluded from studies.

3. Generation of *C. novyi* Spores and *C. novyi-NT*

C. novyi (Type A, ATCC #19402) is a biosafety level 2, spore-forming, Gram-positive anaerobe. The bacterial spores are minimally toxic because *C. novyi* is exquisitely sensitive to oxygen, and unable to germinate in healthy animals (Diaz *et al.*, 2005). Therefore, the spore form is routinely used for systemic administration. Sporulation Medium is made by adding 5 g Na_2HPO_4, 30 g peptone, 0.5 g L-cysteine, 10 g maltose, and 5% wt/vol Difco^{TM} Cooked Meat Medium (BD, Sparks, MD) to 1 l of double-distilled water followed by autoclaving at 121 °C, 15 psi for 30 min. After cooling to room temperature, the medium is filtered using a 0.22-μm Stericup filtration unit (Millipore, Billerica, MA). To generate *C. novyi* spores, Sporulation Medium is inoculated with vegetative bacteria and incubated at 37 °C in an anaerobic chamber (Type A, Coy Labs, Grass Lake, MI) filled with a gas mix containing approximately 5% hydrogen, 5% carbon dioxide, and 90% nitrogen. The spores will settle to the bottom of the culture bottle in a few days, but sporulation should continue for at least 2 weeks for maximum yield. The culture bottle should not be disturbed during the entire sporulation period. Mature spores are purified through two consecutive continuous Percoll (GE Healthcare, Piscataway, NJ) gradients self-generated by centrifugation of 80% Percoll at 12,000*g* for 40 min at 4 °C followed by four washes and resuspensions in phosphate-buffered saline (PBS). Purity of the spore preparations may be determined by examining them under phase contrast microscopy after staining with malachite green and eosin Y, following the procedure described in http://pb.merck.de/servlet/PB/show/1235100/115942en.pdf.

C. novyi-NT is generated by screening for the *C. novyi* clones that have lost the phage episome carrying the lethal *α-toxin* gene (Dang *et al.*, 2001; Eklund *et al.*, 1974). To eliminate the phage, *C. novyi* spores are heated at 70 °C for 15 min, followed by inoculating a Reinforced Clostridial Medium (RCM, BD) agar plate with the heat-treated spores. The inoculated plate is then incubated anaerobically overnight at 37 °C to allow germination and colony formation. Colonies are screened by PCR with

α-toxin-specific primers for *C. novyi* clones cured of the α-*toxin* gene. In our experiment, three such clones were identified, one of which was named *C. novyi-NT*.

4. Preparation of Liposomal Formulations

4.1. Overview

Two liposomal formulations, Doxil and liposomal CPT-11, are used in our study involving liposomase (Cheong *et al.*, 2006). Only the preparation of liposomal CPT-11 is described in this chapter, as Doxil is commercially available (Ortho Biotech, Horsham, PA). However, in principle doxorubicin may be encapsulated following a procedure similar to the one described below (Torchilin and Weissig, 2003). It is useful to think of the preparation of liposomally encapsulated CPT-11 as a two-step process. Liposomes of the desired lipid composition and size containing a specific compound (e.g., ammonium sulfate or manganese sulfate) are first prepared. Following the removal of unencapsulated material, CPT-11 is loaded into these liposomes by using the transmembrane gradient to cause its efficient entrapment within the internal volume of the liposomes.

Numerous methods have been reported for preparing empty liposomes (Torchilin and Weissig, 2003). Generally, lipids are dissolved and mixed in an appropriate solvent, then either dried using rotary evaporation or lyophilized to create a thin film or cake, respectively. These dried lipids are then hydrated using an appropriate buffer to form large multilamellar vesicles (MLVs). MLVs may be further processed and sized down to small unilamellar vesicles (SUVs) using, for example, high-pressure homogenization (e.g., via a French Press), ultrasonication, or extrusion through membranes with defined pore sizes. Of these three methods, extrusion causes the least oxidative damage to lipids and offers the greatest control over the size of liposomes (Olson *et al.*, 1979). At this point, liposome size may be characterized by electron microscopy or quasielastic light scattering (QELS) (Kolchens *et al.*, 1993). Size is of paramount importance because it directly affects the ability of the liposomes to perfuse leaky tumor vasculature via the EPR effect (Yuan *et al.*, 1995). It is desirable for this purpose to size liposomes down to a hydrodynamic radius of between 100 and 200 nm, as larger-sized particles have been shown to be inefficient in penetrating the fenestrations of tumor vessels (Yuan *et al.*, 1995).

The second step is to load CPT-11 into the liposomes. One method is to hydrate the lipids using a buffer in which the drug has been dissolved at high concentrations. This "brute force" method is relatively inefficient (10–30%) for getting polar drugs into liposomes, resulting in a large proportion of the drug remaining unencapsulated (Bachmann *et al.*, 1992). Other methods have sought to increase the encapsulation efficiency by evaporating an excess of volatile solvent (e.g., ether) from a mixture of inverted micelles: small aqueous droplets containing the drug which are stabilized by a phospholipid monolayer. As the solvent is evaporated, the lipids spontaneously self-assemble into complete bilayers surrounding the drug-containing aqueous phase, resulting in relatively high loading efficiencies (>50%) (Szoka and Papahadjopoulos, 1978). One can achieve even more impressive efficiencies approaching 100% for certain drugs by using chemical gradients to passively load the drug (Abraham *et al.*, 2002, 2004). In essence, the objective is to create conditions within the liposome which differ from its external environment such that the drug, upon entering a liposome, becomes charged or otherwise sequestered and unable to return to the external environment. Drugs such as doxorubicin and CPT-11 can become differentially charged on their amine groups and are therefore especially suitable for gradient loading methods (Messerer *et al.*, 2004).

4.2. Liposome preparation

1. A mixture of HEPC:Chol:DSPE-PEG_{2000} at a molar ratio of 50:45:5 in chloroform is placed in a round-bottom flask (ChemGlass 14/20 joint). Our targeted final lipid concentration is 100 m*M* after hydration in step 4.
2. The mixture is dried by rotary evaporation at room temperature for 1–2 h to create a thin lipid film. Drying is continued under high vacuum for 1–2 h to get rid of trace chloroform.
3. While waiting for the previous step, a sonicating water bath (Branson Model 5510 Ultrasonic Cleaner, Kell-Strom, Wethersfield, CT) is equilibrated to 65 °C, and 20 ml of hydration buffer (300 m*M* $MnSO_4$) is preheated in a 50 ml conical tube placed in the 65 °C water bath.
4. Ten ml of preheated hydration buffer is added to the thin lipid film and bath-sonicated for 1 h, maintaining the temperature at 65 °C.
5. While waiting, the extrusion apparatus Lipex Thermobarrel Extruder (Northern Lipids, Vancouver, BC, Canada) is assembled. For the extrusion filters, two Whatman Nuclepore Track-Etch (0.1 μm) membranes are stacked on top of each other with their shiny sides up. The assembled extruder is attached to the 65 °C circulating water bath and N_2 gas tank. The thermobarrel is equilibrated for 5 min.
6. Clean Tygon Tubing (#3603 3/8″ ID 5/8″ OD) is attached to the Exit Port on the Baseplate, and the other end of the tubing is put into a 15-ml conical tube to collect extruded material.
7. The extruder is primed by running 10 ml hydration buffer through it, and the liposomes are extruded 10 times.

8. The liposomes are syringe-filtered (0.45 μm pore size) into a dialysis cassette (Slide-a-Lyzer, Pierce, Rockford, IL) and dialyzed at 4 °C for 24 h against 1 l of dialysis buffer (300 mM sucrose) with buffer changes at 6 and 12 h to remove unencapsulated $MnSO_4$. The mean size of the SUVs is 100.2 nm (polydispersity index = 0.129) as determined by QELS, using a Malvern Zetasizer 3000 (Malvern, Worcestershire, UK).

4.3. Drug loading

Dialysis in the previous step creates a transmembrane Mn^{2+} gradient (high [Mn^{2+}] in the liposomal interior, low in the external milieu). To the liposomal exterior, we add drug, acid, EDTA, and finally calcimycin. Essentially, calcimycin is an ionophore which facilitates the passive exchange of divalent cations for protons across lipid membranes. In this manner, internal Mn^{2+} is exchanged for two external protons, dramatically lowering the pH of the liposomal interior relative to the external environment. Because CPT-11 is preferentially charged at lower pH, it is unable to leave the acidic interior when it enters, resulting in the gradual accumulation of liposomally entrapped CPT-11 over time. Mn^{2+} pumped out in exchange is chelated by EDTA, thus maintaining the Mn^{2+} gradient and keeping the transaction thermodynamically favorable:

1. 2× Drug Mix is prepared by solubilizing the following ingredients in dialysis buffer: 33.3 mM CPT-11 (3:1 lipid:drug molar ratio), 60 mM EDTA, and 40 mM MES.
2. The desired volume of 2× Drug Mix is transferred into a 50-ml tube and an equal volume of empty liposomes is added. This mixture is incubated at 65 °C for 10 min in a water bath.
3. A 100 mM calcimycin stock solution is added at a molar ratio of 0.0002:1 calcimycin:lipid. This works out to 1 μl calcimycin per 10 ml liposome/drug mixture. The solution is incubated at 65 °C for 1 h in a water bath.
4. The liposomes are syringe-filtered (0.45 μm pore size) into a dialysis cassette (Slide-a-Lyzer) and dialyzed at 4 °C for 24 h against 1 l of dialysis buffer (300 mM sucrose) three times at 0, 6, and 12 h to remove unencapsulated material.
5. Liposome-encapsulated drug is stored in glass tubes at 4 °C. Encapsulation efficiency is typically >99% as determined by disruption of liposomes with 1-butanol and fluorometric measurement (excitation at 390 nm, emission at 460 nm) using a fluorescence plate reader (Fluostar Galaxy, BMG LabTech, GmbH). Concentrations are derived by reference to a CPT-11 standard curve.

5. Combination Therapy with *C. novyi-NT* Spores and Liposomes

When treating animals with more than one therapeutic agent, the timing of injection for each agent becomes critical for achieving an optimal therapeutic effect. Maximum liposomal drug release requires established *C. novyi-NT* infection and expression of liposomase. Confoundingly, infection also increases intratumoral pressure due to the inflammatory response, thus obstructing the subsequent delivery of liposomes. An optimal treatment schedule would require injection of liposomes before such inflammation develops. This window of opportunity for injecting liposomes is empirically determined to be between 0 and 16 h after the injection of *C. novyi-NT* spores. Mice are administered *C. novyi-NT* spores by a bolus tail vein injection of 300 million spores suspended in 0.2 ml PBS, followed 16 h later by an injection of 0.1 ml of liposomal drug (10 mg/kg Doxil or 25 mg/kg liposomal CPT-11) also via the tail vein. Solid tumors often comprise a well-perfused periphery and a necrotic core resulting from insufficient blood supply. The necrotic tumor core provides a perfect "breeding ground" for *C. novyi-NT* spores to germinate and proliferate. Spores start to germinate as soon as they enter these areas and elicit a host response manifested as central hemorrhagic necrosis in the infected tumor. The animals treated with spores plus liposomes may appear lethargic and rough-furred about 8–12 h after treatment with liposomes. We typically observe in all mice undergoing the combination treatment, hemorrhagic necrosis expanding over the entire tumor, turning it black. This process takes approximately 3–4 days. During this period, we aggressively hydrate each animal as a supportive measure by subcutaneous injection of 2–3 ml of saline twice daily. Within a week, the necrotic tissue forms a hard scab over the original tumor site and eventually falls off. In our previous study, approximately 70% of animals were cured with a single dose of combination therapy. In comparison, no cures were observed in the controls (Cheong *et al.*, 2006).

6. Purification and Identification of Liposomase

6.1. Biochemical measurement of liposomase activity

The method which we use to measure liposomase activity derives from the well-known principle that fluorescent molecules (such as doxorubicin and CPT-11) will self-quench at high concentrations but increase dramatically in fluorescence as they get diluted into a larger volume (Schlesinger and Saito, 2006). For this reason, the fluorescence of liposomally entrapped doxorubicin and CPT-11 is negligible, but increases over several orders of magnitude once released into the external environment. This time-dependent dequenching is measured by assaying reactions (100 μl sample + 5 μl Doxil in a 96-well plate or 50 μl sample + 2 μl Doxil in a 384-well plate) in a BMG Fluostar Galaxy using excitation at 470 nm and emission at 590 nm. All measurements are performed at 37 °C. A typical readout is shown in Fig. 13.1. Liposome-disrupting activity is defined as the maximum slope of the release curve.

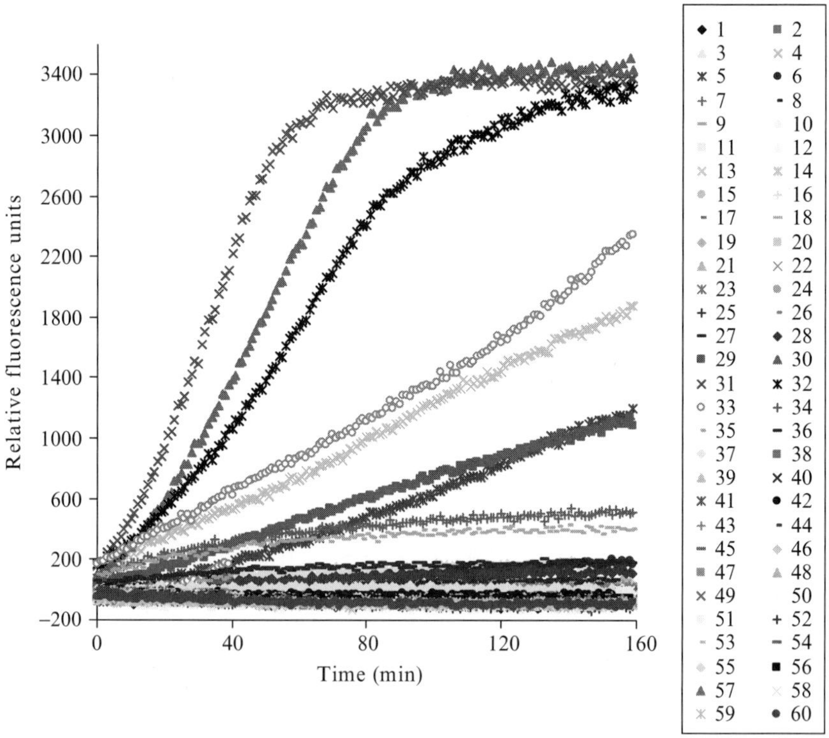

Figure 13.1 Liposome disruption assay. The disruption of liposomal doxorubicin (Doxil) was measured using the dramatic increase in fluorescence as doxorubicin was released from encapsulation. The curves in the graph represent 60 individual fractions from a single anion exchange chromatography run during the biochemical purification of liposomase. Peak activity is observed in fractions 30–32. (See Color Insert.)

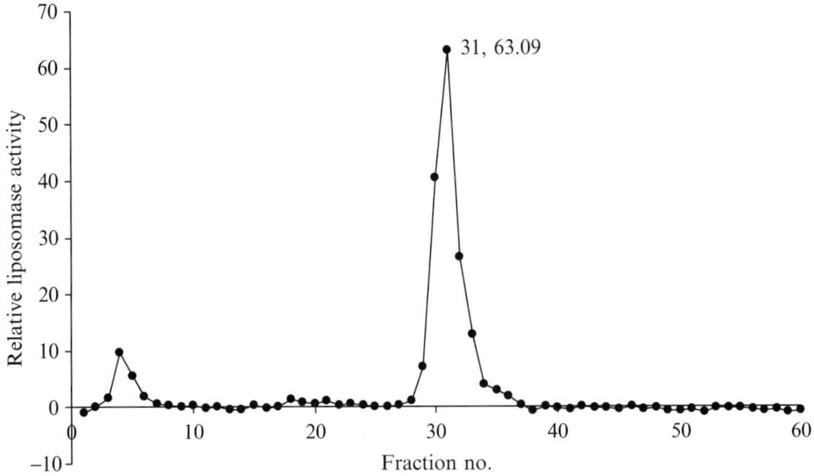

Figure 13.2 Anion exchange chromatography. Medium conditioned by late log *C. novyi-NT* growth was precipitated with ammonium sulfate and resolubilized in a Tris–HCl buffer. This solution was then loaded onto an anion exchange column and eluted using a salt gradient. Relative liposomase activity was calculated by taking the maximum slope of the liposome disruption curves (see Fig. 13.1) and the results were plotted on this graph.

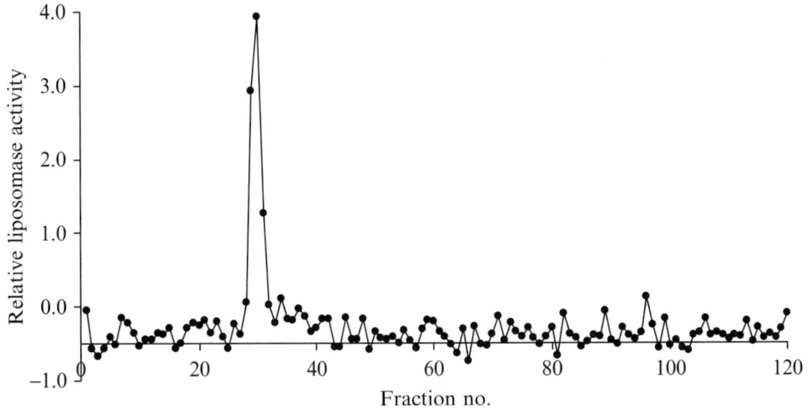

Figure 13.3 Size exclusion chromatography. The two most active fractions (30 and 31) from anion exchange chromatography (see Fig. 13.2) were pooled and further fractionated using size exclusion chromatography. The liposomase activities of the resulting fractions are shown.

6.2. Purification of liposomase activity

The liposome-disrupting factor secreted by *C. novyi-NT* is enriched and purified using the scheme described below. Typical results are shown in Figs. 13.2 and 13.3:

1. Five hundred million *C. novyi-NT* spores are first inoculated into 20 ml Sporulation Medium and then incubated overnight at 37 °C in an anaerobic chamber (Type A, Coy Labs). One milliliter of this starter culture is used to inoculate 100 ml Sporulation Medium that has been pre-equilibrated in the anaerobic chamber. This culture is grown to late log phase ($OD_{600} \sim 0.8$) and then carefully decanted to remove meat particles from the Sporulation Medium. The resulting suspension is then centrifuged at 5000*g* to remove bacterial forms.
2. The supernatant from the previous step is precipitated using 40% saturated ammonium sulfate at 4 °C for 1 h, and centrifuged at 5000*g*. The supernatant is discarded and the pellet is solubilized in TN buffer (100 m*M* Tris–HCl, pH 7.5, 0.1 *M* NaCl). This solution is then syringe-filtered and used for subsequent steps.
3. All chromatography is performed at room temperature on an ÄKTA Purifier FPLC system (Amersham Biosciences, Piscataway, NJ). The filtered sample is loaded onto an anion exchange column (mono-Q 5/50 GL, Amersham) equilibrated in TN buffer. The proteins are eluted using a linear gradient of 10 column volumes formed from TN buffer and 100 m*M* Tris–HCl (pH 7.5), 1 *M* NaCl. The fractions collected are assayed for liposome-disrupting activity as described earlier (Fig. 13.2). The active fractions are highly correlated with an absorption peak (280 nm) eluting at a conductivity of 45 mS/cm as the concentration of NaCl increases.
4. The two most active fractions from the mono-Q column are then pooled and loaded onto a size exclusion column (HiLoad 16/60 Superdex 200, Amersham) equilibrated in 100 m*M* Tris–HCl (pH 7.5), 0.5 *M* NaCl. The column is isocratically eluted with the same buffer over 1.5 column volumes. The fractions are assayed for liposome-disrupting activity as earlier described (Fig. 13.3).
5. Proteins in the two most active fractions are separated by electrophoresis through an SDS-polyacrylamide gel and silver-stained using a Silver-SNAP Stain Kit II (Pierce, Rockford, IL). The single dominant band (apparent molecular weight of ~45 kD) is excised for analysis by LC/MS/MS. Using the *C. novyi-NT* genome as a reference, we identify the 45-kD polypeptide as a putative neutral lipase (NCBI Locus Tag NT01CX_2047), subsequently named liposomase (Cheong *et al.*, 2006). Biochemical and biophysical analyses confirm that liposomase is sufficient for causing the release of drugs from liposomes.

7. Conclusion and Future Perspectives

Human cancer is a heterogeneous genetic disease (Vogelstein and Kinzler, 2004). Cancer genomes, like viral genomes, are unstable and constantly evolve under selective pressure when metastasized to hostile environments or exposed to therapeutic challenges (Lengauer *et al.*, 1997). This instability is often responsible for treatment relapses. Molecularly targeted therapy can discriminate more efficiently between cancer and normal cells, thus minimizing collateral damage, but will still be ineffective against cancer cells that have evolved to become resistant (Kobayashi *et al.*, 2005). Similarly, resistance to conventional chemotherapeutic agents may develop unless doses large enough to quickly eradicate cancer cells can be administered. However, this often cannot be achieved because of the nonspecific nature of the cytotoxic drugs. Targeting these drugs selectively to the cancer site using liposomes represents a viable approach to solve the problem (Torchilin, 2005). The efficacy of liposomal drugs is dictated by their circulation time and release of the payload at the correct target site. In this chapter, we have described a tumor-targeting bacterium that secretes liposomase, a protein capable of facilitating liposomal drug release, thus enhancing the efficacy of liposomal drugs. We have also described in detail the purification of the enzyme. These efforts may lead to therapeutic options in addition to those based on therapeutic bacterial infections. For example, liposomase could be conjugated to tumor-specific antibodies or its gene used for gene therapy. Because a large variety of therapeutic agents can be formulated as liposomal drugs, liposomase offers a number of possibilities for the selective delivery of drugs to tumors.

REFERENCES

Abraham, S. A., Edwards, K., Karlsson, G., MacIntosh, S., Mayer, L. D., McKenzie, C., and Bally, M. B. (2002). Formation of transition metal–doxorubicin complexes inside liposomes. *Biochim. Biophys. Acta* **1565,** 41–54.

Abraham, S. A., McKenzie, C., Masin, D., Ng, R., Harasym, T. O., Mayer, L. D., and Bally, M. B. (2004). *In vitro* and *in vivo* characterization of doxorubicin and vincristine coencapsulated within liposomes through use of transition metal ion complexation and pH gradient loading. *Clin. Cancer Res.* **10,** 728–738.

Allen, T. M., and Cleland, L. G. (1980). Serum-induced leakage of liposome contents. *Biochim. Biophys. Acta* **597,** 418–426.

Allgayer, H., and Fulda, S. (2008). An introduction to molecular targeted therapy of cancer. *Adv. Med. Sci.* **53,** 130–138.

Bachmann, D., Brandl, M., and Gregoriadis, G. (1992). Preparation of liposomes using a Mini-Lab 8.30 H high-pressure homogenizer. *Int. J. Pharm.* **91,** 69–74.

Bettegowda, C., Dang, L. H., Abrams, R., Huso, D. L., Dillehay, L., Cheong, I., Agrawal, N., Borzillary, S., McCaffery, J. M., Watson, E. L., Lin, K. S., Bunz, F., *et al.*

(2003). Overcoming the hypoxic barrier to radiation therapy with anaerobic bacteria. *Proc. Natl. Acad. Sci. USA* **100,** 15083–15088.

Brown, J. M. (2007). Tumor hypoxia in cancer therapy. *Methods Enzymol.* **435,** 297–321.

Cheong, I., Huang, X., Bettegowda, C., Diaz, L. A. Jr., Kinzler, K. W., Zhou, S., and Vogelstein, B. (2006). A bacterial protein enhances the release and efficacy of liposomal cancer drugs. *Science* **314,** 1308–1311.

Coley, W. B. (1991). The treatment of malignant tumors by repeated inoculations of erysipelas. With a report of ten original cases. 1893. *Clin. Orthop. Relat. Res.* **262,** 3–11.

Dang, L. H., Bettegowda, C., Huso, D. L., Kinzler, K. W., and Vogelstein, B. (2001). Combination bacteriolytic therapy for the treatment of experimental tumors. *Proc. Natl. Acad. Sci. USA* **27,** 27.

Dang, L. H., Bettegowda, C., Agrawal, N., Cheong, I., Huso, D., Frost, P., Loganzo, F., Greenberger, L., Barkoczy, J., Pettit, G. R., Smith, A. B. III, Gurulingappa, H., *et al.* (2004). Targeting vascular and avascular compartments of tumors with *C. novyi-NT* and anti-microtubule agents. *Cancer Biol. Ther.* **3,** 326–337.

Diaz, L. A. Jr., Cheong, I., Foss, C. A., Zhang, X., Peters, B. A., Agrawal, N., Bettegowda, C., Karim, B., Liu, G., Khan, K., Huang, X., Kohli, M., *et al.* (2005). Pharmacologic and toxicologic evaluation of *C. novyi-NT* spores. *Toxicol. Sci.* **88,** 562–575.

Drummond, D. C., Meyer, O., Hong, K., Kirpotin, D. B., and Papahadjopoulos, D. (1999). Optimizing liposomes for delivery of chemotherapeutic agents to solid tumors. *Pharmacol. Rev.* **51,** 691–743.

Eklund, M. W., Poysky, F. T., Meyers, J. A., and Pelroy, G. A. (1974). Interspecies conversion of *Clostridium botulinum* type C to *Clostridium novyi* type A by bacteriophage. *Science* **186,** 456–458.

Fearon, E. R., Pardoll, D. M., Itaya, T., Golumbek, P., Levitsky, H. I., Simons, J. W., Karasuyama, H., Vogelstein, B., and Frost, P. (1990). Interleukin-2 production by tumor cells bypasses T helper function in the generation of an antitumor response. *Cell* **60,** 397–403.

Horowitz, A. T., Barenholz, Y., and Gabizon, A. A. (1992). *In vitro* cytotoxicity of liposome-encapsulated doxorubicin: Dependence on liposome composition and drug release. *Biochim. Biophys. Acta* **1109,** 203–209.

Kobayashi, S., Boggon, T. J., Dayaram, T., Janne, P. A., Kocher, O., Meyerson, M., Johnson, B. E., Eck, M. J., Tenen, D. G., and Halmos, B. (2005). EGFR mutation and resistance of non-small-cell lung cancer to gefitinib. *N. Engl. J. Med.* **352,** 786–792.

Kolchens, S., Ramaswami, V., Birgenheier, J., Nett, L., and O'Brien, D. F. (1993). Quasi-elastic light scattering determination of the size distribution of extruded vesicles. *Chem. Phys. Lipids* **65,** 1–10.

Lengauer, C., Kinzler, K. W., and Vogelstein, B. (1997). Genetic instability in colorectal cancers. *Nature* **386,** 623–627.

Maeda, H., Fang, J., Inutsuka, T., and Kitamoto, Y. (2003). Vascular permeability enhancement in solid tumor: Various factors, mechanisms involved and its implications. *Int. Immunopharmacol.* **3,** 319–328.

Messerer, C. L., Ramsay, E. C., Waterhouse, D., Ng, R., Simms, E. M., Harasym, N., Tardi, P., Mayer, L. D., and Bally, M. B. (2004). Liposomal irinotecan: Formulation development and therapeutic assessment in murine xenograft models of colorectal cancer. *Clin. Cancer Res.* **10,** 6638–6649.

Mose, J. R., Mose, G., Propst, A., and Heppner, F. (1967a). Oncolysis of malignant tumors by Clostridium strain M 55. *Med. Klin.* **62,** 189–193.

Mose, J. R., Mose, G., Propst, A., and Heppner, F. (1967b). Oncolysis of malignant tumors through the M55 clostridium strain. II. *Med. Klin.* **62,** 220–225.

Olson, F., Hunt, C. A., Szoka, F. C., Vail, W. J., and Papahadjopoulos, D. (1979). Preparation of liposomes of defined size distribution by extrusion through polycarbonate membranes. *Biochim. Biophys. Acta* **557,** 9–23.

Papahadjopoulos, D., Allen, T. M., Gabizon, A., Mayhew, E., Matthay, K., Huang, S. K., Lee, K. D., Woodle, M. C., Lasic, D. D., Redemann, C., and Martin, F. J. (1991). Sterically stabilized liposomes: Improvements in pharmacokinetics and antitumor therapeutic efficacy. *Proc. Natl. Acad. Sci. USA* **88,** 11460–11464.

Schlesinger, P. H., and Saito, M. (2006). The Bax pore in liposomes, biophysics. *Cell Death Differ.* **13,** 1403–1408.

Szoka, F. Jr., and Papahadjopoulos, D. (1978). Procedure for preparation of liposomes with large internal aqueous space and high capture by reverse-phase evaporation. *Proc. Natl. Acad. Sci. USA* **75,** 4194–4198.

Torchilin, V. P. (2005). Recent advances with liposomes as pharmaceutical carriers. *Nat. Rev. Drug Discov.* **4,** 145–160.

Torchilin, V. P., and Weissig, V. (2003). *Liposomes: A Practical Approach*. Oxford University Press, Oxford.

Toso, J. F., Gill, V. J., Hwu, P., Marincola, F. M., Restifo, N. P., Schwartzentruber, D. J., Sherry, R. M., Topalian, S. L., Yang, J. C., Stock, F., Freezer, L. J., Morton, K. E., *et al.* (2002). Phase I study of the intravenous administration of attenuated *Salmonella typhimurium* to patients with metastatic melanoma. *J. Clin. Oncol.* **20,** 142–152.

Vaupel, P., and Harrison, L. (2004). Tumor hypoxia: Causative factors, compensatory mechanisms, and cellular response. *Oncologist* **9**(Suppl. 5), 4–9.

Vogelstein, B., and Kinzler, K. W. (2004). Cancer genes and the pathways they control. *Nat. Med.* **10,** 789–799.

Wei, M. Q., Ellem, K. A., Dunn, P., West, M. J., Bai, C. X., and Vogelstein, B. (2007). Facultative or obligate anaerobic bacteria have the potential for multimodality therapy of solid tumours. *Eur. J. Cancer* **43,** 490–496.

White, S. C., Lorigan, P., Margison, G. P., Margison, J. M., Martin, F., Thatcher, N., Anderson, H., and Ranson, M. (2006). Phase II study of SPI-77 (sterically stabilised liposomal cisplatin) in advanced non-small-cell lung cancer. *Br. J. Cancer* **95,** 822–828.

Yuan, F., Leunig, M., Huang, S. K., Berk, D. A., Papahadjopoulos, D., and Jain, R. K. (1994). Microvascular permeability and interstitial penetration of sterically stabilized (stealth) liposomes in a human tumor xenograft. *Cancer Res.* **54,** 3352–3356.

Yuan, F., Dellian, M., Fukumura, D., Leunig, M., Berk, D. A., Torchilin, V. P., and Jain, R. K. (1995). Vascular permeability in a human tumor xenograft: Molecular size dependence and cutoff size. *Cancer Res.* **55,** 3752–3756.

CHAPTER FOURTEEN

Targeted Lipoplexes for siRNA Delivery

Ana Cardoso,*[,†] Sara Trabulo,*[,†] João Nuno Moreira,[†,‡] Nejat Düzgüneş,[§] *and* Maria C. Pedroso de Lima*[,†]

Contents

* Department of Biochemistry, Faculty of Science and Technology, University of Coimbra, Coimbra, Portugal
† Center for Neuroscience and Cell Biology, University of Coimbra, Coimbra, Portugal
‡ Laboratory of Pharmaceutical Technology, Faculty of Pharmacy, University of Coimbra, Coimbra, Portugal
§ Department of Microbiology, University of the Pacific, Arthur A. Dugoni School of Dentistry, San Francisco, California, USA

Methods in Enzymology, Volume 465
ISSN 0076-6879, DOI: 10.1016/S0076-6879(09)65014-X

Abstract

RNA interference provides a powerful technology for silencing any single protein within a target cell. Therapeutic applications of small interfering RNAs (siRNAs), however, require vehicles for stable and efficient delivery of these nucleic acid molecules, both *in vitro* and *in vivo*. Targeted lipoplexes have been used as a promising system to mediate siRNA delivery and to achieve gene silencing. Electrostatic association of transferrin (Tf) to cationic liposomes enhances the transfection of siRNA. We describe the methods used for the preparation of Tf-lipoplexes and evaluation of their biocompatibility. Approaches to assess the complexation of siRNAs, the ability of Tf-lipoplexes to mediate siRNA protection, and intracellular delivery, as well as to achieve both mRNA and protein knockdown, are also described. We illustrate the efficiency of Tf-lipoplexes in mediating the knockdown of both green fluorescent protein (GFP) and luciferase in cell lines stably expressing these reporter genes.

1. Introduction

The discovery of RNA interference (RNAi) has generated enthusiasm within the scientific community, mainly due to its potential as a promising therapeutic approach for multiple diseases. The mechanism of RNAi, first described by Fire and Mello in 1998, involves a complex set of regulatory pathways of gene expression. The major intervenients in this process are small double-stranded RNA molecules, known as short interfering RNAs (siRNAs), which are responsible for specific binding and cleavage of a target mRNA. Due to its specificity and potency, gene silencing mediated by RNAi has been proposed as a new therapeutic strategy for the treatment of viral infections, cancer, and genetic or acquired neurodegenerative diseases. Nevertheless, despite its huge potential, the inability of siRNA molecules to cross cellular membranes *per se* and their instability in biological fluids are major limitations for the *in vivo* application of RNAi technology.

Due to the highly lipophilic nature of the plasma membrane, which excludes the entry of large hydrophilic molecules, siRNA passage is restricted by the size of the nucleic acid and the anionic charge of the phosphate backbone. Moreover, siRNAs have a half-life of less than 1 h in human plasma and the circulating molecules are quickly degraded or excreted by the kidneys (Santel *et al.*, 2006). All these factors make the systemic application of siRNAs unlikely to succeed, except when very high doses are employed.

To circumvent these limitations, multiple delivery methodologies, including viral and nonviral vectors, have been developed, for the introduction of siRNAs into cells, both *in vitro* and *in vivo*, with different degrees of success. Although viral vectors have proven to be very efficient for DNA transduction, nonviral vectors are the usual choice for siRNA delivery, since these small RNAs cannot be directly delivered by viruses, except through the incorporation of siRNA sequences in the viral genome. On the contrary, nonviral vectors, such as cationic polymers or liposomes, are able to mediate the internalization of siRNAs through the endocytotic pathway and their subsequent release into the cytoplasm, where the RNA-induced silencing complex (RISC) and the target mRNAs are located.

Besides increasing siRNA delivery, most nonviral vectors have several additional advantages, such as the ability to protect the nucleic acids from nuclease-mediated degradation, both outside and inside the cellular milieu, the simplicity and ease of large-scale production, low price, versatility, and reduced immunogenicity (in most cases). Functional devices can be further introduced to help surpass different barriers to the transfection process, such as traversing the plasma membrane, escaping lysosomal degradation, and overcoming the nuclear envelope. These devices include the use of targeting ligands to increase the specificity of cellular uptake and fusogenic lipids or peptides to enhance endosomal release.

One useful cell-targeting ligand is transferrin (Tf), an iron-transporting protein that interacts with receptors ubiquitously expressed in various tissues (Li and Qian, 2002) and overexpressed in several types of tumors. Targeting transferrin receptors has been applied successfully for DNA delivery by using antibodies against these receptors (Shi *et al.*, 2001a,b), by linking Tf to pegylated liposomes (Hatakeyama *et al.*, 2004), or by associating the protein to cationic liposomes through electrostatic interactions (Simões *et al.*, 1998, 1999a,b). In this regard, we and others have demonstrated that the electrostatic association of Tf to lipoplexes significantly enhances transfection efficiency by promoting lipoplex internalization in a large variety of cells, including dividing and nondividing cells, such as neurons and macrophages (Cardoso *et al.*, 2007; da Cruz *et al.*, 2004, 2005; Pedroso de Lima *et al.*, 2001; Simões *et al.*, 1998, 1999a,b; Tros de Ilarduya and Düzgüneş, 2000). In addition, we have reported that Tf can trigger cytoplasmic delivery of the carried nucleic acids through destabilization of the endosomal membrane under acidic conditions, thus further improving the transfection process (da Cruz *et al.*, 2001; Simões *et al.*, 1999a). These observations paved the way to pursue studies to evaluate the potential of Tf-lipoplexes to deliver siRNAs and mediate efficient gene silencing. Several physicochemical parameters, including size, charge density, and colloidal stability, have been shown to affect the biological activity of the lipoplexes. These parameters control the ability of lipoplexes to complex with and protect siRNAs, as well as the extent of lipoplex–cell interactions, which in turn influence cell binding and internalization and, ultimately, transfection efficiency.

In this chapter, we briefly describe some experimental procedures to study the above referred processes, which may help optimize nonviral lipid-based formulations for siRNA delivery and gene silencing.

2. Liposome and Complex Preparation

2.1. Liposome preparation

Cationic liposomes can be purchased from several companies, or easily prepared in the laboratory. A large number of cationic lipids (monovalent and multivalent) are available commercially and can be used alone or in combination with other lipids that contribute not only to the formation of the liposomes, but also to improve their ability to transfect cells. Dioleoyl-phosphatidylethanolamine (DOPE) and cholesterol (Chol) are the primary "helper" lipids that have been more frequently used in conjunction with cationic lipids. In our laboratory, cationic liposome formulations for nucleic acid delivery are usually prepared from 1,2-dioleoyl-3-trimethylammonium-propane (DOTAP) and Chol at a 1:1 molar ratio. Briefly, a mixture of the selected lipids (1.4 μmol of each lipid), obtained from stock solutions in chloroform, is dried under nitrogen to obtain a thin lipid film. According to the procedure described by Campbell (1995), the film is dissolved in 100 μl of ultrapure ethanol and the resulting ethanol solution is injected from a 250 μl Hamilton syringe into 900 μl of HBS buffer (HEPES-buffered saline solution, 20 m*M* HEPES, 100 m*M* NaCl, pH 7.4), previously heated to 40 °C and continuously vortexed. The resulting solution is sonicated for a period of 2–3 min, diluted in HBS to a final DOTAP concentration of 1 mg/ml and stored at 4 °C until use. This procedure allows the rapid preparation of small liposomes (120–150 nm) that are stable for at least 2 weeks and can be used for *in vitro* transfection assays.

Alternatively, the lipid film can be hydrated with 1 ml deionized water, producing multilamellar vesicles (MLVs). To obtain small unilamellar vesicles (SUVs), the MLVs are sonicated briefly, extruded 21 times through polycarbonate membranes of 50 nm pore diameter using a Liposofast device (Avestin, Toronto, Canada), diluted five times with deionized water and filter-sterilized utilizing Millex 0.22 μm pore diameter filters. This method is particularly useful to generate highly concentrated lipid formulations, as required for *in vivo* studies, since it enables the production of liposomes with a homogeneous size distribution, even when the lipid film is hydrated with a very small volume of water or dextrose solution.

Since DOTAP and other cationic lipids do not contain phosphate groups, total lipid concentration may be estimated by measuring the concentration of the co-lipid by the Bartlett method (in the case of phospholipids, such as DOPE) (Düzgüneş, 2003), or cholesterol quantification using

Infinity™ Cholesterol Reagent (Thermo Fisher Scientific). Liposomes should be stored at 4 °C under nitrogen and used within 3 weeks after being prepared.

2.2. Complex preparation

Cationic liposome/siRNA complexes (plain lipoplexes) are usually prepared by mixing 100 μl of HBS, containing an appropriate amount of liposomes (depending on the (+/−) charge ratio required), with 10 μl of buffer containing 50 pmol of siRNAs (Cardoso *et al.*, 2007). The mixture is then incubated for 15 min at room temperature. Alternatively, transferrin-containing lipoplexes (Tf-lipoplexes) are prepared by preincubating cationic liposomes with iron-saturated human transferrin (32 μg/μg of siRNA) for 15 min followed by addition of 10 μl of a siRNA solution and further incubating the mixture for 15 min at room temperature. Complexes are prepared under aseptic conditions (laminar flow cabinet), using sterile solutions, and should be used immediately after their preparation to avoid aggregation.

It is important to emphasize that the mode of lipoplex preparation determines critically their final properties. Therefore, several parameters, such as the structure and size of cationic liposomes, the concentrations of cationic lipid and siRNAs, the ionic strength and temperature of the suspending medium, the order of addition and the mixing rate of the components, as well as the time of complex maturation, should be considered and optimized, taking into consideration the specific intended application.

3. Physicochemical Characterization of the Complexes

3.1. Size measurement

Evaluation of the hydrodynamic size distribution of lipoplexes can be performed by photon correlation spectroscopy (PCS) (Simões *et al.*, 1998). The PCS technique uses autocorrelation spectroscopy of scattered laser light to determine its time-dependent fluctuations resulting from the Brownian motion of particles in suspension. The light intensity scattered at a given angle is detected by a photomultiplier whose output current is passed to an autocorrelator, which analyzes time dependence, determining the rate of diffusion or Brownian motion of the particles, and hence their size. The detection angle is fixed at 90° and this assay can be performed using a Coulter N4 Plus Instrument (Coulter Electronics, Miami, FL).

The analysis is typically performed at 20 °C, in HBS, and complexes are prepared immediately before analysis.

3.2. Zeta-potential determination

Zeta-potential measurements of the different lipid/siRNA lipoplexes can be performed using a Coulter DELSA 440 instrument (Simões *et al.*, 1998). The DELSA 440 is a laser-based multiple-angle particle electrophoresis analyzer that measures the electrophoretic mobility and zeta-potential distribution simultaneously with the hydrodynamic size of particles in suspension. The principle of the electrophoretic light scattering is based on the measurement of the velocity of particles moving under the influence of an electric field. Electrophoretic mobility is detected by the Doppler shift in frequency, $\Delta\upsilon$, using a heterodyne method, and calculated using Eq. (14.1):

$$\Delta\upsilon = U(Ek), \tag{14.1}$$

where U is the electrophoretic mobility, E is the applied electric field, and k is the scattering angle determined by Eq. (14.2):

$$k = (4\pi n/\lambda)\sin(\theta/2), \tag{14.2}$$

where λ is the wavelength of the helium–neon laser light in vacuum, n is the refractive index of the medium, and θ is the angle between the incident laser beam and the detector. From both equations, electrophoretic mobility can be calculated by Eq. (14.3):

$$U = \frac{\Delta\upsilon\lambda}{2nE\sin(\theta/2)}. \tag{14.3}$$

Electrophoretic mobility is related to the zeta-potential by the assumption derived from the Smoluchowski approximation, which assumes that the double-layer thickness is small compared to the colloidal particle diameter, according to Eq. (14.4):

$$U = \frac{\varepsilon\zeta}{\eta}, \tag{14.4}$$

where ζ is the zeta-potential and ε is defined by $\varepsilon = \varepsilon_0 D$, ε_0 being the permittivity of the free space and D the dielectric constant of water, η is the viscosity, this being a valid assumption for aqueous systems of defined electrical conductivity.

Samples of the prepared complexes are placed in the measuring cell, whose position is adjusted to cover a previously determined stationary layer, and an electric current of 3.0 mA is applied. Measurements are recorded and the zeta-potential (ζ) is calculated for each scattering angle (8.6°, 17.1°, 25.6°, and 34.2°).

4. Assessment of siRNA Protection

Two different assays are used currently to evaluate the degree of siRNA protection conferred by its association with cationic liposomes, in the presence or absence of the targeting ligand transferrin.

4.1. PicoGreen intercalation assay

The principle of this assay is based on the properties of PicoGreen (Molecular Probes, Carlsbad, CA), a fluorescent intercalating probe whose fluorescence is dramatically enhanced upon binding to siRNAs. Complex formation of siRNA and cationic liposomes, in the presence or absence of Tf, can be monitored by evaluating the accessibility of the PicoGreen to the siRNAs associated with the cationic liposomes using a SPEX Fluorolog Spectrometer (SPEX Industries, Edison, NJ), equipped with a sample chamber with a magnetic stirring device and a thermostated circulating water bath. Plain or Tf-lipoplexes, containing 400 ng of siRNA in 1 ml HBS, are added to a cuvette containing 1 ml of PicoGreen solution (1:200 dilution from a stock solution provided by Molecular Probes) and incubated for 5 min. The fluorescence of the probe is then detected at 37 °C, for a period of 2 min, at excitation and emission wavelengths of 480 and 520 nm, respectively (Cardoso *et al.*, 2007). The fluorescence scale should be calibrated such that the initial fluorescence of PicoGreen is set as the residual fluorescence. This is obtained by a 1:200 dilution of the stock solution added to a cuvette containing 1 ml of HBS and incubated for 5 min. The value of fluorescence obtained upon addition of 400 ng of siRNA (control) minus residual fluorescence is set as 100%. For each formulation, the amount of siRNA available to interact with the probe is calculated by subtracting the values of residual fluorescence (PicoGreen in the absence of siRNA) from those obtained for each measurement, and expressed as a percentage of the control that contained naked siRNA only, according to Eq. (14.5):

$$\%\,\text{free siRNA} = \frac{\text{PicoGreen fluorescence}_{\text{complexes}}}{\text{PicoGreen fluorescence}_{\text{naked siRNA}}} \times 100. \qquad (14.5)$$

The proper controls should be performed to exclude a possible nonspecific decrease in PicoGreen fluorescence due to the presence of lipid or protein.

Typical results from application of this assay are presented in Fig. 14.1A for plain and Tf-lipoplexes (Cardoso *et al.*, 2007). As expected, the percentage of free siRNA decreases with increasing lipid/siRNA charge ratio and reaches a plateau for highly positive formulations, indicating that all siRNA

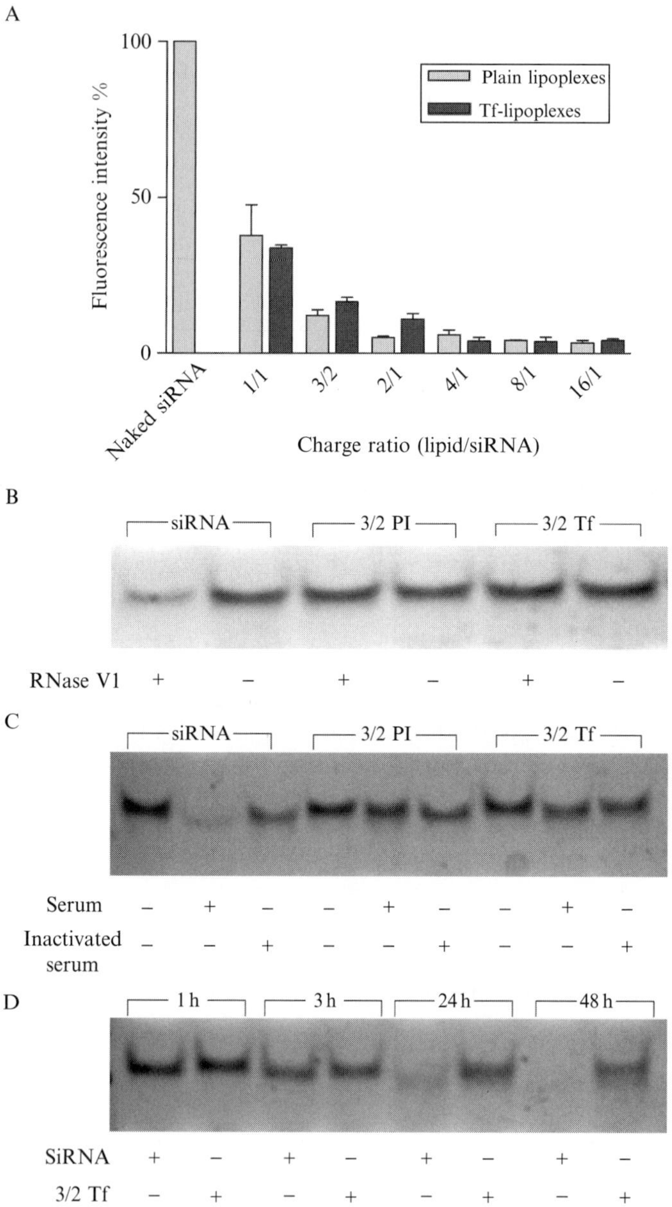

Figure 14.1 Evaluation of siRNA complexation and protection mediated by lipoplexes. (A) To evaluate complex formation of siRNA and cationic liposomes, plain or Tf-lipoplexes, containing 400-ng siRNAs and prepared at different lipid/siRNA charge ratios, are incubated at 37 °C with the fluorescent probe PicoGreen. Relative fluorescence is presented as percentage of control (400-ng naked siRNA). The determined percentage represents the fraction of siRNA in each sample available for PicoGreen

molecules are complexed by the cationic lipid. Moreover, association of Tf to the cationic liposomes is not expected to decrease significantly their ability to complex siRNAs, since no significant increase in the percentage of free siRNAs is usually observed for the same lipid/siRNA charge ratio, in the presence of this protein.

Although the PicoGreen intercalation assay is simple, rapid, and sensitive, it may lead to misinterpretation regarding the access of nucleases to the complexed siRNAs. For example, while certain siRNA duplexes may be accessible to PicoGreen, the latter being a small molecule, they may not be accessible to and degraded by nucleases. A more biologically relevant assay to evaluate the degree of siRNA protection to degradation conferred by cationic liposomes can be performed by following RNAse and serum resistance mediated by the lipoplexes.

4.2. Protection from RNase and serum-mediated degradation

Resistance of complexes to RNase and serum-mediated degradation can be determined by electrophoresis and ethidium bromide (EtBr) staining (Cardoso *et al.*, 2007). Complexes, containing 500 ng of siRNA and prepared in HBS, are incubated for 30 min at 37 °C, in the presence of 0.1 U/μl RNase V1 (Ambion), or for different incubation periods in 50% mouse serum (Jackson ImmunoResearch Laboratories, Inc., West Grove, PA). Parallel experiments are performed by incubating the samples, under the same experimental conditions, in the absence of serum or RNase V1, or in the presence of heat-inactivated mouse serum (where inactivation is achieved by incubating serum at 55 °C for 1 h).

The sensitivity and accuracy of this assay may be limited by the cationic lipid associated to siRNAs and by the presence of high amounts of serum proteins, which may not only prevent EtBr intercalation but also decrease the electrophoretic mobility of the siRNAs. Therefore, it is necessary to recover the siRNA molecules through organic extraction. This is performed by adding 500 μl of Tri-Reagent^{TM} (Sigma) and 100 μl of chloroform to 200 μl of each sample, followed by centrifugation (12,000*g*, 10 min,

interaction. Results are presented as mean values ± SD from triplicates and are representative of three independent experiments. (B–D) Lipoplex-mediated siRNA protection is evaluated following incubation at 37 °C of plain or Tf-lipoplexes, containing 500-ng siRNA and prepared at a 3/2 lipid/siRNA charge ratio, in the presence of RNase V1 or mouse serum. After the indicated incubation periods, siRNA integrity is analyzed by electrophoresis (B). Plain or Tf-lipoplexes are incubated for 30 min at 37 °C in the presence or absence of RNase V1 (0.1 U/μl) (C) or for 4 h in normal or heat-inactivated mouse serum. (D) In parallel experiments, naked siRNAs and Tf-lipoplexes are incubated for 1, 3, 24, and 48 h at 37 °C in the presence or absence of mouse serum. The results are representative of three independent experiments.

4 °C), allowing the siRNAs to be displaced to the aqueous fraction of this mixture while RNAses and other serum proteins will remain in the organic fraction. Following addition of 250 μl of isopropanol to the aqueous fraction, siRNA precipitation is performed overnight at −20 °C. Samples are centrifuged (12,000*g*, 10 min, 4 °C), the supernatant is aspirated and the pellet is washed with 100% ethanol, dried off and resuspended in a 5:1 mixture of RNAse-free water and loading buffer (15% (v/v) Ficol 400, 0.05% (w/v) bromophenol blue, 1% (w/v) SDS, 0.1 M EDTA, pH 7.8). Samples are analyzed in a 15% polyacrylamide gel, applying a voltage of 110 V for 2 h. The gel is stained for 5 min in 1 μg/ml solution of ethidium bromide and the intensity of the siRNA bands is analyzed using imaging software, such as the Versa Doc imaging system (Bio-Rad, Hercules, CA). Figure 14.1 shows the results obtained upon exposure of plain and Tf-lipoplexes to both RNAse V1 (panel B) and serum (panels C and D), when lipoplexes are prepared at a relatively low (albeit positive) lipid/siRNA charge ratio. While naked siRNAs are rapidly degraded in the presence of RNAse and serum, both plain and Tf-lipoplexes are able to efficiently prevent siRNA degradation, even after long incubation periods (Cardoso *et al.*, 2007). Similar results are expected to be obtained for higher lipid/siRNA charge ratios.

5. Assessment of Lipoplex Internalization and Biological Activity *In Vitro*

Different experiments can be performed to clarify the mechanisms by which cationic liposomes deliver siRNAs into cells and contribute to siRNA-mediated gene knockdown. For these studies, it may be useful to employ cell lines stably expressing reporter genes, such as those encoding GFP and luciferase, which allow a rapid readout of both transfection efficiency and gene knockdown using simple assays. In our laboratory, we have used two such cell lines: U-373 cells (glioblastoma-derived cell line) stably expressing GFP and Huh-7 cells (hepatocarcinoma-derived cell line) stably expressing a GFP–luciferase fusion protein (Cardoso *et al.*, 2007). Standard protocols to assess the internalization of plain or Tf-lipoplexes and to evaluate both mRNA and protein silencing mediated by lipoplexes in these cells are outlined below.

5.1. Transfection

All cells lines are maintained at 37 °C, in the appropriate medium and humidified atmosphere containing 5% CO_2. U-373 cells are cultured in DMEM (Dulbecco's modified Eagle's medium) (Sigma), while Huh-7 cells are maintained in DMEM/Ham's F12 (GIBCO, Paisley, Scotland). Both culture media are supplemented with 10% heat-inactivated fetal bovine serum (FBS), 100 μg/ml streptomycin, and 1 U/ml penicillin. In the case of U-373 cells, DMEM should also be supplemented with 24 μl of the selection antibiotic Geneticin® (GIBCO) per ml of medium, to a final concentration of 50 mg/ml.

Twenty-four hours before any experiment, cells are plated at a density of 25,000 cells/cm^2. Immediately before addition of plain or Tf-lipoplexes, cells are washed and the medium is replaced with fresh medium. Lipoplexes, prepared at different lipid/siRNA (+/−) charge ratios and containing siRNAs against the selected target protein, are added to the cells in a total volume of 100 μl to a final siRNA concentration of 100 n*M* (Cardoso *et al.*, 2007). Lipofectamine 2000 (Invitrogen, Carlsbad, CA) diluted in OPTI-MEM (GIBCO) can be used, according to the manufacturer's instructions, as a positive control for siRNA delivery. Parallel studies employing a nonsilencing siRNA sequence (Mut sequence) should always be performed, to ensure the specificity of the selected siRNA and the absence of off-target effects. Following a 4-h incubation, the medium is replaced with fresh medium and the cells are further incubated for different periods of time before evaluation of siRNA delivery and transfection efficiency.

5.2. Epifluorescence and confocal microscopy studies

Analysis of lipoplex internalization and intracellular distribution is usually performed employing fluorescently labeled siRNAs. For this purpose, cells are plated onto 8-well chambered coverslips (Lab-Tek II, Thermo, Waltham, MA) 24 h before lipoplex delivery. Following incubation for 4 h with naked or liposome-associated Cy3-labeled siRNAs, cells are washed with PBS and immediately visualized by confocal microscopy in a Zeiss Axiovert confocal scanning microscope (Zeiss, Jena, Germany), under the 63× oil immersion objective. To reduce the fluorescence background and facilitate observation, a mixture of 20% Cy3-siRNAs and 80% nonfluorescent Mut siRNAs can be used (Cardoso *et al.*, 2007). If 100% Cy3-siRNAs are used too high fluorescence levels are obtained, which may hinder observation. The use of fixing agents, such as paraformaldehyde, should be avoided, since this may lead to artifacts due to cell membrane disruption and subsequent intracellular entry of the probe.

Microscopy assays are also useful to evaluate gene-silencing efficiency mediated by plain or Tf-lipoplexes, following delivery of siRNAs against fluorescent targets, such as GFP (Cardoso *et al.*, 2007). Cells are plated onto 12-well microplates containing glass coverslips 24 h before transfection. Forty-eight hours following delivery of lipoplexes containing anti-GFP siRNAs, the number of GFP-positive cells is determined by epifluorescence microscopy using a Zeiss Axiovert 200 microscope (Zeiss, Thornwood,

NY), equipped with the 20× objective. Briefly, cells are washed with PBS, fixed with 4% paraformaldehyde for 15 min at room temperature, stained with the fluorescent DNA-binding dye Hoechst 33342 (1 μg/ml) for 5 min under dark, rinsed again with PBS, and mounted in Vecta Shield mounting medium (Vecta Laboratories, Inc., Burlingame, CA) for subsequent analysis. For each experimental condition, the number of blue nuclei and green cells is counted in nine separate coverslips (an average of 300 cells per coverslip) and results are expressed as the percentage of GFP-positive cells. It is important to note that all cell counts should be made using the same exposure parameters to allow comparison of the results. Gene-silencing efficiency can be determined by comparing the percentage of GFP-positive cells following lipoplex delivery with that of control cells (Cardoso *et al.*, 2007).

Although this assay allows a simple and direct evaluation of gene-silencing efficiency, the results obtained are mainly qualitative and, therefore, should be further validated employing other techniques.

5.3. Quantification of mRNA silencing by QRT-PCR

QRT-PCR is a useful tool to quantify the decrease of the levels of mRNA of a given target protein upon transfection of siRNA. Twenty-four hours following lipoplex delivery of both target and Mut siRNAs, total RNA is extracted from cells using the RNeasy Mini Kit (Qiagen, Hilden, Germany), according to the manufacturer's recommendations for cultured cells. Briefly, after cell lysis, the total RNA is adsorbed to a silica matrix, washed with the recommended buffers and eluted with 40 μl of RNase-free water by centrifugation. After RNA quantification, cDNA conversion is performed using the Superscript III First Strand Synthesis Kit (Invitrogen, Karlsruhe, Germany), according to the manufacturer's instructions. For each sample, cDNA is produced from 0.5 μg of total RNA in an iQ5 thermocycler (Bio-Rad), by applying the following protocol: 10 min at 25 °C, 30 min at 55 °C, and 5 min at 85 °C. After cDNA synthesis, 30-min incubation with RNase H at 37 °C is performed to remove any remaining RNA contamination. Finally, the cDNA is diluted 1:3 with RNase-free water prior to quantification by QRT-PCR (Cardoso *et al.*, 2007).

Quantitative PCR is performed in an iQ5 thermocycler using 96-well microtiter plates and the iQ SYBR Green Supermix Kit (Bio-Rad). Primers for the selected target gene and housekeeping genes or reference genes (e.g., GADPH and HPRT) can be custom-designed or acquired from Qiagen (QuantiTect Primer). The expression of the reference genes should not vary between samples. Comparison of the target gene with the endogenous reference genes allows the gene expression level of the target gene to be normalized to the amount of total RNA. Normalization corrects for variation in RNA content and any alterations in sample handling.

A master mix is prepared for each primer set, containing a fixed volume of SYBR Green Supermix and the appropriate amount of each primer to yield a final concentration of 150 n*M*. For each reaction, 20 μl of master mix are added to 5 μl of template cDNA. All reactions are performed at least in duplicate (two cDNA reactions per RNA sample) at a final volume of 25 μl per well, using the iQ5 Optical System Software (Bio-Rad). A "no template control" is performed for the target and reference genes, containing all the components of the amplification reaction except for the cDNA template. This control enables detection of contaminants that may be present in the reaction components. The reaction conditions consist of enzyme activation and well-factor determination at 95 °C for 1 min and 30 s, followed by 40 cycles at 95 °C for 10 s (denaturation), 30 s at 55 °C (annealing), and 30 s at 72 °C (elongation). The melting curve protocol should start immediately after amplification and consists of 1-min heating at 55 °C followed by eighty 10-s steps, with a 0.5 °C increase in temperature at each step. This protocol is crucial to evaluate the efficiency of amplification of each sample and determine the integrity and purity of the cDNA. Threshold values for threshold cycle (C_t) determination are generated automatically by the iQ5 Optical System Software.

Target gene expression in cells transfected with the target siRNAs is compared with that in cells treated with Mut siRNAs to determine gene knockdown (Cardoso *et al.*, 2007). Gene expression should be similar in both nontransfected cells and cells transfected with the Mut siRNAs. Any differences in gene expression between these samples indicate the presence of nonspecific effects. The percentage of gene knockdown can be determined by the $\Delta\Delta C_t$ method, using one of the reference genes (HK gene), according to Eqs. (14.6)–(14.9):

$$\Delta C_{\text{t(target siRNA)}} = C_{\text{t(target gene)}} - C_{\text{t(HK gene)}}, \tag{14.6}$$

$$\Delta C_{\text{t(Mut siRNA)}} = C_{\text{t(target gene)}} - C_{\text{t(HK gene)}}, \tag{14.7}$$

$$\Delta\Delta C_t = \Delta C_{\text{t(target siRNA)}} - \Delta C_{\text{t(Mut siRNA)}}, \tag{14.8}$$

$$\%\text{knockdown} = 100\% - (2^{-\Delta\Delta C_t} \times 100). \tag{14.9}$$

The $\Delta\Delta C_t$ value indicates the changes in RNA transcription caused by treatment with target siRNAs. When positive, the $\Delta\Delta C_t$ value indicates downregulation of the target mRNA. To apply this method, the selected housekeeping genes should have amplification efficiencies similar to the target gene. The amplification efficiencies can be determined according to Eq. (14.10):

$$E = 10^{(-1/S)} - 1, \tag{14.10}$$

where S is the slope of the standard curve obtained for each gene. To compare the amplification efficiencies of the target and reference genes, the C_t values of the target gene are subtracted from the C_t values of the reference gene. The difference in C_t values is then plotted against the logarithm of the total RNA amount. If the slope of the resulting straight line is less than 0.1, amplification efficiencies are comparable.

It is important to note that, although mRNA silencing can be observed following transfection, it is usually an early event that precedes protein knockdown and, therefore, should be evaluated 12–48 h following transfection. Moreover, mRNA silencing may not be translated into protein knockdown, making this assay insufficient *per se* to demonstrate a therapeutic effect. Other techniques including Western blot and FACS analysis should always be used in parallel and, therefore, considered as complementary rather than alternatives.

5.4. Determination of GFP silencing by flow cytometry

Flow cytometry simultaneously measures and analyzes multiple physical characteristics of cells, such as relative particle size, relative granularity or internal complexity, and relative fluorescence intensity, as they flow in a fluid stream through a laser beam. A system that records how the cell scatters incident laser light and emits fluorescence is used to determine the different properties of the cells. An important feature of flow cytometry is that it measures fluorescence per cell, in contrast to other techniques, in which the light absorption and transmission at specific wavelengths are measured for an entire population of cells (Mano *et al.*, 2005; Trabulo *et al.*, 2008).

Typically, for flow cytometry analysis of GFP expression, 48-h post-transfection the cells are washed once with PBS and detached with trypsin (5 min at 37 °C). The cells are then further washed, resuspended in PBS, and analyzed immediately. Flow cytometry analysis is performed in live cells and can be performed using a Becton Dickinson FACSCalibur flow cytometer. Live cells are gated by forward/side scattering from a total of a minimum of 25,000 events. Data are finally analyzed using appropriate software, to determine the level of GFP knockdown achieved following transfection.

5.5. Quantification of protein knockdown by Western blot

Since different proteins have different turnover times, to detect the maximum protein knockdown Western blot experiments are performed 48–72 h after transfection (Cardoso *et al.*, 2007). Briefly, protein extracts are obtained from cells transfected in 6-well microplates using a lysis buffer (50 m*M* NaCl, 50 m*M* EDTA, 1% Triton X-100) containing a protease inhibitor cocktail (Sigma), 10 μg/ml DTT, and 1 m*M* PMSF. This procedure should be performed at 4 °C to minimize protein degradation. Protein content is determined using the Bio-Rad protein quantification kit, and 20 μg of total protein are resuspended in protein loading buffer (20% glycerol, 10% SDS, and 0.1% bromophenol blue), followed by incubation for 2 min at 95 °C, and loaded onto a 10% polyacrylamide gel. After electrophoresis, the proteins are blotted onto a PVDF membrane according to standard protocols. After blocking in 5% nonfat milk, the membrane is incubated with the appropriate primary antibody (at a pre-established dilution) overnight at 4 °C, and with the appropriate secondary antibody (suggested dilution 1:2000) for 2 h at room temperature. Equal protein loading is shown by reprobing the membrane with an anti-α-tubulin antibody (1:10,000) (Sigma) and with the appropriate secondary antibody. After this incubation period, the membrane is washed several times with saline buffer (TBS/T: 25 m*M* Tris–HCl, 150 m*M* NaCl, 0.1% Tween, and 5 mg/ml nonfat powder milk), and incubated with ECF (alkaline phosphatase substrate; 20 μl of ECF/cm^2 of membrane) for 5 min at room temperature. The membrane is then submitted to fluorescence detection at 570 nm using a Storm-860 (Molecular Dynamics, CA). The analysis of band intensity is performed using the Quantity One software (Bio-Rad). The target band intensity of all samples is normalized for individual α-tubulin band intensities to correct for variation in total protein content, and any differences in sample handling. Results are usually expressed as percentage of protein levels with respect to nontransfected controls. According to results obtained in our laboratory, which are presented in Fig. 14.2A and B, Tf-lipoplexes are able to mediate significant silencing of GFP in U-373 cells. The maximum GFP knockdown (50%) in this cell line is obtained 48 h after transfection (Cardoso *et al.*, 2007). It is interesting to note that the same formulation leads to a much higher luciferase silencing in Huh-7 cells, showing that transfection efficiency and gene silencing are dependent not only on the vector *per se*, but also on the type of cells and selected target.

5.6. Determination of luciferase silencing

The luciferase assay provides a simple and highly sensitive method to study gene expression, in which luciferase is used as a reporter protein for measuring promoter activity or transfection efficiency. In this assay, firefly luciferase catalyzes luciferin oxidation in the presence of ATP, and light is generated in the reaction (Cardoso *et al.*, 2007; Simões *et al.*, 1998, 1999a).

Luciferase expression in cell lysates is evaluated by measuring light production by luciferase in a luminometer, for example, LMaxTMII384 (Molecular Devices). Forty-eight hours post-transfection, the cells are washed twice with PBS and lysis buffer (Reporter Lysis Buffer, Promega) is added to each well. The luciferase substrate, luciferin, dissolved in assay buffer containing ATP (Luciferase Assay System, Promega), is then added

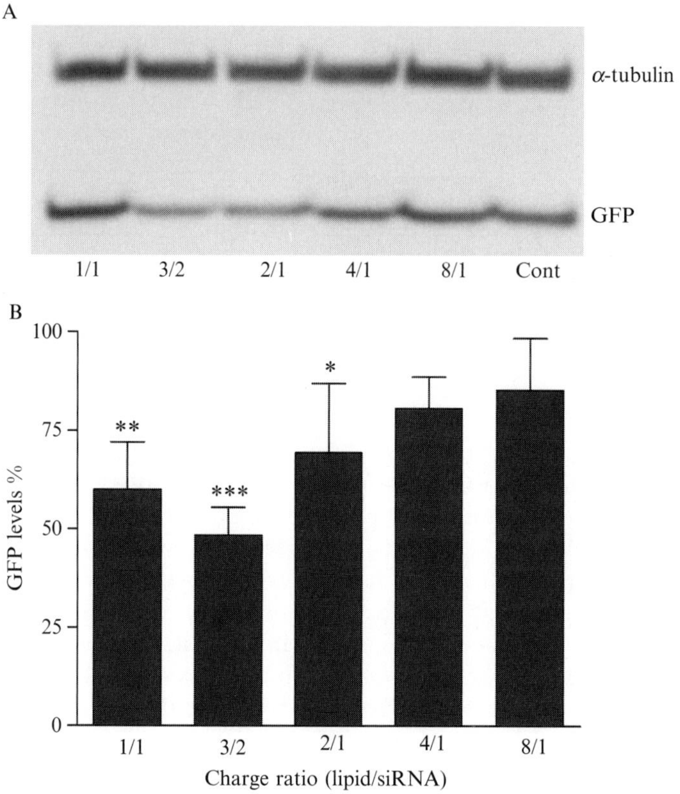

Figure 14.2 Western blot quantification of Tf-lipoplex-mediated GFP knockdown in U-373 cells. Tf-lipoplexes, containing 50 pmol of anti-GFP siRNAs and prepared at the indicated lipid/siRNA charge ratios, are added to U-373 cells to a final siRNA concentration of 100 n*M*. After incubation for 4 h at 37 °C, the medium is replaced with fresh medium and the cells are further incubated for 48 h at the same temperature. GFP and α-tubulin levels are determined by immunoblotting of total protein extracts using specific primary antibodies. (A) Representative gel showing GFP protein levels 48 h following Tf-lipoplex delivery. (B) Quantification of GFP silencing observed in (A) corrected for individual α-tubulin signal intensity. Results are expressed as percentage of GFP expression levels in untreated controls. $^{*}p < 0.05$, $^{**}p < 0.01$, and $^{***}p < 0.001$ compared to control.

to the cell lysates in an opaque multiwell plate. The protein content of the lysates can be measured, using the DC Protein Assay reagent (Bio-Rad, Hercules, CA) and bovine serum albumin as the standard. The results are expressed as RLU of luciferase per mg of total cell protein. Figure 14.3 shows typical results for protein knockdown achieved upon transfection of luciferase-expressing Huh-7 cells with Tf-lipoplexes containing antiluciferase siRNAs (Dharmacon, Lafayette, CO). As observed, the most pronounced

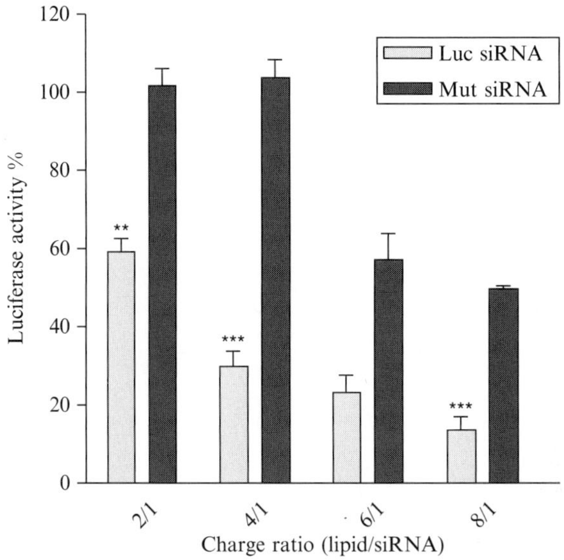

Figure 14.3 Effect of lipid/siRNA charge ratio of Tf-lipoplexes on luciferase silencing in Huh-7 cells. Tf-lipoplexes, containing 50 pmol of antiluciferase (Luc) or nonsilencing (Mut) siRNAs and prepared at the indicated lipid/siRNA charge ratios, are added to Huh-7 cells to a final siRNA concentration of 100 n*M*. After a 4-h incubation at 37 °C, the medium is replaced and the cells are further incubated for 48 h at the same temperature. Luciferase activity is measured in each sample as relative light units (RLU) and calculated as percentage of activity in control (nontransfected) cells. Mean RLU ± SD are obtained from triplicates and are representative of three independent experiments. $^{**}p < 0.01$ and $^{***}p < 0.001$ compared to cells treated with Mut siRNA.

gene silencing is achieved when Tf-lipoplexes are prepared at a 4/1 charge ratio (Cardoso *et al.*, 2007). The optimal charge ratio of Tf-lipoplexes to be used may vary depending on the cell line, which may be partially attributed to differences in the metabolic activity of the cells, and thus in their endocytotic capacity. It is also important to mention that nonspecific silencing may occur when employing high charge ratios of Tf-lipoplexes, pointing out the importance of using control siRNA sequences to ensure the specificity of gene silencing mediated by the delivery system.

6. Cell Viability Studies

Transfection of certain cell types by liposome formulations may lead to toxicity, which can be a limitation to their application *in vitro* and *in vivo*. Therefore, evaluation of the biological activity of lipid-based systems (such as Tf-lipoplexes) for gene delivery should be accompanied by studies on their biocompatibility. These studies can be performed using the protocols described below.

6.1. Alamar Blue

Cell viability under the different experimental conditions can be assessed by a modified Alamar Blue assay (Cardoso *et al.*, 2007; da Cruz *et al.*, 2004; Konopka *et al.*, 1996). This assay is simple to use, since no additional reagents or manipulations are required. Alamar Blue is nontoxic to the cells and to the user and, because no cell lysis is required, the cells can still be used in other experiments or submitted to further analysis. Forty-seven hours post-transfection, the cells are incubated with DMEM containing 10% (v/v) Alamar Blue dye (Sigma). After a 1-h incubation at 37 °C, the absorbance of the medium is measured at 570 and 600 nm. If cells are grown in 24- or 48-well plates, the incubation medium is transferred to 96-well plates and the absorbance is read in a microplate reader. If color development is not sufficient, the incubation time can be increased. Cell viability is calculated, as a percentage of the nontransfected control cells, according to Eq. (14.11):

$$\text{Cell viability (\% of control)} = \left(\frac{A_{570} - A_{600}}{A'_{570} - A'_{600}}\right) \times 100 \qquad (14.11)$$

where A_{570} and A_{600} are the absorbances of the samples, and A'_{570} and A'_{600} are those of control cells, at the indicated wavelengths.

6.2. LDH assay

The integrity of the plasma membrane is also an important parameter to take into consideration when addressing biocompatibility studies. The lactate dehydrogenase (LDH) assay evaluates the release of this enzyme from damaged cells into the extracellular medium, by monitoring the enzyme-dependent conversion of NADH to NAD^+ at 340 nm (Cardoso *et al.*, 2007). Briefly, cells are plated onto 12-well microplates 24 h before transfection and the incubation medium, corresponding to the extracellular fraction, is collected from each well 24 or 48 h following lipoplex delivery. An equal volume of lysis buffer (10 m*M* HEPES, 0.01% Triton X-100, pH 7.4) is then added to each well, the cells are scraped, and the lysate (intracellular fraction) is stored at −80 °C for at least 2 h or until the day of the experiment. All samples of both fractions are centrifuged (12,000*g*, 10 min, 4 °C) to remove cellular debris before measuring NADH conversion. The NADH solution (81.3 m*M* Tris, 203.2 m*M* NaCl, 0.244 m*M* NADH, pH 7.2) and pyruvate solution (81.3 m*M* Tris, 203.2 m*M* NaCl,

9.76 m*M* pyruvate, pH 7.2) should be prepared fresh on the day of the experiment. For each sample, the optical density is monitored at 340 nm, for a period of 2 min, with an integration period of 0.5 s, and the slope of the graph is determined. The percentage of extracellular LDH is determined according to Eq. (14.12):

$$\%\,\text{LDH released} = \frac{\text{slope extracellular LDH}}{\text{slope extracellular LDH} + \text{slope intracellular LDH}}. \tag{14.12}$$

Results are expressed as the percentage of the total LDH activity, and high values indicate loss of membrane integrity and, therefore, high toxicity.

6.3. Evaluation of nuclear fragmentation and cell apoptosis

Lipoplex cytotoxicity may be accompanied by nuclear fragmentation, an event related to apoptotic cell death that can be observed by labeling cells with the DNA-binding fluorescent dye Hoechst 33342 (Cardoso *et al.*, 2007). Briefly, cells are plated onto 12-well microplates (with glass coverslips placed in each well) 24 h before transfection, and 24 h after lipoplex delivery the cells are washed with PBS, fixed with 4% paraformaldehyde for 15 min at room temperature, stained with Hoechst 33342 (1 μg/ml), rinsed again with PBS, and mounted in Vecta Shield mounting medium for subsequent analysis. Stained nuclei are visualized by epifluorescence microscopy using a Zeiss Axiovert 200 microscope, equipped with a 20× objective. At least 200 cells per field should be counted and counts should be made in at least three separate coverslips per experimental condition, without knowledge of treatment history. Cells with condensed or fragmented nuclei are considered apoptotic, while cells with large oval and homogeneous nuclei are considered viable. Results are expressed as the percentage of apoptotic cells with respect to control (nontransfected cells).

7. Concluding Remarks

Although significant progress has been made regarding the safety and stability of siRNA molecules, a major limitation for the *in vivo* application of RNAi technology concerns the inability of siRNAs to cross cellular membranes and reach the cytoplasm, where the gene-silencing machinery is located. Nonviral lipid-based vectors, such as those generated from association of Tf to cationic liposomes, can be used to circumvent this limitation, by promoting siRNA internalization and intracellular release, thus leading to enhanced gene-silencing efficiency. To achieve this goal, it is important to ensure that the developed vectors are able to protect the siRNAs from nuclease-mediated degradation and to promote their efficient intracellular delivery, without causing cytotoxicity. Moreover, when evaluating gene-silencing efficiency, it is advisable to assess both mRNA and protein knock-down using different techniques and the appropriate controls. Such studies are crucial to reach conclusions regarding the potential efficacy of the carrier system for clinical application of RNAi.

REFERENCES

Campbell, M. J. (1995). Lipofection reagents prepared by a simple ethanol injection technique. *Biotechniques* **18,** 1027–1032.

Cardoso, A. L., Simões, S., de Almeida, L. P., Pelisek, J., Culmsee, C., Wagner, E., and Pedroso de Lima, M. C. (2007). siRNA delivery by a transferrin-associated lipid-based vector: A non-viral strategy to mediate gene silencing. *J. Gene Med.* **9,** 170–183.

da Cruz, M. T., Simões, S., Pires, P. P., Nir, S., and de Lima, M. C. (2001). Kinetic analysis of the initial steps involved in lipoplex–cell interactions: Effect of various factors that influence transfection activity. *Biochim. Biophys. Acta* **1510,** 136–151.

da Cruz, M. T., Simões, S., and de Lima, M. C. (2004). Improving lipoplex-mediated gene transfer into C6 glioma cells and primary neurons. *Exp. Neurol.* **187,** 65–75.

da Cruz, M. T., Cardoso, A. L., de Almeida, L. P., Simões, S., and de Lima, M. C. (2005). Tf-lipoplex-mediated NGF gene transfer to the CNS: Neuronal protection and recovery in an excitotoxic model of brain injury. *Gene Ther.* **12,** 1242–1252.

Düzgüneş, N. (2003). Preparation and quantitation of small unilamellar liposomes and large unilamellar reverse-phase evaporation liposomes. *Methods Enzymol.* **367,** 23–27.

Hatakeyama, H., Akita, H., Maruyama, K., Suhara, T., and Harashima, H. (2004). Factors governing the *in vivo* tissue uptake of transferrin-coupled polyethylene glycol liposomes *in vivo*. *Int. J. Pharm.* **281,** 25–33.

Konopka, K., Pretzer, E., Felgner, P. L., and Düzgüneş, N. (1996). Human immunodeficiency virus type-1 (HIV-1) infection increases the sensitivity of macrophages and THP-1 cells to cytotoxicity by cationic liposomes. *Biochim. Biophys. Acta* **1312,** 186–196.

Li, H., and Qian, Z. M. (2002). Transferrin/transferrin receptor-mediated drug delivery. *Med. Res. Rev.* **22,** 225–250.

Mano, M., Teodosio, C., Paiva, A., Simões, S., and Pedroso de Lima, M. C. (2005). On the mechanisms of the internalization of S4(13)-PV cell-penetrating peptide. *Biochem. J.* **390,** 603–612.

Pedroso de Lima, M. C., Simões, S., Pires, P., Faneca, H., and Düzgüneş, N. (2001). Cationic lipid–DNA complexes in gene delivery: From biophysics to biological applications. *Adv. Drug Deliv. Rev.* **47,** 277–294.

Santel, A., Aleku, M., Keil, O., Endruschat, J., Esche, V., Fisch, G., Dames, S., Loffler, K., Fechtner, M., Arnold, W., Giese, K., Klippel, A., *et al.* (2006). A novel siRNA-lipoplex technology for RNA interference in the mouse vascular endothelium. *Gene Ther.* **13,** 1222–1234.

Shi, N., Boado, R. J., and Pardridge, W. M. (2001a). Receptor-mediated gene targeting to tissues *in vivo* following intravenous administration of pegylated immunoliposomes. *Pharm. Res.* **18,** 1091–1095.

Shi, N., Zhang, Y., Zhu, C., Boado, R. J., and Pardridge, W. M. (2001b). Brain-specific expression of an exogenous gene after i.v. administration. *Proc. Natl. Acad. Sci. USA* **98,** 12754–12759.

Simões, S., Slepushkin, V., Gaspar, R., de Lima, M. C., and Düzgüneş, N. (1998). Gene delivery by negatively charged ternary complexes of DNA, cationic liposomes and transferrin or fusigenic peptides. *Gene Ther.* **5,** 955–964.

Simões, S., Slepushkin, V., Pires, P., Gaspar, R., de Lima, M. P., and Düzgüneş, N. (1999a). Mechanisms of gene transfer mediated by lipoplexes associated with targeting ligands or pH-sensitive peptides. *Gene Ther.* **6,** 1798–1807.

Simões, S., Slepushkin, V., Pretzer, E., Dazin, P., Gaspar, R., Pedroso de Lima, M. C., and Düzgüneş, N. (1999b). Transfection of human macrophages by lipoplexes via the combined use of transferrin and pH-sensitive peptides. *J. Leukoc. Biol.* **65,** 270–279.

Trabulo, S., Mano, M., Faneca, H., Cardoso, A. L., Duarte, S., Henriques, A., Paiva, A., Gomes, P., Simões, S., and de Lima, M. C. (2008). S4(13)-PV cell penetrating peptide and cationic liposomes act synergistically to mediate intracellular delivery of plasmid DNA. *J. Gene Med.* **10,** 1210–1222.

Tros de Ilarduya, C., and Düzgüneş, N. (2000). Efficient gene transfer by transferrin lipoplexes in the presence of serum. *Biochim. Biophys. Acta* **1463,** 333–342.

CHAPTER FIFTEEN

Mucosal Delivery of Liposome–Chitosan Nanoparticle Complexes

Edison L. S. Carvalho,[‡,1] Ana Grenha,[†,1] Carmen Remuñán-López,* Maria José Alonso,* *and* Begoña Seijo*

Contents

* Department of Pharmaceutical Technology, Faculty of Pharmacy, University of Santiago de Compostela, Santiago de Compostela, Spain
† Centre for Molecular and Structural Biomedicine (CBME), Institute for Biotechnology and Bioengineering (IBB), University of Algarve, Faro, Portugal
‡ Faculty of Pharmacy, Federal University of Rio de Janeiro, Rio de Janeiro, Brazil
1 Both authors contributed equally

Methods in Enzymology, Volume 465
ISSN 0076-6879, DOI: 10.1016/S0076-6879(09)65015-1

Abstract

Designing adequate drug carriers has long been a major challenge for those working in drug delivery. Since drug delivery strategies have evolved for mucosal delivery as the outstanding alternative to parenteral administration, many new drug delivery systems have been developed which evidence promising properties to address specific issues. Colloidal carriers, such as nanoparticles and liposomes, have been referred to as the most valuable approaches, but still have some limitations that can become more inconvenient as a function of the specific characteristics of administration routes. To overcome these limitations, we developed a new drug delivery system that results from the combination of chitosan nanoparticles and liposomes, in an approach of combining their advantages, while avoiding their individual limitations. These lipid/chitosan nanoparticle complexes are, thus, expected to protect the encapsulated drug from harsh environmental conditions, while concomitantly providing its controlled release. To prepare these assemblies, two different strategies have been applied: one focusing on the simple hydration of a previously formed dry lipid film with a suspension of chitosan nanoparticles, and the other relying on the lyophilization of both basic structures (nanoparticles and liposomes) with a subsequent step of hydration with water. The developed systems are able to provide a controlled release of the encapsulated model peptide, insulin, evidencing release profiles that are dependent on their lipid composition. Moreover, satisfactory *in vivo* results have been obtained, confirming the potential of these newly developed drug delivery systems as drug carriers through distinct mucosal routes.

1. Introduction

The efficient delivery of therapeutic peptides and proteins through routes other than the parenteral has been one of the major scientific challenges in drug delivery research. Mucosal administration of these molecules has a number of advantages, and many design strategies have been explored to administer these biomacromolecules by routes such as the oral, pulmonary, and ocular (Jorgensen *et al.*, 2006). The most valuable approach to address this purpose consists of the application of colloidal carriers like nanoparticles and liposomes (de la Fuente *et al.*, 2008).

Polymeric nanoparticles have reduced dimensions that provide them with extremely increased surface-to-volume ratio and surface functionality (Silva *et al.*, 2007). Furthermore, they have been reported to increase drug absorption by reducing the resistance of the epithelium to drug transport in a localized area or by carrying the drug itself across the epithelium (Csaba *et al.*, 2006). In this context, mucoadhesive polymers, such as chitosan, have been proven adequate materials to design suitable nanoparticulate carriers, facilitating their interaction with mucosal surfaces (Takeuchi *et al.*, 2001b). Chitosan is a polysaccharide with reported ability to improve the permeation of macromolecules across epithelial barriers and chitosan nanoparticles (CS-NP) have demonstrated excellent capacity for protein entrapment and to increase their absorption through the nasal, intestinal, and ocular mucosa (Alonso and Sánchez, 2003; de Campos *et al.*, 2001; Fernández-Urrusuno *et al.*, 1999a; Paolicelli *et al.*, 2009). However, one of the major limitations of these nanoparticles is their limited stability in biological fluids, such as the gastrointestinal media (Issa *et al.*, 2005).

Liposomes are versatile structures that enable the protection of the encapsulated material and tend to be relatively innocuous, because they comprise naturally occurring lipids that are metabolized at endogenous level (Jiang *et al.*, 2007; Torchilin, 2005). Their organized structure (an aqueous core enclosed within one or more phospholipid bilayers) permits the association of drugs with both the aqueous and lipid compartments, and drug release can usually be controlled, depending on the bilayer number and composition (Courrier *et al.*, 2002; Kirby and Gregoriadis, 1999). Moreover, their aqueous core ensures the preservation of protein structure and conformation, while the external lipids might help to improve absorption across biological barriers (El-Maghrabya *et al.*, 2008; Fenske *et al.*, 2008; Gregoriadis, 1988). Nevertheless, one of the most reported problems of liposomes is their lack of stability in terms of leakage of the encapsulated drug (Gabizon, 1995). In fact, if liposomes' inner core was solid instead of liquid, leakage would decrease dramatically, since drug release would imply an extra step of release from the solid core, followed by the traditional diffusion across the lipid bilayer (Campbell *et al.*, 2004; Huang *et al.*, 2005).

We have therefore decided to combine the advantages of each of the described colloidal systems under the form of a single and new drug delivery system, which assembles the chitosan nanoparticles in lipid vesicles (liposome–chitosan nanoparticle, L/CS-NP, complexes) (Carvalho *et al.*, 2001). This system should permit an efficient encapsulation of therapeutic macromolecules, ensuring at the same time their stability and, ideally, providing a controlled release. As expected, as the phospholipid bilayer comprises an extra barrier that should be overcome before release, phospholipids provided a controlled release of the encapsulated model protein, insulin (Grenha *et al.*, 2008a), and also provided stability in biological fluids. Moreover, the complexes demonstrated very low toxicity in ocular epithelial cells (Diebold *et al.*, 2007).

Depending on the administration objective (oral, ocular, or lung delivery), different methodologies have been established to prepare the lipid/chitosan nanoparticle complexes, which are described in detail in this chapter.

2. Preparation of Liposome–Chitosan Nanoparticle (L/CS-NP) Complexes

2.1. Preparation of chitosan nanoparticles

CS-NP are prepared according to the procedure developed by our group, based on the ionotropic gelation of CS with tripolyphosphate (TPP), in which the positively charged amino groups of CS interact with the negatively charged TPP (Calvo *et al.*, 1997). To do so, CS (hydrochloride salt, Protasan Cl 110 or Cl 213, FMC Biopolymer, Norway) is dissolved in purified water to obtain solutions of 1 or 2 mg/ml, and the final CS/TPP mass ratio is adjusted to 6:1, by using TPP (Sigma Chemicals, USA) aqueous solutions of 0.42 and 0.84 mg/ml, respectively, for each concentration of chitosan. The spontaneous formation of nanoparticles occurs upon incorporation of 1.2 ml of the TPP solution in 3 ml of the CS solution, under mild magnetic stirring (Plate A-13 Serie D, SBS, USA) at room temperature. Insulin-loaded CS-NP are obtained following dissolution of the protein (bovine insulin, Sigma Chemicals, USA) in NaOH 0.01 *M* (0.9 mg insulin/0.6 ml NaOH or 3 mg insulin/0.6 ml NaOH) and its subsequent incorporation in the TPP solution (0.6 ml TPP solution + 0.6 ml insulin solution). When insulin-loaded nanoparticles are produced, the TPP solution is prepared at double concentration as compared with that of the unloaded particles, to achieve the same final concentration as used upon mixing with the insulin solution. The insulin concentration in the TPP solution is calculated to obtain CS-NP with a theoretical content of 30% or 50% (w/w) insulin with respect to CS.

CS-NP are afterwards concentrated by centrifugation at 16,000×*g* on a 10 μl glycerol (Sigma Chemicals, USA) layer for 30 min at 15 °C (Beckman Avanti 30, Beckman, USA). The supernatants are discarded and nanoparticles (sediment) are resuspended in 100 μl of purified water. When the final goal is lung administration, the application of the described small scale (3 ml CS + 1.2 ml TPP) to produce CS-NP is restricted to nanoparticles characterization, while the production of nanoparticles for the assembly of lipidic complexes is achieved by means of a large-scale production. In this case, preparation of CS-NP involves adding 12 ml of the TPP solution to 30 ml of the CS solution (10-fold scaling-up) and maintaining the stirring conditions. Centrifugation is performed at 10,000×*g* on a 100 μl glycerol layer for 40 min at 15 °C and the particles are resuspended in 300 μl of purified water. In all other applications, low-scale production is used.

Recommendation. The stirring conditions (speed and type of vial, most importantly) used in the moment of nanoparticles' assembly and the conditions of centrifugation (speed, duration, and amount of glycerol layer) are the most important in obtaining suitable nanocarriers both in the moment of formation and in that of resuspension. Therefore, a correct optimization of the process depending on the involved materials, vials, etc., is advised.

Another important detail is the type of chitosan to be used in the production of the particles. As chitosan exists under many different chemical structures (base, type of salt, molecular weight) and the chitosan type of structure determines the nanoparticles preparation variables, it is thus very important to make a complete optimization of the process of obtaining adequate nanoparticles, concerning CS and TPP concentrations, CS/TPP mass ratio, time and speed of centrifugation, volume of resuspension, etc.

2.2. Preparation of liposomes

Lipid vesicles (empty liposomes) are produced by the technique of hydration of a dry lipid film (Bangham *et al.*, 1965; Beaulac *et al.*, 1999; Marier *et al.*, 2002), using water as the hydrating solution. In some cases, extrusion is applied upon obtaining multilamellar liposomes, to achieve size homogenization and liposomes with approximately the same size as the chitosan nanoparticles. The extrusion process (Lipex Biomembrane, Inc., Vancouver, Canada) is performed under a nitrogen pressure of 100–500 lb/in^2 at 60 °C, and liposomes are extruded five times through polycarbonate membranes of 0.8 μm pore size and, consecutively, an additional five times through pores of 0.4 μm, until the desired vesicle size of around 0.4 μm is obtained. Various lipid mixtures are selected, according to the corresponding L/CS-NP complexes to be prepared, using different combinations of dipalmitoylphosphatidylcholine (DPPC), disteroylphosphatidylcholine (DSPC), dipalmitoylphosphatidylserine (DPPS), dimyristoylphosphatidylglycerol (DMPG) (Lipoid GMBH, Germany) and cholesterol (Sigma Chemicals, USA). In this manner, as shown in Table 15.1, three different liposomal formulations are produced for oral and ocular administration, using the following lipid combinations and molar ratios: Formulation L1—DSPC:DPPS:Chol (6/0.1/4 molar ratio); L2—DPPC: Chol (6/4 molar ratio); L3—DSPC:Chol (6/4 molar ratio); while two formulations are prepared as controls for pulmonary administration; L4—DPPC: DMPG (10/1 molar ratio) and L5—DPPC.

The selection of these combinations of lipids, in the case of the oral delivery, was driven by the necessity of having gastrointestinal resistant vesicles, since DSPC and DPPC possess saturated, relatively long fatty acid chains and high phase-transition temperatures, making them stable in acidic and intestinal media (Aramaki *et al.*, 1993). Moreover, the inclusion of negatively charged lipids aims to favor the interaction with chitosan nanoparticles and cholesterol is included in the formulations to increase the stability and the rigidity of the system (Kirby and Gregoriadis, 1999; Takeuchi *et al.*, 2001a). Respect to the lipids applied in the formulations intended for lung delivery, DPPC and DMPG are chosen as they are endogenous to the lung and principal constituents of the pulmonary

Table 15.1 Lipid molar ratio used in the preparation of liposomes (L) and respective liposome/chitosan nanoparticles (L/CS-NP) complexes

Formulation			Molar ratio				
L	Complexes	Delivery route	DSPC	DPPS	DPPC	DMPG	Cholesterol
L1	L1/CS-NP	Oral/ocular	6	0.1	–	–	4
L2	L2/CS-NP	Oral/ocular	6	–	–	–	4
L3	L3/CS-NP	Oral/ocular	–	–	6	–	4
L4	L4/CS-NP	Pulmonary	–	–	10	1	–
L5	L5/CS-NP	Pulmonary	–	–	10	–	–

Abbreviations: DPPC, dipalmitoylphosphatidylcholine; DSPC, disteroylphosphatidylcholine; DPPS, dipalmitoylphosphatidylserine; DMPG, dimyristoylphosphatidylglycerol.

surfactant. Furthermore, the molar ratios correspond approximately to the phospholipid proportions in the referred surfactant (McAllister *et al.*, 1996; Wright and Clements, 1987) or in this manner, the overall *in vivo* lung environment is simulated. It is important to mention that, in some cases, liposomes are used as controls of the L/CS-NP complexes, while in other occasions liposomes are produced as a part of the technology applied to produce the lipid/chitosan nanoparticles complexes, as occurs in the complexes formulated by lyophilization.

To produce the liposomes, the referred mixtures of lipids are dissolved in 20 ml of chloroform in a round-bottom flask, to a final lipid concentration of 0.3 and 0.12 m*M* for cholesterol, when applicable (formulations for oral and ocular delivery: L1, L2, and L3). Afterwards, 50 g of glass beads are added to increase the surface available for the formation of the dry lipid film. Subsequently, the organic solvent is evaporated under a nitrogen stream in a rotary evaporator (Büchi® R-114, Büchi, Switzerland) for a period of 3 h, at temperatures between 55 and 60 °C and a homogeneous film is formed. Nitrogen is used to ensure removing all the traces of organic solvent. The resulting thin film is then hydrated for 30 min, using 10 ml of water, previously heated to a temperature above that of phase transition and the obtained vesicles are filtered under vacuum to separate them from the glass beads. Formulations prepared for oral and ocular delivery are subsequently extruded according to the procedure described above. All the formulations of liposomes are stored at 4 °C until use.

The applied temperatures of evaporation are established to be above the phase-transition temperature of the lipids or lipid mixtures, so that more flexible vesicles can be obtained (Rodríguez and Xamaní, 2003). These transition temperatures can be easily found in the literature and correspond, for example, to 41 °C for DPPC and 58 °C for DSPC (Delattre *et al.*, 1993).

Recommendation. The phase-transition temperature of lipid mixtures is different from that of the individual lipids. Therefore, when using lipid mixtures, there is the need to find the transition temperature of the specific mixture and, in the absence of that information, the highest transition temperature of the lipids applied in the mixture should be considered, all the work being performed above that temperature.

2.3. Preparation of L/CS-NP complexes by hydration

The preparation of L/CS-NP complexes by the method of hydration of a lipid film consists of adding a suspension of previously prepared CS-NP to a dry lipid film. As can be observed in Table 15.1, each of the presented lipidic combinations is used to produce the liposomal vesicles and the corresponding L/CS-NP complexes. However, instead of hydrating the dry lipid film with water, as for the liposomes, the assembly of L/CS-NP complexes is achieved by using a suspension of chitosan nanoparticles (unloaded or insulin-loaded) as hydrating phase (volume of 10 ml). Complexes for oral or ocular administration are produced with lipid/nanoparticles mass ratio of 2/1, while those intended for lung administration count with a mass ratio of 3/1. Figure 15.1 displays a schematic demonstration of the methodology of the assembly of the L/CS-NP complexes. In the case of the complexes for lung delivery, DPPC or a mixture of DPPC and DMPG (10:1 molar ratio) are dissolved in 20 ml of chloroform, and the same procedure and characteristics described before for the preparation of liposomes are applied until obtaining a dry lipid film. This film is then hydrated for 30 min with a suspension of the CS-NP (unloaded or insulin-loaded), forming the L/CS-NP complexes (3/1, w/w). Immediately afterwards, the complexes are filtered under vacuum to allow their separation from the glass beads.

To provide an efficient pulmonary administration and simultaneously, increase the long-term stability of the formulation, the complexes are submitted to a further step of spray-drying to encapsulate these structures in dry powders with adequate properties for systemic lung delivery. In brief,

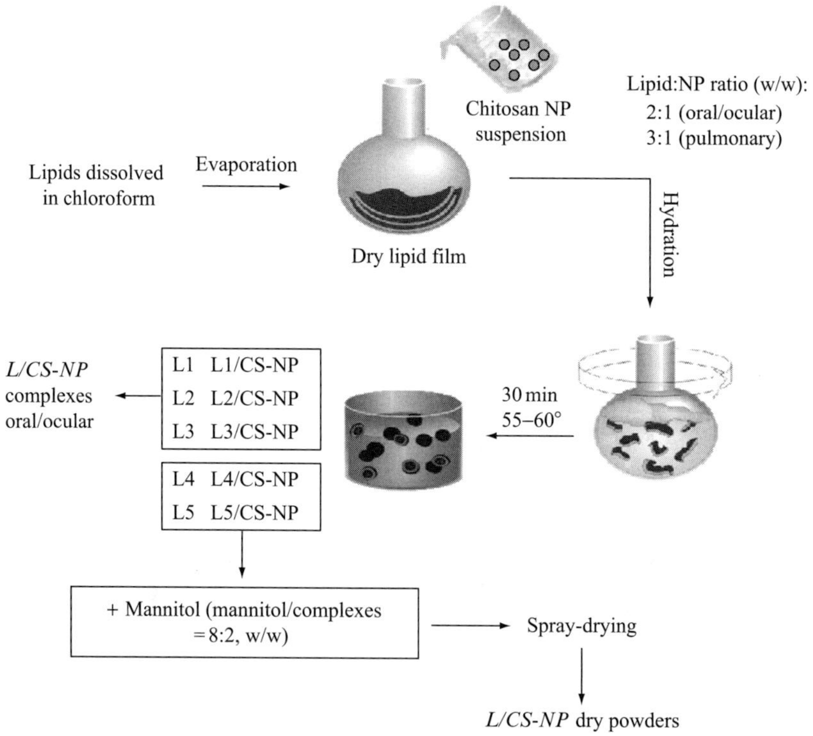

Figure 15.1 Schematic preparation of liposome/chitosan nanoparticles (L/CS-NP) complexes by hydratation.

a suspension of complexes is mixed with mannitol [mannitol/complexes = 80/20 (w/w), final solids content 2.1%], which acts as inert carrier, and this mixture is spray-dried (Büchi® Mini Spray Drier B-290, Büchi, Switzerland) at an inlet temperature of 160 °C. For further details, please consult the original paper (Grenha *et al.*, 2008a).

Complexes for oral and ocular delivery are prepared in a similar manner, except for the absence of the process of spray-drying and the application of different combinations of lipids, as observed in Table 15.1.

As referred for the production of liposomes, the temperatures used for evaporation are, in all cases, adjusted to ensure that the complexes are assembled at temperatures above the phase-transition temperature of all phospholipids of the formulation, enabling the production of more flexible structures (Rodríguez and Xamaní, 2003). The selection of phospholipids to be included in the lipid film surrounding the inner core composed of nanoparticles, which can have very different properties, especially concerning surface charges, dictates the type of covering, which can result either complete or only partial. Therefore, as demonstrated in Fig. 15.2, we propose the production of different structures, depending on the phospholipids composing the film.

Recommendation. Different mass ratios between lipids and nanoparticles should be tested and optimized for each case and for each lipid composition, since the interaction of the various combinations of lipids with the nanoparticles might be different and determine the production of complexes with different properties.

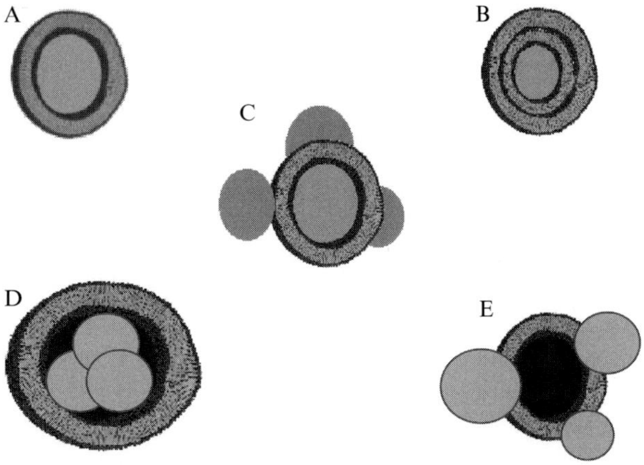

Figure 15.2 Hypothetical structures corresponding to complexes obtained by the method of hydration of a lipid film: (A) complete coating with one lipid layer; (B) complete coating with more than one lipid layer; (C) partial coating of the nanoparticles; (D) coating of several particles within the same complex; and (E) absence of coating, two separated systems coexisting (nanoparticles and liposomes).

2.4. Preparation of L/CS-NP complexes by lyophilization

The preparation of L/CS-NP complexes by the method of lyophilization involves the separate preparation of the basic systems, chitosan nanoparticles, and lipid vesicles (liposomes), with a subsequent mixing of both structures in a suspension.

Upon mixing aliquots of chitosan nanoparticles and phospholipid vesicles suspensions, a procedure of freeze-drying is applied in the presence of 5% trehalose, which acts as a cryoprotectant. The trehalose solution is prepared in double concentration to obtain the final concentration of 5% upon mixing with the suspension of the basic systems.

Two different lipid/nanoparticles mass ratios are used to prepare the complexes by this method: 1/1 and 2/1. In this chapter, only results of the ratio 2/1 are shown. The obtained mixtures are freeze-dried (Labconco freeze dryer, USA) under the following conditions: a primary drying step of 48 h at −30 °C and a secondary drying step until the temperature gradually ascends to +20 °C. After obtaining the lyophilized powder, the L/CS-NP complexes are prepared by hydration of the powder with purified water, under vigorous vortexing. As shown in Fig. 15.3, the complexes obtained by this method are the result of a "sandwich" structure which is formed by the drying of the liposomes intercalated with the nanoparticles, assembling in the form of complexes upon hydration, which obliges the lipids to rearrange into bilayers surrounding the inner core composed of nanoparticles. Albumin labeled with fluorescein isothyocyanate (FITC-BSA) is, in some cases, associated to chitosan nanoparticles, acting as a marker molecule.

Recommendation. One important detail in the freeze-drying step is the type of cryoprotectant to be used in the production of the systems. In previous studies, our group tested different cryoprotectants, such as dextran, sucrose, glucose, and trehalose, all applied to optimize the freeze-drying of chitosan nanoparticles (Fernández-Urrusuno *et al.*, 1999b). Specifically concerning the preparation of L/CS-NP complexes, glucose and trehalose are tested, and trehalose demonstrates to be the most adequate to preserve the integrity of freeze-dried L/CS-NP complexes. The selection of the cryoprotectant has demonstrated to play a major role in the final properties of the developed systems and, thus, this optimization is highly recommended.

3. Characterization of L/CS-NP Complexes

3.1. Morphological examination

The morphological examination of the complexes is conducted by optical (Olympus BH-2, Japan) and transmission electron microscopy (TEM) (CM12 Philips, The Netherlands). For TEM observation, samples are

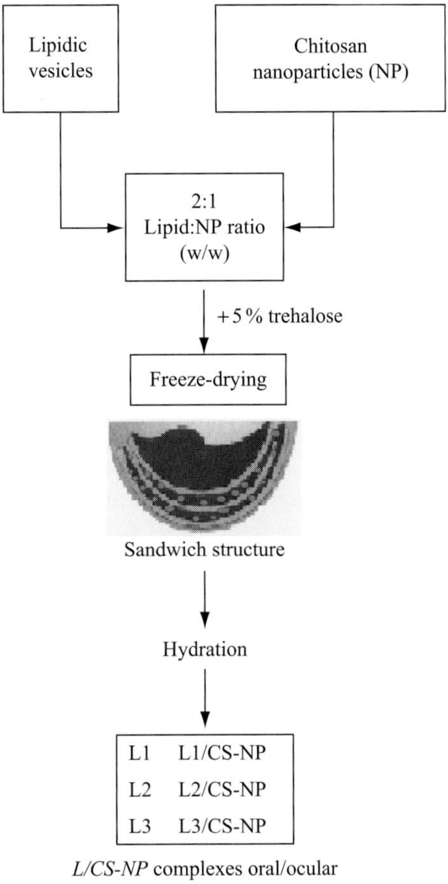

Figure 15.3 Schematic preparation of liposome/chitosan nanoparticles (L/CS-NP) complexes by lyophilization.

mounted on copper grids previously covered with Formvar® films. To obtain adequate samples for viewing, three different steps should be optimized: sample addition to the grid, staining process, and final washing. All these steps should be performed with drops of approximately 10 μl. Initially, the 10 μl sample should be placed in contact with the grid surface for a predetermined period, which is usually of 10 s. Afterwards, the drop is dried using a little piece of filter paper. The second step consists of the sample staining with 10 μl of a solution of 2% (w/v) tungstophosphoric acid (Sigma Chemicals, Germany) that contacts with the sample deposited on the grid for other 10 s. Finally, after drying the staining material, a subsequent step of washing with 10 μl of water is advised to remove the excess of tungstophosphoric acid (10 s once again). Upon the final drying of water, the grid should be placed on a proper grid storage box and stored in a dessicator, for a minimum of 12 h, until use. An optimization of observations can be performed by changing the contact time of the samples with the grid and of the staining material with the samples, and also the time of washing (for instance, washing for 20 s instead of 10 s). Moreover, more than one washing step can be performed if desired.

3.2. Size measurements

The size of formulations (liposomal vesicles and complexes) developed for oral or ocular delivery is determined by photon correlation spectroscopy using a Zetasizer® 3000 HS (Malvern, UK). In this case, the samples should simply be diluted to the appropriate concentration (setup by the equipment) with water and placed in an adequate cell for the measurement. Those formulations developed for lung delivery, because they have not been extruded, and thus, present a larger size, are analyzed for their size using a Coulter counter (Coulter® Multisizer II, Coulter Electronics, UK), equipped with a tube with an orifice aperture of 50 μm. Each particle produces a voltage alteration when passing through the orifice, according to its volume, which is transformed in a size value. To perform the measurements, 20 μl of the complexes' or liposomes' suspension are dispersed in 100 ml of the electrolyte Isoton II (filtered, phosphate-buffered saline solution, PBS). The instrument is previously calibrated using Isoton II and monodisperse latex microspheres of 13 μm, both supplied by Coulter. Nanoparticles' size was, in all cases, determined by photon correlation spectroscopy, using the Zetasizer® 3000 HS and following the procedure described above for sample preparation.

3.3. Zeta-potential measurements

Zeta-potential measurements are performed by laser Doppler anemometry, using the Zetasizer® 3000 HS (Malvern, UK). Samples are diluted with KCl 0.1 mM to an adequate concentration, which is set by the equipment, and placed in the electrophoretic cell, where a potential of $\pm$150 mV is established. The KCl solution, as well as water used to clean the cell between each measurement, must be filtered (0.22 μm) before dilution. Samples are prepared to approximately 5 ml and are introduced in the cell using a syringe.

3.4. Surface characterization

The complexes' surface is analyzed to determine whether the lipids are completely or only partially coating chitosan nanoparticles, using two different and complimentary techniques: X-ray photoelectron spectroscopy (XPS; VG Escalab 250 iXL ESCA, VG Scientific, UK) and static time-of-flight secondary ion mass spectrometry (TOF-SIMS; TOF-SIMS IV, Ion-Tof GmbH, Germany). CS-NP, DPPC and DPPC–DMPG control vesicles, and L/CS-NP complexes (DPPC/CS-NP and DPPC–DMPG/CS-NP) are placed on polished monocrystalline silicon wafers, which are used as sample holders.

XPS measurements are carried out using nonmonochromatic Al-Kα radiation ($h\nu = 1486.62$ eV) and photoelectrons are collected from a take-off angle of 90° relative to the sample surface. Measurements are performed in a constant analyzer energy (CAE) mode with a 100-eV pass energy for survey spectra and 20-eV pass energy for high-resolution spectra. Charge referencing is done by setting the lower binding energy C_{1s} photopeak at 285.0 eV C_{1s} hydrocarbon peak. The high-resolution spectra fitting is based on "Chi-squared" algorithm used to determine the goodness of a peak fit. The chemical functional groups identity is obtained from the high-resolution peak analysis of carbon-1s (C_{1s}), oxygen-1s (O_{1s}), and nitroge-1s (N_{1s}) envelopes. The experimental conditions (X-ray source, power, and analysis area) are kept constant for each analysis.

For TOF-SIMS analyses, a pulsed Gallium primary ion beam ($^{69}Ga^{+}$) generated with a liquid metal ion gun working at 15 kV is used to bombard the samples with 45° incidence respect to the sample surface. The obtained secondary ions are extracted with a 10-kV voltage and their time of flight from the sample to the detector is measured in a reflectron mass spectrometer. Electron flood gun charge compensation is necessary during measurements. A raster size of 500×500 μm is used and at least three different spots are analyzed under the "static" condition with ion doses of about $\approx 10^{12}$ ions/cm^2. The calibration of the mass spectra in the positive mode is based on hydrocarbon peaks such as CH_2^+, CH_3^+, $C_2H_2^+$, and $C_3H_5^+$. The experimental conditions (ion type, beam voltage, and primary ion dose) are maintained constant for each experiment and for compared spectra.

3.5. *In vitro* release studies

To estimate the release rate of insulin from CS-NP and L/CS-NP complexes, suspensions of these systems are incubated under dynamic conditions at 37 °C in PBS, pH 7.4 (which simulates the lung lining fluid) or simulated gastric fluid (US Pharmacopoeia). Briefly, 0.1 ml of each formulation are incubated with 0.9 ml of gastric medium or 0.5 ml with 4.5 ml of PBS, pH 7.4. At appropriate time intervals, individual samples are filtered with a low protein binding filter (0.22 μm Millex® GV, Millipore Ibérica, Spain) and the amount of insulin released is evaluated by the MicroBCA protein assay (Pierce, USA), measuring the absorbances by spectrophotometry (Shimadzu UV-Visible Spectrophotometer UV-1603, Japan) at 562 nm. A calibration curve is made at each time interval using unloaded nanoparticles and L/CS-NP complexes.

3.6. Physical stability in simulated lacrimal fluid

The physical stability of each formulation of L/CS-NP complexes developed for ocular delivery (L1, L2, and L3) is assessed in simulated lacrimal fluid composed of 0.18% KCl, 0.63% NaCl, 0.006% $CaCl_2 \cdot 2H_2O$, and 0.01% $MgCl_2 \cdot 6H_2O$ in distilled water, pH 7.4 ± 0.1 (van Haeringen, 1981). An aliquot of each diluted suspension of complexes is incubated in the fluid at 37 °C, for 2 h, after which the complexes' size is measured by photon correlation spectroscopy using the Zetasizer® 3000 HS.

3.7. Evaluation of the cytotoxicity of the complexes using the IOBA-NHC cell line

The cytotoxicity of L/CS-NP complexes (L1, L2, and L3) is determined upon incubation with IOBA-NHC cells, an immortalized epithelial cell line from normal human conjunctiva. To do so, the 2,3-bis[2-methoxy-4-nitro-5-sulfophenyl]-2*H*-tetrazolium-5-carboxyalinide (XTT) test (XTT assay kit, Sigma Chemicals, USA) is applied, measuring the production of yellow formazan crystals upon cleavage of XTT by the mitochondrial dehydrogenases of viable cells. Briefly, cells are seeded on 96-well plates at a density of 4×10^4 cells/well and incubated until confluency (approximately 20 h) with the appropriate culture medium (DMEM-F12 supplemented with 10% fetal bovine serum, 5000 U/ml penicillin, 5 mg/ml streptomycin, 2 ng/ml human EGF, 1 μg/ml bovine insulin, 0.1 μg/ml cholera toxin, and 0.5 μg/ml hydrocortisone). Afterwards, the cells are washed with supplement-free culture medium for 1 h and, subsequently, 30 μl of CS-NP or L/CS-NP (0.25, 0.5, and 1 mg/ml) suspensions are added over the cells. After 15-, 30-, or 60-min incubation, the cells are washed three times with PBS, pH 5.0, containing 0.27% glucose. The cells are then incubated with 20 μl of XTT solution composed of 1 mg/ml XTT in 100 μl of phenol red-free RPMI culture medium. A solution of 0.005% benzalkonium chloride in DMEM/F-12 culture medium is used as a positive control. The cytotoxicity of the different L/CS-NP complexes tested is expressed as viability, calculated using the following formula:

$$\text{Cell viability}\,(\%) = 100 - [(OD_{test}/OD_c) \times 100] \qquad (15.1)$$

where OD_{test} is the optical density of those wells exposed to CS-NP or L/CS-NP complexes' suspensions and OD_c is the optical density of those wells treated with supplement-free DMEM/F-12 medium.

3.8. Cell uptake studies

To evaluate the ability of L/CS-NP complexes to cross the plasma membrane and enter the cells, IOBA-NHC cells and primary cultures (PCs) of human conjunctival epithelium are exposed to the three L/CS-NP formulations developed for ocular delivery. Confluent monolayers of IOBA-NHC cells and 3-week-old PCs are washed out for 1 h with supplement-free DMEM/F-12 culture medium and then incubated for 15, 30, or 60 min with 0.25, 0.5, or 1 mg/ml of the L/CS-NP formulations. L/CS-NP complexes' uptake by the cells is analyzed immediately after incubation and after a 24-h recovery period in supplemented culture medium.

After incubation, cells are washed three times with PBS, pH 5.0, containing 0.27% glucose and, subsequently, with cold PBS, pH 7.4. They are then fixed in cold methanol (-20 °C) for 10 min and cold acetone for 3 s, followed by incubation with PBS, pH 7.4, containing 0.27% glucose and 0.2% Tween-20® for 30 min. Cells are then counterstained with tetramethylrhodamine isothiocyanate (TRITC)-conjugated phalloidin (1:200) to identify the actin component of the cytoskeleton. Preparations are mounted and examined with a confocal laser scanning microscope (Carl Zeiss LSM310, Jena, Germany) equipped with a krypton–argon laser. FITC and TRITC are excited with a 488- and 543-nm emission laser beam, respectively. Controls include culture medium-treated IOBA-NHC cells and PCs. To confirm the intracellular LCS-NP complexes' location, stacks of serial 10-μm optical sections are captured along the *Z*-axis, first using the 488-nm laser for green FITC emissions and then the 543-nm laser for red TRITC emissions. *Z*-series images are then projected as profiles superimposed with green and red images.

3.9. *In vivo* hypoglycemic effect: Oral administration

The *in vivo* evaluation of the therapeutic response to insulin upon oral administration of the L/CS-NP complexes is carried out using normal male Wistar rats (200–210 g), fasted overnight before the administration, but with free access to water. The rats were previously acclimatized in the place where the experiments are performed and the baseline blood glucose is determined. Only rats with normal blood glucose are used in the experiment. Four experimental groups are established, each composed of seven animals, which correspond to four different formulations to be administered: (1) control water, (2) insulin control solution in PBS, pH 7.4, (3) insulin-loaded CS-NP, and (4) insulin-loaded L/CS-NP complexes. All the formulations are administered in a single dose directly to the stomachs of conscious rats, by means of a glass syringe fitted to a gastric cannula. In all cases, the required dose of insulin (10 UI/kg) is administered to the rats in a total volume of 1 ml. Blood samples are collected from the tail vein at different times after dosing (1, 2, 4, 6, 10, and 24 h) and glycemic levels are determined in the samples by the glucose-oxidase method (Glucose GOD-PAD kit, Spinreact, Spain). The serum glucose level at time zero is taken as a 100% glucose level.

3.10. *In vivo* tolerance assay: Ocular administration

Female albino New Zealand rabbits weighing 2.0–2.5 kg are used to study the acute ocular tolerance to CS-NP, L1/CS-NP, and L3/CS-NP complexes. Only the concentration of the L/CS NP complexes evidencing the best *in vitro* results (0.5 mg/ml) is studied. Rabbits are randomly divided into three groups of five animals, each receiving 30 μl of one of the three formulations in the right eye every 30 min for 6 h. The contralateral eye is used as control and receives no treatment. The animals are then euthanized by air embolism after being deeply anesthetized with an intramuscular overdose of anesthetic and paralyzing mixture of xylazine (20 mg/kg) and ketamine (200 mg/kg). Eyeball and lid tissues are removed and fixed in Davison's fixative solution composed of one part glacial acetic acid, two parts 37% formaldehyde, three parts 95% ethanol, and three parts distilled water. Following fixation, they are embedded in paraffin for pathology. Eyeball and lid sections (6 μm) are evaluated in a masked fashion according to the following criteria: alteration in any of the ocular surface epithelia (cornea, limbus, and conjunctiva), edema in lid tissues, presence of inflammatory cells (eosinophils, neutrophils, mast cells, and lymphocytes), and any other abnormality.

4. Results

CS-NP (unloaded and insulin-loaded) developed to assemble the L/CS-NP complexes present sizes around 400 nm and strong positive zeta-potentials (from +34 to +44 mV), and insulin is associated with high efficiencies (68–96%).

4.1. Pulmonary route

Control vesicles and L/CS-NP complexes composed of DPPC or DPPC-DMPG present sizes around 2 μm. As expected, given the inherent charges of the phospholipids, the physicochemical properties of the complexes are influenced by the phospholipid composition. In this respect, systems comprising DPPC present a neutral or close to neutral charge (−7 to 0 mV), while those with DMPG present a strong negative charge (−54 to −36 mV).

As compared to chitosan nanoparticles, that display a burst release of insulin in 15 min, the L/CS-NP complexes provide a controlled release of the peptide, that is more effective for the complexes containing the mixture DPPC–DMPG. In a general manner, the controlled release of the complexes, as compared to the nanoparticles, is explained by the presence of the phospholipid external layer that represents an extra barrier for the drug diffusing from the nanoparticles before releasing completely from the drug delivery system. The ability of the complexes composed of the mixture of phospholipids that present an overall negative charge, to provide a further control over drug release, is attributed to the fact that the interaction with the positively charged nanoparticles is favored, providing a complete lipid coating of the nanoparticles. In fact, the occurrence of a complete coating could be confirmed by the surface analysis of the complexes by the techniques of XPS and TOF-SIMS, which have also demonstrated the partial coating of complexes formulated with only DPPC. While complexes containing DPPC and DMPG display on their surface only chemical signals attributed to phospholipids, those composed only of DPPC, display some signals assigned to the chitosan nanoparticles, demonstrating that only a partial lipidic coating is achieved in the latter case (Grenha *et al.*, 2008b). In this manner, the major role of electrostatic interactions as driving forces to control the organization of the lipid/nanoparticle assemblies is clearly evident.

It is important to mention that mannitol microspheres developed to act as carriers of the insulin-loaded L/CS-NP complexes to the lung present adequate aerodynamic properties for efficient delivery of the complexes to the alveolar zone, where systemic absorption of the model peptide is expected to occur. The referred microspheres, that are composed of 20% complexes and 80% mannitol, display aerodynamic diameters (obtained with a TSI Aerosizer LD, equipped with an Aerodisperser; Amherst Process instruments, INC; Amherst, Ma, USA) between 2.1 and 2.7 μm and apparent tap densities as low as 0.4–0.5 g/cm^3, depending on the formulation of complexes to be encapsulated. Moreover, the L/CS-NP complexes are easily recovered from the mannitol microspheres by incubation in physiological medium simulating the lung lining fluid (PBS, pH 7.4), without significant alterations in both the morphology (assessed by TEM) and physicochemical characteristics (size and zeta potential). In addition, the microencapsulation process does not have any effect on the insulin release profile (Grenha *et al.*, 2008a). A previous work from our group concerning the microencapsulation of noncomplexed insulin-loaded chitosan nanoparticles as a strategy in lung protein administration, has proven successful following intratracheal administration to rats (Al-Qadi *et al.*, 2009), and the *in vivo* evaluation of the presently developed complexes is currently under way.

4.2. Oral route: L/CS-NP complexes prepared by hydration

The L/CS-NP complexes present sizes within 1.5 and 3 μm. L1/CS-NP and L2/CS-NP formulations have a negative zeta-potential of −17.7 and −3.1 mV, respectively, while L3/CS-NP complexes are positively charged (+15.9 mV).

Figure 15.4 displays the insulin release profile from NP and L/CS-NP formulations in artificial gastric juice. As expected, CS-NP release their payload

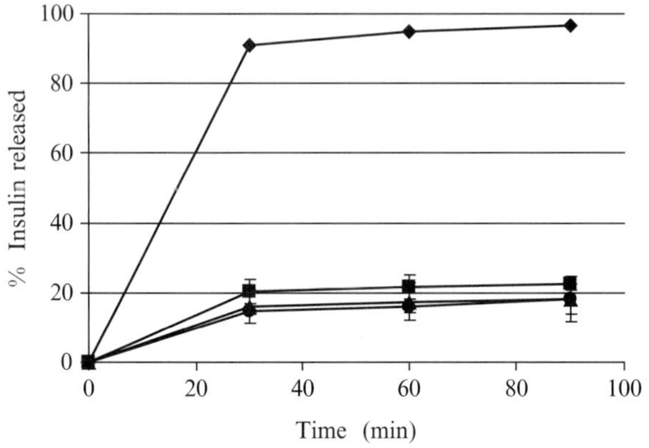

Figure 15.4 Insulin release profile from (◆) chitosan nanoparticles, (■) L1/CS-NP, (▲) L2/CS-NP, and (●) L3/CS-NP complexes prepared by the hydration methodology, in artificial gastric juice.

very rapidly, due to chitosan behavior in acidic pH (dissolution), but all L/CS-NP complexes' formulations exhibit the ability to control the release of the peptide in a similar manner, which is justified by the presence of the lipidic coating, as was confirmed in the complexes designed for lung administration.

These formulations of complexes are administered *in vivo* to rats and, as depicted in Fig. 15.5, different lipidic compositions reflect different hypoglycemic responses. The formulation containing a negatively charged phospholipid (DPPS) (L1/CS-NP) exhibits a more pronounced and rapid reduction in plasma glucose levels ($p < 0.05$), which decreases to about 50% of the baseline, remaining unaltered for at least 24 h. This profile suggests that, as also indicated by the zeta-potential data and as confirmed by the XPS and TOF-SIMS analysis performed in the complexes for lung delivery, a complete lipidic coating of the NP is achieved, constituting an extra barrier that has to be overcome before complete release. The behavior of these complexes is noticeably different than that of the other two formulations (L2/CS-NP and L3/CS-NP), although those also registered significant differences as compared to the profiles displayed by the insulin solution and CS-NP. As expected, the administration of an insulin solution does not lead to significant hypoglycemic effect, given the rapid degradation in the gastric fluid prior to absorption.

4.3. Oral route: L/CS-NP complexes prepared by lyophilization

CS-NP applied to prepare L/CS-NP complexes by lyophilization have the same characteristics as those used in the hydration method. Liposomes display sizes around 300–360 nm, with neutral or negative charge, the latter

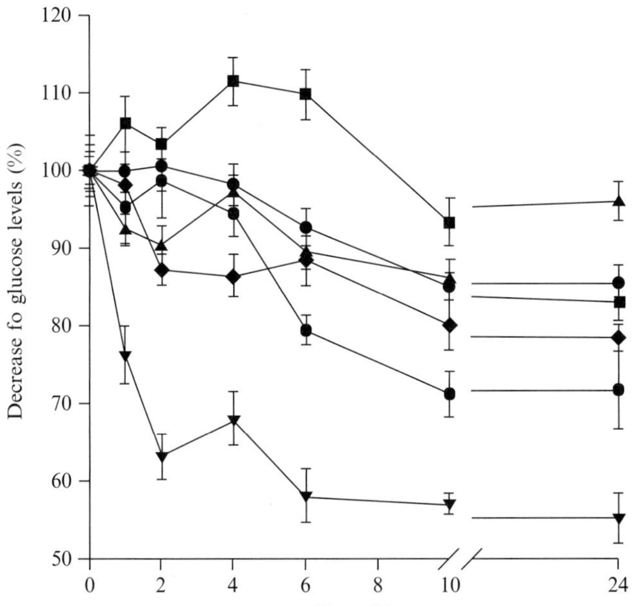

Figure 15.5 Hypoglycemic effect following oral administration of (●) water, (▲) insulin solution, (■) CS-NP, (▼) L1/CS-NP complexes, (◆) L2/CS-NP complexes, and (⬢) L3/CS-NP complexes to rats (insulin doses = 10 UI/kg, data represent mean ± SEM, $n = 7$).

reflecting the presence of DPPS. The size and zeta-potential of the produced L/CS-NP complexes reflect the interaction of lipids with CS-NP. The most important data on these characteristics are that L1/CS-NP complexes have a zeta-potential of +31 mV, while the other two are close to neutral.

Again (as reported for the first method), in contrast with the CS-NP behavior, all L/CS-NP complexes exhibit the ability to control insulin release a long time (Fig. 15.6), the profile varying according to the composition of vesicles used to form the complexes. The formulation with the negatively charged surface (L1/CS-NP) displays the more rapid release, while the other two formulations provide a more controlled release of the peptide.

Zeta-potential data previously suggested that the incorporation of the negatively charged phospholipid in the complexes (L1/CS-NP) does not favor the lipidic coating of the nanoparticles (as happened in the hydration method), because the complexes' zeta-potential in this case is very similar to that presented by the nanoparticles. The release profile confirms this, because this formulation is in fact less effective in controlling drug release. This behavior seems to indicate that the formation of the complexes by the method of lyophilization is very different from that of hydration, probably as a result of the much lower time for interaction between the different parts (lipids and nanoparticles).

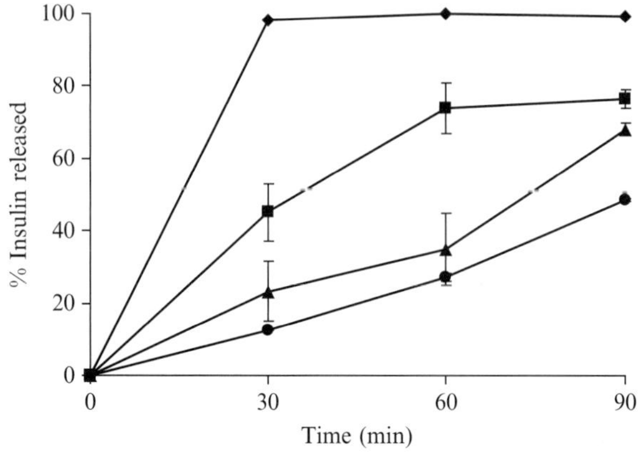

Figure 15.6 Insulin release profiles from (◆) CS-NP, (■) L1/CS-NP, (▲) L2/CS-NP, and (●) L3/CS-NP complexes prepared by the lyophilization method, in artificial gastric juice.

The described *in vitro* behavior is in perfect agreement with the *in vivo* results. Upon oral administration of the different formulations to rats, a different pattern of change in serum glucose levels is observed in each case (Fig. 15.7), also as a function of time (6 h: L2/CS-NP and L3/CS-NP complexes are the more effective; 24 h: L1/CS-NP provokes a long-lasting hypoglycemic effect, decreasing the glucose level to about 70% of basal).

These studies have shown that the complexes formed by both methodologies (hydration and lyophilization) present different structures that are affected by the system composition and by the applied methodology.

4.4. Ocular route

To evaluate the application of the L/CS-NP complexes as drug vehicles by the ocular route, the complexes are prepared by lyophilization. The systems are loaded with FITC-BSA with the goal of assessing the penetration of the complexes in the ocular epithelium, their cytotoxicity, and *in vivo* tolerance. FITC-BSA-loaded CS-NP present sizes around 400 nm, while liposomes vary between 290 and 420 nm, as a function of their lipid composition, resulting in L/CS-NP complexes between 400 and 750 nm. The complexes' zeta-potential varied within +5.8 and +14.7 mV. All the formulations remained stable in simulated lacrimal fluid for at least 2 h and were present inside the cells as early as 15 min after incubation, and more clearly at 30 min. The systems exhibit negligible toxicity *in vitro* and a good tolerance *in vivo*. These results point out L/CS-NP complexes as potentially useful drug delivery carriers through the ocular mucosa. For further details, please consult the original paper (Diebold *et al.*, 2007).

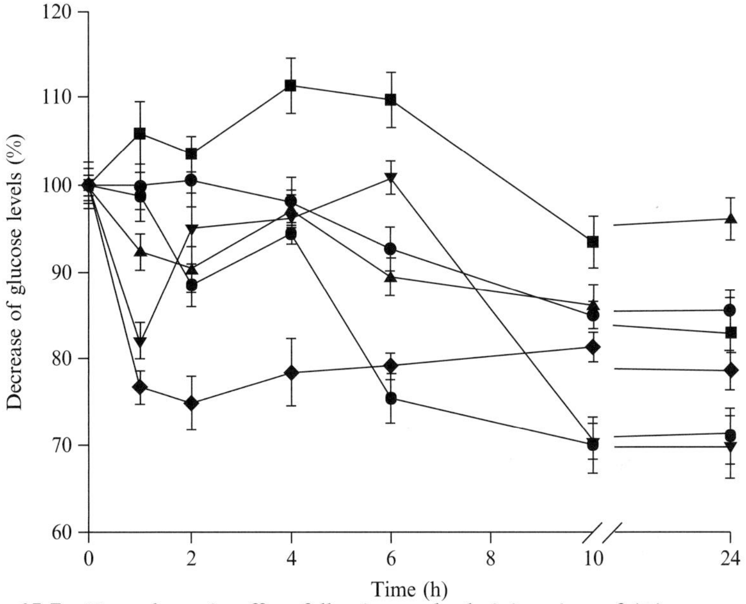

Figure 15.7 Hypoglycemic effect following oral administration of (●) water, (▲) insulin solution, (■) CS-NP, (▼) L1/CS-NP, (◆) L2/CS-NP, and (⬢) L3/CS-NP complexes to rats (insulin doses = 10 UI/kg, data represent the mean ± SEM, n = 7).

5. Conclusions and Prospects

We have developed attractive and simple methodologies to efficiently associate/incorporate chitosan nanoparticles within lipid vesicles. By means of a lyophilization procedure or by performing the hydration of a dry lipid film, we obtain new assemblies (L/CS-NP complexes) that can display different properties, as a function of the lipid composition applied in their formation and the method used for their preparation. As expected, these complexes permit the efficient encapsulation of therapeutic molecules like insulin, offer great stability in biological fluids, and provide a controlled release of their payload. Furthermore, their *in vivo* behavior can be considered very promising. The oral administration of these newly developed complexes loaded with insulin demonstrates their ability to provide an important reduction in plasma glucose levels. Furthermore, these complexes exhibit negligible *in vitro* toxicity in conjunctival epithelial cells and a good tolerance *in vivo*, corroborating their potential as drug carriers to the epithelial surfaces [i.e., ocular and lung (alveolar) surface].

Our strategy can be applied to other naturally occurring polymeric nanoparticles that, in this manner, should benefit from the existence of an extra lipid barrier, being an excellent means of controlling the release pattern. This structure is favored by the interaction between nanoparticles and phospholipids, since it has been reported that the solid cores have a strong ordering effect on the phospholipid molecules.

ACKNOWLEDGMENT

The authors acknowledge funding from the Spanish Ministry of Education and Science and from IBB/CBME, LA, FEDER/POCI 2010.

REFERENCES

Al-Qadi, S., Grenha, A., Seijo, B., and Remuñan-Lopez, C. (2009). Drug and vaccine delivery systems: Microencapsulated nanoparticle-based delivery system for pulmonary administration of therapeutic macromolecules: *in vivo* study in rats. In *Proceedings of the 2nd ESF/UB (European School in Naomedicine)*, Lisboa, Portugal, 85.

Alonso, M. J., and Sánchez, A. (2003). The potential of chitosan in ocular drug delivery. *J. Pharm. Pharmacol.* **55,** 1451–1463.

Aramaki, Y., Tomizawa, H., Hara, T., Yachi, K., Kikuchi, H., and Tsuchiya, S. (1993). Stability of liposomes *in vitro* and their uptake by rat Peyer's patches following oral administration. *Pharm. Res.* **10,** 1228–1231.

Bangham, A. D., Standish, M. M., and Watkins, J. C. (1965). Diffusion of univalent ions across the lamellae of swollen phospholipids. *J. Mol. Biol.* **13,** 238–252.

Beaulac, C., Sachetelli, S., and Lagacé, J. (1999). Aerosolization of low phase transition temperature liposomal tobramycin as dry powder in an animal model of chronic pulmonary infection caused by *Pseudomonas aeruginosa*. *J. Drug Target.* **7,** 33–41.

Calvo, P., Remuñan-Lopez, C., Vila-Jato, J. L., and Alonso, M. J. (1997). Novel hydrophilic chitosan–polyethylene oxide nanoparticles as protein carriers. *J. Appl. Polym. Sci.* **63,** 125–132.

Campbell, A., Taylor, P., Cayre, O. J., and Paunov, V. N. (2004). Preparation of aqueous gel beads coated by lipid bilayers. *Chem. Commun.* **21,** 2378–2379.

Carvalho, E. L. S., Seijo, B., and Alonso, M. J. (2001). "Formation and characterization of chitosan nanoparticles–phospholipid complexes." In *Proceedings of the 13th International Symposium on Microencapsulation*, Angers, France.

Courrier, H. M., Butz, N., and Vandamme, T. F. (2002). Pulmonary drug delivery systems: Recent developments and prospects. *Crit. Rev. Ther. Drug Carr. Syst.* **19,** 425–498.

Csaba, N., Garcia-Fuentes, M., and Alonso, M. J. (2006). The performance of nanocarriers for transmucosal drug delivery. *Expert Opin. Drug Deliv.* **3,** 463–478.

de Campos, A., Sanchez, A., and Alonso, M. J. (2001). Chitosan nanoparticles: A new vehicle for the improvement of the delivery of drugs to the ocular surface. Application to cyclosporin A. *Int. J. Pharm.* **224,** 159–168.

de la Fuente, M., Csaba, N., Garcia-Fuentes, M., and Alonso, M. J. (2008). Nanoparticles as protein and gene carriers to mucosal surfaces. *Nanomedicine* **3,** 845–857.

Delattre, J., Couvreur, P., Puisieux, F., Philippot, J. R., and Schuber, F. (1993). *Les liposomes: Aspects Technologiques, Biologiques et Pharmacologiques*. INSERM, Paris.

Diebold, Y., Jarrín, M., Sáez, V., Carvalho, E. L. S., Orea, M., Calonge, M., Seijo, B., and Alonso, M. J. (2007). Ocular drug delivery by liposome–chitosan nanoparticle complexes (LCS-NP). *Biomaterials* **28,** 1553–1564.

El-Maghrabya, G. M., Barryc, B. W., and Williams, A. C. (2008). Liposomes and skin: From drug delivery to model membranes. *Eur. J. Pharm. Sci.* **34,** 203–222.

Fenske, D. B., Chonn, A., and Cullis, P. R. (2008). Liposomal nanomedicines: An emerging field. *Toxicol. Pathol.* **36,** 21–29.

Fernández-Urrusuno, R., Calvo, P., Remuñan-Lopez, C., Vila-Jato, J. L., and Alonso, M. J. (1999a). Enhancement of nasal absorption of insulin using chitosan nanoparticles. *Pharm. Res.* **16,** 1576–1581.

Fernández-Urrusuno, R., Romani, D., Calvo, P., Vila-Jato, J. L., and Alonso, M. J. (1999b). Development of a freeze-dried formulation of insulin-loaded chitosan nanoparticles intended for nasal administration. *STP Pharm. Sci.* **9,** 429–436.

Gabizon, A. A. (1995). Liposome circulation time and tumor targeting: Implications for cancer chemotherapy. *Adv. Drug Deliv. Rev.* **16,** 285–294.

Gregoriadis, G. (1988). *Liposome as Drug Carrier: Recent Trends and Progress.* John Wiley & Sons, Chichester.

Grenha, A., Seijo, B., Carvalho, E. L. S., and Remuñan-Lopez, C. (2008a). Microspheres containing lipid/chitosan nanoparticles complexes for pulmonary delivery of therapeutic proteins. *Eur. J. Pharm. Biopharm.* **69,** 83–93.

Grenha, A., Seijo, B., Serra, C., and Remuñán-López, C. (2008b). Surface characterisation of lipid/chitosan nanoparticles assemblies, using X-ray photoelectron spectroscopy and time-of-flight secondary ion mass spectrometry. *J. Nanosci. Nanotechnol.* **8,** 358–365.

Huang, Y. Z., Gao, J. Q., Liang, W. Q., and Nakagawa, S. (2005). Preparation and characterization of liposomes encapsulating chitosan nanoparticles. *Biol. Pharm. Bull.* **28,** 387–390.

Issa, M., Koping-Hoggard, M., and Artursson, P. (2005). Chitosan and the mucosal delivery of biotechnology drugs. *Drug Discov. Today: Technol.* **2,** 1–6.

Jiang, W., Kim, B. Y., Rutka, J. T., and Chan, W. C. (2007). Advances and challenges of nanotechnology-based drug delivery systems. *Expert Opin. Drug Deliv.* **4,** 621–633.

Jorgensen, L., Moeller, E. H., van de Weert, M., Nielsen, H. M., and Frokjaer, S. (2006). Preparing and evaluating delivery systems for proteins. *Eur. J. Pharm. Sci.* **29,** 174–182.

Kirby, C. J., and Gregoriadis, G. (1999). Liposomes. *In* "Encyclopedia of Controlled Drug Delivery," (E. Mathiowitz, ed.), pp. 461–492. John Wiley & Sons, New York.

Marier, J.-F., Lavigne, J., and Ducharme, M. P. (2002). Pharmacokinetics and efficacies of liposomal and conventional formulations of tobramycin after intratracheal administration in rats with pulmonary *Burkholderia cepacia* infection. *Antimicrob. Agents Chemother.* **46,** 3776–3781.

McAllister, S. M., Alpar, H. O., Teitelbaum, Z., and Bennett, D. B. (1996). Do interactions with phospholipids contribute to the prolonged retention of polypeptides within the lung? *Adv. Drug Deliv. Rev.* **19,** 89–110.

Paolicelli, P., de la Fuente, M., Sanchez, A., Seijo, B., and Alonso, M. J. (2009). Chitosan nanoparticles for drug delivery to the eye. *Expert Opin. Drug Deliv.* **6,** 239–253.

Rodríguez, R., and Xamaní, M. (2003). Liposomes prepared by high-pressure homogenizers. *In* "Liposomes," (N. Düzgünes, ed.), Vol. 367, pp. 28–45. Elsevier Academic Press, San Diego, CA.

Silva, G. A., Coutinho, O. P., Ducheyne, P., Shapiro, I. M., and Reis, R. L. (2007). Starch-based microparticles as carriers for the release of active platelet-derived growth factor. *Tissue Eng.* **13,** 1259–1268.

Takeuchi, H., Kojima, H., Yamamoto, H., and Kawashima, Y. (2001a). Evaluation of circulation profiles of liposomes coated with hydrophilic polymers having different molecular weights in rats. *J. Control. Release* **75,** 83–91.

Takeuchi, H., Yamamoto, H., and Kawashima, Y. (2001b). Mucoadhesive nanoparticulate systems for peptide drug delivery. *Adv. Drug Deliv. Rev.* **47,** 39–54.

Torchilin, V. P. (2005). Recent advances with liposomes as pharmaceutical carriers. *Nat. Rev. Drug Discov.* **4,** 145–160.
van Haeringen, N. J. (1981). Clinical biochemistry of tears. *Surv. Ophthalmol.* **26,** 84–96.
Wright, J. R., and Clements, J. A. (1987). Metabolism and turnover of lung surfactant. *Am. Rev. Resp. Dis.* **135,** 426–444.

CHAPTER SIXTEEN

Antiangiogenic Photodynamic Therapy with Targeted Liposomes

Naoto Oku *and* Takayuki Ishii

Contents

Department of Medical Biochemistry, University of Shizuoka School of Pharmaceutical Sciences, Shizuoka, Japan

Methods in Enzymology, Volume 465
ISSN 0076-6879, DOI: 10.1016/S0076-6879(09)65016-3

Abstract

Antiangiogenic photodynamic therapy (PDT) is a promising modality for cancer treatment, since it causes efficient cutoff of oxygen and nutrients to the tumor cells and thus indirectly eradicates the tumor cells. For the improvement of therapeutic efficacy of antiangiogenic PDT by using a photosensitizer benzoporphyrin derivative monoacid ring A (BPD-MA) in a liposomal formulation, we endowed the liposomes with an active-targeting probe, Ala-Pro-Arg-Pro-Gly (APRPG), a peptide specific for angiogenic endothelial cells. APRPG-PEG-modified liposomal BPD-MA (APRPG-PEG-LipBPD-MA) accumulated in tumor tissues to a similar extent as PEG-LipBPD-MA at 3-h postinjection. In contrast, APRPG-PEG-LipBPD-MA strongly suppressed tumor growth by PDT treatment, but PEG-LipBPD-MA did not. This finding suggests that antiangiogenic PDT with targeted liposomes is an efficient modality for tumor treatment, whereas PEG-modified nontargeted liposomes are not suitable as a carrier of photosensitizers. The reason for the observed ineffectiveness of PEG-LipBPD-MA is as follows: In the case of PDT, the amount of photosensitizer bound to or taken up into the target cells during the time interval between injection of the agent and laser irradiation is critical, rather than the total amount of photosensitizer in tumor tissue. Therefore, active-targeting technology is quite useful for antiangiogenic PDT.

1. Introduction

Photodynamic therapy (PDT) is a promising cancer treatment without severe side effects, since a photosensitizer such as porphyrin, chlorine, or phthalocyanine derivatives generates active oxygen species at the site that is irradiated with tissue-penetrating laser light (Huang *et al.*, 2008; Ol'shevskaya *et al.*, 2009; Pavani *et al.*, 2009; Vittar *et al.*, 2008). Benzoporphyrin derivative monoacid ring A (BPD-MA; Fig. 16.1), also known as Verteporfin™ or Visudyne™, is one of the second-generation photosensitizers used in PDT, which was developed for the treatment of choroidal neovascularization of age-related macular degeneration (Shah *et al.*, 2009). Since BPD-MA is hydrophobic and liposomal formulations are used clinically, liposome-delivered BPD-MA would be applicable as a targeted drug delivery system (DDS), as its use would reduce the injected dose and enhance the therapeutic efficacy.

Antiangiogenic therapy is thought to be desirable for cancer treatment, since angiogenesis is a crucial event for solid tumor growth and maintenance (Pastorino *et al.*, 2003). In fact, antiangiogenic chemotherapy (which we have termed "antineovascular therapy") with the use of anticancer drugs encapsulated in neovessel-targeted liposomes suppresses tumor growth quite efficiently in tumor-bearing model mice (Shimizu and Oku, 2004).

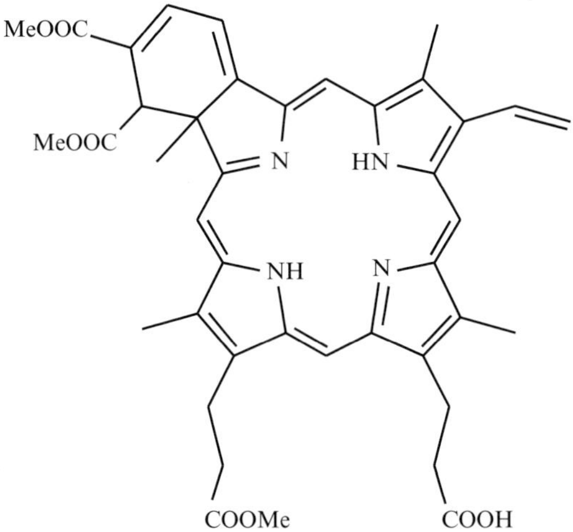

Figure 16.1 Structure of benzoporphyrin derivative monoacid ring A (BPD-MA).

This therapy effectively induced apoptosis of tumor cells through damage to angiogenic endothelial cells. Moreover, this therapy is efficient against drug-resistant tumors, since angiogenic endothelial cells do not acquire drug resistance. For this purpose, we used a 5-mer probe peptide, Ala-Pro-Arg-Pro-Gly (APRPG), which binds specifically to tumor angiogenic sites (Oku *et al.*, 2002), for the modification of liposomes.

Like antiangiogenic chemotherapy, antiangiogenic PDT would be expected to cause efficient tumor regression due to complete cutoff of oxygen and nutrients to the tumor cells (Chen *et al.*, 2008). We established previously a rather stable liposomal BPD-MA, composed of dipalmitoylphosphatidylcholine (DPPC), palmitoyloleoylphosphatidylcholine (POPC), cholesterol, dipalmitoylphosphatidylglycerol (DPPG), and BPD-MA (10/10/10/2.5/0.3 as molar ratio; Ichikawa *et al.*, 2004a). Optimal antiangiogenic PDT scheduling was also established, that is, laser irradiation at 15 min after the injection of the liposomal BPD-MA. In this scheduling, liposomal BPD-MA is abundant near the angiogenic endothelium (Dolmans *et al.*, 2002b; Kurohane *et al.*, 2001). In fact, PDT with this scheduling suppresses tumor growth more efficiently than conventional PDT, that is, laser irradiation at 3 h postinjection, due to hemostasis at the laser-irradiation site (Kurohane *et al.*, 2001). In clinical usage, 3-h PDT has been traditionally performed, since the concentration of BPD-MA in tumor tissue is higher than in normal tissue at 3 h after the injection.

Since the singlet oxygen generated by BPD-MA during irradiation with laser light has a very short half-life, BPD-MA liposomes should be highly abundant around target cells, or should be specifically bound to or taken up by them (Osaki *et al.*, 2006). In this chapter, we describe the use of three different kinds of liposomal BPD-MA for PDT. The first kind is polyethylene glycol (PEG)-coated liposomal BPD-MA, used because the long-circulating characteristic of liposomes achieved by PEG coating is known to cause passive accumulation of liposomal drugs in tumor tissues of tumor-bearing animals (Ishida *et al.*, 2002). In fact, doxorubicin-encapsulated PEG-coated liposomes, Doxil™, are clinically used for the treatment of certain cancer patients (Green and Rose, 2006). In this study, the BPD-MA-encapsulating liposomes were modified with distearoylphosphatidylethanolamine (DSPE)-PEG, and used for damaging tumor cells rather than angiogenic endothelial cells after the treatment with laser irradiation. The second kind is polycation-coated liposomal BPD-MA, chosen since cationic liposomes associate with cells more strongly than neutral or anionic liposomes due to electrostatic interactions, and intravenously (i.v.) injected liposomes interact first with the vessel endothelium in addition to blood cells. The third kind is a liposome that actively targets angiogenic endothelial cells; that is, BPD-MA-containing liposomes are modified with a neovessel-targeting probe, DSPE-PEG-APRPG. In this case, we expect that tumor neovessel endothelial cells would be damaged specifically by irradiation with the laser light.

2. PDT with PEG-Coated Liposomal BPD-MA

2.1. Preparation and characterization of PEG-coated liposomes with encapsulated BPD-MA

DPPC, POPC, cholesterol, DPPG, and BPD-MA (10/10/10/2.5/0.3 as a molar ratio) without (Cont-LipBPD-MA) or with PEG-DSPE (PEG-LipBPD-MA, lipids/PEG-DSPE = 20/1) dissolved in chloroform are dried under reduced pressure, and stored *in vacuo* for at least 1 h. The thin lipid film is hydrated with saline and frozen and thawed for 3 cycles by using liquid nitrogen. Then, the liposomal suspension is sonicated for 15 min at 60 °C in a bath-type sonicator. Finally, the liposomes are sized to 100 nm in diameter by extrusion through a polycarbonate membrane filter, using a syringe extruder (Avestin, Toronto or Avanti Polar Lipids, Alabaster, AL). During preparation of the BPD-MA liposomes, the samples are protected from light as much as possible.

The particle size and ζ-potential of the BPD-MA liposomes are determined by use of an electrophoretic light-scattering spectrophotometer. Typical values are, respectively, 131 nm and 7.4 mV for Cont-LipBPD-MA, and 151 nm and −7.7 mV for PEG-LipBPD-MA.

2.2. Biodistribution of liposomal BPD-MA in tumor tissue assessed by HPLC

Seven days after implantation of Meth-A sarcoma cells (1×10^6 cells/0.2 ml) into the left posterior flank of 5-week-old male BALB/c mice (Japan SLC, Shizuoka, Japan), the tumor-bearing mice are injected intravenously with liposomal BPD-MA (2 mg/kg as BPD-MA). The mice are sacrificed 3-h postinjection under anesthesia with diethyl ether and the tumor is excised from each mouse and homogenized in acetate-buffered saline (pH 5.0). Then, BPD-MA is extracted with ethyl acetate three times. After evaporation of the solvent, the sample is dried completely *in vacuo* overnight and the BPD-MA is dissolved in DMSO. The amount of BPD-MA is analyzed with an HPLC equipped with an ultrasphere C-8 column (Beckman). The mobile phase for the HPLC analysis is composed of 0.08 *M* $(NH_4)_2SO_4$, acetonitrile, tetrahydrofuran, and acetic acid (52:28:28:5 as a volume ratio).

In a typical experiment, the accumulated BPD-MA in the tumor tissue was 0.04% of the injected dose/100 mg tissue for Cont-LipBPD-MA and 0.24% for PEG-LipBPD-MA (Ichikawa *et al.*, 2004a), indicating that PEGylation of liposomes actually greatly enhanced the tumor accumulation of BPD-MA. This finding was expected, because the angiogenic vasculature in tumor tissue is quite leaky, and thus long-circulating liposomes accumulate easily in the interstitial tissues of the tumor due to the enhanced permeability and retention (EPR) effect.

2.3. Therapeutic efficacy of PEG-coated liposomal BPD-MA after PDT

Meth-A sarcoma-bearing mice ($n = 9$ or 10) are injected intravenously with liposomal BPD-MA (0.5 mg/kg as BPD-MA, instead of 2 mg/kg) at day 7 after tumor implantation. Then, the tumor site is irradiated with 689-nm laser light (150 J/cm^2, 0.25 W) at 3-h postinjection by use of a diode-laser system, SP689 (Suzuki Motor Co., Ltd, Japan). The control group is injected intravenously with saline without laser irradiation. The size of the tumor and body weight of each mouse are monitored thereafter. Two bisecting diameters of each tumor are measured with slide calipers to determine the tumor volume, and calculation is performed by using the formula $0.4(a \times b^2)$, where a is the largest, and b, the smallest, diameter. The tumor volume thus calculated correlates well with the actual tumor weight ($r = 0.980$).

The result was unexpected and quite interesting; that is, tumor growth suppression was obvious for the Cont-LipBPD-MA-injected group after treatment with PDT but not significant for the PEG-LipBPD-MA-injected group, although the total amount of BPD-MA in the tumor tissue was about sixfold higher for the latter than for the former, as described above (Ichikawa *et al.*, 2004a). Corresponding to the tumor growth suppression, the mean lifetime of tumor-bearing mice ($n = 5$) was 30.6, 37.0, and 29.4 days for nontreatment, Cont-LipBPD-MA-PDT, and PEG-LipBPD-MA-PDT groups, respectively (Ichikawa *et al.*, 2004a).

This evidence suggests that PEG in PEG liposomes prevents the direct interaction of liposomes with tumor cells, with the result being that the singlet oxygen generated in the liposomes by laser irradiation cannot damage the tumor cells. On the contrary, the noncoated liposome may more easily make contact with tumor cells or angiogenic endothelial cells, and thus deliver BPD-MA to the cells.

PEG liposomes encapsulating anticancer drugs are useful for passive targeting, and the drugs accumulated in tumor tissue act for a longer period time on the tumor cells. In contrast, only the photosensitizers bound to or taken up by tumor cells at 3-h postinjection can damage tumor cells in PDT. We previously observed that glucuronate-modified liposomal BPD-MA strongly suppressed tumor growth in tumor-bearing mice after laser irradiation at 5-h postinjection (Oku *et al.*, 1997). In this case, the enhanced therapeutic efficacy may be explained by the possibility that 5 h is enough time to allow delivery of BPD-MA to tumor cells and/or that glucuronate modification does not prevent the interaction between liposomes and tumor cells.

3. PDT with Polycation-Coated Liposomal BPD-MA

3.1. Preparation and characterization of polycation liposomes with encapsulated BPD-MA

To construct a novel drug delivery carrier that possesses high therapeutic efficacy at a low dosage, we prepared polyethylenimine-modified liposomes (polycation liposome, PCL) and examined the entrapment of BPD-MA for antiangiogenic PDT. For the modification of the liposomal surface with polycations, first cetylated polyethylenimine (cetyl-PEI) is synthesized (Oku *et al.*, 2001; Yamazaki *et al.*, 2000). In brief, PEI (average Mr. = 1800) is stirred at 65 °C in EtOH for 30 min with refluxing and N_2 bubbling, and to the solution is added 1-bromohexadecane (3.44 *M*) and triethylamine (7.17 *M*). After reaction for 6–9 h, the solution is evaporated. The grafting reaction is performed under a N_2 atmosphere. After the reaction, the solvent is removed by evaporation, and the compound is then dissolved in 20% EtOH aqueous solution. Ungrafted compounds are removed over a 2-day period by an ultrafiltration technique using 10 volumes of 20% EtOH aqueous solution (Takeuchi *et al.*, 2003a). Finally, the unfiltered solution is lyophilized. The resulting solid is soluble in $CHCl_3$ and partly soluble in H_2O, enabling the preparation of the PCLs. The stoichiometric percentage of cetyl groups conjugated, as determined by ^1H NMR spectra, was 24% per total PEI amino groups.

Cetyl-PEI-based PCLs were originally developed for gene delivery (Matsuura *et al.*, 2003). The PCL interacts with various cells quite efficiently through electrostatic interaction, is endocytosed, and releases the entrapped materials from endosomes into the cytosol or nucleus due to a proton sponge effect (Arote *et al.*, 2008). The BPD-MA liposome (LipBPD-MA) consists of DPPC, cholesterol, DPPG, and BPD-MA (20/10/5/0.3, respectively, as a molar ratio) (Takeuchi *et al.*, 2003a). On the other hand, the BPD-MA PCL (PCLipBPD-MA) additionally has a 1.75 molar ratio of cetyl-PEI (total lipids/cetyl-PEI = 20/1). Lipids, BPD-MA, and cetyl-PEI dissolved in $CHCl_3$ are evaporated to prepare a thin lipid film, which is then hydrated with PBS. The liposomal solution is frozen and thawed for 3 cycles by using liquid nitrogen. Further preparation methods are similar to those described above. The respective liposomal size and ζ-potential were 103 nm and +5.0 mV for LipBPD-MA, and 114 nm and +32.5 mV for PCLipBPD-MA.

3.2. *In vitro* study using PCL-encapsulated BPD-MA

Human vascular endothelial cell line ECV304 cells (1×10^5 cells) are seeded in 35-mm cell culture dishes in 199 medium supplemented with 10% FBS and incubated in a 5% CO_2 incubator for 48 h at 37 °C. The medium is changed to 990-μl 199 medium with or without the addition of 10% FBS, LipBPD-MA, or PCLipBPD-MA (10 μl; 30–500 ng/ml as the final concentration of BPD-MA) and incubated for 60 min. PDT treatment is performed with irradiation from the top side of the cell culture dish by using a diode-laser system, SP689. The cells that are incubated with liposomal photosensitizers are exposed to the laser light with 2.0 J/cm^2 of fluence (0.25 W, 76.9 s) at 689 nm. At 24 h after the PDT treatment, the cells are washed with PBS and the number of viable cells are determined by the crystal violet dye assay. In brief, the cells are incubated for 10 min in 0.5% crystal violet dissolved in MeOH/H_2O = 1/4 (v/v) followed by the removal of the redundant dye solution by washing with PBS. The dishes are completely dried and 33% AcOH aqueous solution (1 ml) is added to elute the dye from the stained cells. The percentage of surviving cells is quantified spectroscopically by the absorbance at 630 nm, and the parameters are normalized spectroscopically to a set of control cells (without laser exposure or without addition of the liposomal photosensitizer).

The cells that had been treated with PCLipBPD-MA were destroyed efficiently by PDT treatment in comparison to those treated with an equal concentration of control LipBPD-MA in the absence of serum; that is, the phototoxicity in the presence of PCLipBPD-MA was approximately twice that of the BPD-MA liposomes (LD_{50} = 248 ng/ml for LipBPD-MA and 146 ng/ml for PCLipBPD-MA; Takeuchi *et al.*, 2003b). For evaluation of possible direct cytotoxicity, a large amount of PCLipBPD-MA or LipBPD-MA (500 ng/ml as BPD-MA concentration) was applied to ECV304 cells without subsequent laser exposure. The viability of the cells was not affected. Cytotoxicity was also not observed by laser exposure alone (2 J/cm^2).

The phototoxicity of BPD-MA in the normal liposome or PCL formulation in the presence of serum was also examined. The phototoxicity of LipBPD-MA was reduced to 19%, whereas that of PCLipBPD-MA dropped to only 47% (Takeuchi *et al.*, 2003b), suggesting that the inhibitory effect of serum against PDT is overcome by the PCL formulation to some extent.

The cellular uptake of BPD-MA is examined by using ECV304 cells. The cells (3×10^5 cells) are seeded in 60-mm cell culture dishes and incubated in a 5% CO_2 incubator for 48 h at 37 °C. The medium in the cell culture dish is changed to fresh 199 medium supplemented with 10% FBS (2.97 ml). PCLipBPD-MA or LipBPD-MA (30 μl) is added (100 ng/ml as final concentration of BPD-MA), and incubation is continued for 60 min. After the cells are washed with ice-cold PBS, they are dissolved in 0.1% SDS-containing 5 m*M* Tris buffer (1.2 ml), and the amount of BPD-MA is determined fluorometrically with an excitation wavelength of 450 nm and emission at 619 nm, a fluorescence spectrophotometer. Cell protein concentration is determined with the BCA protein assay (Bio-Rad).

Liposomes modified with PEI showed enhanced cellular uptake of BPD-MA, approximately 1.5-fold in the absence of serum and 3.1-fold in the presence of 10% FBS. Furthermore, inhibition of BPD-MA uptake by FBS was reduced by using PCLs; that is, there was an 81.5% inhibition for LipBPD-MA and 61.5% inhibition for PCLipBPD-MA in comparison to the level in the absence of serum (Takeuchi *et al.*, 2003b). Suppression of BPD-MA uptake in the presence of serum may be due to the presence of albumin or other serum proteins. Even so, in the case of PCLs, the electrostatic interaction between the polycations and the plasma membrane of vascular endothelial cells may overcome the suppression of interaction to some extent. Additionally, a remarkable reduction in uptake was observed for both LipBPD-MA and PCLipBPD-MA at the incubation temperature of 20 and 4 °C, suggesting that the BPD-MA was mainly taken up by the cells via the endocytotic pathway.

3.3. Therapeutic efficacy of PCL-encapsulated BPD-MA after PDT

It would be expected that polycations would adhere to the anionic plasma membrane by electrostatic interaction due to the concentrated positive charge by polymerization of amino groups. Therefore, *in vivo* angiogenic vessel-targeted PDT treatment using PCLipBPD-MA is investigated.

The biodistribution of BPD-MA after the injection of LipBPD-MA or PCLipBPD-MA containing [^{14}C]BPD-MA is determined. At day 7 following implantation of post-Meth-A sarcoma, the tumor-bearing mice are i.v. injected with 0.25 mg/kg of [^{14}C]BPD-MA liposomes or [^{14}C]BPD-MA PCLs. After 15 min, the mice are sacrificed under pentobarbital anesthesia, and the plasma and the tissues are collected. These samples are weighed and solubilized in 1 ml of Solvable for 2 days at 50 °C. After bleaching of the solution with H_2O_2 and isopropanol (0.5 ml each) overnight at 50 °C, Hionic-Fluor is added and the radioactivity is measured with a liquid scintillation counter. The biodistribution results are summarized in Fig. 16.2A. Remarkable accumulation of [^{14}C]BPD-MA is detected in the lungs after the injection of PCLipBPD-MA, whereas accumulation of the photosensitizer in the tumor is not significant after the injection of these liposomes.

A therapy-related experiment is performed in a similar manner as described above. Five-week-old male BALB/c mice are implanted subcutaneously (s.c.) into their left posterior flank with 1×10^6 cells/0.2 ml of Meth-A sarcoma cells, and then they are randomly divided into three groups (control and two treatment groups, each $n = 5$). The control group consists of mice that receive an i.v. injection of saline (0.25 ml) without laser exposure. PDT treatment is performed 15-min postinjection of PCLipBPD-MA or LipBPD-MA (0.25 mg/kg of BPD-MA). Figure 16.2B shows the tumor volume change after the PDT treatment. The results indicate that a significant suppression of tumor growth is afforded by the PCLipBPD-MA, even though the tumor accumulation of BPD-MA is rather low compared with that found for LipBPD-MA. It is possible that the remarkable lung infiltration by the [^{14}C]BPD-MA PCLs is due to strong interaction of PCL with pre-existing vessels. However, the toxicity of PCL in the mice is negligible. So, we suggest that the liposomal photosensitizer is rapidly taken up into the surrounding vascular endothelial cells in the microvasculature due to the proximate distance between the PCL and plasma membrane. We consider that this phenomenon also occurs in the angiogenic vasculature at the periphery of the tumor tissue and that photodestruction of vascular endothelial cells and subsequent vascular occlusion occurs there by PDT treatment. This phenomenon is opposite to the case of PEG-LipBPD-MA. PEG-LipBPD-MA accumulates more in the tumor than does the control liposomes; however, the PDT effect is almost negligible due to the lack of liposomal interaction with endothelial or tumor cells. In the case of PCLipBPD-MA, the accumulation is less than that for the control liposomes; however, the interaction of PCLipBPD-MA with endothelial cells will be strong enough to lead to the high accumulation of BPD-MA in these cells.

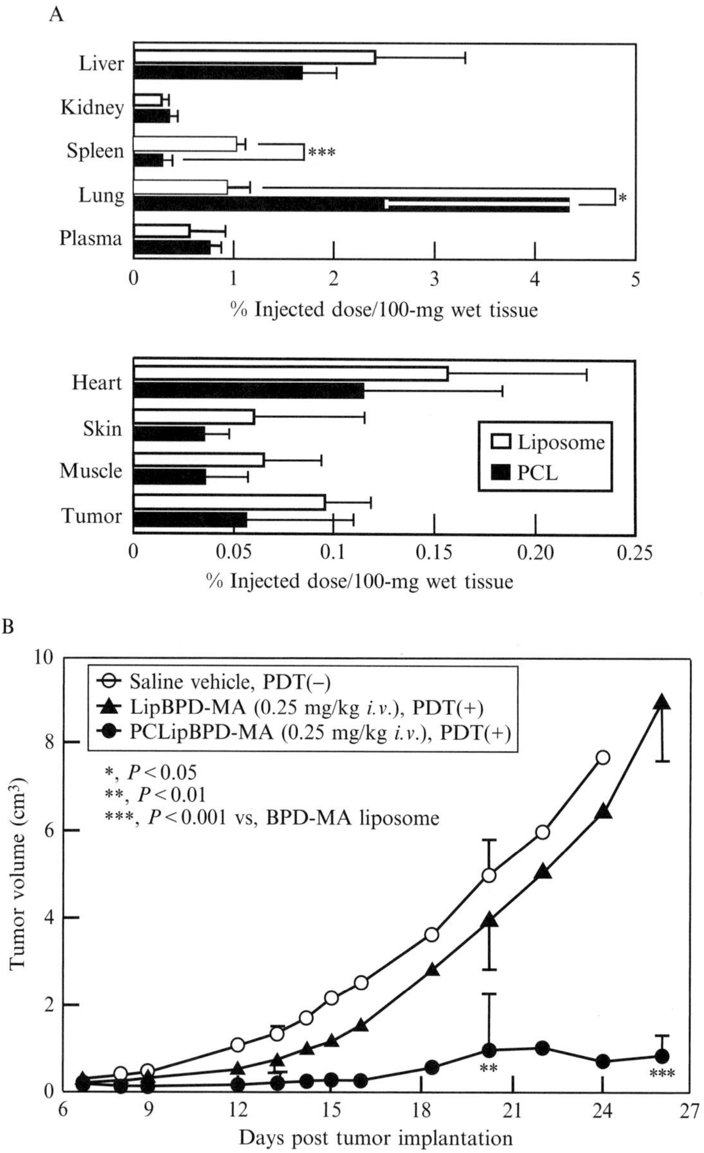

Figure 16.2 Biodistribution of BPD-MA 15-min postinjection with LipBPD-MA or PCLipBPD-MA in tumor-bearing mice, and the therapeutic efficacies. (A) BALB/c mice were implanted s.c. into the left posterior flank with 1×10^6 Meth-A sarcoma cells. At day 7 post-tumor implantation, [^{14}C]BPD-MA entrapped in nonmodified control liposomes (LipBPD-MA) or in polycationic liposomes (PCLipBPD-MA, 0.25 mg/kg in terms of BPD-MA) was i.v. injected. At 15 min after the injection, the mice were sacrificed under pentobarbital anesthesia, and the radioactivity in the plasma and other tissues was determined. Data points represent the mean ± SD ($n = 5$).

3.4. Apoptosis of tumor cells after PDT treatment with PCLipBPD-MA

For clarification of the destruction of the angiogenic vasculature by PDT with PCLipBPD-MA, immunofluorescence double-staining for CD31/PECAM-1 and fragmented DNA by the TUNEL method is performed on tumor specimens. At 24-h post-PDT treatment, tumor tissues are resected, embedded in Tissue-Tek OCT Compound (Sakura Finetek USA), and frozen in -80 °C acetone. The frozen tissues are sectioned at a 5-μm thickness and mounted on MAS-coated glass slides. The specimens are fixed in 4% paraformaldehyde for 15 min at room temperature after tissue air-drying for 1 h, and washed three times with PBS. Protein blocking is performed for 10 min at room temperature by using 1% BSA-containing PBS, and the specimens are then washed three times with PBS.

TUNEL assays are performed with an ApopTag Fluorescein *In Situ* Apoptosis Detection Kit (Intergen) according to the directions supplied with the kit. After tailing of the nicked DNA with digoxygenin-conjugated dNTP by TdT enzyme, the specimens are incubated with biotin-conjugated rat antimouse CD31/platelet–endothelial cell adhesion molecule-1 (PECAM-1) monoclonal antibody (BD Biosciences), diluted at 1:100, for 1 h at room temperature and subsequently washed three times with PBS. Immunofluorescence double staining is performed by using streptavidin–Alexa Fluor 594 (Molecular Probes) and antidigoxygenin fluorescein in the order given. The specimens are finally washed with PBS and coverslips are mounted on the glass slides with Prolong Antifade (Molecular Probes). Immunofluorescent CD31 and DNA fragments are detected under a confocal laser-scanning microscope. Alexa Fluor 594 and fluorescein are excited by using a helium–neon laser at 588 nm and an argon ion laser at 458 nm, respectively. All lines are passed through a white transmission filter.

The results of CD31/TUNEL immunofluorescence double staining of tumor tissue sections following PDT treatment are shown in Fig. 16.3. No TUNEL-positive cells are observed in the saline-injected control mice (Fig. 16.3A). On the contrary, some degree of apoptotic tumor cell death is detected in the PDT-treated mice given LipBPD-MA, although the number of CD31-positive endothelial cells is not much decreased by the

Significant differences: $*p < 0.05$; $***p < 0.001$. (B) LipBPD-MA, PCLipBPD-MA BPD-MA (0.25 mg/kg BPD-MA) or saline was i.v. injected at day 7 postimplantation of Meth-A sarcoma cells. At 15 min after the injection, the BPD-MA-treated mice were exposed to laser light (689 nm, 150 J/cm^2) under pentobarbital anesthesia. The tumor volume was monitored every day. Datum points represent the mean $\pm$ SD ($n = 5$), and SD bars are shown only for the points for days 20 and 26 for the sake of graphic clarity. Significant difference between the two PDT-treated groups: $**p < 0.01$; $***p < 0.001$.

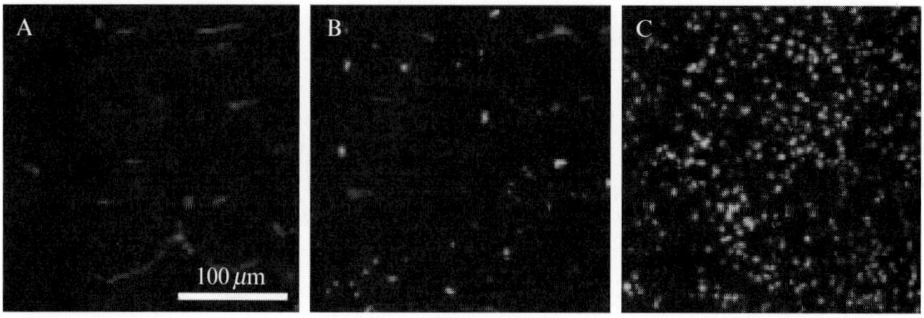

Figure 16.3 Immunofluorescence double staining of vascular endothelial cells and apoptotic cells in tumor tissue sections after PDT. Saline (A), LipBPD-MA (B), or PCLipBPD-MA (0.25 mg/kg BPD-MA) (C) was injected i.v. into tumor-bearing BALB/c mice. At 15-min postinjection of liposomal, the photosensitizer BPD-MA-treated mice were exposed to laser light (689 nm, 150 J/cm^2) under pentobarbital anesthesia. At 24-h postlaser exposure, tumors were resected and immunofluorescently double-stained with red and green fluorescence, indicating CD31-positive vascular endothelial cells (Alexa Fluor 594) and TUNEL-positive apoptotic cells (fluorescein), respectively. (See Color Insert.)

PDT treatment (Fig. 16.3B). Interestingly, the dramatic disappearance of CD31-derived red fluorescence and induction of remarkable apoptosis of tumor cells is simultaneously observed in the PDT-treated mice injected with PCLipBPD-MA (Fig. 16.3C). These results indicate that destruction of angiogenic vessels and subsequent tumor cell apoptosis occurs following PCLipBPD-MA-mediated PDT treatment.

4. PDT with Tumor Angiogenic Vessel-Targeted Liposomal BPD-MA

4.1. Preparation and characterization of neovessel-targeted liposomes with encapsulated BPD-MA

As shown above, PEGylation of liposomes encapsulating BPD-MA (PEG-LipBPD-MA) enhances the passive accumulation of the drug in tumor tissues at 3 h after administration, but does not enhance the therapeutic efficacy after PDT. The reason is speculated to be that PEG-LipBPD-MA accumulates in the interstitial space of tumor tissues and is not taken up effectively into the tumor cells before the laser irradiation at 3 h after administration (Ichikawa *et al.*, 2004b). Therefore, the singlet oxygen generated by laser irradiation of BPD-MA may not damage the cells around the liposomes, since the half-life of active oxygen is too short for the active oxygen to pass from inside of the liposomes to the cells.

On the contrary, PCLipBPD-MA, which has been designed to enhance the interaction of liposomes with cell membranes, shows higher PDT efficacy than control LipBPD-MA, suggesting the importance of the interaction of liposomes with target cells. To enhance the target specificity of BPD-MA liposomes, we modify them with a targeting probe specific for angiogenic endothelial cells. For the study, we use APRPG-PEG-DSPE for the modification of BPD-MA-containing liposomes (APRPG-PEG-LipBPD-MA), since the APRPG peptide is known to be specific for angiogenic endothelium and PEG is known to reduce the reticuloendothelial system (RES) trapping of liposomes (Maeda *et al.*, 2006). We have previously determined the agglutinability of APRPG-PEG-LipBPD-MA in the presence of 50% serum, and do not observe any increase in turbidity (Ichikawa *et al.*, 2005).

Cont-LipBPD-MA and APRPG-PEG-LipBPD-MA are prepared according to procedures similar to those used for preparing PEG-LipBPD-MA. APRPG-PEG-LipBPD-MA consists of DPPC, POPC, cholesterol, DPPG, and BPD-MA (10/10/10/2.5/0.3 as a molar ratio) with APRPG-PEG-DSPE (Lipids/APRPG-PEG-DSPE = 20/1). The particle size and ζ-potential of the APRPG-PEG liposomes containing BPD-MA are 136 nm and −5.9 mV, respectively (Ichikawa *et al.*, 2005).

4.2. Biodistribution of APRPG-PEG-liposomal BPD-MA in tumor tissue

The biodistribution of APRPG-PEG-liposomal BPD-MA is assessed by using HPLC, according to the method used for PEG-liposomal BPD-MA. The accumulation of BPD-MA in the tumor was 0.17% of the injected dose/100 mg, a value is similar to that for PEG-LipBPD-MA (0.24% injected dose/100 mg; Ichikawa *et al.*, 2005).

4.3. Therapeutic efficacy of APRPG-PEG liposomes after PDT

Data on PDT with APRPG-PEG-liposomal BPD-MA are obtained according to the method used to obtain the data on PEG-liposomal BPD-MA. Targeting of the angiogenic vasculature of tumors is promising for cancer treatment. In terms of PDT scheduling, we showed that laser irradiation at a time as early as 15 min after the injection of the photosensitizer was antiangiogenic (Kurohane *et al.*, 2001). Actually, Dolmans *et al.* (2002a,b) demonstrated that a photosensitizer is distributed to vascular endothelial cells to a greater extent at 15 min after the injection than at 4-h postinjection, as determined by intravital microscopy. However, we apply laser irradiation 3-h postinjection of APRPG-PEG-LipBPD-MA, since the liposomes will be actively targeting these cells and taken up by neovessels more at 3-h postinjection than at 15 min.

The data on PDT with APRPG-PEG-LipBPD-MA as well as that with PEG-LipBPD-MA are shown in Fig. 16.4. PDT mediated by APRPG-PEG-LipBPD-MA significantly suppressed tumor growth and prolonged the lifespan of the tumor-bearing mice. On the contrary, consistent with the above results, PEG-LipBPD-MA showed only a little suppression of tumor growth and no increase in the survival time of the tumor-bearing mice.

In conclusion, antiangiogenic PDT treatment using BPD-MA entrapped in neovessel-targeted liposomes results in strong tumor destruction with a low dose of the photosensitizer. The mechanism underlying photodestruction of the tumor is considered to be as follows: APRPG-PEG-LipBPD-MA interacts with the angiogenic vascular endothelial cells and is subsequently taken up into the cells via the endocytotic pathway. Then, photodestruction of angiogenic vessels occurs during the PDT treatment. The loss of oxygen and nutrient supplements delivered via angiogenic vessels thus results in the apoptosis of the tumor cells.

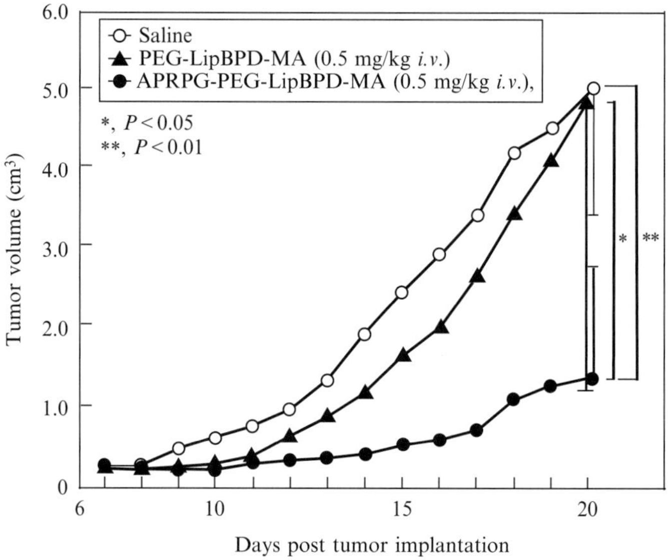

Figure 16.4 Tumor growth suppression after PDT treatment with neovessel-targeted liposomal BPD-MA. BALB/c mice were implanted s.c. into their left posterior flank with 1×10^6 Meth-A sarcoma cells. At day 7 after tumor implantation, saline (open circles), PEG-LipBPD-MA (closed triangles), or APRPG-PEG-LipBPD-MA (closed circles) was intravenously injected. At 3 h after the injection, the liposomal BPD-MA (0.5 mg/kg as BPD-MA)-treated mice were exposed to the laser light (689 nm, 150 J/cm^2) under pentobarbital anesthesia. Tumor volume was monitored thereafter. Datum points represent the mean ± SD ($n = 9$ or 10), and SD bars are shown only at day 20 for the sake of graphic clarity. Significant difference: $^{\star}p < 0.05$, $^{\star\star}p < 0.01$ for bracketed comparisons.

5. Usefulness of Antiangiogenic PDT with Neovessel-Targeted Liposomes

5.1. Antiangiogenic PDT scheduling with nontargeted liposomal BPD-MA

In our early study, we developed antiangiogenic PDT scheduling with PEG-uncoated nontargeted liposomal BPD-MA (Kurohane *et al.*, 2001). Laser irradiation at 15-min postinjection of 2 mg/kg liposomal BPD-MA (15-min PDT) causes complete blocking of blood flow in the neovasculature. In contrast, the irradiation at 3-h postinjection of the same liposomal BPD-MA does not inhibit the blood flow completely. This suggests that tumor growth is strongly inhibited after the 15-min PDT with the liposomal BPD-MA. We speculate the reason why the 15-min PDT is useful is that a high dose of liposomal BPD-MA around the neovessels generates active oxygen that damages the angiogenic endothelial cells. The damage to blood cells also possibly enhances hemostasis (Fig. 16.5A). In the case of 3-h PDT, the amount of BPD-AM in the vicinity of the endothelial cells is not high enough to cause complete damage to the neovessels (Fig. 16.5B). Antiangiogenic PDT scheduling does not require specific targeting probes, but in that case a rather high dose of photosensitizers is needed; and there would be the occurrence of unfavorable damage to blood cells.

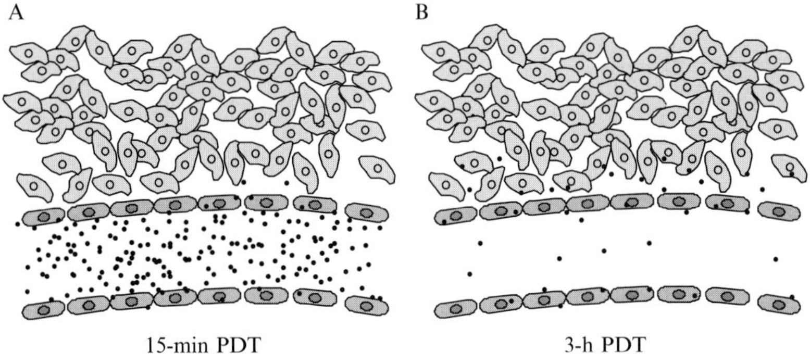

Figure 16.5 Possible scheme of intratumoral distribution of nontargeted liposomes containing BPD-MA and their efficacy after PDT. (A) At 15-min postinjection of nonmodified liposomes (shown as small dots), they are abundant in the bloodstream and may damage endothelial and blood cells during laser irradiation. This damage causes efficient hemostasis and subsequent tumor cell apoptosis due to lack of oxygen and nutrients. (B) At 3-h postinjection, some of the nonmodified liposomes interact with endothelial and tumor cells, and cause damage to these cells during the laser irradiation. However, complete tumor cell killing is not expected. In our experiment, 2 mg/kg BPD-MA was used.

5.2. Antiangiogenic PDT scheduling with neovessel-targeted liposomal BPD-MA

In contrast to the use of nontargeted liposomal BPD-MA in antiangiogenic PDT, neovessel-targeted liposomal BPD-MA would reduce the injected dose and unfavorable damage to blood cells, as neovessel-targeted liposomes would be expected to deliver the encapsulated agents to the angiogenic endothelial cells quite efficiently. Furthermore, these liposomes would be expected to bind to or be internalized into the cells. Therefore, active oxygen generated by PDT would efficiently damage the cells. In the case of PDT, the total amount of photosensitizer in the tumor tissue is not the important factor, but rather the amount of it bound to or taken up by target cells during the time interval between the injection of the photosensitizer and laser irradiation is critical. Therefore, active-targeting technology is quite useful, especially targeting of angiogenic endothelial cells rather than tumor cells, since the damage to angiogenic endothelial cells would eradicate tumor cells through the cutoff of oxygen (Fig. 16.6B). The ineffectiveness of PEG-LipBPD-MA for 3-h PDT, shown above, may be explained as follows: PEG-modified liposomes were present in the interstitial tissue and produced singlet oxygen there, but were not destructive, because they did not interact strongly with either endothelial or tumor cells (Fig. 16.6A).

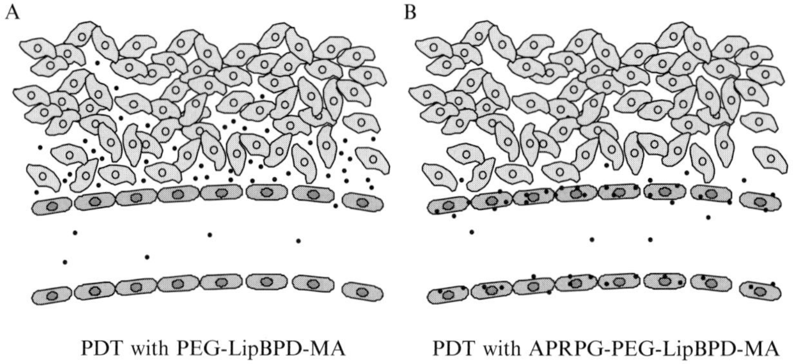

Figure 16.6 Possible scheme of intratumoral distribution of PEG and APRPG-PEG liposomes containing BPD-MA and their efficacy after PDT. (A) At 3-h postinjection of PEG liposomes (shown as small dots), these liposomes accumulate in the interstitial spaces of the tumor tissue by passive targeting. However, since PEG prevents the interaction of these liposomes with tumor cells, active oxygen generated by the laser irradiation does not induce tumor cell killing. (B) APRPG-PEG liposomes specifically interact with and are taken up into angiogenic endothelial cells. Therefore, PDT treatment causes efficient destruction of the angiogenic vasculature that causes apoptosis of tumor cells indirectly and efficiently. In our experiment, 0.25–0.5 mg/kg BPD-MA was used.

6. Concluding Remarks

PEG liposomes are widely used in the case of chemotherapy, because, in this case, the slow and sustained release of chemotherapeutic agents at the tumor site is favorable. Taken together, the available data indicate that active targeting, not passive targeting, is useful for antiangiogenic PDT, since the photosensitizers not only would be delivered to the target tissue but also would be bound and taken up by the target cells in a short period of time. Furthermore, the findings described here indicate that antiangiogenic PDT is a promising modality for cancer treatment.

REFERENCES

Arote, R. B., Hwang, S. K., Yoo, M. K., Jere, D., Jiang, H. L., Kim, Y. K., Choi, Y. J., Nah, J. W., Cho, M. H., and Cho, C. S. (2008). Biodegradable poly(ester amine) based on glycerol dimethacrylate and polyethylenimine as a gene carrier. *J. Gene Med.* **10,** 1223–1235.

Chen, B., Crane, C., He, C., Gondek, D., Agharkar, P., Savellano, M. D., Hoopes, P. J., and Pogue, B. W. (2008). Disparity between prostate tumor interior versus peripheral vasculature in response to verteporfin-mediated vascular-targeting therapy. *Int. J. Cancer* **123,** 695–701.

Dolmans, D. E., Kadambi, A., Hill, J. S., Waters, C. A., Robinson, B. C., Walker, J. P., Fukumura, D., and Jain, R. K. (2002a). Vascular accumulation of a novel photosensitizer, MV6401, causes selective thrombosis in tumor vessels after photodynamic therapy. *Cancer Res.* **62,** 2151–2156.

Dolmans, D. E., Kadambi, A., Hill, J. S., Flores, K. R., Gerber, J. N., Walker, J. P., Borel Rinkes, I. H., Jain, R. K., and Fukumura, D. (2002b). Targeting tumor vasculature and cancer cells in orthotopic breast tumor by fractionated photosensitizer dosing photodynamic therapy. *Cancer Res.* **62,** 4289–4294.

Green, A. E., and Rose, P. G. (2006). Pegylated liposomal doxorubicin in ovarian cancer. *Int. J. Nanomed.* **1,** 229–239.

Huang, Z., Xu, H., Meyers, A. D., Musani, A. I., Wang, L., Tagg, R., Barqawi, A. B., and Chen, Y. K. (2008). Photodynamic therapy for treatment of solid tumors—Potential and technical challenges. *Technol. Cancer Res. Treat.* **7,** 309–320.

Ichikawa, K., Takeuchi, Y., Yonezawa, S., Hikita, T., Kurohane, K., Namba, Y., and Oku, N. (2004a). Antiangiogenic photodynamic therapy (PDT) using Visudyne causes effective suppression of tumor growth. *Cancer Lett.* **205,** 39–48.

Ichikawa, K., Hikita, T., Maeda, N., Takeuchi, Y., Namba, Y., and Oku, N. (2004b). PEGylation of liposome decreases the susceptibility of liposomal drug in cancer photodynamic therapy. *Biol. Pharm. Bull.* **27,** 443–444.

Ichikawa, K., Hikita, T., Maeda, N., Yonezawa, S., Takeuchi, Y., Asai, T., Namba, Y., and Oku, N. (2005). Antiangiogenic photodynamic therapy (PDT) by using long-circulating liposomes modified with peptide specific to angiogenic vessels. *Biochim. Biophys. Acta* **1669,** 69–74.

Ishida, T., Harashima, H., and Kiwada, H. (2002). Liposome clearance. *Biosci. Rep.* **22,** 197–224.

Kurohane, K., Tominaga, A., Sato, K., North, J. R., Namba, Y., and Oku, N. (2001). Photodynamic therapy targeted to tumor-induced angiogenic vessels. *Cancer Lett.* **167,** 49–56.

Maeda, N., Miyazawa, S., Shimizu, K., Asai, T., Yonezawa, S., Kitazawa, S., Namba, Y., Tsukada, H., and Oku, N. (2006). Enhancement of anticancer activity in antineovascular therapy is based on the intratumoral distribution of the active targeting carrier for anticancer drugs. *Biol. Pharm. Bull.* **29,** 1936–1940.

Matsuura, M., Yamazaki, Y., Sugiyama, M., Kondo, M., Ori, H., Nango, M., and Oku, N. (2003). Polycation liposome-mediated gene transfer *in vivo*. *Biochim. Biophys. Acta* **1612,** 136–143.

Oku, N., Saito, N., Namba, Y., Tsukada, H., Dolphin, D., and Okada, S. (1997). Application of long-circulating liposomes to cancer photodynamic therapy. *Biol. Pharm. Bull.* **20,** 670–673.

Oku, N., Yamazaki, Y., Matsuura, M., Sugiyama, M., Hasegawa, M., and Nango, M. (2001). A novel non-viral gene transfer system, polycation liposomes. *Adv. Drug Deliv. Rev.* **52,** 209–218.

Oku, N., Asai, T., Watanabe, K., Kuromi, K., Nagatsuka, M., Kurohane, K., Kikkawa, H., Ogino, K., Tanaka, M., Ishikawa, D., Tsukada, H., Momose, M., *et al.* (2002). Anti-neovascular therapy using novel peptides homing to angiogenic vessels. *Oncogene* **21,** 2662–2669.

Ol'shevskaya, V. A., Nikitina, R. G., Savchenko, A. N., Malshakova, M. V., Vinogradov, A. M., Golovina, G. V., Belykh, D. V., Kutchin, A. V., Kaplan, M. A., Kalinin, V. N., Kuzmin, V. A., and Shtil, A. A. (2009). Novel boronated chlorin e6-based photosensitizers: Synthesis, binding to albumin and antitumour efficacy. *Bioorg. Med. Chem.* **17,** 1297–1306.

Osaki, T., Hoshino, S., Hoshino, Y., Takagi, S., Okumura, M., Kadosawa, T., and Fujinaga, T. (2006). Clinical pharmacokinetics of anti-angiogenic photodynamic therapy with benzoporphyrin derivative monoacid ring-A in dogs having naturally occurring neoplasms. *J. Vet. Med. A Physiol. Pathol. Clin. Med.* **53,** 108–112.

Pastorino, F., Brignole, C., Marimpietri, D., Cilli, M., Gambini, C., Ribatti, D., Longhi, R., Allen, T. M., Corti, A., and Ponzoni, M. (2003). Vascular damage and anti-angiogenic effects of tumor vessel-targeted liposomal chemotherapy. *Cancer Res.* **63,** 7400–7409.

Pavani, C., Uchoa, A. F., Oliveira, C. S., Iamamoto, Y., and Baptista, M. S. (2009). Effect of zinc insertion and hydrophobicity on the membrane interactions and PDT activity of porphyrin photosensitizers. *Photochem. Photobiol. Sci.* **8,** 233–240.

Shah, G. K., Sang, D. N., and Hughes, M. S. (2009). Verteporfin combination regimens in the treatment of neovascular age-related macular degeneration. *Retina* **29,** 133–148.

Shimizu, K., and Oku, N. (2004). Cancer anti-angiogenic therapy. *Biol. Pharm. Bull.* **27,** 599–605.

Takeuchi, Y., Kurohane, K., Ichikawa, K., Yonezawa, S., Nango, M., and Oku, N. (2003a). Induction of intensive tumor suppression by anti-angiogenic photodynamic therapy using polycation-modified liposomal photosensitizer. *Cancer* **97,** 2027–2034.

Takeuchi, Y., Kurohane, K., Ichikawa, K., Yonezawa, S., Ori, H., Koishi, T., Nango, M., and Oku, N. (2003b). Polycation liposome enhances the endocytic uptake of photosensitizer into cells in the presence of serum. *Bioconjug. Chem.* **14,** 790–796.

Vittar, N. B., Prucca, C. G., Strassert, C., Awruch, J., and Rivarola, V. A. (2008). Cellular inactivation and antitumor efficacy of a new zinc phthalocyanine with potential use in photodynamic therapy. *Int. J. Biochem. Cell Biol.* **40,** 2192–2205.

Yamazaki, Y., Nango, M., Matsuura, M., Hasegawa, Y., Hasegawa, M., and Oku, N. (2000). Polycation liposomes, a novel nonviral gene transfer system, constructed from cetylated polyethylenimine. *Gene Ther.* **7,** 1148–1155.

CHAPTER SEVENTEEN

Controlling the *In Vivo* Activity of Wnt Liposomes

L. Zhao,[1] S. M. Rooker,[1] N. Morrell,[1] P. Leucht,[1] D. Simanovskii,[2] *and* J. A. Helms[1]

Contents

Abstract

Liposomes offer a method of delivering small molecules, nucleic acids, and proteins to sites within the body. Typically, bioactive materials are encapsulated within the liposomal aqueous core and liposomal phase transition is elicited by pH or temperature changes. We developed a new class of liposomes for the *in vivo* delivery of lipid-modified proteins. First, we show that the inclusion of a chromophore into the liposomal or vesosomal membrane renders these lipid vesicles extremely sensitive to very small (μJ) changes in energy. Next, we demonstrate that the lipid-modified Wnt protein is not encapsulated within a liposome but rather is tethered to the exoliposomal surface in an active configuration. When applied to intact skin, chromophore-modified liposomes do not penetrate past the corneal layer of the epidermis, but remain localized to the site of application. Injury to the epidermis allows rapid penetration of liposomes into the dermis, which suggests that mild forms of dermabrasion will greatly enhance transdermal delivery of liposome-packaged molecules. Finally, we demonstrate that topical application of Wnt3a liposomes rapidly stimulates proliferation of cells in the corneal layer, resulting in a thicker, more fibrillous epidermis.

[1] Division of Plastic and Reconstructive Surgery, Department of Surgery, Stanford University School of Medicine, Stanford, California, USA
[2] Hansen Experimental Physics Laboratory (HEPL), Stanford University School of Medicine, Stanford, California, USA

Methods in Enzymology, Volume 465
ISSN 0076-6879, DOI: 10.1016/S0076-6879(09)65017-5

1. Introduction

The physical and chemical properties of liposomes have made them popular vehicles for the delivery of drugs, nucleic acids, and small molecules. Liposomes appear to be particularly well suited for the transdermal delivery of agents to treat dermatologic conditions (de Leeuw *et al.*, 2009). The penetration of most molecules through the epidermis is limited by the corneal layer, which consists of apoptotic keratinocytes organized into keratin-rich, hydrophobic envelopes, and lipid bilayers with hydrophilic regions in between. As a consequence of its hydrophobic nature, the corneal layer allows the preferential penetration of lipid-soluble, hydrophobic molecules over hydrophilic molecules (Bos and Meinardi, 2000). Strongly hydrophobic compounds are hampered from passing through the corneal layer by the hydrophilic regions; consequently, the optimal penetration of molecules through the epidermis requires a careful balance between hydrophobicity and hydrophilicity (reviewed by de Leeuw *et al.*, 2009). Liposomal penetration through the skin is enhanced by dermabrasion, which removes the epidermis but leaves the underlying dermis intact (Friedman and Lippitz, 2009). Even if a helpful drug can pass through the abraded epidermis, the next challenge is to package the reagent in such a way as to avoid degradation in the protease-rich injury site (Toriseva and Kahari, 2009). The packaging method must also not impede the biological activity of the intended drug.

Here, we investigated the transdermal delivery of a lipid-modified protein to reduce scarring and stimulate hair regeneration. We describe the engineering of liposomes containing Wnt3a, then show that treatment with Wnt liposomes stimulates activity in the dermis.

Wnt proteins act as long-range, concentration-dependent morphogens, so-called because of their abilities to specify patterns of cellular responses based on their tissue distribution (Cayuso and Marti, 2005; Logan and Nusse, 2004). Wnt signals are involved in stem cell self-renewal and proliferation (Nusse *et al.*, 2008) and in adult tissue regeneration (Leucht *et al.*, 2008), and thus may be particularly relevant in dermatologic regenerative medicine. For example, Wnt signals are essential for hair (DasGupta and Fuchs, 1999; Narhi *et al.*, 2008) and skin (Jia *et al.*, 2008; Nguyen *et al.*, 2006; Ohtola *et al.*, 2008) development, Wnt signaling is activated after skin wounding (Morrell *et al.*, 2008; Widelitz, 2008), and there is a growing literature demonstrating that Wnt signals are necessary for skin wound repair (Fathke *et al.*, 2006; Ito *et al.*, 2007; Ohtola *et al.*, 2008).

Lipid modification of the Wnt protein is required for *in vivo* activity (Mann and Beachy, 2004; Nusse, 2003; Willert *et al.*, 2003). The fatty acid tail appears to be required for cellular secretion and transport (Kadowaki *et al.*, 1996; van den Heuvel *et al.*, 1993), and may allow the Wnt/Wg protein to remain associated with a lipid bilayer during intercellular signaling (Zhai *et al.*, 2004). Furthermore, this membrane association may facilitate localization of the Wnt protein at its target receptors, allowing the protein to reach a threshold level required for biological activity (Miura *et al.*, 2006). For instance, there is evidence that the *Drosophila* Wnt homolog, Wingless, is transported across long distances in a membrane-bound form (Greco *et al.*, 2001). This theoretical transport mechanism hints at an appealing opportunity: if the endogenous method of lipidated Wnt transport involves association with some sort of lipid raft (Eaton, 2006; Nusse, 2003; Panakova *et al.*, 2005), then perhaps liposomal packaging can be used for the delivery of purified Wnt protein *in vivo*.

Liposomes can form single (unilamellar) or multiple (multilamellar) lipid bilayers that surround an aqueous core. The amphiphathic nature of liposomes allows a hydrophilic molecule to be encapsulated in the aqueous core (de Leeuw *et al.*, 2009). For example, considerable time and energy has been invested in adjusting the physical properties of liposomes, such as lipid bilayer composition, to improve both drug encapsulation and drug release in response to mild hyperthermic conditions (Kakinuma *et al.*, 1996). In an effort to prolong their circulating half-life, "stealth" liposomes have been developed (Cattel *et al.*, 2003). The addition of polyethylene glycol to the exoliposomal surface appears to allow these stealth liposomes to avoid detection by the reticuloendothelial system (Immordino *et al.*, 2006).

We postulated that the amphiphathic arrangement of liposomes could be exploited for the *in vivo* delivery of hydrophobic molecules. We reasoned that the lipid-modified nature of Wnt (Willert *et al.*, 2003) and Hedgehog (Pepinsky *et al.*, 1998) proteins would effectively tether the proteins to the exoliposomal surface of the liposome. This postulated mechanism coupled with dermabrasion, would facilitate the delivery of these molecules to the dermis. While other studies have tracked the distribution of fluorescent molecules after dermabrasion (Watt and Collins, 2008), our study is the first to combine dermabrasion with the liposomal delivery of a lipid-modified protein.

Perhaps the single greatest challenge in using liposomes as drug delivery vehicles is controlling when and where the liposomal "payload" is released. In an effort to restrict the site of biological activity, thermosensitive liposomes and localized heating have been employed. Heating causes the lipid bilayer to become more permeable, and agents contained in the aqueous core are released at the site. Heating can be achieved using a waterbath (Kono, 2001) or lasers (Ebrahim *et al.*, 2005; Kim *et al.*, 2007).

In previous work we demonstrated that laser sources can trigger phase transition of liposomes *in vivo*, without causing harm to cells in the heated area (Kim *et al.*, 2007). In this chapter we describe a method to enhance this phase transition while simultaneously reducing the risk of unintended thermal damage to adjacent cells and tissues. This is accomplished by the addition of chromophore-modified lipids into the liposomal membrane. Due to their lipophilic nature, 1,1′-dioctadecyl-3,3,3′,3′-tetramethylindocarbocyanine perchlorate (DiI) and 1,1′-dioctadecyl-3,3,3′,3′-tetramethyl-indotricarbocyanine iodide (DiR) can be directly incorporated into lipid vesicles (Gullapalli *et al.*, 2008). We reasoned that the addition of chromophores to a liposome would effectively create a larger margin of thermal safety *in vivo*: in effect, the thermal energy delivered to tissues would be preferentially absorbed by the chromophore rather than by the cells. This preferential absorption could then lead to liposome phase transition without inadvertent thermal injury to surrounding tissues. We describe the results of these experiments and our conclusions, which illustrate a unique and precedent approach to the *in vivo* delivery of Wnt and other lipid-modified proteins and their selective activation via external energy sources.

2. Materials, Methods, and Results

Liposomes are prepared using 1,2-dimyristoyl-*sn*-glycero-3-phosphocholine (DMPC) in chloroform or 1,2-dihexadecanoyl-*sn*-glycero-3-phosphocholine (DPPC) in chloroform, both from Avanti Polar Lipids, Inc. (Alabaster, AL). D-Luciferin is obtained from Sigma-Aldrich (St. Louis, MO). Vybrant® DiI cell-labeling solution (1 m*M*) and "DiR" $DiIC_{18}$ (7) are purchased from Invitrogen Molecular Probes (Eugene, OR). Indocyanine green (ICG) (100 mg) is obtained from TCI America (Portland, OR).

Purified mouse Wnt3a protein in 1% CHAPS + 0.5 *M* NaCl in 1× PBS (20 μg/ml) is obtained from R. Nusse (Willert *et al.*, 2003) and is added to liposomes as described previously (Morrell *et al.*, 2008). The lipid-modified Wnt preferentially associates with the lipid fraction during preparation (Morrell *et al.*, 2008). An LSL cell-based reporter assay is then used to determine the biological activity of Wnt3a liposomes. LSL cells (ATCC Global Bioresource Center, Manassas, VA) are stably transfected with a construct containing three copies of the TCF-binding site driving luciferase expression in response to exogenous Wnt signals (Ishitani *et al.*, 1999). LSL cells also constitutively express β-galactosidase. Therefore, relative luciferase activity can be normalized against cell number. Both luciferase and β-galactosidase are measured using the Dual-Light® Combined Reporter Gene Assay System (Applied Biosystems, Foster City, CA) using a luminometer (Kim *et al.*, 2007).

2.1. Manufacture and functional characterization of the phase transition of chromophore-modified liposomes

Thermosensitive liposomes are fabricated in such a way that the phospholipid membrane changes from a gel phase to a liquid-crystalline phase once the critical temperature is reached (Kuo *et al.*, 2000; Needham and

Dewhirst, 2001). While this method allows liposomes to release their content upon heating, it also has a distinct disadvantage: both the liposome and the surrounding tissue are subjected to the same heating. This issue becomes especially important when the liposome drug target (e.g., tumor) is located deep within the body; in these cases the external temperature must be very high in order to generate hyperthermic conditions at the preferred site of drug release. Since many cellular components, including DNA and vital enzymes, are sensitive to elevated temperatures (Kassahn *et al.*, 2009), even mildly hyperthermic conditions required to activate liposomal drug release can be potentially detrimental to cells and tissues.

To avoid these potential complications, we reasoned that the incorporation of chromophores into the liposomal membrane would provide a thermal margin of safety. In effect, the chromophores would act as an "antennae" for thermal energy, selectively absorbing and concentrating thermal energy at the liposomal surface. This selective absorption by the chromophores reduces the amount of heat required to activate liposomal phase transition and spares adjacent cells from thermal damage.

Chromophore-modified liposomes are prepared by adding DiI or DiR at 1.5–2% molar amounts to 14 μmol of DMPC (unless otherwise noted). The chromophore and lipid mixture are dried under nitrogen gas in a 10-ml round bottom flask, wrapped in tin foil to preserve the integrity of the chromophore. After drying, 1.0 ml 1× PBS is added to the flask and the mixture is vortexed vigorously until no film remains on the sides of the flask. The mixture is extruded 30 times through a 100-nm polycarbonate membrane positioned in a thermo-barrel extruder (Avanti Polar Lipids, Inc.) to create the chromophore-labeled liposomes. The liposomes are centrifuged at 14,000 rpm for 30 min at 4 °C, after which the supernatant is removed and the pellet is resuspended in 0.5 ml of 1× DMEM. These chromophore-modified liposomes are centrifuged are protected from light by storage in foil-covered eppendorf tubes at 4 °C.

In a separate series of experiments volumetric liposomes are generated, by encapsulating ICG into the liposomal (aqueous) core. Lipid drying is conducted as described above. After drying, 0.284 μmol of ICG (2% of lipid concentration) is dissolved in 1 ml of 1× PBS and added to the lipid-lined flask. The solution is vortexed vigorously until no lipid is observed on the flask, and the lipid-ICG solution mixture is extruded as described above, followed by centrifugation and resuspension in 0.5 ml of 1× DMEM. Volumetric (ICG) liposomes are stored as described above.

The phase transition of both chromophore-modified liposomes and volumetric liposomes can be induced through several methods of heating. For these experiments we use a titanium:sapphire (Ti:S; Spectra Physics Tsunami) laser source. A similar laser source has been employed by our group to elicit liposomal phase transition *in vivo* (Kim *et al.*, 2007).

After testing a variety of conditions we found that a 1-ns pulse duration, 10 J/cm^2 and adjustable light range from a Ti:S laser was sufficient to activate chromophore-modified and volumetric liposome phase transition. We used an acoustic signal, generated when the liposomes rupture (Wu *et al.*, 2008), as a proxy for identifying the point of liposomal phase transition. The acoustic signal intensity, as determined by hydrophone recording, increased in proportion to the energy of the laser (Fig. 17.1A). DiR liposomes exhibited the highest acoustic signal whereas volumetric ICG liposomes produced the lowest acoustic signal when exposed to 350 μJ of energy from a Ti:S laser (Fig. 17.1B).

For all three types of chromophore-modified liposomes, we demonstrated that the force of liposomal rupture increased as the energy of the laser increased. ICG and DiR/DiI chromophores are excited at different wavelengths; therefore not all chromophore-modified liposomes responded to a specified laser intensity in the same manner. For example, DiR-modified liposomes showed the largest increase in rupture intensity relative to baseline (Fig. 17.1B) when compared to ICG- and DiI-modified liposomes (Fig. 17.1B). Thus, our data demonstrated that DiR-modified liposomes were more sensitive to laser heating than other forms of chromophore-modified liposomes. These results also raised the possibility that combinations of chromophore-modified liposomes might be useful for sustained and/or sequential drug delivery: by varying the intensity of an external energy source, the selective rupture of one class of chromophore-modified liposomes might be achieved while the integrity of another class of liposomes is maintained.

2.2. Using vesosomes as delivery vehicles for liposomes

The inclusion of chromophores into a liposomal membrane is a useful method to enhance phase transition and the release of molecules from the aqueous liposomal core. For hydrophobic molecules that are tethered to the liposomal membrane, however, activity is not dependent on phase transition (Morrell *et al.*, 2008). Rather, molecules positioned on the exoliposomal surface are already accessible to cells in the environment (Morrell *et al.*, 2008). Therefore, liposomes in which the "payload" is already positioned in an active configuration (i.e., on the external surface of the liposome) must be contained within some other vesicle first. In a second step, their activity can be triggered by release from this vesicle. A vesosome is a larger version of a liposome that has room in its aqueous core to harbor liposomes (Kisak *et al.*, 2004). We tested the idea that Wnt liposomes could be packaged into chromophore-modified vesosomes and that these vesosomes could be induced to undergo a phase transition and release the active Wnt liposomes.

As a proof-of-principle, we first generated luciferin-containing liposomes (Kim *et al.*, 2007) and then packaged these liposomes into vesosomes.

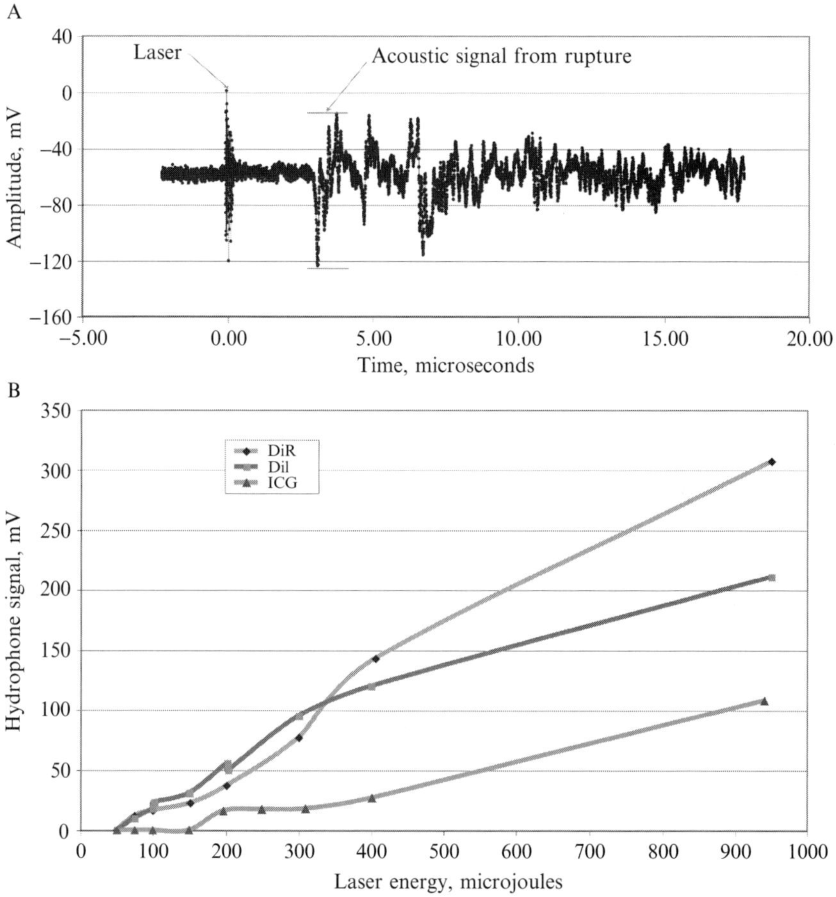

Figure 17.1 Liposome membrane phase transition is induced by laser heating. (A) The acoustic signal from the hydrophone measures pressure fluctuations associated with liposomal membrane phase transition during laser heating. The laser pulse is seen at time = 0.00, and the phase transition is measured microseconds later with a sudden increase in amplitude of the acoustic signal caused by liposomal membrane disruption. (B) The amplitude of the hydrophone signal at the time of liposomal membrane phase transition is compared with the energy of the laser for liposomes modified with DiR, DiI, and ICG chromophores. In all groups, the intensity of the acoustic signal increases as the energy of the laser is increased. DiR-modified liposomes show the greatest change in acoustic signal amplitude as laser energy increases and ICG volumetric liposomes show the least change as laser energy increases.

By encapsulating luciferin in the core of the liposome, we could follow the phase transition and stability of the liposome/vesosome combination.

Liposomes that contain a 0.3% solution of D-luciferin are generated as described (Kim *et al.*, 2007). After preparation, free luciferin is removed by dialysis in 1× PBS overnight. The luciferin liposomes are suspended in 1× DMEM, stored at 4 °C, and remain stable for up to 3 weeks (data not shown). Vesosomes are made by drying 15 μmol of DPPC in chloroform with nitrogen gas and then redissolving the lipids in 0.5 ml of HEPES buffered saline solution; ethanol is added to a final concentration of 3 *M*. The solution is allowed to sit for 2 h and is then transferred to a 15-ml Falcon tube containing 3 ml 1× DMEM. This mixture is centrifuged three times at 3000 rpm for 5 min. Excess ethanol, which is contained in the supernatant, is removed and 185 μl of luciferin liposomes are added, following which, the solution is briefly vortexed. The mixture is then heated to 46 °C for 20 min under agitation and centrifuged at 2500 rpm for 5 min. The supernatant is removed and the remaining pellet is resuspended in 0.5 ml 1× DMEM. As before, chromophore-modified liposomes and vesosomes are protected from light.

We tested the ability of luciferin liposomes to undergo a selective phase transition in response to external energy (Fig. 17.2). For all of these experiments, an 805-nm laser with a 1-ns exposure is used, generating $\sim 10^{-13}$ μJ. As expected, luciferin liposomes that lacked a chromophore did not undergo phase transition in response to laser exposure: we found no significant differences in luminescence between luciferin liposomes exposed to the laser and those that were not (p-value $= 0.05$; Fig. 17.2A).

The inclusion of a chromophore in the liposomal membrane, however, rapidly and reliably induced a liposomal phase transition. When exposed to $\sim 10^{-13}$ μJ, the DiR-modified liposomes showed an increase in luminescence, compared to unheated DiR-modified luciferin liposomes (p-value $= 0.008$; Fig. 17.2B). These results demonstrate that the inclusion of a chromophore into a liposomal membrane dramatically enhances the phase transition in response to very small doses of energy.

The stability of the vesosomes is tested by encapsulating luciferin-containing liposomes in the vesosomal core. In this configuration, the stability of both the vesosomal and liposomal membranes are tested. We found no significant increases in luminescence of these vesosomes compared to unheated controls (p-value $= 0.04$; Fig. 17.2C). These data confirm that our vesosomal preparation provided a stable method for packaging liposomes.

In the third series of experiments, the permeability of the vesosomal membrane alone is tested. To do this, DiR-modified luciferin liposomes are packaged into vesosomes whose membranes lacked any chromophores, and as before, the vesosomes are exposed to $\sim 10^{-13}$ μJ of laser energy. Compared to unheated control, the heated samples showed only a slight increase in luminescence (p-value $= 0.02$; Fig. 17.2D). These data show that only chromophore-modified liposomes underwent a phase transition in response to $\sim 10^{-13}$ μJ; there was, however, a small amount of leakage of the hydrophilic luciferin out of the vesosomal membrane.

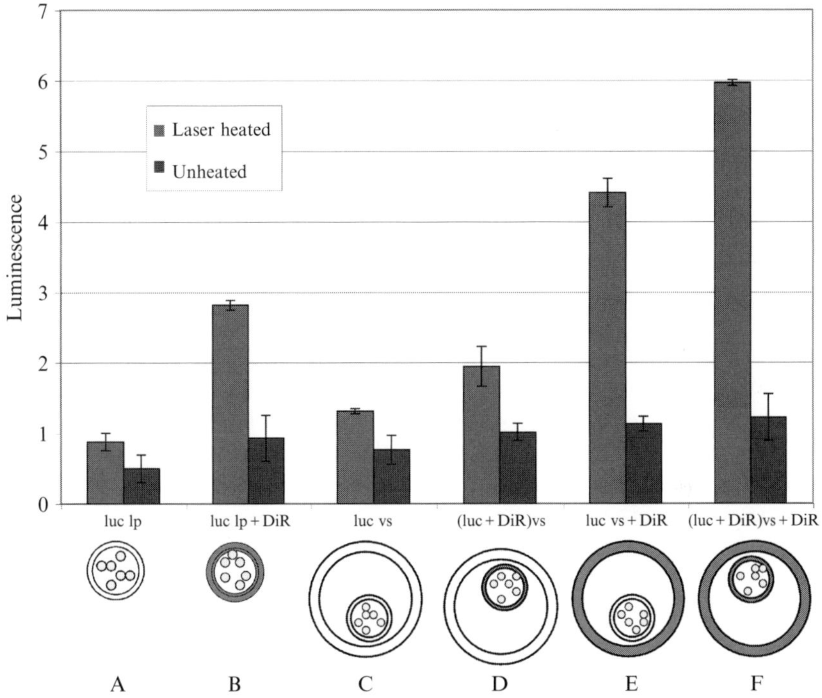

Figure 17.2 The inclusion of a chromophore in the membrane allows targeted activation of liposomal and vesosomal phase transitions. 4T1 cells, which are stably transfected with the firefly luciferase gene (Thorne *et al.*, 2006), were maintained near confluence; various compositions of liposomes and vesosomes were added to the cells, following which the cells and liposome/vesosomes were exposed to an 805-nm laser for 1 ns. Luciferase activity was measured by a luminometer. (A) In the absence of a chromophore, heating of luciferin-containing liposomes did not elicit a significant change in luciferase activity. (B) In the presence of the DiR chromophore, heated luciferin liposomes exhibit significantly greater luminescence than unheated luciferin-containing liposomes. (C) In the absence of a chromophore, heating of vesosomes containing luciferin liposomes produced a modest change in luciferase activity relative to unheated controls. (D) Inclusion of the DiR chromophore in the liposomal surface resulted in moderately higher luminescence compared to unheated controls. (E,F) The inclusion of a DiR chromophore in the vesosomal membrane resulted in higher luminescence compared to unheated controls. Including DiR chromophores in both liposomal and vesosomal membranes resulted in the largest increase in luminescence, compared to unheated controls. Luc lp, luciferin liposomes; luc lp + DiR, luciferin liposomes with DiR; luc vs, luciferin vesosomes; (luc + DiR)vs, luciferin liposomes with DiR in vesosomes; luc vs + DiR, luciferin liposomes in vesosomes with DiR; and (luc + DiR) vs + DiR, luciferin liposomes with DiR in vesosomes with DiR. Error bars indicate standard deviation. (See Color Insert.)

In the final set of experiments, the ability of laser heating to trigger the release of liposomes from the vesosomal core is tested. As before, luciferin liposomes are placed into the vesosomal core. In one case (Fig. 17.2E), the vesosomal membrane alone contains a chromophore; in the other case, both the liposomal and vesosomal membranes contain chromophores (Fig. 17.2F). In both cases we found that $\sim 10^{-13}$ μJ was sufficient to induce a phase transition, which was reported by an increase in luminescence after laser heating (p-value $= 0.0001$, Fig. 17.2E; p-value $= 0.001$, Fig. 17.2F). In the latter case, the inclusion of a chromophore into both membranes enhanced the increase in luminescence after exposure to the laser.

Our experiments with DiR-modified vesosomes demonstrate that vesosomes can be manufactured in such a way as to encapsulate liposomes. The inclusion of a chromophore and brief exposure to laser heating can selectively trigger the release of these liposomes from their vesosomal compartment. This method of release can be used for both hydrophilic and hydrophobic liposomal molecules, since hydrophilic molecules can be contained in the aqueous core of the vesosome and hydrophobic molecules can be incorporated into the liposomal membrane. We envision such a method could be employed for the simultaneous delivery of a hydrophilic Wnt agonist and the hydrophobic Wnt protein.

In the next series of experiments, the extent to which heat treatment affects Wnt liposome activity is tested. Wnt liposomes are fabricated as described previously (Morrell *et al.*, 2008) and then heated for 1 min in a 41 °C incubator. This duration and extent of heating produced no significant change in Wnt activity (Fig. 17.3). Thus, the activity of Wnt liposomes does not depend upon liposomes undergoing a phase transition. These data confirm that, rather than being contained in the aqueous core, Wnt proteins are positioned on the external liposomal surface.

2.3. DiI liposome penetration and Wnt3a liposome *in vivo* activity experiment

In the next series of experiments, how the penetration of liposomes into intact and abraded skin is investigated. For these purposes we generated DiI-modified liposomes as described above. The skin injury model consists of a 4-mm diameter injury; the edges of the abraded skin are prevented from retracting by the placement of an O-ring sutured to the intact adjacent epidermis. DiI-modified liposomes are prepared as described above. Between 1 and 2 μl of DiI-modified liposomes are topically administered immediately after wounding, and the treated skin is collected 20 h postsurgery. Shaved, intact skin is treated with 1–2 μL of DiI-modified liposomes served as a control. Hoechst 33342 stain is added to visualize cell nuclei, and the liposomes are viewed under fluorescent light (excitation wavelength 360-340 nm).

We discovered that when applied to intact skin, DiI-modified liposomes remained concentrated at the epidermis (Fig. 17.4A). When the tissue was viewed using differential interference contrast (DIC) imaging, the extent of

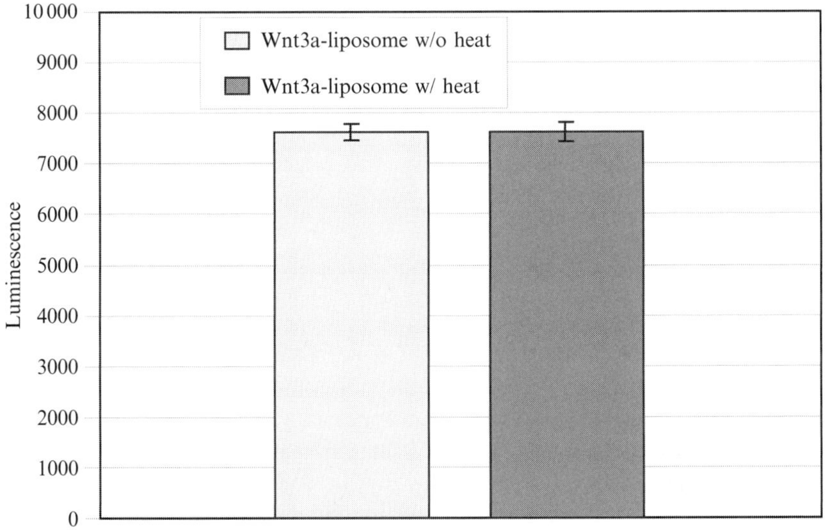

Figure 17.3 The activity of Wnt liposomes is not dependent upon phase transition. LSL cells were treated with Wnt liposomes, and then subjected to heating; controls were maintained at 37 °C, below the critical temperature for phase transition. Heated and unheated Wnt liposomes exhibited similar levels of activity as measured by luminescence. These data demonstrate that the Wnt protein is maintained in an active configuration that does not require liposomal phase transition, and that elevated temperatures do not decrease the activity of the protein. Error bars indicate standard deviation. (See Color Insert.)

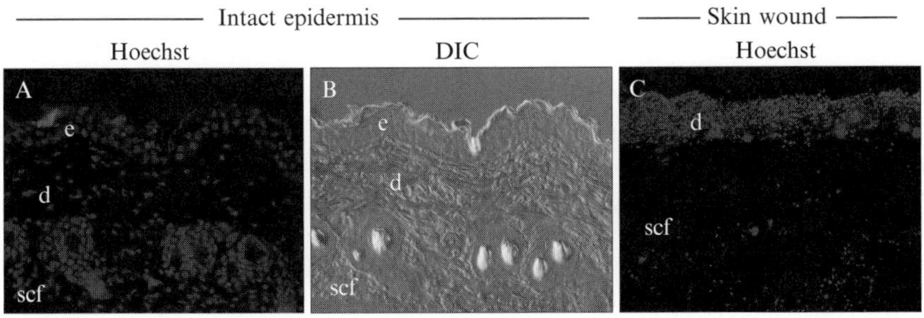

Figure 17.4 Liposomal penetration into skin can be monitored with the inclusion of a chromophore in the membrane. Intact skin was treated with DiI-modified liposomes, and then visualized under appropriate conditions. (A) Topically applied DiI liposomes (red) do not penetrate through the corneal layer of the epidermis; rather, they accumulate within the epidermis and are maintained at the site of application for at least 20 h. (B) Histology of the skin including the epidermis, dermis and subcutaneous fat shown by DIC microscopy. (C) After injury, DiI liposomes (red) rapidly penetrated into the dermis and the subcutaneous fat layer. e, epidermis; d, dermis; scf, subcutaneous fat. (See Color Insert.)

penetration could be confirmed as extending no further than the corneal layer (Fig. 17.4B). This finding is in keeping with published reports (Bos and Meinardi, 2000). When the skin was wounded, however, DiI-modified liposomes readily penetrated into the dermis and subcutaneous fat (Fig. 17.4C). The liposomes were still detectable within the wound site 20 h after initial delivery. These data demonstrate that the corneal layer of the epidermis is a barrier to liposome penetration, and that wounding the skin permits liposome penetration as deep as the subcutaneous fat layer, even if only a single application is used.

In the last series of experiments, the effects of Wnt liposomes on intact skin are tested. Wnt liposomes are prepared as previously described (Morrell *et al.*, 2008). All liposome preparation materials, including flasks, syringes, and extrusion apparatus are thoroughly washed in concentrated acid detergent, 70% ethanol and deionized water. Briefly, 14 μmol of DMPC are dried in chloroform with nitrogen gas in a 10-ml round bottom flask, and the lipid film is dried overnight in vacuum. PBS is used to dilute stock Wnt3a protein to 1.3 μg/ml. One milliliter of 1.3 μg/ml Wnt (or 1× PBS for PBS liposomes) is added to the lipid-lined flask. The lipid-lined flask is vortexed vigorously until the solution is cloudy, and no lipid remains on the sides of the flask. The cloudy protein-lipid solution is extruded 30 times through a 100-nm polycarbonate membrane in a thermo-barrel extruder (Avanti Polar Lipids, Inc.) held at 30–32 °C. After extrusion, the liposomes are centrifuged at 14,000 rpm for 30 min at 4 °C to pellet the Wnt3a liposomes from free Wnt3a protein in solution. The supernatant containing free Wnt3a protein is aspirated and the Wnt3a liposome pellet is resuspended in 0.5 ml 1× DMEM. The liposomes are stored at 4 °C when not in use.

To quantify Wnt liposome activity, a LSL cell-based luciferase assay is conducted as previously described (Morrell *et al.*, 2008). Mouse LSL cells (25,000 cells/well) are plated in 1× DMEM, 10% FBS, 1% penicillin/streptomycin for overnight incubation at 37 °C and 5% CO_2. The cells are then treated with varying volumes (5, 10, and 15 μl, each done in triplicate) of purified Wnt3a (3.25 μg/ml) and Wnt3a liposomes. After treatment, the cells are incubated overnight again. PBS liposomes and 1× DMEM treatments are used as baseline controls, and the well volume does not exceed 150 μl. Wnt3a activity is measured by the Dual-Light® Combined Reporter Gene Assay System and relative luciferase units are normalized over β-galactosidase activity.

Axin2LacZ/+ reporter mice (Jho *et al.*, 2002) are used to visualize Wnt signaling *in vivo*. In these reporter mice, Wnt signaling leads to the activation of LacZ, which subsequently results in β-galactosidase production. Xgal staining can then be used to visualize this enzyme production.

In our experiments, Wnt and PBS liposomes are topically administered to shaved intact skin on the back of Axin2LacZ/+ mice. Tissues are collected after 5 days; the collected tissue is fixed in 0.4% paraformaldehyde overnight before being infused with 30% sucrose for 24 h. We then embed the samples in OCT medium and cryosection (12 μm) on Superfrost-plus slides (Fisher Scientific, Pittsburgh, PA). Xgal staining is performed as previously described (Brugmann *et al.*, 2007).

We evaluated the baseline level of Wnt signaling in intact skin. Untreated, intact skin from Axin2LacZ/+ mice exhibited Xgal staining around the hair follicles (data not shown). Intact Axin2LacZ/+ skin treated with PBS liposomes showed the same intensity and pattern of Xgal staining around the hair follicles (Fig. 17.5A). Intact Axin2LacZ/+ skin treated with Wnt liposomes showed a robust increase in Xgal staining in the epidermis (Fig. 17.5B). This increase in Wnt signaling was only detectable in the epidermis, coincident with the location of liposome congregation in the epidermis as visualized by DiI liposomes (Fig. 17.4). Together, these data demonstrate that a single application of Wnt liposomes to intact skin dramatically increases Wnt signaling in the epidermis. Liposomes do not penetrate further than the corneal layer, indicating that the extent of Wnt signal amplification can be effectively limited to the outer layers of skin.

We evaluated the physiological response to enhanced Wnt signaling in the epidermis. Intact skin treated with PBS liposomes showed a normal organization in the epidermis (Fig. 17.5C). In contrast, the epidermal layer of liposomal Wnt-treated skin was substantially thicker (Fig. 17.5D). This thicker epidermal layer also exhibited greater cell proliferation (Fig. 17.5F) compared to PBS liposome treated skin (Fig. 17.5E). Finally, Wnt liposome treated skin displayed substantially more fibrillous matrix than skin treated with PBS liposomes (Fig. 17.5G and H).

3. Concluding Remarks

The therapeutic potential of Wnt proteins for the treatment of dermatologic conditions is enormous. A growing literature implicates Wnt signaling in the development and homeostasis of skin and hair (reviewed by Watt and Collins, 2008), and in the role of Wnt signaling in skin wound healing (Fathke *et al.*, 2006; Ito *et al.*, 2007). Purified Wnt protein, however, does not show sufficient biological activity to induce physiological changes *in vivo*. The incorporation of Wnt protein into liposomes provides a biomimetic method for exogenous Wnt delivery that results in enhanced Wnt responsiveness in tissues, and a subsequent physiological response in keeping with the role of Wnt signaling in epidermal stem cell self-renewal and proliferation (Jia *et al.*, 2008; Watt and Collins, 2008).

The development of a Wnt liposome delivery system with robust *in vivo* activity is a critical step toward actualizing Wnts as therapeutic agents.

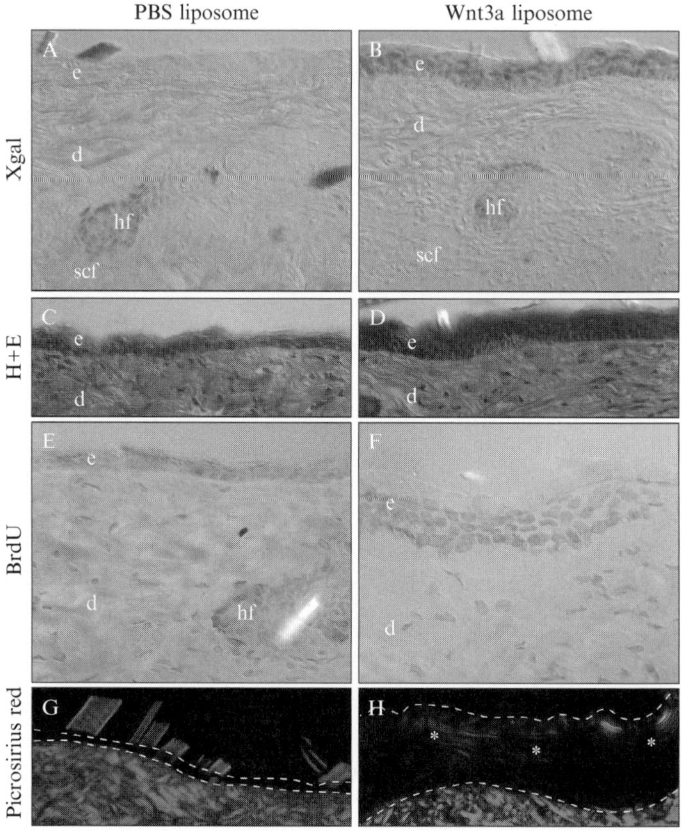

Figure 17.5 Topical application of Wnt liposomes to intact epidermis elicits a biological response, stimulates cell proliferation, and enhances birefringent fibril composition in the epidermis. Equal volumes of PBS or Wnt liposomes were topically applied onto the intact skin of Axin2LacZ/+ reporter mice. (A) On postapplication day 5, baseline Xgal staining was detectable around the hair follicles; no obvious Xgal staining was detectable in the epidermis. This pattern of Xgal staining was indistinguishable from untreated control skin. (B) In contrast, treatment with Wnt liposomes elicited robust Xgal staining in the epidermis; Xgal staining in the deeper hair follicles was unchanged from (A). (C) Histological tissue sections reveal the thin, flattened epidermis present in PBS-treated skin. (D) Treatment of intact skin with Wnt liposomes elicits a dramatic thickening of the epidermis. (E) BrdU labeling indicates proliferation of cells around the hair follicle, and minimal proliferation in the corneal layer of the epidermis. (F) In contrast, liposomal Wnt treatment of intact skin resulted in substantially more BrdU-positive cells in the epidermis. Cell proliferation around the hair follicle was unchanged. (G) When visualized under polarized light, picrosirius red staining highlights birefringent fibrils; note the conspicuous lack of picrosirius red staining in the mature, thin epidermis. (H) Liposomal Wnt treatment produced a thicker epidermis, and picrosirius red staining indicated a greater amount of fibrillous matrix within that thicker epidermis (asterisks). (See Color Insert.)

We also show that the addition of chromophores to the liposomal and vesosomal surfaces offers a novel and highly sensitive method for inducing the phase transition of liposomes. The employment of a laser to induce heating is also an innovative advancement in drug delivery because heating can be produced with precise spatiotemporal control, and with only brief (ns) exposure times. This selective heating method significantly improves the thermal margin of safety and allows targeted liposomal activation without causing undue disruption to cells in the target vicinity. Future experiments will be directed toward optimizing the stability and controlling the activity of Wnt liposomes *in vivo*.

REFERENCES

Bos, J. D., and Meinardi, M. M. (2000). The 500 Dalton rule for the skin penetration of chemical compounds and drugs. *Exp. Dermatol.* **9,** 165–169.

Brugmann, S. A., Goodnough, L. H., Gregorieff, A., Leucht, P., ten Berge, D., Fuerer, C., Clevers, H., Nusse, R., and Helms, J. A. (2007). Wnt signaling mediates regional specification in the vertebrate face. *Development* **134,** 3283–3295.

Cattel, L., Ceruti, M., and Dosio, F. (2003). From conventional to stealth liposomes: A new frontier in cancer chemotherapy. *Tumori* **89,** 237–249.

Cayuso, J., and Marti, E. (2005). Morphogens in motion: Growth control of the neural tube. *J. Neurobiol.* **64,** 376–387.

DasGupta, R., and Fuchs, E. (1999). Multiple roles for activated LEF/TCF transcription complexes during hair follicle development and differentiation. *Development* **126,** 4557–4568.

de Leeuw, J., de Vijlder, H. C., Bjerring, P., and Neumann, H. A. (2009). Liposomes in dermatology today. *J. Eur. Acad. Dermatol. Venereol.* **23,** 505–516.

Eaton, S. (2006). Release and trafficking of lipid-linked morphogens. *Curr. Opin. Genet. Dev.* **16,** 17–22.

Ebrahim, S., Peyman, G. A., and Lee, P. J. (2005). Applications of liposomes in ophthalmology. *Surv. Ophthalmol.* **50,** 167–182.

Fathke, C., Wilson, L., Shah, K., Kim, B., Hocking, A., Moon, R., and Isik, F. (2006). Wnt signaling induces epithelial differentiation during cutaneous wound healing. *BMC Cell Biol.* **7,** 4.

Friedman, S., and Lippitz, J. (2009). Chemical peels, dermabrasion, and laser therapy. *Dis. Mon.* **55,** 223–235.

Greco, V., Hannus, M., and Eaton, S. (2001). Argosomes: A potential vehicle for the spread of morphogens through epithelia. *Cell* **106,** 633–645.

Gullapalli, R. R., Demirel, M. C., and Butler, P. J. (2008). Molecular dynamics simulations of DiI-C18(3) in a DPPC lipid bilayer. *Phys. Chem. Chem. Phys.* **10,** 3548–3560.

Immordino, M. L., Dosio, F., and Cattel, L. (2006). Stealth liposomes: Review of the basic science, rationale, and clinical applications, existing and potential. *Int. J. Nanomedicine* **1,** 297–315.

Ishitani, T., Ninomiya-Tsuji, J., Nagai, S., Nishita, M., Meneghini, M., Barker, N., Waterman, M., Bowerman, B., Clevers, H., and Shibuya, H. (1999). The TAK1-NLK-MAPK-related pathway antagonizes signalling between beta-catenin and transcription factor TCF. *Nature* **399,** 798–802.

Ito, M., Yang, Z., Andl, T., Cui, C., Kim, N., Millar, S. E., and Cotsarelis, G. (2007). Wnt-dependent *de novo* hair follicle regeneration in adult mouse skin after wounding. *Nature* **447,** 316–320.

Jho, E. H., Zhang, T., Domon, C., Joo, C. K., Freund, J. N., and Costantini, F. (2002). Wnt/beta-catenin/Tcf signaling induces the transcription of Axin2, a negative regulator of the signaling pathway. *Mol. Cell. Biol.* **22,** 1172–1183.

Jia, L., Zhou, J., Peng, S., Li, J., Cao, Y., and Duan, E. (2008). Effects of Wnt3a on proliferation and differentiation of human epidermal stem cells. *Biochem. Biophys. Res. Commun.* **368,** 483–488.

Kadowaki, T., Wilder, E., Klingensmith, J., Zachary, K., and Perrimon, N. (1996). The segment polarity gene porcupine encodes a putative multitransmembrane protein involved in Wingless processing. *Genes Dev.* **10,** 3116–3128.

Kakinuma, K., Tanaka, R., Takahashi, H., Sekihara, Y., Watanabe, M., and Kuroki, M. (1996). Drug delivery to the brain using thermosensitive liposome and local hyperthermia. *Int. J. Hyperthermia* **12,** 157–165.

Kassahn, K. S., Crozier, R. H., Portner, H. O., and Caley, M. J. (2009). Animal performance and stress: Responses and tolerance limits at different levels of biological organisation. *Biol. Rev. Camb. Philos. Soc.* **84,** 277–292.

Kim, J. B., Leucht, P., Morrell, N. T., Schwettman, H. A., and Helms, J. A. (2007). Visualizing *in vivo* liposomal drug delivery in real-time. *J. Drug Target.* **15,** 632–639.

Kisak, E. T., Coldren, B., Evans, C. A., Boyer, C., and Zasadzinski, J. A. (2004). The vesosome—A multicompartment drug delivery vehicle. *Curr. Med. Chem.* **11,** 199–219.

Kono, K. (2001). Thermosensitive polymer-modified liposomes. *Adv. Drug Deliv. Rev.* **53,** 307–319.

Kuo, Y. H., Lin, C. H., Hwang, S. Y., Shen, Y. C., Lee, Y. L., and Shyh-Yuan, L. (2000). A novel cytotoxic C-methylated biflavone from the stem of *Cephalotaxus wilsoniana*. *Chem. Pharm. Bull. (Tokyo)* **48,** 440–441.

Leucht, P., Minear, S., Ten Berge, D., Nusse, R., and Helms, J. A. (2008). Translating insights from development into regenerative medicine: The function of Wnts in bone biology. *Semin. Cell Dev. Biol.* **19**(5)**,** 434–443.

Logan, C. Y., and Nusse, R. (2004). The Wnt signaling pathway in development and disease. *Annu. Rev. Cell Dev. Biol.* **20,** 781–810.

Mann, R. K., and Beachy, P. A. (2004). Novel lipid modifications of secreted protein signals. *Annu. Rev. Biochem.* **73,** 891–923.

Miura, G. I., Buglino, J., Alvarado, D., Lemmon, M. A., Resh, M. D., and Treisman, J. E. (2006). Palmitoylation of the EGFR ligand Spitz by Rasp increases Spitz activity by restricting its diffusion. *Dev. Cell* **10,** 167–176.

Morrell, N. T., Leucht, P., Zhao, L., Kim, J. B., ten Berge, D., Ponnusamy, K., Carre, A. L., Dudek, H., Zachlederova, M., McElhaney, M., Brunton, S., Gunzner, J., *et al.* (2008). Liposomal packaging generates Wnt protein with *in vivo* biological activity. *PLoS ONE* **3,** e2930.

Narhi, K., Jarvinen, E., Birchmeier, W., Taketo, M. M., Mikkola, M. L., and Thesleff, I. (2008). Sustained epithelial beta-catenin activity induces precocious hair development but disrupts hair follicle down-growth and hair shaft formation. *Development* **135,** 1019–1028.

Needham, D., and Dewhirst, M. W. (2001). The development and testing of a new temperature-sensitive drug delivery system for the treatment of solid tumors. *Adv. Drug Deliv. Rev.* **53,** 285–305.

Nguyen, H., Rendl, M., and Fuchs, E. (2006). Tcf3 governs stem cell features and represses cell fate determination in skin. *Cell* **127,** 171–183.

Nusse, R. (2003). Wnts and Hedgehogs: Lipid-modified proteins and similarities in signaling mechanisms at the cell surface. *Development* **130,** 5297–5305.

Nusse, R., Fuerer, C., Ching, W., Harnish, K., Logan, C., Zeng, A., ten Berge, D., and Kalani, Y. (2008). Wnt signaling and stem cell control. *Cold Spring Harb. Symp. Quant. Biol.* **73,** 59–66.

Ohtola, J., Myers, J., Akhtar-Zaidi, B., Zuzindlak, D., Sandesara, P., Yeh, K., Mackem, S., and Atit, R. (2008). β-Catenin has sequential roles in the survival and specification of ventral dermis. *Development* **135,** 2321–2329.

Panakova, D., Sprong, H., Marois, E., Thiele, C., and Eaton, S. (2005). Lipoprotein particles are required for Hedgehog and Wingless signalling. *Nature* **435,** 58–65.

Pepinsky, R. B., Zeng, C., Wen, D., Rayhorn, P., Baker, D. P., Williams, K. P., Bixler, S. A., Ambrose, C. M., Garber, E. A., Miatkowski, K., *et al.* (1998). Identification of a palmitic acid-modified form of human Sonic hedgehog. *J. Biol. Chem.* **273,** 14037–14045.

Thorne, S. H., Negrin, R. S., and Contag, C. H. (2006). Synergistic antitumor effects of immune cell-viral biotherapy. *Science* **311,** 1780–1784.

Toriseva, M., and Kahari, V. M. (2009). Proteinases in cutaneous wound healing. *Cell. Mol. Life Sci.* **66,** 203–224.

van den Heuvel, M., Harryman-Samos, C., Klingensmith, J., Perrimon, N., and Nusse, R. (1993). Mutations in the segment polarity genes wingless and porcupine impair secretion of the wingless protein. *EMBO J.* **12,** 5293–5302.

Watt, F. M., and Collins, C. A. (2008). Role of beta-catenin in epidermal stem cell expansion, lineage selection, and cancer. *Cold Spring Harb. Symp. Quant. Biol.* **73,** 503–512.

Widelitz, R. B. (2008). Wnt signaling in skin organogenesis. *Organogenesis* **4,** 123–133.

Willert, K., Brown, J. D., Danenberg, E., Duncan, A. W., Weissman, I. L., Reya, T., Yates, J. R. III., and Nusse, R. (2003). Wnt proteins are lipid-modified and can act as stem cell growth factors. *Nature* **423,** 448–452.

Wu, G., Mikhailovsky, A., Khant, H. A., Fu, C., Chiu, W., and Zasadzinski, J. A. (2008). Remotely triggered liposome release by near-infrared light absorption via hollow gold nanoshells. *J. Am. Chem. Soc.* **130,** 8175–8177.

Zhai, L., Chaturvedi, D., and Cumberledge, S. (2004). Drosophila wnt-1 undergoes a hydrophobic modification and is targeted to lipid rafts, a process that requires porcupine. *J. Biol. Chem.* **279,** 33220–33227.

CHAPTER EIGHTEEN

Convection-Enhanced Delivery of Liposomes to Primate Brain

Michal T. Krauze, John Forsayeth, Dali Yin, *and* Krystof S. Bankiewicz

Contents

Abstract

Direct delivery of therapeutic agents to the human central nervous system remains an inadequately studied field. Our group has extensively studied and refined a powerful method for distributing various macromolecules and nanoparticles into the parenchyma by means of a procedure called convection-enhanced delivery (CED). First, we developed an improved design of infusion cannula that greatly decreased the likelihood of reflux of infusate up the outside of the cannula. Second, we began to use liposomes loaded with the MRI contrast reagent, Gadoteridol (Gd), to track infusions into brain parenchyma in real time. This innovation generated a wealth of quantitative and qualitative data that in turn drove further improvements in CED. In this chapter, we review many of the recently devised methods needed to ensure controlled distribution of therapeutic agents in the brain.

Department of Neurological Surgery, University of California San Francisco, San Francisco, California, USA

Methods in Enzymology, Volume 465
ISSN 0076-6879, DOI: 10.1016/S0076-6879(09)65018-7

1. Introduction

Drug delivery to human central nervous system (CNS) has long posed a major challenge. Despite recent advances in pharmacology, the blood–brain barrier (BBB) remains the limiting factor in delivery of therapeutics to the CNS. Less than 1% of systemically administered drugs reach the brain, regardless of the physical and chemical properties of the drug, and regardless of manipulations to increase capillary permeability. Various other methods have been employed to increase CNS uptake of intravenously applied agents with either poor outcomes or inability to translate into *in vivo* applications (Hsieh *et al.*, 2005; Weyerbrock *et al.*, 2003; Zhang and Pardridge, 2005). Introduction of convection-enhanced delivery (CED) as new way of bypassing the BBB improved direct delivery to CNS drastically (Bobo *et al.*, 1994). CED uses a pressure gradient established at the tip of an infusion catheter to initiate bulk flow that forces infusate through the space between brain cells (i.e., the extracellular space). The pressurized infusate then engages the perivascular space, and distribution is significantly aided by the pulsation of blood vessels. We have shown previously that pulse pressure appears to be the main driver of CED (Hadaczek *et al.*, 2006). As a result, infusate is distributed evenly, and at higher concentrations over a larger area, than in the absence of CED, that is, by diffusion alone. Initial experimental applications for CED were limited by extremely long delivery times, reflux along the outside of the cannula and variable distribution within various parts of the CNS (Krauze *et al.*, 2005a; Laske *et al.*, 1997; Lonser *et al.*, 2002). Development of a reflux-free stepped cannula has almost entirely eliminated the risk of backflow and consequently also improved infusion rates (Krauze *et al.*, 2005a). Another advance in CED was real-time imaging by MRI during infusion (Saito *et al.*, 2005) in which Gadoteridol-loaded liposomes (GDL) were included in the infusate. This approach allowed us to monitor infusion of liposomal drugs into brain tumors (Dickinson *et al.*, 2008) and viral gene therapy vectors into parenchyma (Fiandaca *et al.*, 2008a). Moreover, because infusions could be visualized, we were able to define quantitative relationships between infusate volume (Vi) and subsequent distribution (Vd) for both white and gray matter (Krauze *et al.*, 2005b). Initial safety studies suggest that the infusion of GDL into the primate brain is safe and may be considered for clinical application (Krauze *et al.*, 2008). We envisage the clinical use of this approach in brain diseases where efficacious agents have been identified but suffer major disadvantages in terms of brain penetration or poor pharmacokinetics. We have, for example, shown that intratumoral CED of a liposomal formulation of CPT-11 resulted in a dramatic increase in tissue half-life, considerably reduced systemic toxicity, and more profound antitumor efficacy than seen with free CPT-11 (Krauze *et al.*, 2007). Although continued refinements in CED may be expected, our work over the past decade has resulted in a new paradigm for direct parenchymal delivery that may find increasing application in the treatment of currently intractable diseases like Alzheimer's and Parkinson's disease, brain tumors and movement disorders. Here we describe in detail our methods that are essential to achieve optimal CED, especially with respect to the primate CNS.

2. Liposome Preparation

Separate liposomes are prepared for detection either by MRI or by histology. Liposomes containing the MRI contrast agent were composed of 1,2-dioleoyl-*sn*-glycero-3-phosphocholine (DOPC)/cholesterol/1,2-distearoyl-*sn*-glycero-3-[methoxy(polyethylene glycol)-2000] (PEG-DSG) with a molar ratio of 3:2:0.3. DOPC is purchased from Avanti Polar Lipids (Alabaster, AL), PEG-DSG from NOF Corporation (Tokyo, Japan), and cholesterol from Calbiochem (San Diego, CA). The lipids are dissolved in chloroform/methanol (90:10, v/v), and the solvent is removed by rotary evaporation, resulting in a thin lipid film. The lipid film is dissolved in ethanol and heated to 60 °C. A commercial United States Pharmacopeia solution of 0.5 *M* Gadoteridol (10-(2-hydroxy-propyl)-1,4,7,10-tetraazacyclododecane-1,4,7-triacetic acid) (Prohance; Bracco Diagnostics, Princeton, NJ) is heated to 60 °C and injected rapidly into the ethanol/lipid solution. Unilamellar liposomes are formed by extrusion (Lipex; Northern Lipids, Vancouver, Canada) by 15 passes through double-stacked polycarbonate membranes (Whatman Nucleopore, Clifton, NJ) with a pore size of 100 nm, resulting in a liposome diameter of 124 ± 24.4 nm as determined by quasi-elastic light scattering (N4Plus particle size analyzer; Beckman Coulter, Fullerton, LA). Unencapsulated Gadoteridol is removed with a Sephadex G-75 (Sigma, St. Louis, MO) size-exclusion column eluted with pH 6.5 HEPES-buffered saline (5 m*M* HEPES, 135 m*M* NaCl, pH adjusted with NaOH). Liposomes loaded with rhodamine for histological studies are formulated with the same lipid composition and preparation method as the Gadoteridol-containing liposomes, except that the lipids are hydrated directly with 20 m*M* sulforhodamine B (Sigma) in pH 6.5 HEPES-buffered saline by six successive cycles of rapid freezing and thawing rather than by ethanol injection. The sulforhodamine B liposomes had a diameter of 90 ± 30 nm (used alone for histological analysis) or 115 ± 40.1 nm (used for coinfusion with GDL in the MRI-monitoring study). For the preparation of liposomes containing a DiI-DS fluorescent probe, 1,1′dioctadecyl-3,3,3′,3′-tetramethylindocarbocyanine-5,5′ disulfonic acid ($DiIC_{18}(3)$-DS) (DiI-DS; Molecular Probes, Eugene, OR) is added to the lipid solution at a concentration of 0.2 mol% of the total lipid. DiI-DS liposomes had a diameter of 110 ± 40 nm. For further detail regarding formulation of liposomes loaded with chemotherapeutic agents, refer to Krauze *et al.* (2007).

During our studies it became apparent that, even when using the same method (CED), the volumes of distribution for different compounds were inconsistent. For simple infusates, CED distribution was significantly increased if the infusate was more hydrophilic or had weaker tissue affinity (Saito *et al.*, 2006a). Encapsulation of tissue-affinitive molecules by liposomes significantly increased their tissue distribution. However, it appeared that liposomal surface properties (cationic versus neutral liposome, surface charge, and percentage of PEGylation) affected parenchymal volume of distribution (Vd). We found that PEGylation, which provided steric stabilization and reduced surface charge, yielded the greatest Vd when compared with volume of infusion (Vi) of all other liposomal formulations (Saito *et al.*, 2006a).

3. Quantification of Liposome-Entrapped Gadoteridol by Magnetic Resonance Imaging

To determine the total concentration of Gadoteridol infused into the brain, we first used nuclear MR relaxivity measurements to calculate the concentration of Gadoteridol entrapped in the liposomes. The relationship between the change in the intrinsic relaxation rate imposed by a paramagnetic agent (ΔDR1), also known as "T1 shortening," and the concentration of the agent is defined by the equation: $\Delta DR1 = r[\text{agent}]$, where r is relaxivity of the paramagnetic agent and $\Delta DR1 = (1/T1_{\text{observed}} - 1/T1_{\text{intrinsic}})$. Since Gadoteridol was encapsulated within the liposome, we corrected for the change in the observed T1 imposed by the lipid by measuring the T1 of solubilized liposomes, with and without Gadoteridol, by means of an iterative inversion-recovery MRI sequence on a 2-T Brucker Omega scanner (Brucker Medical, Karlsruhe, Germany). The relaxivity of Gadoteridol had been empirically derived previously on the same system and was known to have a value of 4.07 $mM^{-1}\ s^{-1}$. The concentration of Gadoteridol encapsulated within the liposomes was then calculated with the following equation: $[\text{Gadoteridol}] = [(1/T1_{\text{wGado}}) - (1/T1_{\text{w/oGado}})]/4.07$, where $T1_{\text{wGado}}$ is the T1 of the Gadoteridol-filled liposomes and $T1_{\text{w/oGado}}$ is the T1 of empty liposomes (with no Gadoteridol). The total Gadoteridol dose delivered could then be determined directly from the concentration of liposomes infused into each brain region. Local dose distribution within the tissue was established by coinfused liposomes containing a quantitative histological indicator. Gadoteridol shortens the T1 of water with which it interacts; thus, local enzymatic and/or thermal processes that alter the permeability or size of the liposomes may cause a T1 change that is independent of the local concentration of liposomes (Rubesova *et al.*, 2002; Saito *et al.*, 2005).

4. Experimental Subjects

Animal protocols are reviewed and approved by the Institutional Animal Care and Use Committees at the University of California San Francisco (San Francisco, CA). Adult male Cynomolgus monkeys (*Macaca fascicularis*, $n = 6$, 3–10 kg) are individually housed in stainless steel cages. Each animal room is maintained on a 12-h light/dark cycle and room temperature ranges between 64 and 84 °F. Prior to assignment to the study, all animals underwent at least a 31-day quarantine period mandated by the Centers for Disease Control and Prevention (Atlanta, GA). All animals undergoing CED were surgically managed as described in detail elsewhere (Bankiewicz *et al.*, 2000; Saito *et al.*, 2005).

5. Infusion Catheter Design and Infusion Procedure

Cannula design has been one of the most neglected features of brain delivery protocols (Fig. 18.1). Reflux is defined as the phenomenon of movement of infusate back up the outside of the cannula rather than into the tissue (Fig. 18.1A and B). Although earlier studies showed that smaller cannula diameters permit better delivery, the crucial problem of reflux was either not assessed or not measurable. In our early studies, we confirmed that smaller cannula diameters allowed faster delivery rates but the smallest available catheter needles were associated with increasing rates of reflux when the rate of infusion exceeded 0.5 μl/min (Krauze *et al.*, 2005a), clearly a significant problem when infusing large volumes. Recently (Krauze *et al.*, 2005a), we have been able to increase the CED infusion rate to 5 μl/min without reflux by means of an innovative cannula design (Fig. 18.1C–F). The early metal cannula has been replaced by one made of silica that also features sharp transitions in outer diameter that prevent reflux (Fig. 18.2) (Eberling *et al.*, 2008). Pushing flow rates significantly above 5 μl/min, however, can induce reflux even with this catheter, as we have shown in canine and primate studies (Varenika *et al.*, 2008).

The animal infusion system consists of three components: (i) sterile infusion stepped silica cannula, (ii) sterile Teflon loading line (0.03 in. i.d.) containing the liposomes, and (iii) a nonsterile Teflon infusion line attached to a 1-ml Hamilton Syringe containing olive oil. The infusion cannula consists of fused silica (o.d., 0.03 in.; i.d., 0.06 in.; Polymicro Technologies, Phoenix, AZ)

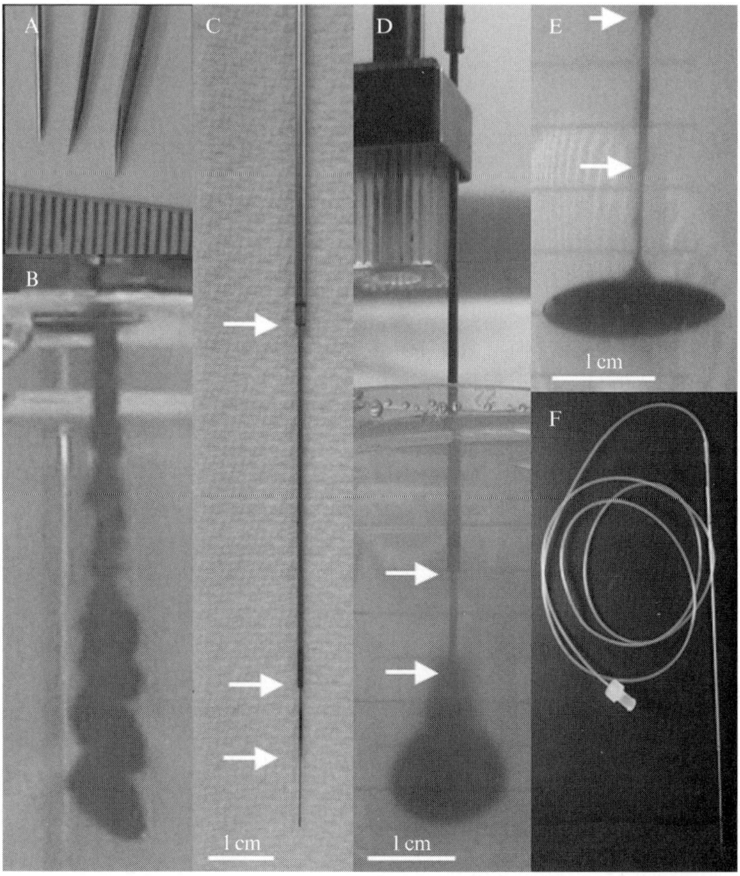

Figure 18.1 Test infusions in agarose gels. (A) Standard stainless steel needles with corresponding millimeter scale on bottom used in initial CED pilot studies. (B) CED of Trypan blue in agarose gel at 5 μl/min through standard 23-gauge stainless steel needle showing reflux along the needle shaft. (C) Initial stepped stainless steel cannula. (D) CED of Trypan blue in agarose gel at 5 μl/min through stepped cannula. (E) CED of Trypan blue at greater than 20 μl/min showing discoid distribution pattern. (F) Stepped cannula complete with silicon tubing and connector for the loading line.

fitted with a smaller diameter fused silica (o.d., 0.016 in.; i.d., 0.008 in.) and placed in Teflon tubing (0.03 in. i.d.; Upchurch Scientific, Seattle, WA), such that the distal tip of the silica extends approximately 15 mm out of the tubing (Fig. 18.3A). The needle/silica is secured into the tubing with superglue and the system is checked for leaks prior to use. At the proximal end of the tubing, a Tefzel fitting and ferrule are connected to the adjacent loading line. Loading and infusion lines consist of 50-cm sections of Teflon tubing (o.d., 0.062 in.; i.d., 0.03 in.) fitted with Tefzel 1/16-in. ferrules, unions, and male Luer-lock

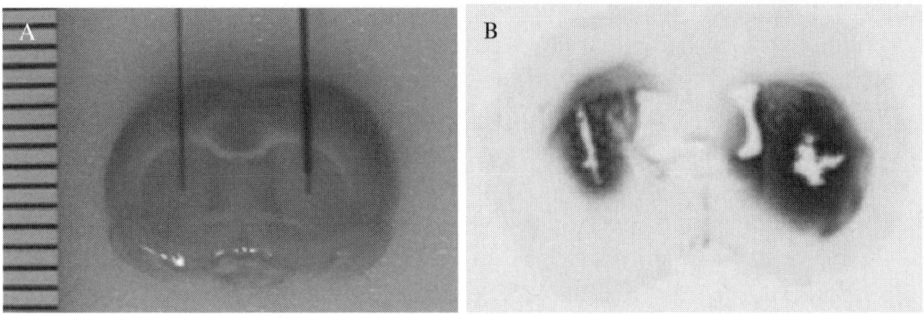

Figure 18.2 Stepped versus simple cannulae in rat brain. (A) Rodent brain showing simple fused silica cannula on the left and the stepped cannula used currently in all our studies consisting of a larger diameter fused silica glued over a smaller diameter fused silica creating a crucial step (with corresponding millimeter scale on the left border). (B) Sample infusion of 5 μl Trypan blue into rodent striatum through simple fused silica on the left and stepped cannula on the right showing decreased distribution on the right due to reflux.

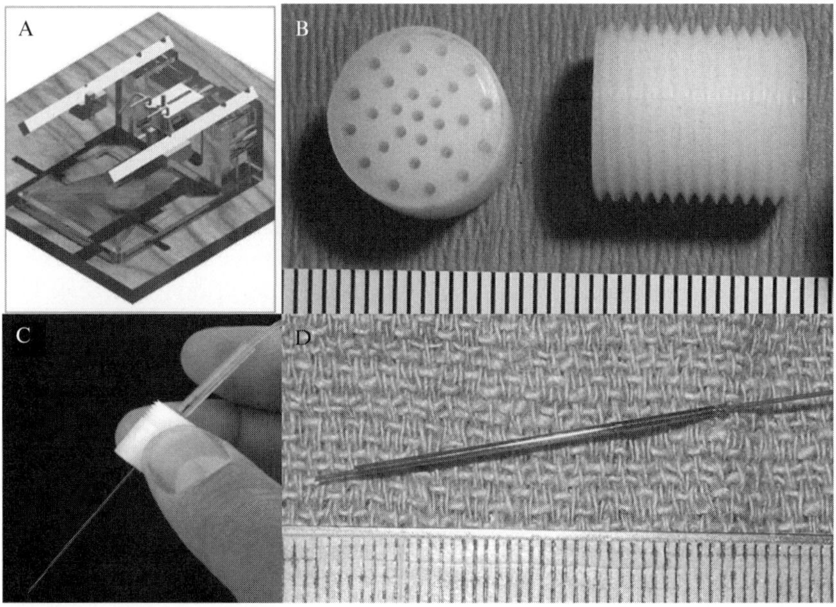

Figure 18.3 Infusion instrumentation. (A) Stereotactic frame used to secure primate head during infusion and MRI. (B) Detailed view of guide block that is secured onto animal's cranium with dental acrylic after baseline MRI for exact placement coordinates. Scale at bottom shows mm gradations. (C) Image of stepped cannula used in ongoing studies with plastic guide block pieces and silicone tubing. (D) Glued fused silica consisting of outer and inner parts with corresponding mm scale.

adapters (Upchurch Scientific, Oak Harbor, WA) at the distal ends. The sterile loading lines accommodate up to a 1-ml volume and are primed with olive oil prior to use. After connecting the entire system, all air is removed from the loading line and proper flow from the tip of the catheter is confirmed (Saito *et al.*, 2005). We have found that a constant delivery rate of 0.1 μl/min via a rate-controllable microinfusion pump (Bioanalytical Systems, Lafayette, IN) before starting *in vivo* delivery, prevents clogging at the catheter tip, a risk that increases with smaller tip diameters. Each customized stepped cannula is cut to a specified length and stereotactically guided to its target. In earlier studies, we used simple burr-holes created in the skull to access the brain. More recently, guide cannula pieces were secured to the skull with dental acrylic, and the tops of the guide-cannula assemblies were capped with stylet screws for simple access during the later infusion procedure (Fig. 18.3D).

Two to four weeks after surgical recovery from guide-cannula placement, CED procedures are performed to infuse GDL into each target site during MRI procedures. Briefly, under isoflurane anesthesia, the animal's head is placed in an MRI-compatible stereotactic frame and a baseline MRI scan is performed (Fig. 18.3C). Vital signs including heart rate and pO_2 are monitored during the procedure. Up to three stepped cannulas connected aseptically to microinfusion pump systems are introduced into the brain through the implanted guide blocks (Fig. 18.3). The length measured from the distal tip of the cannula to the step is of crucial significance for reflux-free CED. We have used longer distance in the past, but our results suggested an increased risk of reflux beyond 3 mm (Fig. 18.3B). In smaller animals, such as rodents, the tip should be shortened to 1 mm (D. Yin and K. Bankiewicz, unpublished data). After secure placement of cannulae, the animal's head is repositioned in the MRI gantry and CED procedures are initiated while MRI data are continuously acquired. Once the cannulae are inserted and the infusion commences, SPGR scans are taken consecutively throughout the infusion (Fig. 18.4). The scan-time is dependent on the number of slices needed to cover the extent of infusion and ranges from 9 min 44 s to 11 min 53 s. After turning off delivery pumps at the end of each infusion session, each catheter remains in place for another 5 min to allow the built-up tissue pressure to abate. Earlier rodent studies have shown that immediate removal of catheter after finishing CED leads to backflow of infusate to the surface of the brain. After completion of the CED/MRI procedure, each cannula is removed from the brain and the respective guide-cannula entry site is cleansed with alcohol. The animal is removed from the MRI scanner and monitored for full recovery from anesthesia. Each animal receives up to three infusion procedures into the same anatomic location, at least 8 weeks apart. In early studies, we also determined the retention of liposomes in primate CNS. MRI was acquired 24 and 48 h after the first infusion. We found only a weak signal after 24 h and complete resolution after 48 h (Saito *et al.*, 2005).

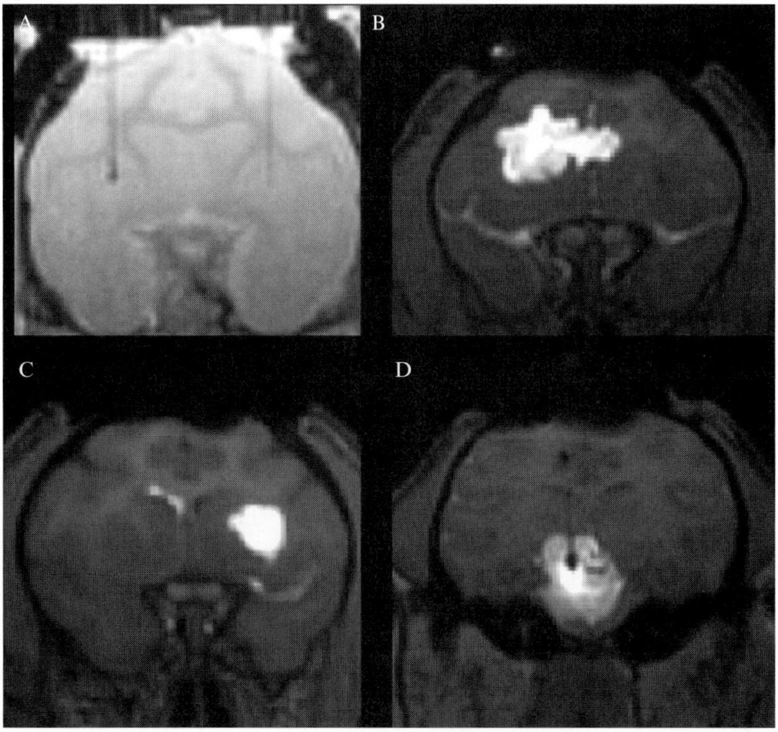

Figure 18.4 MRI-monitored infusion of GDL to primate brain. (A) Confirmation of stepped cannula placement in bilateral primate putamen prior to infusion initiation. (B) CED to primate striatum showing distribution into contralateral hemisphere along white matter tracts of the corpus callosum (infusion volume exceeding 500 μl). (C) CED to primate putamen with perivascular transport of liposomes along medial cerebral artery on the infused side to insular cortex (see also part B). Concomitant infusion of corona radiata is seen as trace signal in contralateral hemisphere of putamen infusion. (D) CED to primate brainstem at with visualization of infusion cannula.

6. Distribution of Liposomes Within Anatomic Structures of the Primate Brain

Although the primate brain is much smaller than the human brain, it is organized similarly into anatomically distinct domains that offer both barriers and conduits to infusate. To explore pathways of distribution within the primate brain, three anatomical structures are chosen for CED: putamen, corona radiata, and brainstem. Infusions in the putamen represent an important step toward therapeutic applications as it represents the main side of dopamine action that is severely disrupted in Parkinson's disease.

Putaminal infusions showed that optimal distribution within the putamen is highly dependent on catheter placement. Lateral positioning of the infusion catheter in the putamen leads to enhanced trafficking of liposomes by the perivascular pump of the lateral striatal arteries (Fig. 18.4C) (Krauze *et al.*, 2005c). The perivascular pump was previously known to function only in the context of the transport of physiologic nutrients in the brain. In our experiments, the perivascular pump was found to be a powerful transport mechanism for liposomes that followed large CNS vessels up to the insular cortex of the temporal lobe (see also Fig. 18.5B). Catheter placement in the putamen close to the white matter led to an "escape" of distribution into the white matter. We found that the increased cellularity of the putamen creates a denser environment and that CED infusions distribute along the path of least resistance that is much lower in the less cellular white matter tracts adjacent to the putamen (Krauze *et al.*, 2005c, 2008). We are currently trying to define a putaminal locus that allows maximal distribution in this anatomic structure without risking untoward distribution (manuscript in preparation). CED in corona radiata showed excellent distribution along white matter tracts (Fig. 18.4B). We did not observe any distribution into other anatomic structures when infusing into corona radiata. We saw that with higher infusion volumes (>400 μl) distribution extended along white matter tracts into the contralateral noninfused hemisphere. Similarly, brainstem infusions were without complications when using MRI guidance (Fig. 18.4D), and distribution within this deep brain structure was homogenous, permeating mainly rostral and caudal regions.

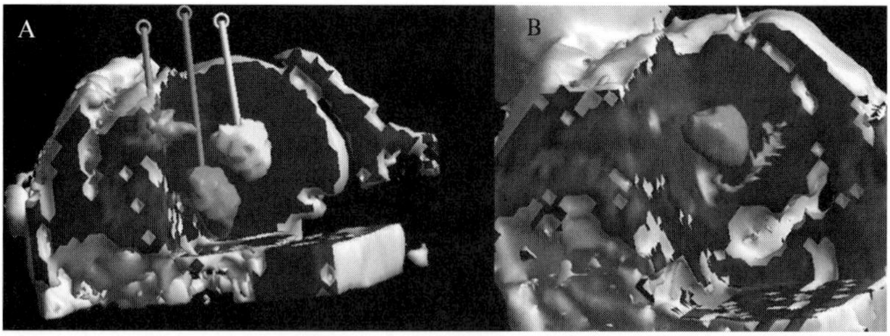

Figure 18.5 Three-dimensional reconstructions of infusions. (A) Three-dimensional reconstruction of all three anatomic infusion locations allowing exact volumetric calculations based on MR data (frontolateral view with partial digital subtraction for better visualization). (B) Digital reconstruction of putamen infusion with perivascular transport of liposomes along large brain arteries (frontolateral view with digital subtraction).

7. Volumetric Calculations of Liposomal Distribution in Primate CNS

MRI-monitored CED allows precise calculation of the volume of distribution (Vd) in any given structure of the brain for any volume infused (Vi). In order to validate this system we compared data from histology (fluorescent liposomes) and MRI (GDL). For volumetric quantification from fluorescent images, primate brains are sectioned on a cryostat. Sequential 40-μm sections at 400-μm intervals are obtained. Rhodamine fluorescence is visualized via an ultraviolet light source, and a charged-coupled device camera with a fixed aperture is used to capture the images. The volume of fluorescent liposome distribution is analyzed on a Macintosh-based image analysis system (NIH Image 1.62; NIH, Bethesda, MD) as described previously (Hamilton *et al.*, 2001). For volumetric quantification from MR images, the volume of liposomal distribution within each infused brain region is quantified with BrainLab® software (BrainLab, Heimstetten, Germany) (Fig. 18.5). MR images acquired during the infusion procedure are correlated with Vi in each series. BrainLab software reads all data specifications from MR images. After the pixel threshold value for liposomal signal is defined, the software calculates the signal above a defined threshold value and establishes the volume of distribution from primate brain (Fig. 18.5A). This allows volume of distribution to be determined at any given time point and can be reconstructed as a three-dimensional image. The ability to accurately calculate distribution *in vivo* during MRI showed that volume of distribution plateaus when leakage to cortical surface or drainage into ventricle is detected despite continuous infusion (Fig. 18.5B) (Varenika *et al.*, 2008). This fact is of particular importance for future clinical applications of CED.

8. Future and Outlook of CED to Brain

The growing amount of potential therapeutic options for CNS pathologies clearly warrants a refined direct delivery system to the CNS. Our group has continuously refined and improved CED in the recent years. The system that we described here in detail represents an efficient, but still complex method of delivery. Numerous clinical trials have employed at least some variant of CED so far (Gill *et al.*, 2003; Gupta and Sarin, 2002; Kunwar, 2003). The enthusiastic outcomes of some early reports were soon replaced by inexplicable side effects (Sherer *et al.*, 2006). Unfortunately, none of the past and current clinical trials employs any method of imaging during delivery as described here. Hence, clinical failures suspected to have

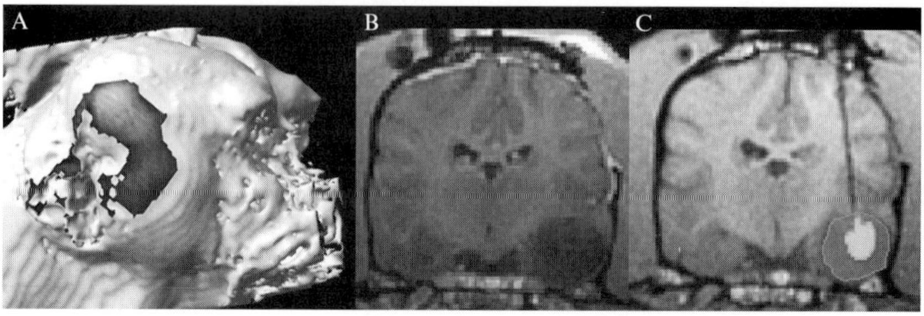

Figure 18.6 Current therapeutic application of CED in spontaneous canine brain tumors. (A) Digital reconstruction of canine head after baseline MRI. (B) Image of large brain tumor located at the base of the left temporal lobe. (C) MR image during infusion of tumor (red delineation with a mix of GDL and chemotherapeutic-loaded liposomes (green delineation)). (See Color Insert.)

been caused by suboptimal delivery are not amenable to review. We have shown in canine tumor studies that a mixture of GDL mixed with chemotherapeutic liposomes is an elegant means of monitoring and targeting delivery in brain tumors (Fig. 18.6A–C) (Dickinson *et al.*, 2008). Earlier rodent studies showed excellent therapeutic properties of chemotherapeutic liposomes (Krauze *et al.*, 2007; Saito *et al.*, 2006b; Yamashita *et al.*, 2007). When compared with the free drug, liposomal drugs are less toxic and have extended half-life due to slow release. The combination of real-time visualization of infusions and enhanced pharmacokinetic properties of antitumor agents has permitted us to localize toxicity within the margins of tumors. Ongoing studies in our laboratory are seeking to establish GDL as a surrogate marker for infusion of viral vectors (Fiandaca *et al.*, 2009) and small molecules. More studies are needed to bring our system into clinical use, but encouraging results are emerging.

ACKNOWLEDGMENT

This work was supported in part by NIH grants (P01).

REFERENCES

Bankiewicz, K. S., Eberling, J. L., Kohutnicka, M., Jagust, W., Pivirotto, P., Bringas, J., Cunningham, J., Budinger, T. F., and Harvey-White, J. (2000). Convection-enhanced delivery of AAV vector in parkinsonian monkeys; *in vivo* detection of gene expression and restoration of dopaminergic function using pro-drug approach. *Exp. Neurol.* **164,** 2–14.

Bobo, R. H., Laske, D. W., Akbasak, A., Morrison, P. F., Dedrick, R. L., and Oldfield, E. H. (1994). Convection-enhanced delivery of macromolecules in the brain. *Proc. Natl. Acad. Sci. USA* **91,** 2076–2080.

Dickinson, P. J., LeCouteur, R. A., Higgins, R. J., Bringas, J. R., Roberts, B., Larson, R. F., Yamashita, Y., Krauze, M., Noble, C. O., Drummond, D., Kirpotin, D. B., Park, J. W., *et al.* (2008). Canine model of convection-enhanced delivery of liposomes containing CPT-11 monitored with real-time magnetic resonance imaging: Laboratory investigation. *J. Neurosurg.* **108**(5), 989–998.

Eberling, J. L., Jagust, W. J., Christine, C. W., Starr, P., Larson, P., Bankiewicz, K. S., and Aminoff, M. J. (2008). Results from a phase I safety trial of *hAADC* gene therapy for Parkinson disease. *Neurology* **70**(21), 1980–1983.

Fiandaca, M., Forsayeth, J., and Bankiewicz, K. (2008). Current status of gene therapy trials for Parkinson's disease. *Exp. Neurol.* **209**(1), 51–57.

Fiandaca, M. S., Varenika, V., Eberling, J., McKnight, T., Bringas, J., Pivirotto, P., Beyer, J., Hadaczek, P., Bowers, W., Park, J., Federoff, H., Forsayeth, J., *et al.* (2009). Real-time MR imaging of adeno-associated viral vector delivery to the primate brain. *Neuroimage* **47** (Suppl. 2), T27–T35.

Gill, S. S., Patel, N. K., Hotton, G. R., O'Sullivan, K., McCarter, R., Bunnage, M., Brooks, D. J., Svendsen, C. N., and Heywood, P. (2003). Direct brain infusion of glial cell line-derived neurotrophic factor in Parkinson disease. *Nat. Med.* **9,** 589–595.

Gupta, T., and Sarin, R. (2002). Poor-prognosis high-grade gliomas: Evolving an evidence-based standard of care. *Lancet Oncol.* **3,** 557–564.

Hadaczek, P., Yamashita, Y., Mirek, H., Tamas, L., Bohn, M. C., Noble, C., Park, J. W., and Bankiewicz, K. (2006). The "perivascular pump" driven by arterial pulsation is a powerful mechanism for the distribution of therapeutic molecules within the brain. *Mol. Ther.* **14,** 69–78.

Hamilton, J. F., Morrison, P. F., Chen, M. Y., Harvey-White, J., Pernaute, R. S., Phillips, H., Oldfield, E., and Bankiewicz, K. S. (2001). Heparin coinfusion during convection-enhanced delivery (CED) increases the distribution of the glial-derived neurotrophic factor (GDNF) ligand family in rat striatum and enhances the pharmacological activity of neurturin. *Exp. Neurol.* **168,** 155–161.

Hsieh, C. H., Chen, Y. F., Chen, F. D., Hwang, J. J., Chen, J. C., Liu, R. S., Kai, J. J., Chang, C. W., and Wang, H. E. (2005). Evaluation of pharmacokinetics of 4-borono-2-(18)F-fluoro-L-phenylalanine for boron neutron capture therapy in a glioma-bearing rat model with hyperosmolar blood-brain barrier disruption. *J. Nucl. Med.* **46,** 1858–1865.

Krauze, M. T., Saito, R., Noble, C., Tamas, M., Bringas, J., Park, J. W., Berger, M. S., and Bankiewicz, K. (2005a). Reflux-free cannula for convection-enhanced high-speed delivery of therapeutic agents. *J. Neurosurg.* **103,** 923–929.

Krauze, M. T., McKnight, T. R., Yamashita, Y., Bringas, J., Noble, C. O., Saito, R., Geletneky, K., Forsayeth, J., Berger, M. S., Jackson, P., Park, J. W., and Bankiewicz, K. S. (2005b). Real-time visualization and characterization of liposomal delivery into the monkey brain by magnetic resonance imaging. *Brain Res. Brain Res. Protoc.* **16,** 20–26.

Krauze, M. T., Saito, R., Noble, C., Bringas, J., Forsayeth, J., McKnight, T. R., Park, J., and Bankiewicz, K. S. (2005c). Effects of the perivascular space on convection-enhanced delivery of liposomes in primate putamen. *Exp. Neurol.* **196,** 104–111.

Krauze, M. T., Noble, C. O., Kawaguchi, T., Drummond, D., Kirpotin, D. B., Yamashita, Y., Kullberg, E., Forsayeth, J., Park, J. W., and Bankiewicz, K. S. (2007). Convection-enhanced delivery of a topoisomerase I inhibitor (nanoliposomal topotecan) and a topoisomerase II inhibitor (pegylated liposomal doxorubicin) in intracranial brain tumor xenografts. *Neuro Oncol.* **9**(1), 20–28.

Krauze, M. T., Vandenberg, S. R., Yamashita, Y., Saito, R., Forsayeth, J., Noble, C., Park, J., and Bankiewicz, K. S. (2008). Safety of real-time convection-enhanced delivery of liposomes to primate brain: A long-term retrospective. *Exp. Neurol.* **210,** 638–644.

Kunwar, S. (2003). Convection enhanced delivery of IL13-PE38QQR for treatment of recurrent malignant glioma: Presentation of interim findings from ongoing phase 1 studies. *Acta Neurochir. Suppl.* **88,** 105–111.

Laske, D. W., Morrison, P. F., Lieberman, D. M., Corthesy, M. E., Reynolds, J. C., Stewart-Henney, P. A., Koong, S. S., Cummins, A., Paik, C. H., and Oldfield, E. H. (1997). Chronic interstitial infusion of protein to primate brain: Determination of drug distribution and clearance with single-photon emission computerized tomography imaging. *J. Neurosurg.* **87,** 586–594.

Lonser, R. R., Walbridge, S., Garmestani, K., Butman, J. A., Walters, H. A., Vortmeyer, A. O., Morrison, P. F., Brechbiel, M. W., and Oldfield, E. H. (2002). Successful and safe perfusion of the primate brainstem: *In vivo* magnetic resonance imaging of macromolecular distribution during infusion. *J. Neurosurg.* **97,** 905–913.

Rubesova, E., Berger, F., Wendland, M. F., Hong, K., Stevens, K. J., Gooding, C. A., and Lang, P. (2002). Gd-labeled liposomes for monitoring liposome-encapsulated chemotherapy: Quantification of regional uptake in tumor and effect on drug delivery. *Acad. Radiol.* **9**(Suppl. 2), S525–S527.

Saito, R., Krauze, M. T., Bringas, J. R., Noble, C., McKnight, T. R., Jackson, P., Wendland, M. F., Mamot, C., Drummond, D. C., Kirpotin, D. B., Hong, K., Berger, M. S., *et al.* (2005). Gadolinium-loaded liposomes allow for real-time magnetic resonance imaging of convection-enhanced delivery in the primate brain. *Exp. Neurol.* **196,** 381–389.

Saito, R., Krauze, M. T., Noble, C. O., Tamas, M., Drummond, D. C., Kirpotin, D. B., Berger, M. S., Park, J. W., and Bankiewicz, K. S. (2006a). Tissue affinity of the infusate affects the distribution volume during convection-enhanced delivery into rodent brains: Implications for local drug delivery. *J. Neurosci. Methods* **154**(1–2), 225–232.

Saito, R., Krauze, M. T., Noble, C. O., Drummond, D. C., Kirpotin, D. B., Berger, M. S., Park, J. W., and Bankiewicz, K. S. (2006b). Convection-enhanced delivery of Ls-TPT enables an effective, continuous, low-dose chemotherapy against malignant glioma xenograft model. *Neuro Oncol.* **8,** 205–214.

Sherer, T. B., Fiske, B. K., Svendsen, C. N., Lang, A. E., and Langston, J. W. (2006). Crossroads in GDNF therapy for Parkinson's disease. *Mov. Disord.* **21,** 136–141.

Varenika, V., Dickinson, P., Bringas, J., Lecouteur, R., Higgins, R., Park, J., Fiandaca, M., Berger, M., Sampson, J., and Bankiewicz, K. (2008). Detection of infusate leakage in the brain using real-time imaging of convection-enhanced delivery. *J. Neurosurg.* **109,** 874–880.

Weyerbrock, A., Walbridge, S., Pluta, R. M., Saavedra, J. E., Keefer, L. K., and Oldfield, E. H. (2003). Selective opening of the blood-tumor barrier by a nitric oxide donor and long-term survival in rats with C6 gliomas. *J. Neurosurg.* **99,** 728–737.

Yamashita, Y., Krauze, M. T., Kawaguchi, T., Noble, C. O., Drummond, D. C., Park, J. W., and Bankiewicz, K. S. (2007). Convection-enhanced delivery of a topoisomerase I inhibitor (nanoliposomal topotecan) and a topoisomerase II inhibitor (pegylated liposomal doxorubicin) in intracranial brain tumor xenografts. *Neuro Oncol.* **9,** 20–28.

Zhang, Y., and Pardridge, W. M. (2005). Delivery of beta-galactosidase to mouse brain via the blood-brain barrier transferrin receptor. *J. Pharmacol. Exp. Ther.* **313,** 1075–1081.

CHAPTER NINETEEN

Hemoglobin-Vesicles as an Artificial Oxygen Carrier

Hiromi Sakai, Keitaro Sou, *and* Eishun Tsuchida

Contents

Abstract

Hemoglobin-vesicles (HbV) or liposome-encapsulated hemoglobin (LEH) are artificial oxygen carriers that mimic the cellular structure of RBCs. In contrast to other liposomal products containing antifungal or anticancer drugs, one injection of HbV in place of a blood transfusion is estimated as equivalent to a massive dose, such as several hundred milliliters or a few liters of normal blood contents. The fluid must therefore contain a sufficient amount of Hb, the binding site of oxygen, to carry oxygen like blood. Encapsulation of Hb can shield various toxic effects of molecular Hbs. On the other hand, the liposomal structure, surface property, and the balance between the stability for storage and blood circulation and instability for the prompt degradation in the reticuloendothelial system must be considered to establish an optimal transfusion alternative.

Research Institute for Science and Engineering, Waseda University, 3-4-1 Okubo, Shinjuku, Tokyo, Japan, and Waseda Bioscience Research Institute in Singapore, 11 Biopolis Way, Helios, Singapore

Methods in Enzymology, Volume 465
ISSN 0076-6879, DOI: 10.1016/S0076-6879(09)65019-9

1. Introduction: Encapsulated Hemoglobin as an Artificial Oxygen Carrier

Hemoglobin (Hb) (MW 64,500) is the most abundant protein in blood. It contains four subunits with a heme on each, where oxygen binds and dissociates reversibly. About two million Hb molecules are encapsulated in one red blood cell (RBC). The intracellular Hb concentration is as high as 35 g/dL, which makes the Hb concentration of blood as high as 12–15 g/dL. After the discovery of blood type antigens on the outer surface of RBCs by Landsteiner in 1901, the idea was born to use purified Hb as an oxygen-carrying fluid that is free of any blood type. Nevertheless, it was unsuccessful because of various toxic effects. In spite of its abundance in blood, Hb becomes toxic when it is released from RBCs. Dissociation of tetramer Hb subunits into two dimers, $\alpha_2\beta_2 \rightarrow 2\alpha\beta$, induces renal toxicity. Entrapment of gaseous messenger molecules (NO and CO) induces vasoconstriction, hypertension, reduction of blood flow and tissue oxygenation at microcirculatory levels (Sakai *et al.*, 2000c, 2008a,b), neurological disturbances, and malfunction of esophageal motor functions (Murray *et al.*, 1995). Some chemically modified Hb solutions (cell-free Hb-based oxygen carriers) are now confronting difficulties in clinical trials (Chang, 2004; Natanson *et al.*, 2008). These side effects of molecular Hb imply the importance of the cellular structure.

Hb encapsulation was first performed by Chang in the 1950s (Chang, 1991), using a polymer membrane. Some Japanese groups also tested Hb encapsulation with gelatin, gum Arabic, silicone, etc. Nevertheless, it was extremely difficult to regulate the particle size to be appropriate for blood flow in the capillaries and to obtain sufficient biocompatibility. After Bangham and Horne (1964) reported that phospholipids assemble to form vesicles in aqueous media, and that they encapsulate water-soluble materials in their inner aqueous interior, it seemed reasonable to use such vesicles for Hb encapsulation. Djordjevich and Miller (1977) prepared a liposome-encapsulated Hb (LEH) composed of phospholipids, cholesterol, fatty acid, etc. Since then, many groups have tested encapsulated Hbs using liposomes (Djordjevich and Ivankovich, 1988; Gaber *et al.*, 1983; Hayward *et al.*, 1985; Hunt *et al.*, 1985; Jopski *et al.*, 1989; Kato *et al.*, 1984; Liu and Yonetani, 1994) (Table 19.1). Some failed initially, and some are progressing with the aim of clinical usage. The Naval Research Laboratory presented remarkable progress of LEH (Rudolph *et al.*, 1991), but it suspended development about 10 years ago. What we call Hb-vesicles (HbV) with high-efficiency production processes and improved properties have been established by our group, based on nanotechnologies of molecular assembly and precise analyses of pharmacological and physiological aspects (Tsuchida, 1998).

Table 19.1 Trials of liposome encapsulated Hb

References	Lipid composition	Characteristic preparation methods
Djordjevich and Ivankovich (1988)	L-α-Phosphatidylcholine/ cholesterol/palmitic acid	Sonication
Gaber *et al.* (1983), Farmer and Gaber (1987)	EYPC/cholesterol/bovine brain phosphatidylserine; DMPC/cholesterol/ dicetylphosphate	Extrusion
Hunt *et al.* (1985)	EYPC/cholesterol/DPPA/ α-tocopherol	Reverse phase evaporation and extrusion
Beissinger *et al.* (1986)	HSPC/cholesterol/ dicetylphosphate or DMPG	Microfluidizer
Rabinovici *et al.* (1993), Rudolph *et al.* (1988)	HSPC/cholesterol/DMPG/ α-tocopherol; trehalose is added to store LEH as a lyophilized powder	Bovine Hb; thin film hydration and emulsification
Jopski *et al.* (1989)	EYL/PS (EYPA)	Detergent dialysis
Sato *et al.* (1992), Sakai *et al.* (1992), Akama *et al.* (2000)	DODPC/cholesterol/ octadecadienoic acid; gamma-ray polymerization	HbCO, extrusion method
Hayward *et al.* (1985)	Diacetylene phospholipid/ cholesterol; UV-irradiation for polymerization	HbCO, sonication
Kato *et al.* (1984)	EYL/carboxymethyl chitin	Reverse phase evaporation
Yoshioka (1991), Takahashi (1995)	HSPC/cholesterol/myristic acid/α-tocopherol/DPPE-PEG	Microfluidizer
Mobed and Chang (1991)	HSPC/DMPG/ α-tocopherol/ carboxymethyl chitin	Reverse phase evaporation
Liu and Yonetani (1994)	EYL/cholesterol/ dicetylphosphate/ α-tocopherol	Freeze–thaw method
Phillips *et al.* (1999)	DSPC/cholesterl/PEG_{5000}-DSPE/a-tocopherol	α-Cross-linked human Hb; microfluidizer
Sakai *et al.* (1996), Takeoka *et al.* (1996)	DPPC/cholesterol/DPPG or palmitic acid	HbCO, extrusion

(*continued*)

Table 19.1 (*continued*)

References	Lipid composition	Characteristic preparation methods
Sakai *et al.* (1997)	DPPC/cholesterol/DPPG/ DSPE-PEG$_{5000}$	HbCO, extrusion
Sou *et al.* (2003), Sakai *et al.* (2002a)	DPPC/cholesterol/DHSG/ DSPE-PEG$_{5000}$	HbCO, extrusion
Li *et al.* (2005)	DMPC/cholesterol/DMPG/ DSPE-PEG$_{2000}$/actin	Extrusion
Pape *et al.* (2008)	HSPC/cholesterol/stearic acid/DSPE-PEG$_{5000}$	Lipid paste rapid dispersion

DMPC: 1,2-dimyristoyl-*sn*-glycero-3-phosphatidylcholine; EYPC: egg yolk phosphatidylcholine; DPPA: 1,2-dipalmitoyl-*sn*-glycero-3-phosphatidic acid; HSPC: hydrogenated soy phosphatidylcholine; DMPG: 1,2-dimyristoyl-*sn*-glycero-3-phosphatidylglycerol; EYL: egg yolk lecithin; PS: phosphatidylserine; DODPC: 1,2-dioctadecadienoyl-*sn*-glycero-3-phosphatidylcholine; DPPE: 1,2-dipalmitoyl-*sn*-glycero-3-phosphatidylethanolamine; DSPE: 1,2-distearoyl-*sn*-glycero-3-phosphatidylethanolamine; DPPC: 1,2-dipalmitoyl-*sn*-glycero-3-phosphatidylcholine; DHSG: 1,5-*O*-dihexadecyl-*N*-succinyl-L-glutamate; HbCO: carbonylhemoglobin.

The major expected advantages of an artificial oxygen carrier are the absence of any pathogen, its stability for long-term storage, and its sufficient oxygen transporting capability. The important points of HbV or LEH to be considered are the following:

i. Source of Hb, virus inactivation, and virus removal procedure
ii. Improvement of encapsulation efficiency without Hb denaturation
iii. Sufficient stability for storage and blood circulation
iv. Blood compatibility and prompt metabolism in the reticuloendothelial system
v. Physicochemical properties comparable with those of blood.

The last point differs completely from other liposomal products. It is necessary to design not only a particle but also its suspension because we intend to prepare a substitute for blood, which has 12–15 g/dL hemoglobin, 40–50% hematocrit, and 5–7 g/dL plasma proteins, and unique rheological and osmotic properties.

2. Encapsulation of Concentrated Hb in Liposomes

The performance of LEH depends on the weight ratio of Hb to lipid ([Hb]/[Lipids]): to carry more Hb with fewer vehicles made of lipids. This value is improved by lowering the number of bilayer membranes (lamellarity) of the vesicle, and raising the concentration of Hb in the interior of the vesicle ([Hb] < 40 g/dL). Table 19.1 shows that many groups add cholesterol and negatively charged lipids to phospholipids, which is plausible because cholesterol not only improves membrane stability but it also reduces the curvature for LUV. Addition of a small amount of negatively charged lipid increases the repulsive force between the lipid membranes and reduces the lamellarity in addition to controlling the zeta potential for blood compatibility (Sou and Tsuchida, 2008). Saturated phospholipids (e.g., HSPC, DSPC, and DPPC) are preferred to unsaturated lipids (e.g., EYL and soy phosphatidylcholines) because of the synergistic, facilitated oxidation of both unsaturated lipids and Hb and physical instability (Szebeni *et al.*, 1984, 1985), but cholesterol lowers such Hb denaturation to some degree.

To encapsulate Hb solution in liposomes, some groups have used detergent dialysis and reverse phase evaporation methods. However, it is difficult to guarantee the absence of a detergent and an organic solvent in the final product that would result in a massive dosage. A simple mixing of a concentrated Hb solution and lipids (as a powder, paste, or thin film on the wall of a flask) produces multilamellar vesicles with a wide particle size distribution, necessitating size regulation by a homogenizer (microfluidizer) (Beissinger *et al.*, 1986; Vidal-Naquet *et al.*, 1989; Vivier *et al.*, 1992), or an extrusion method, which would be more practical. In the case of a microfluidizer, the solution temperature increases by a strong shear force. Denaturation of the components and wide particle distribution must be prevented. Use of an extrusion method can avoid such difficulties; but conditions for efficient filter permeation must be established.

We studied optimal conditions for Hb encapsulation using the extrusion method, and considered the behavior of the Hb and lipid assemblies as a kind of polymer electrolyte (Takeoka *et al.*, 1993, 1994a,b, 1996). The maximum [Hb]/[Lipid] ratio is obtainable at ca. pH 7, which would relate to the isoelectric point (p*I*) of Hb. The Hb molecule is negatively charged when the pH is greater than 7.0, and the electrostatic repulsion between Hb and the negatively charged bilayer membrane results in lower encapsulation efficiency. Lower pH, however, is expected to enhance Hb denaturation through strong interaction with the lipid bilayer membrane and metHb formation. Therefore, a physiological pH, 7.0–7.4, would be optimal. In addition, the higher ionic strength shields repulsion between the negatively charged lipid bilayer membranes and thereby increases the lamellarity.

The lamellarity decreases concomitantly with increased microviscosity of the lipid bilayer membrane (decreased lipid mobility). Multilamellar vesicles are converted to smaller vesicles with a smaller lamellarity during the extrusion procedure. When the membrane fluidity is high, deformation of the vesicles during extrusion occurs more easily even for multilamellar vesicles, resulting in larger lamellarity in the final vesicles. Therefore, a phospholipid with a higher phase transition temperature that typically forms a membrane with lower membrane fluidity is preferred. However, these lipids would make extrusion more difficult, because a higher shear rate (high extrusion pressure) is required. Based on this reasoning, mixed lipids contain 1,2-dipalmitoyl-*sn*-glycero-3-phosphatidylcholine (DPPC) as the main component. The encapsulation efficiency of the Hb solution in a size-regulated phospholipid vesicle has been improved using an extrusion method (Sakai *et al.*, 1996; Sou *et al.*, 2003). Mixed lipids—DPPC, cholesterol, 1,5-*O*-dihexadecyl-*N*-succinyl-L-glutamate (DHSG), and DSPE-$PEG_{5,000}$ at a molar ratio of 5:5:1:0.033—are hydrated using a NaOH solution (7.6 m*M*) to obtain a polydispersed multilamellar vesicle dispersion (50 nm–30 μm diameter). The polydispersed vesicles are converted to smaller vesicles with average diameter of ca. 500 nm by repeating two cycles of freezing by liquid nitrogen and thawing at 40 °C, and finally the suspension is freeze-dried to obtain a lyophilized powder. It is then rehydrated into a concentrated Hb solution (40 g/dL), retaining the size and size distribution of the original vesicles. The resulting vesicle dispersion permeates smoothly through the membrane filters during extrusion (Extruder, Northern Lipids Inc.; Filters, Fuji Film Microfilter, Filter pore size: 3.0, 0.45, 0.3, and 0.22 μm). The average permeation rate of the frozen–thawed vesicles is ca. 30 times higher than that of the simple hydrated vesicles. During extrusion, the Hb solution is encapsulated into reconstructed vesicles with 250 ± 20 nm diameter. The [Hb]/[lipid] ratio reaches 1.7–1.8. This improvement of the Hb encapsulation procedure is a breakthrough for HbV scalability. The main physicochemical characteristics of HbV and their analytical methods are presented in Table 19.2.

3. Source of Hb and Its Purification

Hb is the most abundant protein in blood. It is easily obtained from human or animal blood by washing RBCs to remove plasma proteins, hemolysis by adding pure water, removal of the cell membrane components by ultrafiltration, and dialysis, or ion chromatography (DeLoach *et al.*, 1986; Palmer *et al.*, 2009; Sheffield *et al.*, 1987; Winslow and Chapman, 1994). The significantly high dosage as a transfusion alternative requires high purity, absence of pathogen (sterility assurance, virus-free), and the abundance of Hb as a starting material (Sakai *et al.*, 2004d). Possible sources of Hb would be (i) outdated human donated blood, (ii) bovine Hb, because blood is abundant in the cattle industry (Sakai *et al.*, 2002b), and (iii) recombinant Hb. Because of some patients' religious beliefs, human-derived or cow-derived Hb would be unacceptable. The absence of prion contamination must be verified in the case of bovine Hb. The influence of

Table 19.2 Physicochemical characteristics of Hb vesicles (HbV) developed at Waseda University

Parameter		Analytical method (references)
Particle diameter	250–280 nm	Light scattering, Beckman Coulter submicron particle analyzer (Sakai *et al.*, 1996; Sato *et al.*, 2009)
P_{50}	25–28 Torr (pH 7.4)	Hemox Analyer (37 °C, pH 7.4) (Sakai *et al.*, 2000a)
[Hb]	10 g/dL	A modified cyanomet Hb method
[heme] (iron)	6.2 m*M*	Atomic absorption, Inductively coupled plasma spectrometry
[metHb]	<3%	Soret band absorption (Sakai *et al.*, 1996)
[HbCO]	<2%	Soret band absorption (Sakai *et al.*, 1996)
[Lipids]	5.3–6.3 g/dL	Choline oxidase DAOS method, molybdenum blue method, cholesterol oxidase method (Sakai *et al.*, 1996; Sou *et al.*, 2003)
Occupied volume of particles	40%	Ultracentrifugation (50,000×*g*, 30 min) and supernatant volume measurement
Colloid osmotic pressure	20 Torr in 5% HSA	Colloid Osmometer; Wescor Model 4420 (Sakai *et al.*, 2000a)
Lipid composition	DPPC/cholesterol/DHSG/DSPE-PEG_{5000}	(Sakai *et al.*, 2002a; Sou *et al.*, 2003)
ζ potential	−18.7	Zetasizer ([NaCl] = 20 m*M*, pH 7.4) (Sou and Tsuchida, 2008)
Viscosity	3.8 cP	Anton Parr Rheometer M301 (268 s^{-1}, 25 °C) (Sakai *et al.*, 2007)
Lamellarity	Nearly 1	Theoretical calculation (Takeoka *et al.*, 1996); small angled X-ray scattering (Sato *et al.*, 2009)
Sterility, [LPS]	<0.2 EU/mL	A modified limulus amebocyte lysate test with kinetic-turbidimetric gel clotting analysis and pretreatment of surfactant. (Sakai *et al.*, 2004d)
Stability for storage at room temperature	>2 years, purged with N_2	Measurements of particle size distribution, turbidity, P_{50}, hemolysis, [metHb], lipid decomposition (Sakai *et al.*, 2000b)
Circulation half-life	35 h (rats)	Biodistribution of ^{99m}Tc-, ^{3}H-, and ^{125}I-labeled HbV (Sou *et al.*, 2005; Taguchi *et al.*, 2009a,b)

immunoglobulin G antibovine Hb has to be examined carefully (Hamilton and Kickler, 2007). A commercially available powdered Hb reagent for laboratory use is not recommended for *in vivo* use, because some molecules are already oxidized and a regeneration procedure would be required. Sterility (absence of lipopolysaccharide contamination) also presents a problem.

In Japan, the research and development of blood substitutes was initiated because of the beneficial use of outdated blood. The primary advantage of artificial O_2 carriers is expected to present no fear of infectious disease derived from human blood. The donated blood is inspected strictly by the nucleic acid amplification test (NAT) for possible detection of human immunodeficiency virus, hepatitis B virus, and hepatitis C virus. It is necessary, however, to introduce procedures to inactivate and remove viruses in the process of Hb purification from outdated RBCs to guarantee sterility, in light of the unforgettable tragedy of the HIV transmission caused by the distribution of nonpasteurized, plasma-derived products. In our purification process, virus inactivation was performed by pasteurization at 60 °C for 10 h, which is the same condition for the pasteurization of human serum albumin (HSA) (Fukutomi *et al.*, 2002; Sakai *et al.*, 1993). This process can be used because carbonylhemoglobin (HbCO) is stable under these conditions. Thermograms of HbCO indicate a denaturation temperature at 78 °C, which is much higher than that for oxyhemoglobin (64 °C) (Sakai *et al.*, 2002b).

The virus inactivation efficiency must be evaluated using spike tests (Abe *et al.*, 2001; Huang *et al.*, 2002). The Hb solution spiked with vesicular stomatitis virus (VSV) is treated at 60 °C for 1 h under a CO atmosphere. VSV is inactivated by $>6.0\ \log_{10}$ without metHb formation and denaturation. Some protein bands other than Hb disappear on sodium dodecyl sulfate–polyacrylamide gel electrophoresis (SDS–PAGE) and isoelectric focusing after heat treatment. During pasteurization, all other proteins are denatured and precipitated. Consequently, we obtain an ultrapure Hb solution. This high purity is crucial for preventing membrane plugging during subsequent ultrafiltration to remove viruses. The US FDA requires two orthogonal steps, not only of virus inactivation but also of virus removal.

We tested ultrafiltration of the HbCO solution to remove viruses using PLANOVA™-35N and -15N (P35N, P15N, Bemberg Microporous Membrane, BMM; Asahi Kasei Medical Co. Ltd.) (Naito *et al.*, 2002). The virus removal mechanism is size exclusion through the capillary pores, and by depth filtration. The unit membrane, which has a network structure of capillaries and voids, is accumulated to form 150 layers. P35N and P15N have respective mean pore sizes of 35 and 15 nm. P35N is suitable for removing lipid-enveloped viruses of 40–100 nm such as HIV and HCV. P15N is useful to remove naked capsid viruses smaller than

40 nm, such as parvoviruses. The permeation flux is sufficiently high because of the absence of denaturation of HbCO. Denatured and insoluble proteins cause plugging of filter pores. A high removal efficiency of a bacteriophage, $\phi \times 174$, (>7.7 log) was confirmed (Naito *et al.*, 2002). In addition, P15N is effective for the process of virus removal from the Hb solution. We also confirmed the effectiveness of other virus removal ultrafiltration systems such as Viresolve (Millipore Corp.).

The obtained purified HbCO solution can be concentrated to greater than 40 g/dL using a tangential flow ultrafiltration process (e.g., Millipore Biomax V screen, cut off Mw: 8 kDa). After regulation of the electrolyte concentrations, this is supplied for the encapsulation procedure. The ligand of the resulting HbV, CO, is photodissociated and converted to O_2 by illuminating the liquid membrane of the HbV suspension with visible light under flowing O_2 (Chung *et al.*, 1995). A recent new idea is intravenous injection of CO-bound Hb-based O_2 carriers. Actually, CO is dissociated unexpectedly rapidly in the bloodstream while showing some cytoprotective effects (Sakai *et al.*, 2009c; Vandegriff *et al.*, 2008).

An alternative purification method would be to preserve the well-organized, but unstable, enzymatic systems that are originally present in RBCs, aiming at the prolonged stability of the ferrous state of Hb (Ogata *et al.*, 1997). The enzymatic system can be preserved in part with the compensation of insufficient virus removal or inactivation, but this cannot guarantee the safety of the resulting Hb-based oxygen carriers. One advantage of HbV is that any reagent can be coencapsulated in the vesicles. It has been confirmed that coencapsulation of the appropriate amount of a reductant, such as glutathione or homocysteine, and active oxygen scavengers, such as catalase, retards metHb formation (Sakai *et al.*, 2000b, 2004c; Takeoka *et al.*, 1997, 2002; Teramura *et al.*, 2003). It is also interesting to coencapsulate L-tyrosine to establish an artificial catalase system by the combination of metHb (Atoji *et al.*, 2006). Our recent interpretation, however, is that metHb formation might not present a serious problem in an emergency situation, because HbV would be infused as an interim measure until blood transfusion in a clinical setting.

4. Regulation of Oxygen Affinity

The O_2 affinity of purified Hb (expressed as P_{50}, O_2 tension at which Hb is half-saturated with O_2) is about 14 Torr in phosphate buffered saline. Hb strongly binds O_2 and does not release O_2 at 40 Torr (partial pressure of mixed venous blood). Historically, it has been regarded that the O_2 affinity should be regulated similar to that of RBC—about 25–30 Torr—using an allosteric effector or by direct chemical modification of the Hb molecules.

Theoretically, this allows sufficient O_2 unloading during blood microcirculation, as might be evaluated according to the arteriovenous difference in the levels of O_2 saturation, in accordance to an O_2 equilibrium curve. The P_{50} of Hb in HbV is expected to be equivalent to that of human RBCs, that is, 28 Torr, or higher if this theory is correct. Pyridoxal 5′-phosphate (PLP) is coencapsulated in HbV as an allosteric effector to regulate P_{50} (Sakai *et al.*, 2000a; Wang *et al.*, 1992). The main binding site of PLP is the N-termini of the α- and β-chains, and β-82 lysine within the β-cleft, which is part of the binding site of the natural allosteric effector, 2,3-diphosphoglyceric acid (2,3-DPG). The bound PLP retards dissociation of the ionic linkage between the β-chains of Hb during conversion of deoxy to oxyHb in the same manner as 2,3-DPG. Consequently, the O_2 affinity of Hb decreases in the presence of PLP. The P_{50} of HbV can be regulated to 5–150 Torr by coencapsulating the appropriate amount of PLP or inositol hexaphosphate (IHP) as an allosteric effector (Takahashi, 1995; Wang *et al.*, 1992). PLP equimolar to Hb (PLP/Hb = 1:1 by mol) is coencapsulated, and P_{50} is regulated to 18 Torr. When the molar ratio PLP/Hb is 3:1, P_{50} is regulated to 32 Torr. Furthermore, IHP strongly binds Hb; the resulting oxygen dissociation curve is distorted, and the Hill number becomes nearly 1. When bovine Hb is encapsulated in liposomes, Cl^- can regulate P_{50} (Sakai *et al.*, 2002b). It is 16 Torr without Cl^-, but it becomes 28 Torr with 100 mM Cl^-.

The O_2 affinities of HbV can be regulated easily without changing other physical parameters, but chemical structures determine O_2 affinities in the case of the other modified Hb solutions. For this reason, regulation is difficult. The appropriate O_2 affinities for O_2 carriers have not yet been decided completely. Easy regulation of the O_2 affinity, however, might be useful to meet the requirements of clinical indications such as oxygenation of ischemic tissues (Cabrales *et al.*, 2005; Contaldo *et al.*, 2003; Sakai and Tsuchida, 2007).

5. Structural Stabilization of Liposome-Encapsulated Hb

Hb autoxidizes to form metHb and loses its O_2-binding ability during storage, as well as during blood circulation. Therefore, prevention of metHb formation is necessary. Deoxygenated Hb and HbCO are stable, and can be stored in a liquid state (Kerwin *et al.*, 1999). Hb oxidation requires bound O_2 and depends on the O_2 partial pressure. Therefore, complete depletion of O_2 can prevent metHb formation. For LEH or HbV, not only the encapsulated Hb, but also the membrane structure must be physically stabilized, because liposomes, as molecular assemblies, have generally been characterized as structurally unstable. The US Naval

Research Laboratory tested the addition of disaccharides such as trehalose and maltose as cryoprotectants and lyoprotectants to LEH for its preservation as a powder, without causing hemolysis after rehydration (Rudolph, 1988). Furthermore, stabilization methods using polymer chains have been suggested, such as carboxymethylchitin, poly(ethylene glycol) (PEG), and actin (Kato *et al.*, 1984; Li *et al.*, 2005; Mobed and Chang, 1991; Yoshioka, 1991). Polymerization of phospholipids containing diacetylene groups was studied by Hayward *et al.* (1985). Liposomes encapsulating HbCO were irradiated by UV rays to obtain polymerized liposomes. The absorption spectrum of the resulting suspension differs from that of Hb, indicating that the color would not be the same as that of blood. Our group extensively studied the polymerization of phospholipids containing two dienoyl groups (1,2-dioctadecadienoyl-*sn*-glycero-3-phosphatidylcholine; DODPC) (Hosoi *et al.*, 1997; Sakai *et al.*, 1992; Sato *et al.*, 1992). For example, gamma-ray irradiation induces radiolysis of water molecules and generates OH radicals that initiate intermolecular polymerization of dienoyl groups in DODPC. This method produces remarkably stable liposomes encapsulating Hb, which are resistant to freeze-thawing, freeze-drying, and rehydration. A salient disadvantage, however, is that the polymerized liposomes are so stable that they are not degraded easily in macrophages, even 30 days after injection (Akama *et al.*, 2000). Polymerized lipids would be inappropriate for intravenous injection because of the difficulty in excretion. Subsequently, selection of appropriate lipids (phospholipid/cholesterol/negatively charged lipid/ PEG-conjugated phospholipid) and the composition are important to enhance the stability of nonpolymerized liposomes.

Surface modification of phospholipid vesicles with PEG-conjugated lipid is a well-known means to prolong the circulation time of the vesicles *in vivo* for drug delivery systems (Klibanov *et al.*, 1990; Papahadjopoulos *et al.*, 1991). The surface of HbV can be modified with PEG chains to improve the dispersion state of the vesicles when mixed with blood components (Yoshioka, 1991). PEG-modified HbV has shown improved blood circulation and tissue oxygenation attributable to the prevention of HbV aggregate formation and viscosity elevation (Sakai *et al.*, 1997, 1998), and prolonged circulation *in vivo* (Phillips *et al.*, 1999; Sou *et al.*, 2005). We studied the possibility of long-term preservation of HbV through the combination of two technologies: surface modification of HbV with PEG chains, and complete deoxygenation during storage for 2 years (Sakai *et al.*, 2000b). Samples stored at 4 and 23 °C show a stable dispersion state for 2 years, but the sample stored at 40 °C exhibits precipitation and decomposition of the vesicular components, decreased pH, and 4% leakage of the total Hb after 1 year. The PEG chains on the vesicle surface stabilize the dispersion, and prevent aggregation and fusion because of their stearic hindrance. The original metHb content (ca. 3%) before the preservation decreases gradually to less than 1% in all samples after 1 month because of the presence of homocysteine inside the vesicles that consume residual O_2 (thiol groups in homocysteines react with oxygen to generate disulfide and active oxygen species), and gradually reduce the trace amount of metHb. The rate of metHb formation is strongly dependent on the O_2 partial pressure; no increase in the metHb formation is observed because of the intrinsic stability of the deoxygenated Hb. These observations suggest that the HbV suspension can be stored at room temperature for at least 2 years, which would enable stockpiling of HbV for any emergency. Hemolysis (rupture of liposomes and release of Hb) must be prevented both *in vitro* and *in vivo*. In comparison to RBCs, HbV are highly resistant to hypotonic shock, freeze-thawing, and enzymatic attack by phospholipase A_2 (Sakai *et al.*, 2009b) because of the larger curvature of liposomes (i.e., larger surface volume ratio, or larger amount of lipids compared to RBCs), and the PEG chains that prevent intervesicular access and enzymatic attack.

Biodistribution of HbV can be examined using ^{99m}Tc-conjugated homocysteine or glutathione containing HbV (Phillips *et al.*, 1999; Sou *et al.*, 2005) and HbV containing ^{125}I-labeled Hb (Taguchi *et al.*, 2009a). These experiments show that HbV are finally captured by macrophages, mainly in the spleen and liver. Electron microscopic observation can detect the presence of Hb-encapsulating particles in the phagosomes of macrophages because of the high densities of protein and electrons (derived from Fe) in the particles such as RBCs. The HbV particles disappear in 1 week (Sakai *et al.*, 2001). This rapid disappearance is in contrast to the polymerized liposomes using DODPC (Akama *et al.*, 2000), which remained in the liver and spleen even after 30 days. Immunohistochemical staining with antihuman Hb antibody and antimethoxy-PEG indicates that Hb and PEG of HbV disappear in 2 weeks (Sakai *et al.*, 2001, 2004b, 2009a). It was shown recently that ^{125}I-labeled Hb and ^{3}H-labeled cholesterol in HbV have identical blood clearance, indicating that HbV retains its integrity in the bloodstream, and distributes to the reticuloendothelial system together. However, ^{125}I mainly appears in urine, and ^{3}H in feces, showing different metabolic routes in the macrophages (Taguchi *et al.*, 2009b).

Taking the points described above into consideration, it is important to consider the instability of particles after their capture by macrophages and their prompt excretion, in addition to sufficient stability during storage for years, and during blood circulation for a week.

6. Blood Compatibility of LEH and HbV

A so-called injection reaction, or pseudo-allergy, resulting from complement activation after injection of a small amount of liposome is well known, giving rise to anaphylatoxins, which trigger various hypersensitivity reactions (Chonn *et al.*, 1991; Loughrey *et al.*, 1990; Szebeni *et al.*, 2005). Transient thrombocytopenia and pulmonary hypertension in relation to complement activation is an extremely important hematologic effect observed in rodent and porcine models after infusion of LEH (containing DPPG) developed by the US Naval Research Laboratory (Phillips *et al.*, 1997; Rabinovici *et al.*, 1992; Szebeni *et al.*, 1999). Neo red cells (Terumo Corp.) containing stearic acid showed pulmonary hypertension in beagle and porcine models (Pape *et al.*, 2008), but not in monkeys. In our group, exchange transfusion of prototype HbV (containing DPPG, no PEG modification) in anesthetized rats engendered transient thrombocytopenia and slight hypertension (Izumi *et al.*, 1997). The transient reduction in platelet counts and increase of thromboxane B_2 caused by complement-bound liposomes was also associated with sequestration of platelets in the lung and liver (Phillips *et al.*, 1997). In the present formation of HbV, we use a negatively charged lipid (DHSG) instead of DPPG. It does not induce thrombocytopenia or complement activation in animal experiments (Abe *et al.*, 2007; Sou and Tsuchida, 2008), probably because it contains PEGylated lipids and a different type of negatively charged lipid (DHSG), instead of DPPG or a fatty acid.

The *in vitro* human blood compatibility of HbV has been extensively studied (Abe *et al.*, 2006, 2007; Wakamoto *et al.*, 2001, 2005). The present PEG-modified HbV containing DHSG does not affect the extrinsic or intrinsic coagulation activities of human plasma, although HbV-containing DPPG and no PEG modification tends to shorten the intrinsic coagulation time. The kallikrein–kinin cascade of plasma was activated slightly by the prototype DPPG-HbV, but not by the present PEG–DHSG–HbV. The exposure of human platelets to high concentrations of the this HbV (up to 40%) *in vitro* do not cause platelet activation and do not affect adversely the formation and secretion of prothrombotic substances or proinflammatory substances that are triggered by platelet agonists (Ito *et al.*, 2001). These results imply that HbV, at concentrations of up to 40%, do not have aberrant interactions with either unstimulated or agonist-induced platelets. It can be concluded that the PEG–DHSG–HbV described here have higher blood compatibility.

7. Regulation of Osmotic Pressure and Suspension Rheology to Mimic and Overwhelm the Function of Blood

The osmotic pressure of blood comprises crystalloid osmotic pressure and colloid osmotic pressure (COP) components. The former is derived mainly from Na^+, Cl^-, other electrolytes, and small molecules; the latter is derived from plasma proteins, mainly albumin, which is dissolved at about 5 g/dL in blood. Albumin regulates the fluid balance across a semipermeable biomembrane between the interstitial tissue and blood. A solution of Hb has

COP, while the HbV suspension does not, because one HbV particle (ca. 250 nm diameter) contains about 30,000 Hb molecules. In fact, HbV acts as a particle, not as a solute, as do RBCs. Therefore, HbV must be suspended in or coinjected with an aqueous solution of a plasma substitute, because injection of a blood substitute results in substitution of a large amount of blood. This contrasts with the characteristics of other Hb-based O_2 carriers, intramolecular cross-linked Hbs, polymerized Hbs, and polymer conjugated Hbs, which all possess high COP as protein solutions (Sakai *et al.*, 2000a).

Animal tests of HbV suspended in 5% plasma-derived HSA or rHSA showed an O_2 transporting capacity that is comparable to that of blood in extreme hemodilution and resuscitation from hemorrhagic shock studies (Izumi *et al.*, 1997; Sakai *et al.*, 1997, 1998, 1999, 2002a, 2004a, 2009a). As a primer fluid for extracorporeal circulation, HbV suspended in rHSA is effective (Yamazaki *et al.*, 2006). We previously reported that HbV suspended in plasma-derived HSA or rHSA was almost Newtonian: no aggregation was detected microscopically (Sakai *et al.*, 1997, 1998, 2009d,e). In Japan, rHSA was approved for clinical use in 2008, but various plasma substitutes are used worldwide, such as hydroxylethyl starch (HES, MW 70, 130, 200, 670 kDa), dextran (DEX, 70 kDa), and modified fluid gelatin (MFG, 32 kDa). The selection among these plasma substitutes is best determined not only according to their safety and efficacy but also according to their associated cost, experience of clinicians, and customs in different countries. Water-soluble polymers generally interact with particles such as polystyrene beads, liposomes, and RBCs to induce aggregation or flocculation by depletion interaction (Meyuhas *et al.*, 1996; Sakai *et al.*, 2009d; Sato *et al.*, 2009). A depletion layer develops near a particle surface that is in contact with a polymer solution if the loss of the configurational entropy of the coil of the polymer is not balanced by adsorption energy. Within this layer, the polymer concentration is lower than in the bulk phase. Consequently, as particles approach the osmotic pressure difference between the interparticle polymer-poor depletion layer and the bulk phase engenders solvent displacement into the bulk phase and consequent depletion interaction. Because of this interaction, the attractive force of particles tends to minimize the polymer-poor space between the particles, thereby inducing flocculation. Consequently, it is important to determine the compatibility of HbV with these plasma substitutes as water-soluble polymers. With that background, we studied rheological properties of HbV suspended in these plasma substitute solutions using a complex rheometer and a microchannel array (Sakai *et al.*, 2007). The rheological property of an Hb-based O_2 carrier is important because the infusion amount is expected to be considerably large, which might affect the blood viscosity and hemodynamics.

HbV suspended in rHSA is nearly Newtonian. Its viscosity resembles that of blood, and the mixtures with RBCs at various mixing ratios has viscosities of 3–4 cP. Other polymers, HES, DEX, and MFG, induce flocculation of

HbV, possibly by depletion interaction, and render the suspensions as non-Newtonian with the shear-thinning profile. These HbV suspensions have high viscosities and high storage moduli (G') because of the presence of flocculated HbV. On the other hand, HbV suspended in rHSA exhibits a very low G'. The viscosities of HbV suspended in DEX, MFG, and high-molecular-weight HES solutions respond quickly to rapid step changes of shear rates of 0.1–100 s^{-1} and a return to 0.1 s^{-1}, indicating that flocculation formation is both rapid and reversible. Microscopically, the flow pattern of the flocculated HbV perfused through microchannels (4.5 μm deep, 7 μm wide, and 20 cm H_2O applied pressure) shows no plugging. Furthermore, the time required for passage is related directly to the viscosity.

Lower blood viscosity after hemodilution has been regarded as being effective for tissue perfusion. Microcirculatory observations indicate, however, that in some cases lower "plasma viscosity" decreases shear stress on the vascular wall, causing vasoconstriction and reduction of the functional capillary density (Tsai *et al.*, 1998). Therefore, an appropriate viscosity that maintains the normal tissue perfusion level might exist. The large molecular dimension of HbV can yield a transfusion fluid with high viscosity. Liposomes with a large molecular dimension are also effective in reducing vascular permeability and minimizing the reaction with NO and CO as vasorelaxation factors (Sakai *et al.*, 2000c, 2008a,b). Erni and collaborators have shown that HbV with a high O_2-binding affinity (low P_{50}, such as 8–15 Torr) and high viscosity (such as 11 cP) suspended in a high-molecular-weight HES solution is effective for oxygenation of an ischemic skin flap (Contaldo *et al.*, 2003, 2005; Plock *et al.*, 2005, 2007). HbV retains O_2 in the upper arterioles, then perfuses through collateral arteries, and delivers O_2 to the targeted ischemic tissues: a concept of targeted O_2 delivery by a Hb-based O_2 carrier. A high O_2-binding affinity (low P_{50}) is also effective for improving the O_2 saturation of Hb in pulmonary capillaries when exposed to a hypoxic atmosphere or with an impaired lung function. Some plasma substitutes cause flocculation of HbV and hyperviscosity. Hyperviscosity, however, would not necessarily be deleterious in the body and might even be advantageous in some cases (Martini *et al.*, 2006).

The use of HbV provides a unique opportunity to manipulate the suspension rheology, P_{50}, and other physicochemical properties, not only as a transfusion alternative, but also for other clinical applications, including oxygenation of ischemic tissues and *ex vivo* perfusion systems.

8. Concluding Remarks

In spite of the long history of blood substitute development, no material using Hb is clinically available yet, because many investigators have specifically addressed cell-free Hb-based oxygen carriers, which are much easier to produce. Now that the physiological significance of the cellular structure of RBCs has been widely recognized, cellular type HbV or LEH is becoming the next promising blood substitute. The use of HbV presents both advantages and disadvantages in comparison to sophisticated RBC functions. We hope this chapter can provide various useful alternatives to investigators for modification or manipulation of the components, and production of new formulations of LEH.

REFERENCES

Abe, H., Ikebuchi, K., Hirayama, J., Fujihara, M., Takeoka, S., Sakai, H., Tsuchida, E., and Ikeda, H. (2001). Virus inactivation in hemoglobin solution by heat treatment. *Artif. Cells Blood Substit. Immobil. Biotechnol.* **29,** 381–388.

Abe, H., Fujihara, M., Azuma, H., Ikeda, H., Ikebuchi, K., Takeoka, S., Tsuchida, E., and Harashima, H. (2006). Interaction of hemoglobin vesicles, a cellular-type artificial oxygen carrier, with human plasma: Effects on coagulation, kallikrein-kinin, and complement systems. *Artif. Cells Blood Substit. Immobil. Biotechnol.* **34,** 1–10.

Abe, H., Azuma, H., Yamaguchi, M., Fujihara, M., Ikeda, H., Sakai, H., Takeoka, S., and Tsuchida, E. (2007). Effects of hemoglobin vesicles, a liposomal artificial oxygen carrier, on hematological responses, complement and anaphylactic reactions in rats. *Artif. Cells Blood Substit. Immobil. Biotechnol.* **35,** 157–172.

Akama, K., Awai, K., Yano, Y., Tokuyama, S., and Nakano, Y. (2000). *In vitro* and *in vivo* stability of polymerized mixed liposomes coposed of 2, 4-octadecadienoyl groups of phospholipids. *Polym. Adv. Technol.* **11,** 280–287.

Atoji, T., Aihara, M., Sakai, H., Tsuchida, E., and Takeoka, S. (2006). Hemoglobin vesicles containing methemoglobin and L-tyrosine to suppress methemoglobin formation *in vitro* and *in vivo*. *Bioconjug. Chem.* **17,** 1241–1245.

Bangham, A. D., and Horne, R. W. (1964). Negative staining of phospholipids and their structure modification by surface-active agents as observed in the electron microscope. *J. Mol. Biol.* **8,** 660–668.

Beissinger, R. L., Farmer, M. C., and Gossage, J. L. (1986). Liposome-encapsulated hemoglobin as a red cell surrogate. *ASAIO Trans.* **32,** 58–63.

Cabrales, P., Sakai, H., Tsai, A. G., Takeoka, S., Tsuchida, E., and Intaglietta, M. (2005). Oxygen transport by low and normal oxygen affinity hemoglobin vesicles in extreme hemodilution. *Am. J. Physiol. Heart Circ. Physiol.* **288,** H1885–H1892.

Chang, T. M. S. (1991). Blood substitutes based on modified hemoglobin prepared by encapsulation or crosslinking: An overview. *Biomater. Artif. Cells Immobilization Biotechnol.* **20,** 159–182.

Chang, T. M. (2004). Hemoglobin-based red blood cell substitutes. *Artif. Organs* **28,** 789–794.

Chonn, A., Cullis, P. R., and Devine, D. V. (1991). The role of surface charge in the activation of the classical and alternative pathways of complement by liposomes. *J. Immunol.* **146,** 4234–4241.

Chung, J. E., Hamada, K., Sakai, H., Takeoka, S., Nishide, H., and Tsuchida, E. (1995). Ligand exchange reaction of carbonylhemoglobin to oxyhemoglobin in a hemoglobin liquid membrane. *Nippon Kagaku Kaishi* **1995,** 123–127.

Contaldo, C., Schramm, S., Wettstein, R., Sakai, H., Takeoka, S., Tsuchida, E., Leunig, M., Banic, A., and Erni, D. (2003). Improved oxygenation in ischemic hamster flap tissue is correlated with increasing hemodilution with Hb vesicles and their O_2 affinity. *Am. J. Physiol. Heart Circ. Physiol.* **285,** H1140–H1147.

Contaldo, C., Plock, J., Sakai, H., Takeoka, S., Tsuchida, E., Leunig, M., Banic, A., and Erni, D. (2005). New generation of hemoglobin-based oxygen carriers evaluated for oxygenation of critically ischemic hamster flap tissue. *Crit. Care Med.* **33,** 806–812.

DeLoach, J. R., Sheffield, C. L., and Spates, G. E. (1986). A continuous-flow high yield process for preparation of lipid-free hemoglobin. *Anal. Biochem.* **157,** 191–198.

Djordjevich, L., and Ivankovich, A. D. (1988). Liposomes as carriers of hemoglobin. *In* "Liposomes as Drug Carriers," (G. Gregoriadis, ed.), pp. 551–567. John Wiley & Sons, Chapter 39.

Djordjevich, L., and Miller, I. F. (1977). Lipid encapsulated hemoglobin as a synthetic erythrocyte. *Fed. Proc.* **36,** 567.

Farmer, M. C., and Gaber, B. P. (1987). Liposome-encapsulated hemoglobin as an artificial oxygen carrying system. *Methods Enzymol.* **149,** 184–200.

Fukutomi, I., Sakai, H., Takeoka, S., Nishide, H., Tsuchida, E., and Sakai, K. (2002). Carbonylation of oxyhemoglobin solution using a membrane oxygenator. *J. Artif. Organs* **5,** 102–107.

Gaber, B. P., Yager, P., Sheridan, J. P., and Chang, E. L. (1983). Encapsulation of hemoglobin in phospholipid vesicles. *FEBS Lett.* **153,** 285–288.

Hamilton, R. G., and Kickler, T. S. (2007). Bovine hemoglobin (glutamer-250, Hemopure)-specific immunoglobulin G antibody cross-reacts with human hemoglobin but does not lyse red blood cells *in vitro*. *Transfusion* **47,** 723–728.

Hayward, J. A., Levine, D. M., Neufeld, L., Simon, S. R., Johnston, D. S., and Chapman, D. (1985). Polymerized liposomes as stable oxygen-carriers. *FEBS Lett.* **187,** 261–266.

Hosoi, F., Omichi, H., Akama, K., Awai, K., Endo, S., and Nakano, Y. (1997). Radiation-induced polymerization of unsaturated phospholipid mixtures for the synthesis of artificial red cells. *Nucl. Instrum. Methods Phys. Res. B* **131,** 329–334.

Huang, Y., Takeoka, S., Sakai, H., Abe, H., Hirayama, J., Ikebuchi, K., Ikeda, H., and Tsuchida, E. (2002). Complete deoxygenation from a hemoglobin solution by an electrochemical method and heat treatment for virus inactivation. *Biotechnol. Prog.* **18,** 101–107.

Hunt, C. A., Burnette, R. R., MacGregor, R. D., Strubbe, A. E., Lau, D. T., Taylor, N., and Kawada, H. (1985). Synthesis and evaluation of prototypal artificial red cells. *Science* **230,** 1165–1168.

Ito, T., Fujihara, M., Abe, H., Yamaguchi, M., Wakamoto, S., Takeoka, S., Sakai, H., Tsuchida, E., Ikeda, H., and Ikebuchi, K. (2001). Effects of poly(ethyleneglycol)-modified hemoglobin vesicles on N-formyl-methionyl-leucyl-phenylalanine-induced responses of polymorphonuclear neutrophils *in vitro*. *Artif. Cells Blood Substit. Immobil. Biotechnol.* **29,** 427–437.

Izumi, Y., Sakai, H., Takeoka, S., Kose, T., Hamada, K., Yoshizu, A., Horinouchi, H., Kato, R., Nishide, H., Tsuchida, E., and Kobayashi, K. (1997). Evaluation of the capabilities of a hemoglobin vesicle as an artificial oxygen carrier in a rat exchange transfusion model. *ASAIO J.* **43,** 289–297.

Jopski, B., Pirkl, V., Jaroni, H.-W., Schubert, R., and Schmidt, K.-H. (1989). Preparation of hemoglobin-containing liposomes using octyl glucoside and octyltetraoxyethylene. *Biochim. Biophys. Acta* **978,** 79–84.

Kato, A., Arakawa, M., and Kondo, T. (1984). Liposome-type artificial red blood cells stabilized with carboxymethylchitin. *Nippon Kagaku Kaishi* **6,** 987–991 (in Japanese).

Kerwin, B. A., Akers, M. J., Apostol, I., Moore-Einsel, C., Etter, J. E., Hess, E., Lippincott, J., Levine, J., Mathews, A. J., Revilla-Sharp, P., Schubert, R., and

Looker, D. L. (1999). Acute and long-term stability studies of deoxy hemoglobin and characterization of ascorbate-induced modifications. *J. Pharm. Sci.* **88,** 79–88.

Klibanov, A. L., Maruyama, K., Torchilin, V. P., and Huang, L. (1990). Amphipathic polyethylene glycols effectively prolong the circulation time of liposomes. *FEBS Lett.* **268,** 235–237.

Li, S., Nickels, J., and Palmer, A. F. (2005). Liposome-encapsulated actin-hemoglobin (LEAcHb) artificial blood substitutes. *Biomaterials* **26,** 3759–3769.

Liu, L., and Yonetani, T. (1994). Preparation and characterization of liposome-encapsulated haemoglobin by a freeze–thaw method. *J. Microencapsul.* **11,** 409–421.

Loughrey, H. C., Bally, M. B., Reinish, L. W., and Cullis, P. R. (1990). The binding of phosphatidylglycerol liposomes to rat platelets is mediated by complement. *Thromb. Haemost.* **64,** 172–176.

Martini, J., Cabrales, P., Tsai, A. G., and Intaglietta, M. (2006). Mechanotransduction and the homeostatic significance of maintaining blood viscosity in hypotension, hypertension and haemorrhage. *J. Intern. Med.* **259,** 364–372.

Meyuhas, D., Nir, S., and Lichtenberg, D. (1996). Aggregation of phospholipid vesicles by water-soluble polymers. *Biophys. J.* **71,** 2602–2612.

Mobed, M., and Chang, T. M. S. (1991). Preparation and surface characterization of carboxymethylchitin-incorporated submicron bilayer-lipid membrane artificial cells (liposomes) encapsulating hemoglobin. *Biomater. Artif. Cells Immobilization Biotechnol.* **19,** 731–744.

Murray, J. A., Ledlow, A., Launspach, J., Evans, D., Loveday, M., and Conklin, J. L. (1995). The effects of recombinant human hemoglobin on esophageal motor function in humans. *Gastroenterology* **109,** 1241–1248.

Naito, Y., Fukutomi, I., Masada, Y., Sakai, H., Takeoka, S., Tsuchida, E., Abe, H., Hirayama, J., Ikebuchi, K., and Ikeda, H. (2002). Virus removal from hemoglobin solution using Planova membrane. *J. Artif. Organs* **5,** 141–145.

Natanson, C., Kern, S. J., Lurie, P., Banks, S. M., and Wolfe, S. M. (2008). Cell-free hemoglobin-based blood substitutes and risk of myocardial infarction and death: A meta-analysis. *JAMA* **299,** 2304–2312.

Ogata, Y., Goto, H., Kimura, T., and Fukui, H. (1997). Development of neo red cells (NRC) with the enzymatic reduction system of methemoglobin. *Artif. Cells Blood Substit. Immobil. Biotechnol.* **25,** 417–427.

Palmer, A. F., Sun, G., and Harris, D. R. (2009). Tangential flow filtration of hemoglobin. *Biotechnol. Prog.* **25,** 189–199.

Papahadjopoulos, D., Allen, T. M., Gabizon, A., Mayhew, E., Matthay, K., Huang, S. K., Lee, K. D., Woodle, M. C., Lasic, D. D., Redemann, C., and Martin, F. J. (1991). Sterically stabilized liposomes: Improvements in pharmacokinetics and antitumor therapeutic efficacy. *Proc. Natl. Acad. Sci. USA* **88,** 11460–11464.

Pape, A., Kertscho, H., Meier, J., Horn, O., Laout, M., Steche, M., Lossen, M., Theisen, A., Zwissler, B., and Habler, O. (2008). Improved short-term survival with polyethylene glycol modified hemoglobin liposomes in critical normovolemic anemia. *Intensive Care Med.* **34,** 1534–1543.

Phillips, W. T., Klipper, R., Fresne, D., Rudolph, A. S., Javors, M., and Goins, B. (1997). Platelet reactivity with liposome-encapsulated hemoglobin in the rat. *Exp. Hematol.* **25,** 1347–1356.

Phillips, W. T., Klipper, R. W., Awasthi, V. D., Rudolph, A. S., Cliff, R., Kwasiborski, V., and Goins, B. A. (1999). Polyethylene glycol-modified liposome-encapsulated hemoglobin: A long circulating red cell substitute. *J. Pharmacol. Exp. Ther.* **288,** 665–670.

Plock, J. A., Contaldo, C., Sakai, H., Tsuchida, E., Leunig, M., Banic, A., Menger, M. D., and Erni, D. (2005). Is the Hb in Hb vesicles infused for isovolemic hemodilution necessary to improve oxygenation in critically ischemic hamster skin? *Am. J. Physiol. Heart Circ. Physiol.* **289,** H2624–H2631.

Plock, J. A., Tromp, A. E., Contaldo, C., Spanholtz, T., Sinovcic, D., Sakai, H., Tsuchida, E., Leunig, M., Banic, A., and Erni, D. (2007). Hemoglobin vesicles reduce hypoxia-related inflammation in critically ischemic hamster flap tissue. *Crit. Care Med.* **35,** 899–905.

Rabinovici, R., Rudolph, A. S., Ligler, F. S., Smith, E. F. 3rd, and Feuerstein, G. (1992). Biological responses to exchange transfusion with liposome-encapsulated hemoglobin. *Circ. Shock* **37,** 124–133.

Rabinovici, R., Rudolph, A. S., Vernick, J., and Feuerstein, G. (1993). A new salutary resuscitative fluid: Liposome encapsulated hemoglobin/hypertonic saline solution. *J. Trauma* **35,** 121–127.

Rudolph, A. S. (1988). The freeze–dried preservation of liposome encapsulated hemoglobin: A potential blood substitute. *Cryobiology* **25,** 277–284.

Rudolph, A. S., Klipper, R. W., Goins, B., and Phillips, W. T. (1991). *In vivo* biodistribution of a radiolabeled blood substitute: ^{99m}Tc-labeled liposome-encapsulated hemoglobin in an anesthetized rabbit. *Proc. Natl Acad. Sci. USA* **88,** 10976–10980.

Sakai, H., and Tsuchida, E. (2007). Hemoglobin-vesicles for a transfusion alternative and targeted oxygen delivery. *J. Liposome Res.* **17,** 227–235.

Sakai, H., Takeoka, S., Yokohama, H., Nishide, H., and Tsuchida, E. (1992). Encapsulation of Hb into unsaturated lipid vesicles and γ-ray polymerization. *Polym. Adv. Technol.* **3,** 389–394.

Sakai, H., Takeoka, S., Yokohama, H., Seino, Y., Nishide, H., and Tsuchida, E. (1993). Purification of concentrated Hb using organic solvent and heat treatment. *Protein Expr. Purif.* **4,** 563–569.

Sakai, H., Hamada, K., Takeoka, S., Nishide, H., and Tsuchida, E. (1996). Physical properties of hemoglobin vesicles as red cell substitutes. *Biotechnol. Prog.* **12,** 119–125.

Sakai, H., Takeoka, S., Park, S. I., Kose, T., Izumi, Y., Yoshizu, A., Nishide, H., Kobayashi, K., and Tsuchida, E. (1997). Surface-modification of hemoglobin vesicles with poly(ethylene glycol) and effects on aggregation, viscosity, and blood flow during 90%-exchange transfusion in anesthetized rats. *Bioconjug. Chem.* **8,** 23–30.

Sakai, H., Tsai, A. G., Kerger, H., Park, S. I., Takeoka, S., Nishide, H., Tsuchida, E., and Intaglietta, M. (1998). Subcutaneous microvascular responses to hemodilution with red cell substitutes consisting of polyethylene glycol-modified vesicles encapsulating hemoglobin. *J. Biomed. Mater. Res.* **40,** 66–78.

Sakai, H., Tsai, A. G., Rohlfs, R. J., Hara, H., Takeoka, S., Tsuchida, E., and Intaglietta, M. (1999). Microvascular responses to hemodilution with Hb-vesicles as red cell substitutes: Influences of O_2 affinity. *Am. J. Physiol. Heart Circ. Physiol.* **276,** H553–H562.

Sakai, H., Yuasa, M., Onuma, H., Takeoka, S., and Tsuchida, E. (2000a). Synthesis and physicochemical characterization of a series of hemoglobin-based oxygen carriers: Objective comparison between cellular and acellular types. *Bioconjug. Chem.* **11,** 56–64.

Sakai, H., Tomiyama, K., Sou, K., Takeoka, S., and Tsuchida, E. (2000b). Poly(ethylene glycol)-conjugation and deoxygenation enable long-term preservation of hemoglobin-vesicles as oxygen carriers in a liquid state. *Bioconjug. Chem.* **11,** 425–432.

Sakai, H., Hara, H., Yuasa, M., Tsai, A. G., Takeoka, S., Tsuchida, E., and Intaglietta, M. (2000c). Molecular dimensions of Hb-based O_2 carriers determine constriction of resistance arteries and hypertension. *Am. J. Physiol. Heart Circ. Physiol.* **279,** H908–H915.

Sakai, H., Horinouchi, H., Tomiyama, K., Ikeda, E., Takeoka, S., Kobayashi, K., and Tsuchida, E. (2001). Hemoglobin-vesicles as oxygen carriers: Influence on phagocytic activity and histopathological changes in reticuloendothelial system. *Am. J. Pathol.* **159,** 1079–1088.

Sakai, H., Takeoka, S., Wettstein, R., Tsai, A. G., Intaglietta, M., and Tsuchida, E. (2002a). Systemic and microvascular responses to the hemorrhagic shock and resuscitation with Hb-vesicles. *Am. J. Physiol. Heart Circ. Physiol.* **283,** H1191–H1199.

Sakai, H., Masada, Y., Takeoka, S., and Tsuchida, E. (2002b). Characteristics of bovine hemoglobin as a potential source of hemoglobin-vesicles for an artificial oxygen carrier. *J. Biochem.* **131,** 611–617.

Sakai, H., Horinouchi, H., Masada, Y., Yamamoto, M., Takeoka, S., Kobayashi, K., and Tsuchida, E. (2004a). Hemoglobin-vesicles suspended in recombinant human serum albumin for resuscitation from hemorrhagic shock in anesthetized rats. *Crit. Care Med.* **32,** 539–545.

Sakai, H., Masada, Y., Horinouchi, H., Ikeda, H., Takeoka, S., Suematsu, M., Kobayashi, K., and Tsuchida, E. (2004b). Physiologic capacity of reticuloendothelial system for degradation of hemoglobin-vesicles (artificial oxygen carriers) after massive intravenous doses by daily repeated infusions for 14 days. *J. Pharmacol. Exp. Ther.* **311,** 874–884.

Sakai, H., Masada, Y., Onuma, H., Takeoka, S., and Tsuchida, E. (2004c). Reduction of methemoglobin via electron transfer from photoreduced flavin: Restoration of O_2-binding of concentrated hemoglobin solution coencapsulated in phospholipid vesicles. *Bioconjug. Chem.* **15,** 1037–1045.

Sakai, H., Hisamoto, S., Fukutomi, I., Sou, K., Takeoka, S., and Tsuchida, E. (2004d). Detection of lipopolysaccharide in hemoglobin-vesicles by Limulus amebocyte lysate test with kinetic-turbidimetric gel clotting analysis and pretreatment of surfactant. *J. Pharm. Sci.* **93,** 310–321.

Sakai, H., Sato, A., Takeoka, S., and Tsuchida, E. (2007). Rheological property of hemoglobin vesicles (artificial oxygen carriers) suspended in a series of plasma substitute aqueous solutions. *Langmuir* **23,** 8121–8128.

Sakai, H., Sato, A., Masuda, K., Takeoka, S., and Tsuchida, E. (2008a). Encapsulation of concentrated hemoglobin solution in phospholipid vesicles retards the reaction with NO, but not CO, by intracellular diffusion barrier. *J. Biol. Chem.* **283,** 1508–1517.

Sakai, H., Sato, A., Sobolewski, P., Takeoka, S., Frangos, J. A., Kobayashi, K., Intaglietta, M., and Tsuchida, E. (2008b). NO and CO binding profiles of hemoglobin vesicles as artificial oxygen carriers. *Biochim. Biophys. Acta* **1784,** 1441–1447.

Sakai, H., Seishi, Y., Obata, Y., Takeoka, S., Horinouichi, H., Tsuchida, E., and Kobayashi, K. (2009a). Fluid resuscitation with artificial oxygen carriers in hemorrhaged rats: Profiles of hemoglobin-vesicle degradation and hematopoiesis for 14 days. *Shock* **31,** 192–200.

Sakai, H., Okamoto, M., Ikeda, E., Horinouchi, H., Kobayashi, K., and Tsuchida, E. (2009b). Histopathological changes of rat brain after direct injection of hemoglobin vesicles (oxygen carriers) and neurological impact in an intracerebral hemorrhage model. *J. Biomed. Mater. Res. A* **88A,** 34–42.

Sakai, H., Horinouchi, H., Tsuchida, E., and Kobayashi, K. (2009c). Hemoglobin-vesicles and red blood cells as carriers of carbon monoxide prior to oxygen for resuscitation after hemorrhagic shock in a rat model. *Shock* **31,** 507–514.

Sakai, H, Sato, A., Takeoka, S., and Tsuchida, E. (2009d). Mechanism of flocculate formation of phospholipid vesicles suspended in a series of water-soluble biopolymers. *Biomacromolecules* **10,** 2344–2350.

Sakai, H., Sato, A., Okuda, N., Takeoka, S., Maeda, N., and Tsuchida, E. (2009e). Peculiar flow patterns of RBCs suspended in viscous fluids and perfused through a narrow tube (25 μm). *Am. J. Physiol. Heart Circ. Physiol.* **297,** H905–H910.

Sato, T., Kobayashi, K., Sekiguchi, S., and Tsuchida, E. (1992). Characteristics of artificial red cells: Hemoglobin-encapsulated in poly-lipid vesicles. *ASAIO J.* **38,** M580–M584.

Sato, T., Sakai, H., Sou, K., Medebach, M., Glatter, O., and Tsuchida, E. (2009). Static structure and dynamics of hemoglobin vesicle (HbV) developed as a transfusion alternative. *J. Phys. Chem. B* **113,** 8418–8428.

Sheffield, C. L., Spates, G. E., Droleskey, R. E., Green, R., and DeLoach, J. R. (1987). Preparation of lipid-free human hemoglobin by dialysis and ultrafiltration. *Biotechnol. Appl. Biochem.* **9,** 230–238.

Sou, K., and Tsuchida, E. (2008). Electrostatic interactions and complement activation on the surface of phospholipid vesicle containing acidic lipids: Effect of the structure of acidic groups. *Biochim. Biophys. Acta* **1778,** 1035–1041.

Sou, K., Naito, Y., Endo, T., Takeoka, S., and Tsuchida, E. (2003). Effective encapsulation of proteins into size-controlled phospholipid vesicles using freeze–thawing and extrusion. *Biotechnol. Prog.* **19,** 1547–1552.

Sou, K., Klipper, R., Goins, B., Tsuchida, E., and Phillips, W. T. (2005). Circulation kinetics and organ distribution of Hb vesicles developed as a red blood cell substitute. *J. Pharmacol. Exp. Ther.* **312,** 702–709.

Szebeni, J., Breuer, J. H., Szelenyi, J. G., Bathori, G., Lelkes, G., and Hollan, S. R. (1984). Oxidation and denaturation of hemoglobin encapsulated in liposomes. *Biochim. Biophys. Acta* **798,** 60–67.

Szebeni, J., Di Iorio, E. E., Hauser, H., and Winterhalter, K. H. (1985). Encapsulation of hemoglobin in phospholipid liposomes: Characterization and stability. *Biochemistry* **24,** 2827–2832.

Szebeni, J., Fontana, J. L., Wassef, N. M., Mongan, P. D., Morse, D. S., Dobbins, D. E., Stahl, G. L., Bunger, R., and Alving, C. R. (1999). Hemodynamic changes induced by liposomes and liposome-encapsulated hemoglobin in pigs: A model for pseudoallergic cardiopulmonary reactions to liposomes. Role of complement and inhibition by soluble CR1 and anti-C5a antibody. *Circulation* **99,** 2302–2309.

Szebeni, J., Baranyi, L., Sávay, S., Bodó, M., Milosevits, J., Alving, C. R., and Bünger, R. (2005). Complement activation-related cardiac anaphylaxis in pigs: Role of C5a anaphylatoxin and adenosine in liposome-induced abnormalities in ECG and heart function. *Am. J. Physiol. Heart Circ. Physiol.* **290,** H1050–H1058.

Taguchi, K., Maruyama, T., Iwao, Y., Sakai, H., Kobayashi, K., Horinouchi, H., Tsuchida, E., Kai, T., and Otagiri, M. (2009a). Pharmacokinetics of single and repeated injection of hemoglobin-vesicles in hemorrhagic shock rat model. *J. Control. Release* **136,** 232–239.

Taguchi, K., Urata, Y., Anraku, M., Maruyama, T., Watanabe, H., Sakai, H., Horinouchi, H., Kobayashi, K., Tsuchida, E., Kai, T., and Otagiri, M. (2009b). Pharmacokinetic study of enclosed hemoglobin and outer lipid component after the administration of hemoglobin-vesicles as an artificial oxygen carrier. *Drug Dispos. Metabol.* **37,** 1456–1463.

Takahashi, A. (1995). Characterization of neo red cells (NRCs), their function and safety *in vivo* tests. *Artif. Cells Blood Substit. Immobil. Biotechnol.* **23,** 347–354.

Takeoka, S., Sakai, H., Nishide, H., and Tsuchida, E. (1993). Preparation conditions of human hemoglobin-vesicles covered with lipid membranes. *Jpn. J. Artif. Organs* **22,** 566–569.

Takeoka, S., Terase, K., Sakai, H., Yokohama, H., Nishide, H., and Tsuchida, E. (1994a). Interaction between phosphoslipid assemblies and hemoglobin (Hb). *J. Macromol. Sci. Pure Appl. Chem.* **A31,** 97–108.

Takeoka, S., Sakai, H., Terease, K., Nishide, H., and Tsuchida, E. (1994b). Characteristics of Hb-vesicles and encapsulation procedure. *Artif. Cells Blood Substit. Immobil. Biotechnol.* **22,** 861–866.

Takeoka, S., Ohgushi, T., Terase, K., and Tsuchida, E. (1996). Layer-controlled hemoglobin vesicles by interaction of hemoglobin with a phospholipid assembly. *Langmuir* **12,** 1755–1759.

Takeoka, S., Sakai, H., Kose, T., Mano, Y., Seino, Y., Nishide, H., and Tsuchida, E. (1997). Methemoglobin formation in hemoglobin vesicles and reduction by encapsulated thiols. *Bioconjug. Chem.* **8,** 539–544.

Takeoka, S., Teramura, Y., Atoji, T., and Tsuchida, E. (2002). Effect of Hb-encapsulation with vesicles on H_2O_2 reaction and lipid peroxidation. *Bioconjug. Chem.* **13,** 1302–1308.

Teramura, Y., Kanazawa, H., Sakai, H., Takeoka, S., and Tsuchida, E. (2003). The prolonged oxygen-carrying ability of Hb vesicles by coencapsulation of catalase *in vivo*. *Bioconjug. Chem.* **14,** 1171–1176.

Tsai, A. G., Friesenecker, B., McCarthy, M., Sakai, H., and Intaglietta, M. (1998). Plasma viscosity regulates capillary perfusion during extreme hemodilution in hamster skinfold model. *Am. J. Physiol. Heart Circ. Physiol.* **275,** H2170–H2180.

Tsuchida, E. Ed. (1998). *Blood Substitutes: Present and Future Perspectives*. Elsevier, Chichester.

Vandegriff, K. D., Young, M. A., Lohman, J., Bellelli, A., Samaja, M., Malavalli, A., and Winslow, R. M. (2008). CO-MP4, a polyethylene glycol-conjugated haemoglobin derivative and carbon monoxide carrier that reduces myocardial infarct size in rats. *Br. J. Pharmacol.* **154,** 1649–1661.

Vidal-Naquet, A., Gossage, J. L., Sullivan, T. P., Haynes, J. W., Gilruth, B. H., Beissinger, R. L., Sehgal, L. R., and Rosen, A. L. (1989). Liposome-encapsulated hemoglobin as an artificial red blood cell: Characterization and scale-up. *Biomater. Artif. Cells Artif. Organs* **17,** 531–552.

Vivier, A., Vuillemard, J. C., Ackermann, H. W., and Poncelet, D. (1992). Large-scale blood substitute production using a microfluidizer. *Biomater. Artif. Cells Immobilization Biotechnol.* **20,** 377–397.

Wakamoto, S., Fujihara, M., Abe, H., Sakai, H., Takeoka, S., Tsuchida, E., Ikeda, H., and Ikebuchi, K. (2001). Effects of poly(ethyleneglycol)-modified hemoglobin vesicles on agonist-induced platelet aggregation and RANTES release *in vitro*. *Artif. Cells Blood Substit. Immobil. Biotechnol.* **29,** 191–201.

Wakamoto, S., Fujihara, M., Abe, H., Yamaguchi, M., Azuma, H., Ikeda, H., Takeoka, S., and Tsuchida, E. (2005). Effects of hemoglobin vesicles on resting and agonist-stimulated human platelets *in vitro*. *Artif. Cells Blood Substit. Immobil. Biotechnol.* **33,** 101–111.

Wang, L., Morizawa, K., Tokuyama, S., Satoh, T., and Tsuchida, E. (1992). Modulation of oxygen-carrying capacity of artificial red cells (ARC). *Polym. Adv. Technol.* **4,** 8–11.

Winslow, R., and Chapman, K. W. (1994). Pilot-scale preparation of hemoglobin solutions. *Methods Enzymol.* **231,** 3–16.

Yamazaki, M., Aeba, R., Yozu, R., and Kobayashi, K. (2006). Use of hemoglobin vesicles during cardiopulmonary bypass priming prevents neurocognitive decline in rats. *Circulation* **114**(1 Suppl), I220–I225.

Yoshioka, H. (1991). Surface modification of haemoglobin-containing liposomes with poly (ethylene glycol) prevents liposome aggregation in blood plasma. *Biomaterials* **12,** 861–864.

Author Index

Subject Index

C

D

P

Q

R

S

Y

Z

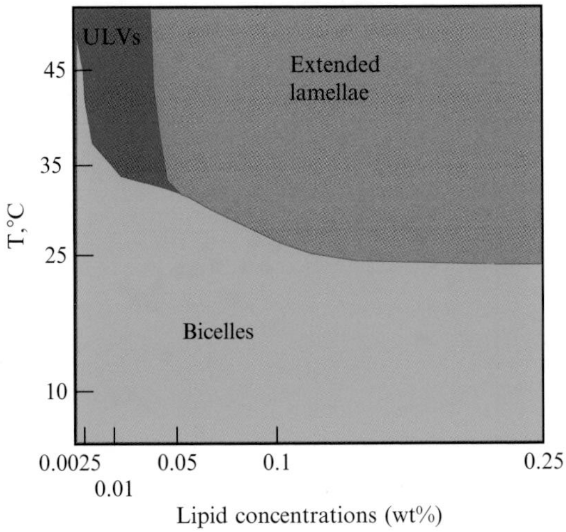

Mu-Ping Nieh *et al.*, Figure 1.2 Structural phase diagram constructed by SANS.

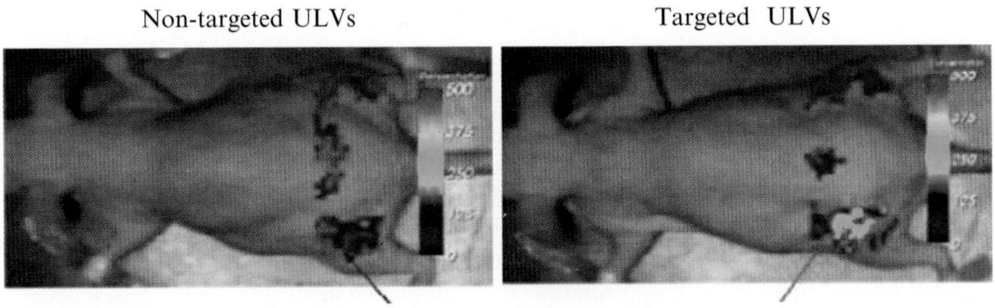

Mu-Ping Nieh *et al.*, Figure 1.8 Imaging payloads delivered to a xenograft tumor using antibody-functionalized ULVs loaded with Gd, Cy5.5, and the C225 antibody. The Cy5.5 signal is predominant in the tumor where functionalized ULVs were used.

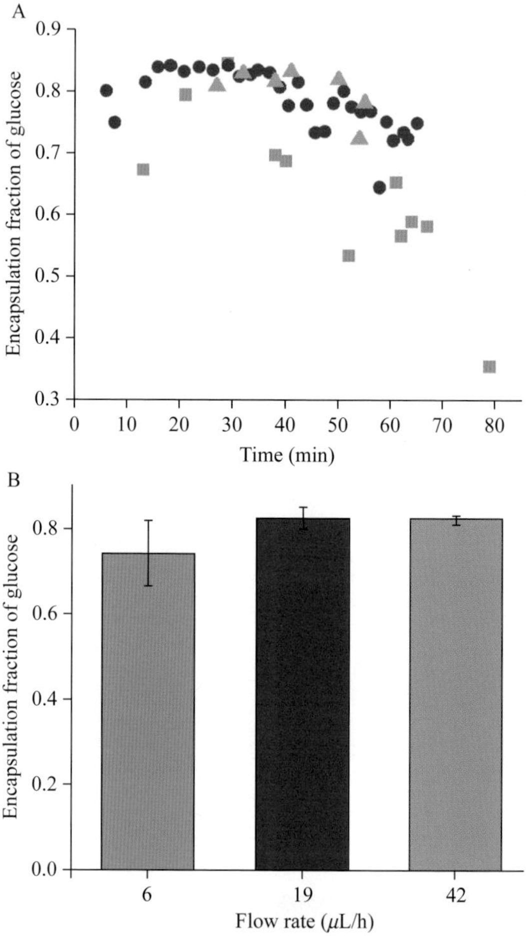

Thomas H. Li *et al.*, Figure 4.6 Experimentally determined glucose encapsulation fraction by the falling vesicle method comparing three different nozzle flow rates, with (A) encapsulation fraction over time and (B) average encapsulation fraction for each flow rate during the first 40 min. Vesicles are formed by jetting sucrose solution through glucose. The lack of a trend suggests that continuous flow through the nozzle does not significantly affect encapsulation fraction in vesicles.

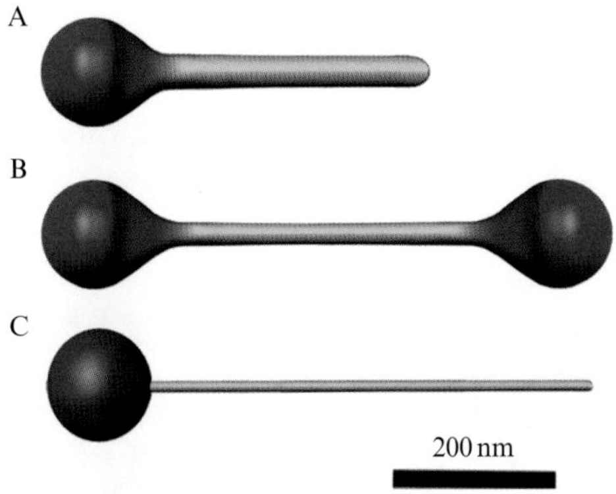

Alexandra Zidovska *et al.*, Figure 6.3 Schematic depiction of the three main types of block liposomes. Different colors represent different membrane Gaussian curvature: positive (red), negative (blue), and zero (yellow). (A) Pear-tube diblock liposome. (B) Pear-tube-pear triblock liposome. (C) Sphere-rod diblock liposome. Reprinted in part from Zidovska *et al.* (2009a) with permission. Copyright 2008, American Chemical Society.

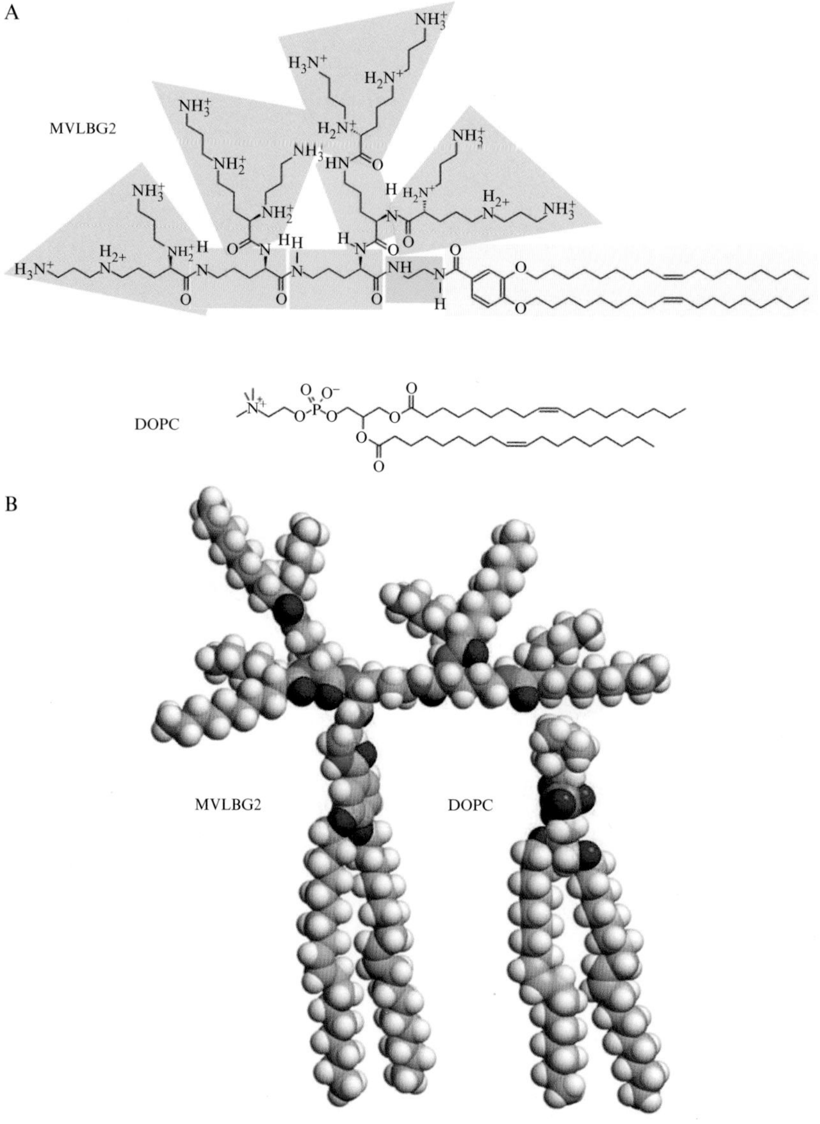

Alexandra Zidovska *et al.*, Figure 6.4 Chemical structures and molecular models of MVLBG2 and DOPC. (A) The chemical structures of MVLBG2 and DOPC. The hydrophibic tail, spacer, ornithine branching units, and charged carboxyspermine moieties of MVLBG2 are underlaid in tan, blue, green, and red, respectively. (B) Space filling molecular models of MVLBG2 and DOPC demonstrating their conical and cylindrical molecular shape, respectively. Reprinted in part from Zidovska *et al.* (2009a) with permission. Copyright 2008, American Chemical Society.

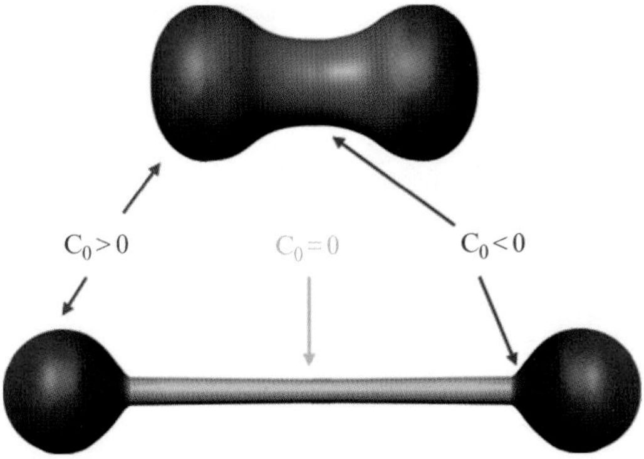

Alexandra Zidovska *et al.*, Figure 6.6 Comparison of the Gaussian curvature of a dumbbell vesicle and a triblock liposome. In the schematic, membrane regions of positive, negative, and zero Gaussian curvature are shown in red, blue, and yellow, respectively. The illustration demonstrates that regions of all three Gaussian curvatures are present in the block liposome, while the dumbbell morphology only contains regions of positive and negative Gaussian curvature.

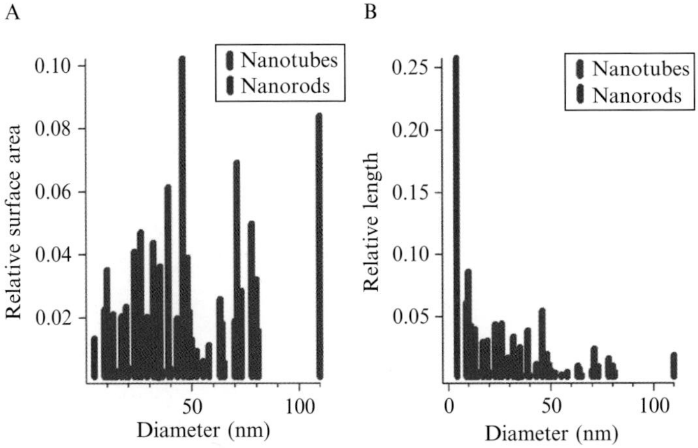

Alexandra Zidovska *et al.*, Figure 6.9 Statistical analysis of the nanotube and nanorod populations. (A) Diameter histogram weighted by the surface area, indicating the relative amount of lipid in the nanorod and nanotube state. (B) Diameter histogram weighted by the length of the structure, highlighting the striking length of nanorods compared to nanotubes. Reprinted in part from Zidovska *et al.* (2009a) with permission. Copyright 2008, American Chemical Society.

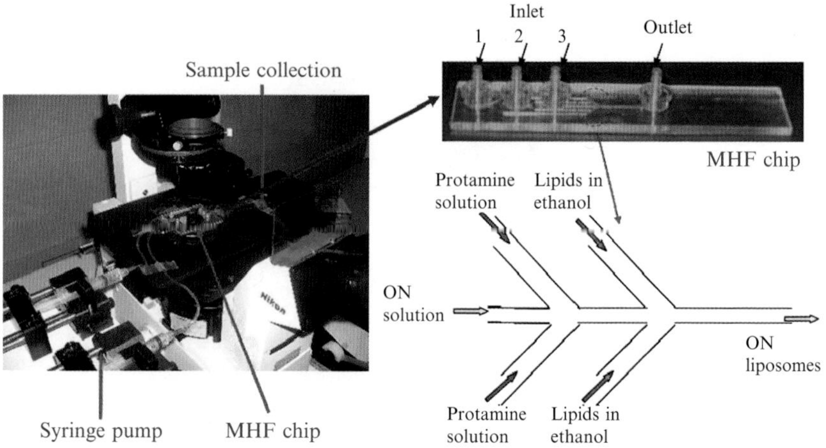

Bo Yu *et al.*, Figure 7.2 Photographs of five-inlet MHF methods to prepare liposomes containing oligonucleotides.

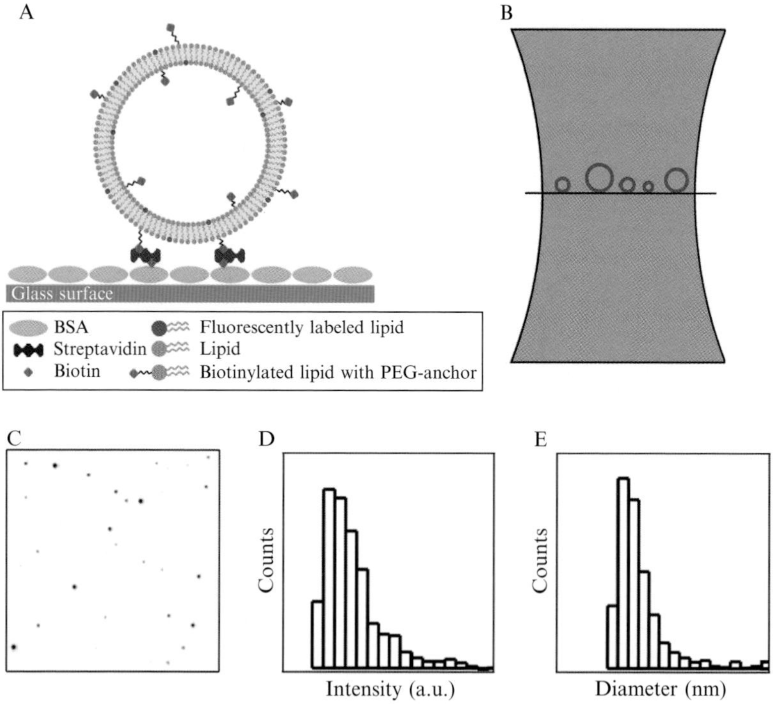

Christina Lohr *et al.*, Figure 8.1 Size distribution of vesicle samples based on single-particle fluorescence measurements. (A) Single fluorescence-labeled vesicles are immobilized at low densities on a protein-functionalized glass substrate. (B) The membrane-labeled vesicles are imaged with wide-field epi-fluorescence microscopy. The large depth of field insures that all vesicles are illuminated with equal efficiency. (C) In a fluorescence micrograph vesicles of different sizes appear as diffraction-limited spots with different total intensity. An intensity histogram of all vesicles (D) can be transformed into a size distribution histogram (E) after a calibration measurement.

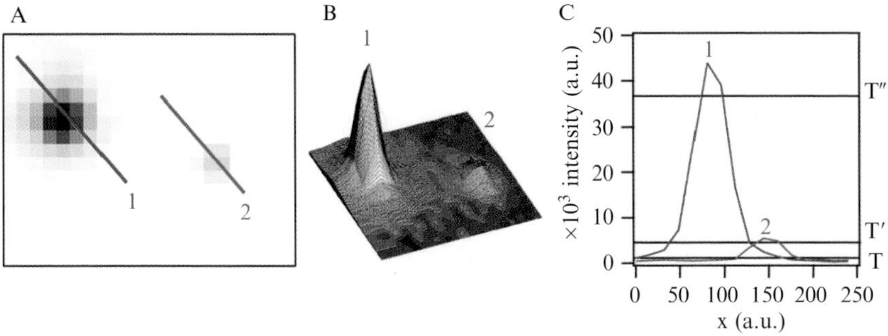

Christina Lohr *et al.*, Figure 8.2 How to threshold the fluorescence image of single vesicles. (A) Two neighboring vesicles from a fluorescent micrograph exhibit different fluorescent intensities, big particle (1, red line) and small particle (2, blue line). (B) A three-dimensional view shows a strong intensity signal for particle 1 and a fluorescence signal that is similar in intensity to the background noise for particle 2. (C) The threshold value regulates the detection of particles. Fluorescence signals below the threshold value will not be detected. Different threshold values (T, T', and T'') might be applied to the image: T'' only allows the detection of particles with a strong fluorescence signal and is therefore not suitable for application. T' includes the detection of the peaks of small particles, but still excludes particles with a weaker signal that are still strong enough to be distinguished from the background. Threshold T allows the cut-off of the background noise without losing particles expressing a weak fluorescence, and represents a suitable value for application to the images.

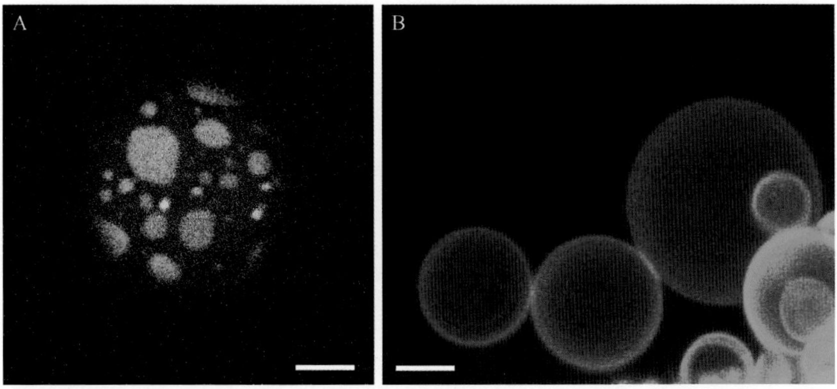

Philippe Méléard *et al.*, Figure 9.4 Fluorescent images of giant unilamellar vesicles composed of (A) DOPC/DPPC (1:1, mol/mol) plus 20 mol% cholesterol, showing coexistence of liquid ordered (red, fluorescent probe naphtopyrene) and liquid disordered (green, fluorescent probe rhodamine-DOPE), and (B) erythrocyte ghosts labeled with the fluorescent probe DiIC18. The GUVs were prepared using the electroformation protocol under physiological conditions (Table 9.2). The bar corresponds to 10 μm.

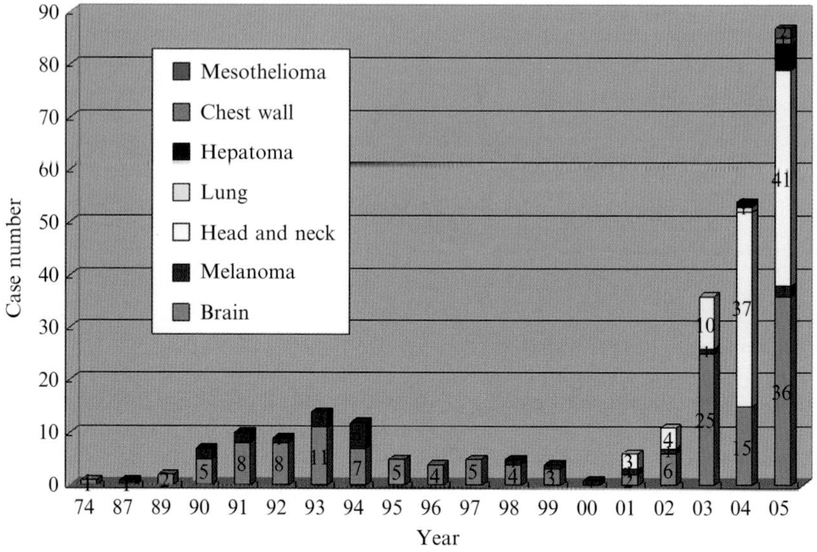

Hiroyuki Nakamura, Figure 10.2 Number of cases for treatment of cancers with BNCT at KUR.

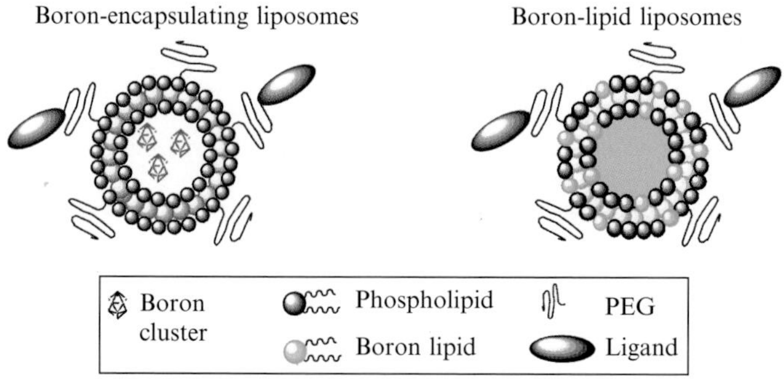

Hiroyuki Nakamura, Figure 10.3 Boron-encapsulating liposomes and boron-lipid liposomes.

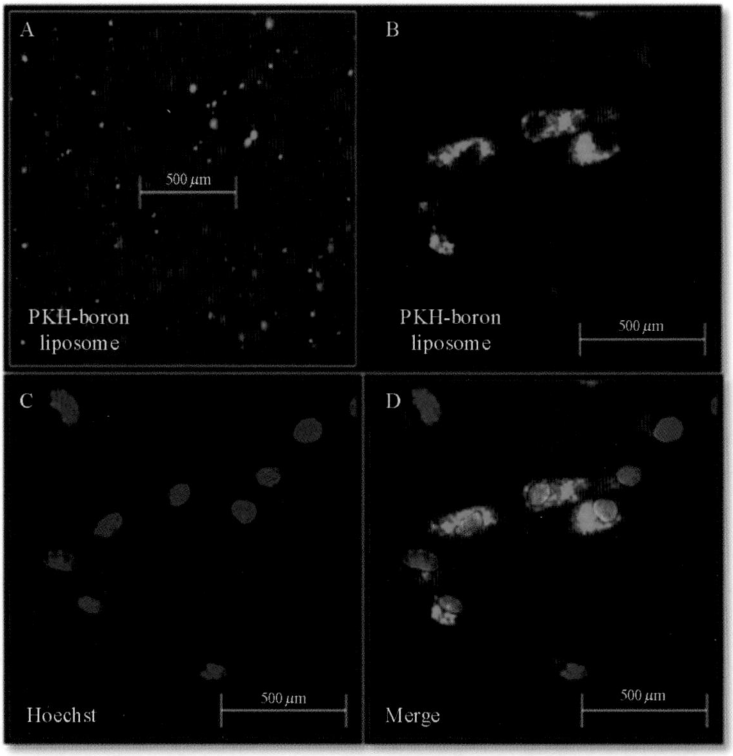

Hiroyuki Nakamura, Figure 10.11 PKH67-labeled boron liposomes and intracellular location of PKH-labeled boron liposomes. (A) PKH67-labeled 25% DSBL liposomes in PBS were visualized in fluorescent microscope. (B) Intracellular (HeLa) location of PKH67-labeled 25% DSBL liposomes were visualized in fluorescent microscope. (C) Hoechst-labeled nucleuses in HeLa were visualized in fluorescent microscope. (D) The merge image of PKH67-labeled 25% DSBL liposomes (B) and Hoechst-labeled nucleus (C) is shown in (D).

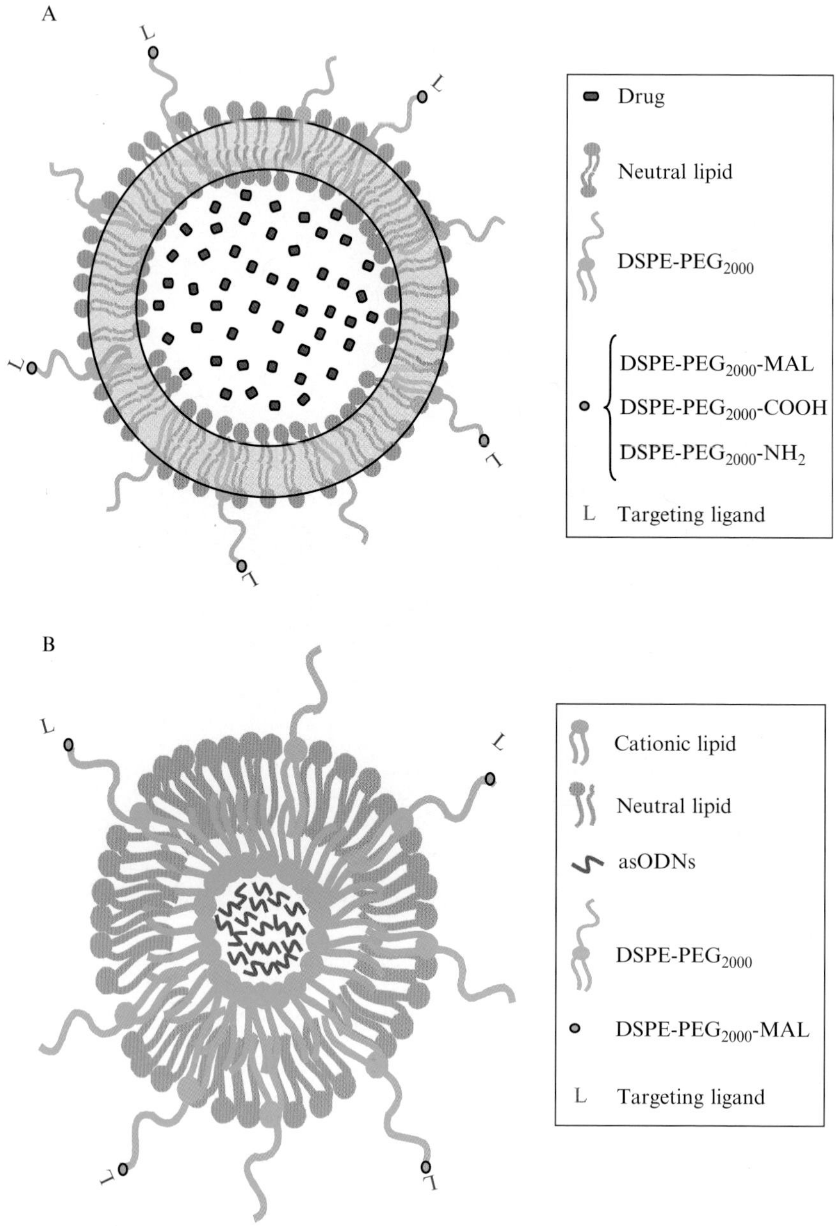

Daniela Di Paolo *et al.*, Scheme 12.1 Representative images of targeted liposomes (A) and targeted asODN-entrapping coated cationic liposomes (B).

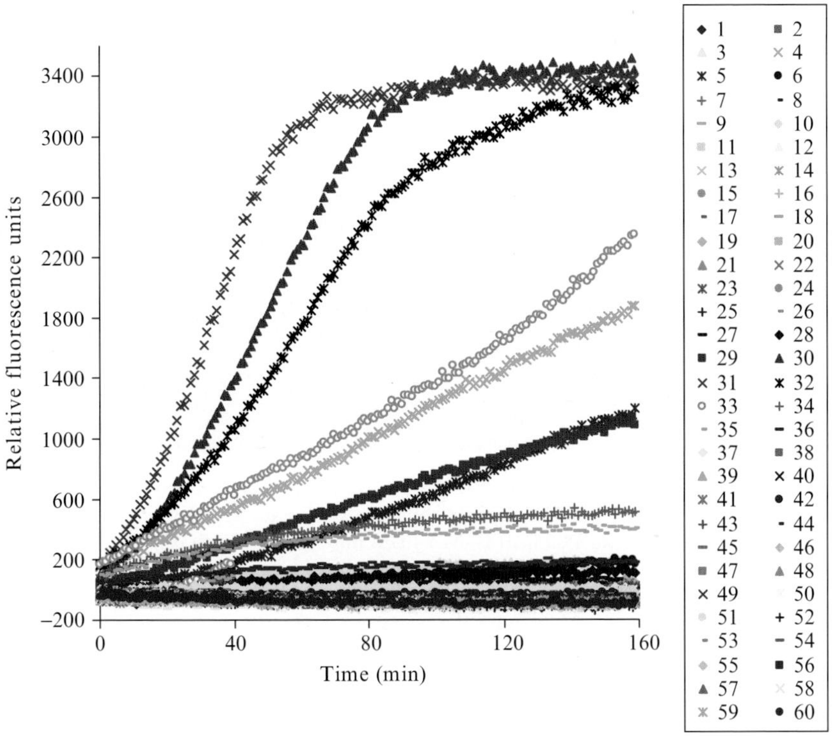

Ian Cheong and Shibin Zhou, Figure 13.1 Liposome disruption assay. The disruption of liposomal doxorubicin (Doxil) was measured using the dramatic increase in fluorescence as doxorubicin was released from encapsulation. The curves in the graph represent 60 individual fractions from a single anion exchange chromatography run during the biochemical purification of liposomase. Peak activity is observed in fractions 30–32.

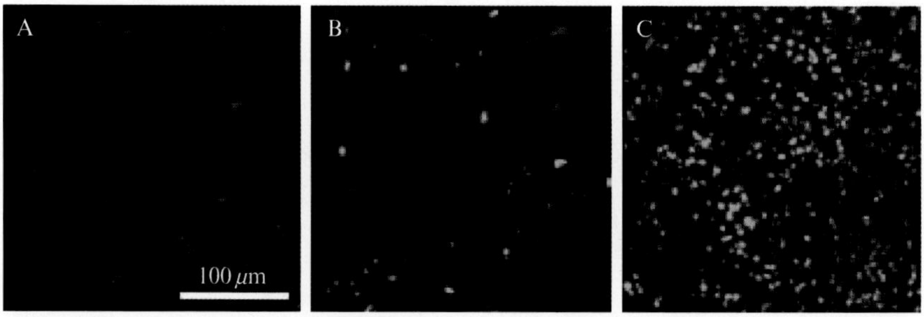

Naoto Oku and Takayuki Ishii, Figure 16.3 Immunofluorescence double staining of vascular endothelial cells and apoptotic cells in tumor tissue sections after PDT. Saline (A), LipBPD-MA (B), or PCLipBPD-MA (0.25 mg/kg BPD-MA) (C) was injected i.v. into tumor-bearing BALB/c mice. At 15-min postinjection of liposomal, the photosensitizer BPD-MA-treated mice were exposed to laser light (689 nm, 150 J/cm^2) under pentobarbital anesthesia. At 24-h postlaser exposure, tumors were resected and immunofluorescently double-stained with red and green fluorescence, indicating CD31-positive vascular endothelial cells (Alexa Fluor 594) and TUNEL-positive apoptotic cells (fluorescein), respectively.

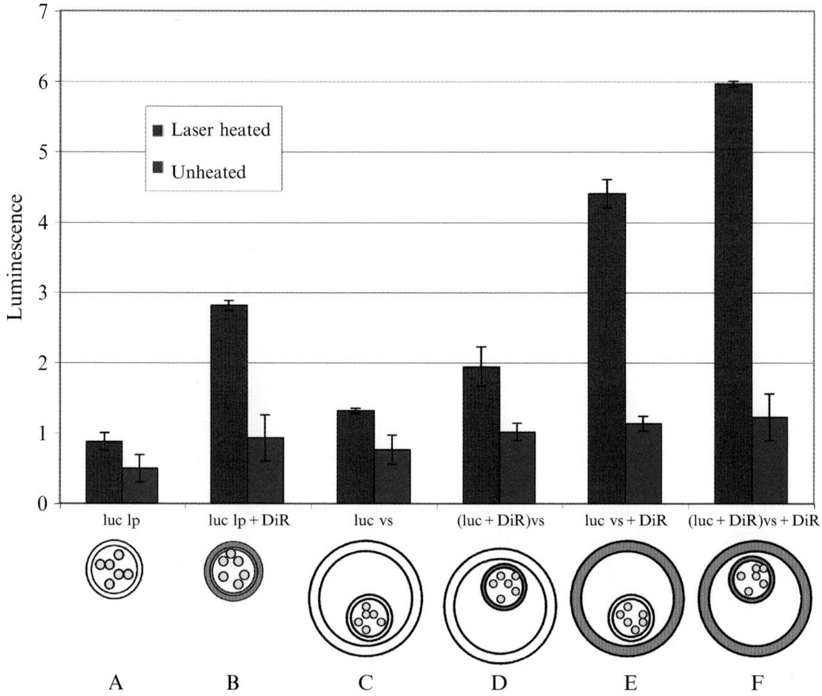

L. Zhao *et al.*, Figure 17.2 The inclusion of a chromophore in the membrane allows targeted activation of liposomal and vesosomal phase transitions. 4T1 cells, which are stably transfected with the firefly luciferase gene (Thorne *et al.*, 2006), were maintained near confluence; various compositions of liposomes and vesosomes were added to the cells, following which the cells and liposome/vesosomes were exposed to an 805-nm laser for 1 ns. Luciferase activity was measured by a luminometer. (A) In the absence of a chromophore, heating of luciferin-containing liposomes did not elicit a significant change in luciferase activity. (B) In the presence of the DiR chromophore, heated luciferin liposomes exhibit significantly greater luminescence than unheated luciferin-containing liposomes. (C) In the absence of a chromophore, heating of vesosomes containing luciferin liposomes produced a modest change in luciferase activity relative to unheated controls. (D) Inclusion of the DiR chromophore in the liposomal surface resulted in moderately higher luminescence compared to unheated controls. (E,F) The inclusion of a DiR chromophore in the vesosomal membrane resulted in higher luminescence compared to unheated controls. Including DiR chromophores in both liposomal and vesosomal membranes resulted in the largest increase in luminescence, compared to unheated controls. Luc lp, luciferin liposomes; luc lp + DiR, luciferin liposomes with DiR; luc vs, luciferin vesosomes; (luc + DiR)vs, luciferin liposomes with DiR in vesosomes; luc vs + DiR, luciferin liposomes in vesosomes with DiR; and (luc + DiR) vs + DiR, luciferin liposomes with DiR in vesosomes with DiR. Error bars indicate standard deviation.

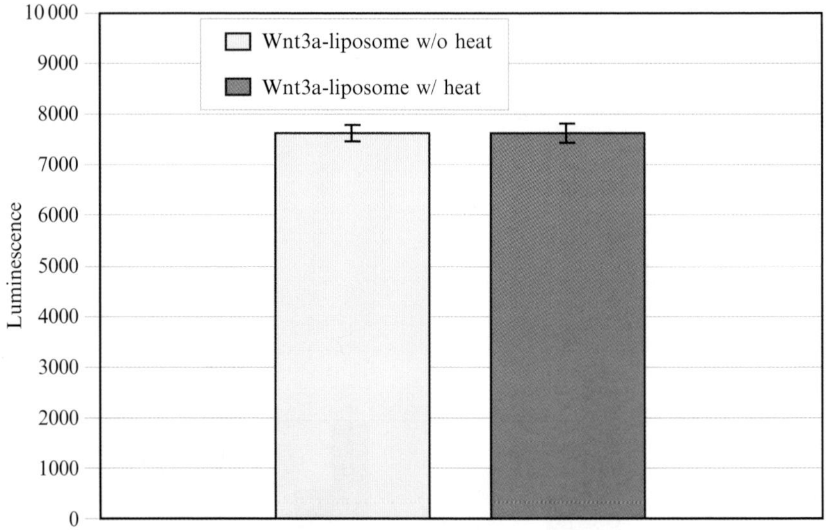

L. Zhao *et al.*, Figure 17.3 The activity of Wnt liposomes is not dependent upon phase transition. LSL cells were treated with Wnt liposomes, and then subjected to heating; controls were maintained at 37 °C, below the critical temperature for phase transition. Heated and unheated Wnt liposomes exhibited similar levels of activity as measured by luminescence. These data demonstrate that the Wnt protein is maintained in an active configuration that does not require liposomal phase transition, and that elevated temperatures do not decrease the activity of the protein. Error bars indicate standard deviation.

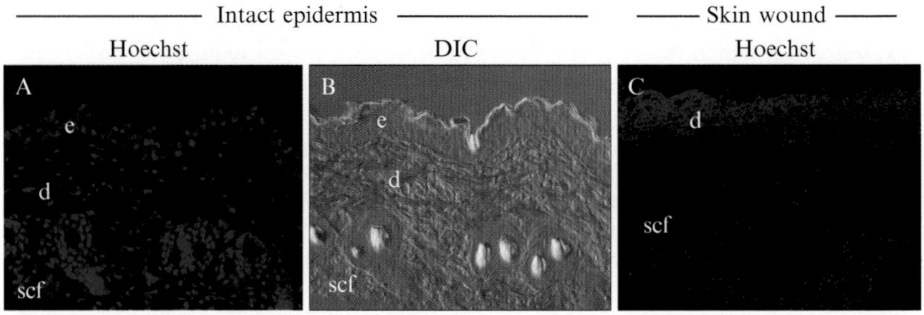

L. Zhao *et al.*, Figure 17.4 Liposomal penetration into skin can be monitored with the inclusion of a chromophore in the membrane. Intact skin was treated with DiI-modified liposomes, and then visualized under appropriate conditions. (A) Topically applied DiI liposomes (red) do not penetrate through the corneal layer of the epidermis; rather, they accumulate within the epidermis and are maintained at the site of application for at least 20 h. (B) Histology of the skin including the epidermis, dermis and subcutaneous fat shown by DIC microscopy. (C) After injury, DiI liposomes (red) rapidly penetrated into the dermis and the subcutaneous fat layer. e, epidermis; d, dermis; scf, subcutaneous fat.

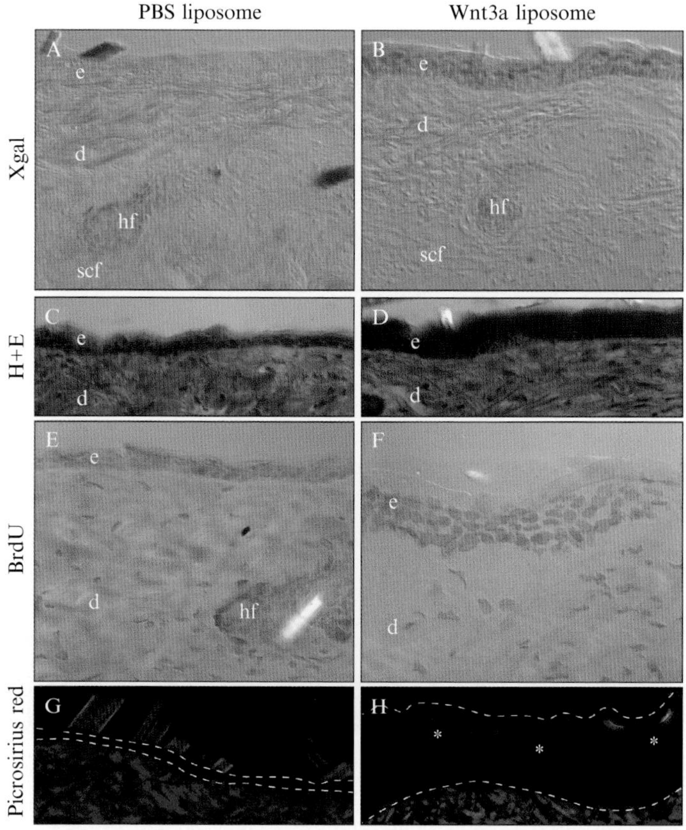

L. Zhao *et al.*, Figure 17.5 Topical application of Wnt liposomes to intact epidermis elicits a biological response, stimulates cell proliferation, and enhances collagen deposition in the epidermis. Equal volumes of PBS or Wnt liposomes were topically applied onto the intact skin of Axin2LacZ/+ reporter mice. (A) On postapplication day 5, baseline Xgal staining was detectable around the hair follicles; no obvious Xgal staining was detectable in the epidermis. This pattern of Xgal staining was indistinguishable from untreated control skin. (B) In contrast, treatment with Wnt liposomes elicited robust Xgal staining in the epidermis; Xgal staining in the deeper hair follicles was unchanged from (A). (C) Histological tissue sections reveal the thin, flattened epidermis present in PBS-treated skin. (D) Treatment of intact skin with Wnt liposomes elicits a dramatic thickening of the epidermis. (E) BrdU labeling indicates proliferation of cells around the hair follicle, and minimal proliferation in the corneal layer of the epidermis. (F) In contrast, liposomal Wnt treatment of intact skin resulted in substantially more BrdU-positive cells in the epidermis. Cell proliferation around the hair follicle was unchanged. (G) When visualized under polarized light, picrosirius red staining highlights collagenous fibrils; note the conspicuous lack of picrosirius red staining in the mature, thin epidermis. (H) Liposomal Wnt treatment produced a thicker epidermis, and picrosirius red staining indicated a greater amount of collagenous matrix within that thicker epidermis (asterisks).

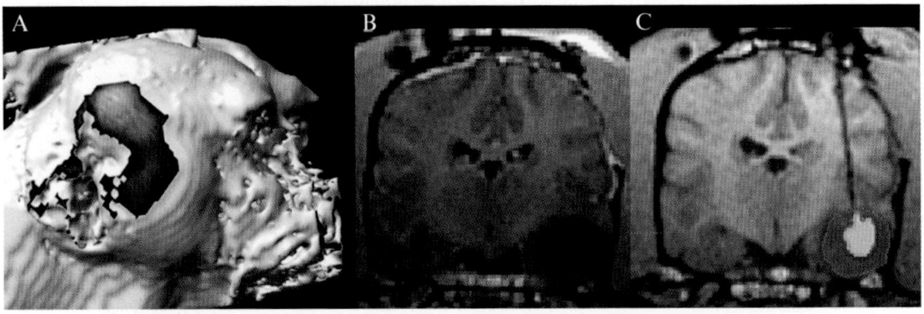

Michal T. Krauze *et al.*, Figure 18.6 Current therapeutic application of CED in spontaneous canine brain tumors. (A) Digital reconstruction of canine head after baseline MRI. (B) Image of large brain tumor located at the base of the left temporal lobe. (C) MR image during infusion of tumor (red delineation with a mix of GDL and chemotherapeutic-loaded liposomes (green delineation)).